T0328373

Marine Mammal Ecotoxicology

Marine Mammal Ecotoxicology

Impacts of Multiple Stressors on Population Health

Edited by

Maria Cristina Fossi

Cristina Panti

Academic Press is an imprint of Elsevier
125 London Wall, London EC2Y 5AS, United Kingdom
525 B Street, Suite 1650, San Diego, CA 92101, United States
50 Hampshire Street, 5th Floor, Cambridge, MA 02139, United States
The Boulevard, Langford Lane, Kidlington, Oxford OX5 1GB, United Kingdom

Library of Congress Cataloging-in-Publication Data
A catalog record for this book is available from the Library of Congress

British Library Cataloguing-in-Publication Data
A catalogue record for this book is available from the British Library

ISBN: 978-0-12-812144-3

Publisher: Andre Gerhard Wolff
Acquisition Editor: Anna Valutkevich
Editorial Project Manager: Pat Gonzalez
Production Project Manager: Sreejith Viswanathan
Cover Designer: Matthew Limbert

Typeset by TNQ Technologies

Contents

Section I
Legacy and Emerging Contaminants in Marine Mammal Populations

Section II
Effects of Toxicological and Cumulative Stress on Marine Mammal Health

List of Contributors

Manuel Arbelo University of Las Palmas de Gran Canaria, Las Palmas, Spain

Matteo Baini University of Siena, Siena, Italy

Jonathan L. Barber Centre for Environment, Fisheries and Aquaculture Science, Lowestoft, United Kingdom

James Barnett University of Exeter, Penryn, United Kingdom

Sarah Baulch Environmental Investigation Agency (EIA), London, United Kingdom

Asunción Borrell University of Barcelona, Barcelona, Spain

Andrew Brownlow SRUC Veterinary Services, Inverness, United Kingdom

Cinzia Centelleghe University of Padua, Padova, Italy

Tracy K. Collier Ocean Associates, Inc., Bainbridge Island, WA, United States

Maria Cristina Fossi University of Siena, Siena, Italy

Krishna Das University of Liege, Liege, Belgium

Nicholas J. Davison SRUC Veterinary Services, Inverness, United Kingdom

Robert Deaville Zoological Society of London, London, United Kingdom

Jean-Pierre Desforges Aarhus University, Roskilde, Denmark

Giovanni Di Guardo University of Teramo, Teramo, Italy

Rune Dietz Aarhus University, Roskilde, Denmark

Renata Emin-Lima Museu Paraense Emílio Goeldi, Coordenação de Zoologia, Setor de Mastozoologia, Grupo de Estudos de Mamíferos Aquáticos da Amazônia (GEMAM) and Programa de Capacitação Institucional, Belém, Brazil

Patricia A. Fair Medical University of South Carolina, Charleston, SC, United States

Antonio Fernandez University of Las Palmas de Gran Canaria, Las Palmas, Spain

Céline A.J. Godard-Codding The Institute of Environmental and Human Health Texas Tech University, Lubbock, TX, United States

Rachel A. Hauser-Davis Escola Nacional de Saúde Pública (ENSP), Centro de Estudos em Saúde do Trabalhador e Ecologia Humana (CESTEH), Fiocruz, Rio de Janeiro, Brazil

Magali Houde Aquatic Contaminants Research Division, Environment and Climate Change Canada, Montreal, QC, Canada

Paul D. Jepson Zoological Society of London, London, United Kingdom

Begoña Jiménez Institute of Organic Chemistry (IQOG-CSIC), Madrid, Spain

Juan José Alava Institute for the Oceans and Fisheries, University of British Columbia, Vancouver, BC, Canada; Fundación Ecuatoriana para el Estudio de Mamíferos Marinos (FEMM), Guayaquil, Ecuador; Ocean Pollution Research Program, Coastal and Ocean Research Institute, Ocean Wise Conservation Association, Vancouver, BC, Canada

Helena A. Kehrig Universidade Estadual do Norte Fluminense, Campos dos Goytacazes, Brazil; Grupo de Estudos de Mamíferos Marinhos da Região dos Lagos (GEMM-Lagos), Rio de Janeiro, Brazil

Ricardo Lavandier Pontifícia Universidade Católica do Rio de Janeiro, Rio de Janeiro, Brazil

Robin J. Law Zoological Society of London, London, United Kingdom; Centre for Environment, Fisheries and Aquaculture Science, Lowestoft, United Kingdom

Kristina Lehnert University of Veterinary Medicine Hannover, Büsum, Germany

Leila S. Lemos Grupo de Estudos de Mamíferos Marinhos da Região dos Lagos (GEMM-Lagos), Rio de Janeiro, Brazil; Escola Nacional de Saúde Pública (ENSP), Centro de Estudos em Saúde do Trabalhador e Ecologia Humana (CESTEH), Fiocruz, Rio de Janeiro, Brazil

Milton Levin University of Connecticut, Storrs, CT, United States

Susan M. Bengtson Nash Griffith University, Brisbane, QLD, Australia

Bjørn M. Jenssen Norwegian University of Science and Technology, Trondheim, Norway; The University Centre in Svalbard, Longyearbyen, Norway

Annalaura Mancia University of Ferrara, Ferrara, Italy

Helene Marsh James Cook University, Townsville, QLD, Australia

Letizia Marsili University of Siena, Siena, Italy

Sandro Mazzariol University of Padua, Padova, Italy

Elizabeth A. McHuron University of California Santa Cruz, Santa Cruz, CA, United States

Isabel Moreira Pontifícia Universidade Católica do Rio de Janeiro, Rio de Janeiro, Brazil

Jailson F. Moura Leibniz Center for Tropical Marine Ecology – ZMT, Systems Ecology Group, Bremen, Germany

Sinéad Murphy Galway-Mayo Institute of Technology, Galway, Ireland; Zoological Society of London, London, United Kingdom

Todd M. O'Hara University of Alaska Fairbanks, Fairbanks, AK, United States

Thomas J. O'Shea U.S. Geological Survey, retired, Glen Haven, CO, United States

Mark P. Simmonds Humane Society International, London, United Kingdom; University of Bristol, School of Veterinary Sciences, Bristol, United Kingdom

Cristina Panti University of Siena, Siena, Italy

Rod Penrose Marine Environmental Monitoring, Cardigan, United Kingdom

Matthew W. Perkins Zoological Society of London, London, United Kingdom

Sarah H. Peterson University of California Santa Cruz, Santa Cruz, CA, United States

Natalia S. Quinete Southeast Environmental Research Center (SERC), Florida International University, Miami, FL, United States

Peter S. Ross Ocean Pollution Research Program, Coastal and Ocean Research Institute, Ocean Wise Conservation Association, Vancouver, BC, Canada

Heli Routti Norwegian Polar Institute, Fram Centre, Tromsø, Norway

Salvatore Siciliano Instituto Oswaldo Cruz/Fiocruz, Laboratório de Enterobactérias (LABENT), Pavilhão Rocha Lima, Rio de Janeiro, Brazil

Ursula Siebert University of Veterinary Medicine Hannover, Büsum, Germany

Eva Sierra University of Las Palmas de Gran Canaria, Las Palmas, Spain

Christian Sonne Aarhus University, Roskilde, Denmark

Noel Y. Takeuchi University of South Florida St. Petersburg and Fish and Wildlife Research Institute, Florida Fish and Wildlife Conservation Commission, St Petersburg, FL, United States

Sabrina Tartu Norwegian Polar Institute, Fram Centre, Tromsø, Norway

Davi C. Tavares Universidade Estadual do Norte Fluminense, Campos dos Goytacazes, Brazil; Grupo de Estudos de Mamíferos Marinhos da Região dos Lagos (GEMM-Lagos), Rio de Janeiro, Brazil

Liesbeth Weijs The University of Queensland, Coopers Plains, QLD, Australia

Foreword

The world's seas and oceans have never been more impacted by human activities than they are today. As the global population has increased over the last 200 years from c. 1 billion to c. 7.6 billion currently, we have dramatically increased the use of the marine environment as a source of food and energy, as a means of transporting huge amounts of goods globally, and as a site for the unprecedented disposal of waste. In addition, we have deliberately, accidentally, or uncaringly let contaminants of all kinds enter inshore and, ultimately, offshore waters, from the equator to both polar seas.

What then is the fate of the denizens of the deep? Those mammalian species high in the marine food web may provide early warning of, and important insights into, damaging effects on marine ecosystems. In the past, efforts have been made to examine individual threats, one at time. However, this is not how marine organisms experience their world day to day. Instead, across their life course, they are simultaneously exposed to multiple stressors, both natural and anthropogenic, that combine to push animals to the limits of their tolerance. It is only by understanding the nature and extent of these integrated adverse impacts that we can be sufficiently well informed to take effective mitigatory action.

This book represents an important step on the path toward more effective action to manage our seas and oceans so that marine mammals, as well as other species, can maintain their populations in sustainable ecosystems. Guided by the editors, the authors explore the effects of exposure to multiple stressors, including anthropogenic pollutants, on marine mammals in different seas and oceans. Currently, both on ecotoxicologic and conservation grounds, it is crucial to determine the status of marine mammal populations and to instigate measures for their conservation. As top predators in complex marine food webs, several marine mammal species are especially vulnerable to environmental contaminant accumulation and to associated detrimental effects on important health endpoints.

Marine mammals, and particularly cetaceans, are widely regarded as reliable environmental sentinels due to their position near the top of the marine food web, their conspicuous nature, and their reliance on marine resources. As their physiology and/or diets are similar to those of humans, they can also potentially provide an early indication of adverse health effects and insights into the toxic mechanisms of diverse, well-known, and emerging hazardous substances. Such toxicologic stress, arising from the bioaccumulation of anthropogenic contaminants, combines with infectious diseases, competition from invasive species, food deprivation, and climate change to pose a growing risk to marine mammal populations worldwide. For this reason, efforts have intensified recently to find ways to use marine mammals as charismatic sentinels of ocean change.

The impacts of multiple stressors are extensively explored in this book. It is well known that the various cetacean species exhibit different home ranges and occupy different habitats. This knowledge can be exploited in hot spot areas where a range of species can serve as sentinels of marine environmental quality at different times during the annual cycle. Marine mammals have relatively long lifespans, thereby permitting the study of chronic diseases including reproductive failure, abnormalities in growth and development, and cancer. As apex predators, some marine mammals feed at or near the top of the food web. Consequently, biomagnification of the levels of anthropogenic contaminants is found in top predators and long-living species' tissues.

One of the main outcomes of this book, which is based on the recent findings in the field of ecotoxicology of marine mammal species worldwide, is that the application of robust examination procedures and biochemical, immunological, and microbiological techniques, combined with pathologic examination and behavioral analysis, has led to the development of health assessment methods at the individual and population levels in wild marine mammals. With these tools in hand, investigators have begun to unravel the relationships between exposures to environmental multiple stressors (e.g., climate change, pollutants, marine litter, pathogens, biotoxins) and to identify a range of disease endpoints that can be used as indicators of ecosystems health.

This book paves the way to the development of a comprehensive approach to the study of the threats to marine mammals combining conservation approaches to new and emerging technologies. It comprises 17 chapters contributed by 59 scientists from 13 countries. Part 1 examines legacy and emerging contaminants in marine mammal populations; Part 2 explores the effects of toxicologic threats and cumulative stress on marine mammal health, while Part 3 considers some implications for monitoring and conservation of marine mammals. A brief overview follows:

SECTION I: LEGACY AND EMERGING CONTAMINANTS IN MARINE MAMMAL POPULATIONS

Four chapters (Chapters 1, 4, 7, and 8) focus primarily on organochlorine contaminants (notably polychlorinated biphenyls and dichlorodiphenyltrichloroethane) that still persist, bioaccumulate, and biomagnify within marine food webs. Many of these legacy pollutants are endocrine disruptors that can potentially exert adverse effects on hormonal and immune functions in humans, laboratory animals, and wildlife. In Chapter 2, McHuron and coauthors, evaluate feeding ecology as a tool for estimating contaminant exposure in coastal mammals. Chapter 3 focuses on an important class of pollutants related to oil exposure, while Chapter 5 addresses a class of emerging pollutants, poly- and perfluoroalkyl compounds. In Chapter 6 the impacts of the marine litter are discussed in several cetacean species.

SECTION II: EFFECTS OF TOXICOLOGIC AND CUMULATIVE STRESS ON MARINE MAMMAL HEALTH

Part 2, the most extensive of the book, is composed of eight chapters focused on the effects of toxicologic and cumulative stress on marine mammal health. Three chapters (Chapters 9–11) review new technologies for monitoring marine mammal health, from biochemical and gene expression biomarkers to omics techniques. In Chapter 12, the authors report a new approach to evaluate the immunotoxic effects of environmental pollutants in marine mammals. Chapter 13 proceeds to describe ecotoxicologic stressors in Arctic marine mammals, with a particular focus on polar bears. The toxicologic risks and considerations associated with lipophilic contaminant burdens of Mysticetes in Antarctic ecosystems are dealt with in Chapter 14. The authors of Chapter 15 report on the emerging pathogens and stress syndromes in cetaceans in European waters, especially cumulative effects. Part 2 is completed with Chapter 16, in which the ecotoxicology of the Sirenia is presented as a case study of cumulative stress in marine mammals.

SECTION III: IMPLICATION FOR MONITORING AND CONSERVATION OF MARINE MAMMALS

The third and final section of the book is composed by one concluding chapter that explores in detail the implications of the previous chapters for monitoring and conservation of marine mammals.

Chapter 17, written by Simmonds, provides a final overview of multiple stressor effects in marine mammals and the implications for conservation and policy. The author states that marine mammals face many threats in the 21st century, and a comprehensive description is provided of these threats and some of the efforts to try to study their combined effects. This draws mainly on work undertaken under the auspices of the International Whaling Commission and, most recently, the US National Academies of Sciences, Engineering, and Medicine.

The consideration of combined or cumulative effects is complex, but progress is being made, leading potentially to improved conservation policy. However, the pathway from better scientific understanding to real conservation action in the field is frequently difficult. In addition, the potential of veterinary sciences, including health studies of wild animal populations, to identify problems is noted, and a cross-fertilization between conventional conservation approaches and wild animal "welfare science" is encouraged.

The challenges inherent in addressing multiple stressors make the need for precaution in marine mammal matters all the greater. However, for decision-makers to properly respond to complexity, they need to have the limitations of science clearly and honestly explained.

In summary, this book provides an important contribution to the investigation of the status of the world's seas and oceans, and the marine mammals that live within them. Hopefully, it will be a useful guide for those working in this research area and help to generate further interest in the practical management and conservation of marine mammals, seas, and oceans.

The editors are grateful to all the authors for their contribution and, also, to the numerous experts who provided reviews of chapters, who made the realization of the book possible. Finally, the editors thank all the authors, who have been extremely patient and generous with their time and energy during the long gestation of the present book. Many thanks also go to the staff of Elsevier Academic Press for their guidance and patience throughout the birth, development, and final production of the book.

Prof. Maria Cristina Fossi, Editor
University of Siena, Siena, Italy

Dr. Cristina Panti, Editor
University of Siena, Siena, Italy

Prof. Michael H. Depledge, CBE DSc FRCP
Chair of Environment and Human Health
European Centre for Environment and
Human Health University of Exeter Medical School
Honorary Professor of Public Policy
University College, London

List of Abbreviations

β-HCH	Beta-hexachlorocyclohexane
2D-PAGE	Two-dimensional polyacrylamide gel electrophoresis
4MBC	4-Methylbenzylidene camphor
Ab	Antibodies
ACCOBAMS	Agreement on the Conservation of Cetaceans in the Black Sea, Mediterranean Sea, and Contiguous Atlantic Area
ACTH	Adrenocorticotropic hormone
ADME	Absorption, distribution, metabolism, and elimination
AFFF	Aqueous firefighting foam
AFS	Antifouling systems
AHR	Aryl hydrocarbon receptor
AI	Adipocyte index
AIDS	Acquired immunodeficiency syndrome
ALDFG	Abandoned, lost, or discarded fishing gear
AN	Androstenedione
APP	Acute phase proteins
ARNT	Aryl hydrocarbon receptor nuclear translocator
ASCOBANS	Agreement on the Conservation of Small Cetaceans of the Baltic, North East Atlantic, Irish, and North Seas
ASVCP	American Society for Veterinary Clinical Pathology
BaP	Benzo-a-pyrene
BB	Barataria Bay
BBzP	Benzyl butyl phthalate
BCI	Body condition index
BFRs	Brominated flame retardants
BMFs	Biomagnification factors
BPA	Bisphenol A
BPMO	Benzo(a)pyrene monooxygenase
CAT	Catalase activity
CCAMLR	Convention for the Conservation of Antarctic Marine Living Resources
CDV	Canine distemper virus
CDW	Converted to dry weight
CE	Central Eastern

CECs	Chemicals/contaminants of emerging concern
CeMV	Cetacean morbillivirus
CHLs	Chlordanes
CNS	Central nervous system
CNV	Copy number variation
COAFG	Commercial and other active fishing gears
COD	Cause of death
COYs	Cubs of the year
CRP	C-reactive protein
CSIA	Compound-specific stable isotope analysis
CSIP	Cetacean Strandings Investigation Programme
CTL	Cytotoxic T lymphocytes
CUPs	Current-use pesticides
CYP	Cytochrome P450
CYP4A	Cytochrome P450 4A
DBDPE	Decabromodiphenylethane
DDE	Dichlorodiphenyldichloroethylene
DDT	Dichlorodiphenyltrichloroethane
Dec 602	Dechlorane 602
Dec 603	Dechlorane 603
DEHP	Bis(2-ethylhexyl) phthalate
DIOP	Diisooctyl phthalate
DMV	Dolphin morbillivirus
DNA	Deoxyribonucleic acid
DNHP	Di-n-hexyl phthalate
DP	Dechlorane plus
DTH	Delayed-type hypersensitivity
DWH	Deepwater Horizon
E2F1	Transcription factor E2F1
EDCs	Endocrine disruptor compounds
EHMC	2-Ethyl-hexyl-4-trimethoxycinnamate
EIDs	Emerging infectious diseases
ENSO	El Niño Southern Oscillation
ERα	Estrogen receptor alpha
ESA	Endangered Species Act
EtFOSA	N-ethyl perfluorooctane sulfonamide
EVOS	Exxon Valdez spill
F-53B	6:2 Chlorinated polyfluorinated ether sulfonate (6:2 Cl-PFAES, with trade name F-53B)
FA	Fatty acid
FABP4	Fatty acid binding protein 4
FAO	Food and Agricultural Organization of the United Nations
FASA	Fatty acid signature analysis

FOSA	Perfluorooctane sulfonamide
FOSE	Sulfonamido ethanol
FT3	Free triiodothyronine
FT4	Free thyroxin
FTOH	Fluorotelomer alcohol
FWMFs	Food web magnification factors
GC–MS	Gas chromatography–mass spectrometry
GLMs	Generalized linear models
GMR	Galapagos Marine Reserve
GNP	Galapagos National Park
GSH	Glutathione
GST	Glutathione-s-transferase
HAHs	Halogenated aromatic hydrocarbons
HBB	Hexabromobenzene
HBCDD/HBCD	Hexabromocyclododecane
HCB	Hexachlorobenzene
HCHs	Hexachlorocyclohexanes
HFRs	Halogenated flame retardants
HGP	Human Genome Project
HgSe	Mercury selenide
HNPs	Halogenated natural products
HOC	Halogenated organic compounds
Hp	Haptoglobin
HPA	Hypothalamic-pituitary-adrenal
HRCGC	High-resolution capillary gas chromatography
HSP70	Heat shock protein 70
ID	Photo identification
IFNγ	Interferon-gamma
Ig	Immunoglobulin
IgA	Immunoglobulin alpha
Igf1	Insulin-like growth factor receptor 1
IgG	Immunoglobulin gamma
IgM	Immunoglobulin mu
IL	Interleukin
Il1b	Interleukin-1 β
IL-4	Interleukin-4
IMMAs	Important Marine Mammal Areas
IOC	Intergovernmental Oceanographic Commission
IUCN	International Union for Conservation of Nature
IWC	International Whaling Commission
KOH	Potassium hydroxide
LC-MS/MS	Liquid chromatography-tandem mass spectrometry
LOAEL	Lowest observed adverse effect level
LOEC	Lowest observed effect concentration

LTT	Lymphocyte proliferation test
MALT	Mucosal-associated lymphoid tissue
MAO	Monoamine oxidase activity
MBP	Monobutyl phthalate
MBzP	Monobenzyl phthalate
MeHg	Methylmercury
MEHP	Mono(2-ethylhexyl)phthalate
MeO-PBDEs	Methoxylated PBDEs
MEPC	Marine Environment Protection Committee
$MeSO_2$	Methylsulfonyl
MHCII	Major histocompatibility complex II
MMPATF	Marine Mammal Protected Areas Task Force
MPAs	Marine Protected Areas
MPO	Myeloperoxidase
MPs	Microplastics
mRNA	Messenger ribonucleic acid
MRSA	Methicillin-resistant *Staphylococcus aureus*
MS	Mass spectrometry
MS	Mississippi Sound
MSF	Methyl sulfone
MSFD	Marine Strategy Framework Directive
MTs	Metallothioneins
NCP	Northern Contaminants Program, Indigenous and Northern Affairs Canada
nGoM	Northern Gulf of Mexico
NIST	National Institute of Standards and Technology
NK	Natural killer
NMDA	N-methyl-D-aspartate
NMMTB	National Marine Mammal Tissue Bank
NOAA	National Oceanic and Atmospheric Administration
Nr3c1	Glucocorticoid receptor
NRDA	Natural Resource Damage Assessment
OA	Okadaic acid
OCDD	Octachlorodibenzodioxin
OCs	Organochlorine contaminants
OCT	Octocrylene
ODE	Old dog encephalitis
OH	Hydroxylated
OT	Organotin
PAEs	Phthalate esters
PBDEs	Polybrominated diphenyl ethers
PBEB	Pentabromoethylbenzene
PBMCs	Peripheral blood mononuclear cells
PBPK	Physiologically based pharmacokinetic

PBT	Persistent, bioaccumulative, and toxic
PBTK	Physiologically based toxicokinetic
PbTx	Brevetoxins
PCBs	Polychlorinated biphenyls
PCDDs	Polychlorinated dibenzo-p-dioxins
PCDFs	Polychlorinated dibenzofurans
PCoD	Population Consequences of Disturbance
PCoMS	Population Consequences of Multiple Stressors
PDMS	Polydimethylsiloxane
PDV	Phocine distemper virus
PFAS	Poly- and perfluoroalkyl substances
PFBA	Perfluorobutyric acid
PFBS	Perfluorobutane sulfonic acid
PFCA	Perfluoroalkyl carboxylate
PFDA	Perfluorodecanoic acid
PFDoA	Perfluorododecanoic acid
PFECHS	Perfluoro-4-ethylcyclohexane sulfonic acid
PFHxA	Perfluorohexanoic acid
PFHxS	Perfluorohexane sulfonic acid
PFNA	Perfluorononanoic acid
PFOA	Perfluorooctanoic acid
PFOS	Perfluorooctane sulfonic acid
PFPIA	Perfluoroalkyl phosphinic acid
PFSA	Perfluoroalkyl sulfonate
PFTriA	Perfluorotridecanoic acid
PFUnA	Perfluorododecanoic acid
PhV-1	Phocine herpesvirus
PMV	Porpoise morbillivirus
POPs	Persistent organic pollutants
PPARα	Peroxisome proliferator-activated receptor alpha
PPARγ	Peroxisome proliferator-activated receptor gamma
PPCPs	Pharmaceuticals and personal care products
PRE	Pregnenolone
PreFOS	Perfluorooctanesulfonate precursor
PT	Phenyltin
PWMV	Pilot whale morbillivirus
PXR	Pregnane X receptor
PYR	Synthetic pyrethroids
QFASA	Quantitative fatty acid signature analysis
qRT-PCR	Quantitative real-time PCR
RI	Reference intervals
RNA-seq	RNA sequencing
ROS	Reactive oxygen species
RQs	Risk quotients

RT-qPCR	Real time-quantitative PCR
SAA	Serum amyloid A
SAGs	Structure activity groups
SAP	Serum amyloid P
SCANS	Small Cetaceans in European Atlantic Waters and the North Sea
SealAV	Seal anellovirus
SNPs	Single-nucleotide polymorphisms
SOD	Superoxide dismutase
SPAMI	Specially Protected Areas and Biological Diversity in the Mediterranean
SPAMI	Specially Protected Areas of Mediterranean Importance
SSPE	Subacute sclerosing panencephalitis
STX	Saxitoxin
TBECH	Tetrabromoethylcyclohexane
TBT	Tributyltin
TDF	Trophic discrimination factor
TEFs	Toxic equivalent factors
TEQs	Toxic equivalents
TetraBHD	2,5,7-Tribromo-4a-bromomethyl-1,1-dimethyl-2,3,4,4a,9,9a-hexahydro-1H-xanthene
Th1	T helper 1
Th2	T helper 2
THg	Total Hg
Thra	Thyroid hormone receptor alpha
TL	Trophic level
TMF	Trophic magnification factor
TNF	Tumor necrosis factor
TPT	Triphenyltin
TriBHD	2,7-dibromo-4a-bromomethyl-1,1-dimethyl-2,3,4,4a,9,9a-hexahydro-1H-xanthene
TT3	Total triiodothyronine
TT4	Total thyroxine
TTR	Transthyretin
UG	Uteroglobin
UMEs	Unusual mortality events
UNEP	United Nation Environmental Program
UNESCO	United Nations Educational, Scientific, and Cultural Organization
USEPA	United States Environmental Protection Agency
WGS	Whole genome sequence
WHO	World Health Organization
WHO-TEQ	World Health Organization-Toxic Equivalents
YRL	Yearlings
ZcAV	Species-specific anellovirus

Section I

Legacy and Emerging Contaminants in Marine Mammal Populations

Chapter 1

Organochlorine Contaminants and Reproductive Implication in Cetaceans: A Case Study of the Common Dolphin

Sinéad Murphy[1,2], Robin J. Law[2,3], Robert Deaville[2], James Barnett[4], Matthew W. Perkins[2], Andrew Brownlow[5], Rod Penrose[6], Nicholas J. Davison[5], Jonathan L. Barber[3], Paul D. Jepson[2]

[1]*Galway-Mayo Institute of Technology, Galway, Ireland;* [2]*Zoological Society of London, London, United Kingdom;* [3]*Centre for Environment, Fisheries and Aquaculture Science, Lowestoft, United Kingdom;* [4]*University of Exeter, Penryn, United Kingdom;* [5]*SRUC Veterinary Services, Inverness, United Kingdom;* [6]*Marine Environmental Monitoring, Cardigan, United Kingdom*

INTRODUCTION

Organochlorines (OCs) are known persistent organic pollutants (POPs) that both bioaccumulate and biomagnify within marine food webs. This chapter will focus on the legacy pollutants polychlorinated biphenyls (PCBs) and dichlorodiphenyltrichloroethane (DDT), which have been reported to have adverse effects on endocrine, reproductive, and immune functions in marine mammals (Hall et al., 2006a; Murphy et al., 2015; O'Shea et al., 1999; Reijnders et al., 1999; Vos et al., 2003). The chapter is divided into three sections, including an introduction to OCs and their effects, a review of the published cases of reproductive failure and dysfunction in cetaceans associated with exposures to OCs, and a new case study exploring the effects of PCBs on reproduction in short-beaked common dolphins (*Delphinus delphis*) in the Northeast Atlantic. Thus, this chapter provides a comprehensive overview of published and new information on OCs and reproductive implications in cetaceans.

PCBs were originally synthesized in the 19th century, but they came into widespread use in the 1930s. They were primarily used in mineral oils used as dielectric fluids in electrical equipment such as transformers and capacitors, but they were also used in a wide variety of other applications, including as flame retardants, in paints and lacquers, hydraulic fluids, caulks, and sealants (Diamond et al., 2010). A number of different products were produced, differing

Marine Mammal Ecotoxicology. https://doi.org/10.1016/B978-0-12-812144-3.00001-2

in the proportion of chlorine incorporated and the range of congeners comprising the product (e.g., Aroclor 1242, 1254 and 1260, manufactured by Monsanto: 42%–60% chlorine). Most of the PCB products were produced in the United States and Europe (Law and Jepson, 2017). In the mid-1960s, they were discovered as contaminants in fish from the River Viskan in Sweden, and later in other wildlife, highlighting their widespread environmental occurrence (Jensen, 1996; Jensen et al., 1969). Subsequently, it was discovered that PCBs are persistent and highly mobile, being transported to the Arctic by a process known as global distillation, or long-range atmospheric transport (Jepson and Law, 2016). Controls on production and use in both the United States and Europe, where over 1 million tons of PCBs were produced, began around the early 1980s.

Initially, marine environmental concentrations of PCBs declined rapidly following the ban in America and Europe in 1979 and 1985, respectively, but in some instances, this decline has stalled (Aguilar and Borrell, 2005; Borrell and Aguilar, 2007; Jepson et al., 2016; Law et al., 2012). Continued input into the marine environment through activities such as dredging of PCB-laden sediment and mariculture, as well as from land-based sources such as leakages from old landfills and PCB-contaminated buildings material are suspected (Jepson et al., 2016; Tornero and Hanke, 2016). Further, as large quantities of PCB-containing equipment still require disposal (CLEEN, 2005) and the half-life of some PCB congeners is up to 100 years (Hickie et al., 2007; Jonsson et al., 2003; Sinkkonen, 2000), the problems associated with these compounds will continue for decades to come. Traditional organochlorine pesticide compounds (e.g., DDT, aldrin, dieldrin) have been restricted or scheduled for elimination under the Stockholm Convention since 2004 (United Nations Environment Programme, 2001), and declines in some of these compounds, notably, DDT, have been reported in some marine mammal populations (e.g., European waters; Aguilar and Borrell, 2005; Borrell and Aguilar, 2007; Law et al., 2012). However, concentrations of DDT have been observed to increase again in tropical wildlife species, where use has been recommitted by the World Health Organization to tackle malaria (Alava et al., 2011).

As top predators with long life spans, cetaceans accumulate high concentrations of OCs in their lipid tissue, as these lipophilic compounds bind to fatty acids in the blubber (90%–95% of total body burden; Aguilar, 1985), and odontocetes through their higher trophic level status have a higher exposure to OCs compared to mysticetes (Houde et al., 2005). Blubber concentrations in male cetaceans increase with age, with reported annual PCB accumulation rates of 1.1 mg/kg in UK harbor porpoises (*Phocoena phocoena*) (Murphy et al., 2015) and 2.96 mg/kg in bottlenose dolphins (*Tursiops truncatus*) from Sarasota Bay, Florida (Hall et al., 2006b). In species such as killer whales (*Orincus orca*), a growth dilution phase was observed after weaning up to around 10 years in age, following which an increase in ΣPCBs was observed in males (Hickie et al., 2007). Once sequestered in blubber tissue, compounds are mobilized during times of fasting/starvation or intense energetic demand. In southern resident killer whales (*Orcinus orca*), concentrations of POPs, possibly mobilized from endogenous lipid stores, were highest and had the greatest potential for toxicity during periods of low prey abundance (Lundin et al., 2016). During pregnancy and lactation, female cetaceans offload a high proportion of their lipophilic pollutant burden to offspring, particularly in lipid-rich milk. In bottlenose dolphins, up to c. 80% of OCs can be offloaded to firstborn calves during the first 7 weeks of lactation (Cockcroft et al., 1989). This is similar to long-finned pilot whales (*Globicephala melas*), where 60%–100% of their pollutant load is offloaded during lactational transfer, compared to just 4%–10% through placental transfer (Borrell et al., 1995), and in striped dolphins (*Stenella coeruleoalba*), 72%–91% of body burdens were offloaded during lactation and only 4%–9% during gestation (Fukushima and Kawai, 1981 reported in Yordy et al., 2010). For all cetacean species studied to date, gestational and lactational transfer of OCs has ranged from 3.5% to 15% and 67.6%–99.9%, respectively (Mongillo et al., 2016). DDTs are more easily transferred than PCBs, as transplacental and lactational transfer is easier for lower chlorinated compounds (Borrell and Aguilar, 2005). It is during these periods of mobilization of OCs that adverse health effects may occur, for example, exerting toxic effects on fetal growth and development.

Both PCBs and DDT have been reported as endocrine disruptors in marine mammals, though the latter has also shown direct lethal effects on wildlife. Endocrine-disrupting chemicals (EDC) are "an exogenous chemical, or mixture of chemicals, that interferes with any aspect of hormone action" (Zoeller et al., 2012). By hormone action, it is inferred to mean hormone receptor activation, i.e., interfering or interacting with the hormone receptors themselves, such as mimicking hormones or blocking hormone receptors, and also delivery (synthesis, release, transport including blood and across membranes, metabolism, or clearance) of hormones to those receptors (Zoeller et al., 2014). Thus, it is any inference with the endocrine system's role in maintaining homeostasis, as well as roles in sexual differentiation, development, metabolism, and stress responses. In marine mammals, it has been hypothesized that PCBs and DDT, or metabolites thereof, bind to hormone receptors and/or hormone carriers, or in some cases break down steroid hormones through increased OC metabolic-induced cytochrome (CYP) enzyme activity (Reijnders, 2003). EDCs differ somewhat from general toxicants as they (e.g., chemicals with hormone-like properties) have the ability to act at low doses, exhibit nonmonotonic dose responses (e.g., U-shaped curves), show varying effects over an individual's lifespan, delayed effects (of sexual dysfunction and physical abnormalities) that are not evident until later in life or until future generations, and have the potential to show combination effects when exposed to multiple pollutants (Bergman et al., 2013; Ingre-Khans et al., 2017).

The EDC effects of PCBs depends on the specific congeners in the compound, as some congeners have a dioxin-like mechanism mode of action, while other congeners and metabolites thereof have estrogenic and/or antiandrogenic or antiestrogenic effects (EEA Technical Report, 2012; Fossi and Marsili, 2003; Letcher et al., 2010). Both *p,p′*-DDT and *o,p′*-DDT, which comprise DDT, promote estrogenic activity but have also been shown to have both antiestrogenic and antiandrogenic effects, while the breakdown product, *p,p′*-DDE, has been shown to exert a wide variety of effects (Fossi and Marsili, 2003). Hence, the toxic effects of OCs can be exerted by the parent compound or by their metabolites, and different cetaceans (and even different populations, depending on previous levels of exposure) have varying capacities to metabolize/biotransform OC compounds depending on the presence and activity of CYP enzymes (Houde et al., 2005; Muir et al., 1996). It is beyond the scope of this chapter to discuss the properties of OCs in detail; however, fully comprehensive reviews are provided elsewhere (Aguilar et al., 1999; Houde et al., 2005; Letcher et al., 2010; O'Shea et al., 1999; Reijnders et al., 1999).

Using data on UK harbor porpoises from Murphy et al. (2015), we assessed the influence of metabolism and the degree of offloading of PCBs in females. PCBs were grouped into five structure activity groups (SAGs) based on their capacity for biotransformation as defined previously by Boon et al. (1994). Seven PCB congeners (PCB118, -138, -149, -153, -170, -180, and -187) contributed to 79% of the ΣPCB content in female harbor porpoises blubber samples (Fig. 1.1). PCB118, PCB170, and PCB180 are dioxin-like PCBs, whereas PCB138 and PCB153 are nondioxin-like, with the latter having estrogen-like activities. The top PCB congeners in the blubber of females were PCBs 153 > 138 > 149 > 180 > 187, accounting for 54% of the ΣPCB concentration. PCB congener profiles did not differ markedly among reproductive groups, and this is due to the persistence of some congeners and the lack of ability by cetaceans to metabolize those congeners. The highly persistent CB-153 ranged from 23% of ΣPCB concentration in pregnant females to 27% in resting females (neonates = 26%, immature = 26%).

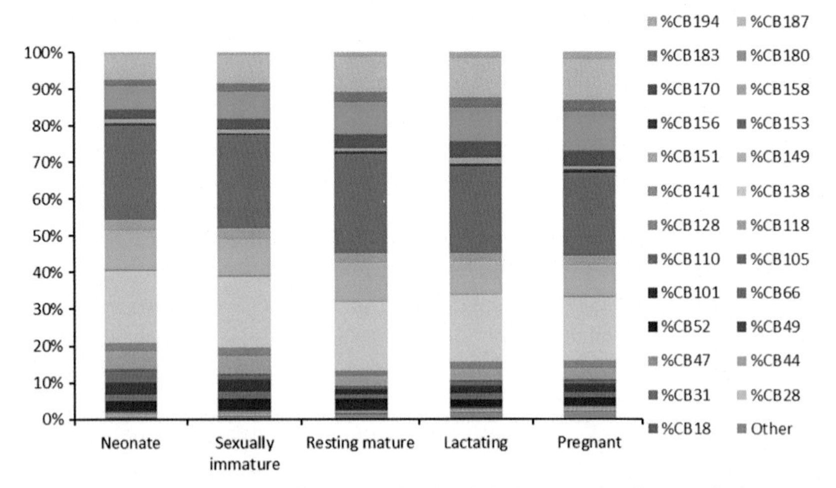

FIGURE 1.1 PCB congener profiles in 278 female UK harbor porpoises by reproductive status. Neonates $n = 14$, sexually immature $n = 145$, resting mature $n = 54$, lactating $n = 28$, and pregnant $n = 37$. *(Raw data taken from Murphy et al. (2015).)*

PCB153, -138, and -149 are SAG 1, 2, and 5 congeners, respectively, and are considered nonbiotransformable in cetaceans (Boon et al., 1994; Yordy et al., 2010). For SAG 3 congeners (PCB28, -31, -66, -105, -118, and -156) that are metabolized by CYP1A1-mediated enzymes in cetaceans (Boon et al., 1994; Yordy et al., 2010), relatively low concentrations were observed within the blubber tissue of all maturity groups. Other cetaceans have also shown an accumulation with age of metabolically refractory PCBs, but not for those PCBs that are subject to metabolism (e.g., bottlenose dolphins; Yordy et al. 2010).

There has been continued debate over the use of thresholds (below which no adverse effect occurs) in human studies, as not all people are equally sensitive to a particular dose, so a graded response is expected (Zoeller et al., 2014). Further, there are so many confounding factors that have to be taken into account in human toxicity studies that dose thresholds are impossible to prove or disprove (Zoeller et al., 2014). Namely, these are accounting for additive, synergistic, or nonadditive effects between pollutants, and also other stressors, which leads to difficulties in identifying chronic and sublethal responses for specific pollutants, particularly where adverse responses show delayed latent effects. Similar problems have been observed in marine mammal toxicity studies, so few attempts have been made at understanding the exposure-response relationship, and for the most part, proposed thresholds have relied on experimental studies on surrogate species. Kannan et al. (2000) proposed a toxicity threshold concentration of 17 mg/kg PCB lw (for Aroclor 1254; or 9 mg/kg for ΣPCBs determined by Jepson et al., 2016) for the onset of physiologic (immunological and reproductive) endpoints in marine mammals, and it was based on observed effects in experimental studies on seals, otters, and mink (Kannan et al., 2000). These endpoints (responses to a hazard) were for sublethal effects, causing morbidity, i.e., impairing the health of the animal through disruption of thyroid hormones, suppression of natural killer cell activity, etc. Higher-level exposure to certain PCBs may result in more severe immunosuppression, increasing risks of opportunistic infections that may ultimately lead to death (Hall et al., 2006a). Other studies that have applied this threshold to cetacean species noted that caution was warranted due to differing sensitivities among species; however, it did provide a benchmark for which to assess if PCB exposure was biologically significant (Hickie et al., 2007; Jepson et al., 2005, 2016; Murphy et al., 2015). It should be noted that threshold levels for adverse health effects during critical periods in development would be lower than those proposed for adults (Borrell and Aguilar, 2005), and Hall et al. (2006b) estimated a threshold of 10 mg/kg for calf survival in bottlenose dolphins from Sarasota Bay.

CASES OF REPRODUCTIVE FAILURE AND DYSFUNCTION ASSOCIATED WITH EXPOSURE TO OCs IN CETACEANS

The effects of EDCs may be expressed as infertility, cancer, or other types of diseases, and if EDCs interfere during critical periods of development, these effects may not be observed until later in life (Bergman et al., 2013). Proposed reproductive toxicity endpoints for EDCs in animals and humans include spontaneous abortions, early puberty, disorders of lactation and ovulation, including polycystic ovarian syndrome and premature ovarian failure, ovarian and mammary gland tumors, and other reproductive tract disorders/diseases including endometriosis and uterine fibroids, as well as leiomyomas and vaginal adenocarcinomas (reviewed in Diamanti-Kandarakis et al., 2009).

Within marine mammals, few cases of reproductive failure and dysfunction have been associated with exposure to OCs, and even fewer in cetaceans. This is primarily due to confounding factors and difficulties in identifying the mechanism or mode of action for specific pollutants. For those cases of reproductive failure and reproductive dysfunction that have been documented, the etiologies of the observed conditions have usually been uncertain (Reijnders, 2003). In cetaceans, observed cases of reproductive failure that have been associated with high exposure to OCs include fetal and/or newborn mortality in harbor porpoises (Murphy et al., 2015) and increased firstborn calf mortality (stillbirths or neonate mortality) in bottlenose dolphins, where PCB-related risk of reproductive failure in primiparous females ranged from 60% to 79% among three populations (Schwacke et al., 2002; Wells et al., 2005). Evidence of reproductive abnormalities and disorders associated with exposure to OCs include the presence of ovarian luteinized cysts in striped dolphins (Munson et al., 1998) and the development of cancer of the reproductive system and hermaphroditism in beluga whales (*Delphinapterus leucas*) (Béland et al., 1993; De Guise et al., 1994a; Martineau et al., 2002).

Beluga whales inhabiting the St. Lawrence Estuary in Canada are one of the most heavily contaminated cetacean populations worldwide, exposed to a variety of chemicals, including PCBs and DDT, as well as polycyclic aromatic hydrocarbons (PAHs) (Martineau, 2012). This threatened (COSEWIC status 2004) population has the highest prevalence of cancer in cetaceans, and it is the second leading cause of death with tumors (primarily of the digestive tract) identified in 27% of sampled adult whales (Martineau et al., 2002; McAloose and Newton, 2009). Putative causes include environmental exposure to carcinogenic substances and/or reduced immune function due to limited genetic diversity (Martineau et al., 2002; Newman and Smith, 2006). Evidence of reproductive disease in females between 1983 and 1999 included three cases of ovarian tumors (see Table 1.1), and adenocarcinomas were observed in mammary gland tissue of a further three individuals and the uterus of another (Martineau et al., 2002). Two hermaphrodite cases were documented out of 94 animals examined. These include an atypical bilateral true hermaphrodite, which was attributed to hormonal disturbance in early pregnancy (De Guise et al., 1994a), and a male pseudohermaphrodite (Reijnders, 2003). A follicular cyst was found on the left ovary of the true hermaphrodite measuring 6 cm × 6 cm × 5 cm (De Guise et al., 1995). Cases of adrenal and thyroid proliferative and degenerative lesions (adenomatous hyperplasia and follicular cysts) were observed (reviewed

TABLE 1.1 Individual Cases of Ovarian Neoplasms in Cetaceans

Ovarian Neoplasms	Cetacean Species	Location	Reference
Dysgerminoma (*germ cell tumor*)	Dusky dolphin	Peru	Van Bressem et al. (2000)
Dysgerminoma[a]	Beluga Whale	St. Lawrence Estuary	De Guise et al. (1994b) and Martineau et al. (2002)
Granulosa cell tumor (*sex cord-stromal tumor*)	Harbor porpoise	German North Sea	Seibel et al. (2012)
Granulosa cell tumors	Beluga whale	St. Lawrence Estuary	Martineau et al. (1988)
Granulosa cell tumor	Beluga whale	St. Lawrence Estuary	De Guise et al. (1994b)
Granulosa cell tumor	Pilot whale	Japan	Benirschke and Marsh (1984)
Granulosa cell tumor	Fin whale	Antarctica	Geraci et al. (1987) and Rewell and Willis (1950)
Granulosa cell tumor	Fin whale	Antarctica, S. Georgia	Geraci et al. (1987) and Rewell and Willis (1950)
Granulosa cell tumor	Blue whale	Antarctica, S. Georgia	Geraci et al. (1987) and Rewell and Willis (1950)
Ovarian carcinoma (granulosa cell tumor?[b])	Fin whale*	Antarctica	Geraci et al. (1987) and Stolk (1950)
Mucinous cystadenoma (*epithelial tumor*)	Blue whale	Antarctic	Geraci et al. (1987) and Rewell and Willis (1950)

[a]*Originally classified as a granulosa cell tumor and reclassified as a dysgerminoma by Martineau et al. (2002).*
[b]*Reclassified as a (possible) granulosa cell tumor in Geraci et al. (1987).*

in Martineau, 2012), as well as incidences of mastitis (De Guise et al., 1995). Evidence of reproductive failure was also apparent in the population, as the proportions of calves and juveniles were lower than less polluted populations off Alaska (Martineau et al., 1987). Compromised milk production (inflammatory change and cancer) was reported in 41% of stranded females (Martineau et al., 1994), which may have impacted fecundity rates. For all cases of reproductive failure and dysfunction, a contaminant-based etiology was proposed (Martineau, 2012; Martineau et al., 2002), although the mechanisms of action were not established. PAHs more so than PCBs may have been involved in the etiology of cancer in the observed cases (Martineau et al., 2002).

Multiple luteinized cysts were observed on the ovaries of four striped dolphins (out of 56 examined) that died during a morbillivirus epizootic (Munson et al., 1998). A mass die-off of more than 1000 striped dolphins occurred in the Mediterranean Sea between 1990 and 1992, and it was suggested that PCBs and other OC pollutants with the potential for immunosuppressive effects may have triggered the event, or enhanced its spread and lethality (Aguilar and Borrell, 1994). It is not known if the presence of ovarian luteinized cysts and lower fecundity, demonstrated by a high number of abortions, were caused by high PCB levels, the morbillivirus infection, or a combination of the two (Munson et al., 1998). Luteinized cysts occur when ovulation is impeded and were potentially caused by the effects of PCBs or morbillivirus on hypothalamic/pituitary function, or PCBs on ovarian responsiveness (Munson et al., 1998).

The Northeast Atlantic harbor porpoise population exhibits a lower pregnancy rate and longer calving interval than other conspecific populations, which could be resulting from PCB-mediated effects (Murphy et al., 2015). A decline in blubber ΣPCBs concentrations was observed in UK porpoises in the mid-1990s; however, this plateaued after 1998 (Jepson et al., 2016; Law et al., 2012), whereas levels of blubber ∑DDT showed a significant decline from the early 1990s onward, reflecting a lack of fresh input into the marine environment of this agricultural pesticide (Law et al., 2012). Murphy et al. (2015) assessed for evidence of reproductive failure and reproductive dysfunction in the population using samples and data collected over a 22-year period and from 329 female UK stranded porpoises. Twenty-five of 127 (19.7%) mature females showed direct observations of reproductive failure including fetal death, spontaneous abortion, dystocia, and stillbirth. Further, 16.5% (21 of 127) of mature females had infections of the reproductive tract or tumors of reproductive tract tissues that could contribute to reproductive failure. These included malignant tumors such as cervix squamous cell carcinoma, benign tumors such as leiomyoma, papilloma-like lesions and vaginal plaques, endometritis, and other infections and inflammations of the reproductive tract, some of which were previously reported as toxicity endpoints in animals and humans (Caserta et al., 2008; Diamanti-Kandarakis et al., 2009; Herbst et al., 1971; Murphy et al., 2015; Padmanabhan et al., 2010; Steinberg et al., 2008). Difficulties arose in showing casual associations between cases of reproductive dysfunction and ΣPCBs concentration due to the female's capability in offloading lipophilic pollutants (Murphy et al., 2015). However, 47% of females had ∑PCB concentrations above Kannan's threshold for the onset of adverse health effects, which included 52% of sexually immature and 53% of resting (not pregnant or lactating) mature individuals. Resting females were more likely to have higher ΣPCB burdens than other maturity groups (apart from sexually immature individuals), and where data were available, these nonoffloading females were previously gravid, which suggests fetal or newborn mortality (Murphy et al., 2015).

Above all, these studies on cetaceans have not shown a cause-effect relationship related to a particular mechanism, nor a causal relationship between an observed disorder and exposure to certain OCs (Reijnders, 2003). A semifield experimental study on harbor seals (*Phoca vitulina*) did show a possible mechanism for reproductive failure following exposure to PCBs. Between the 1950s and 1970s, average pup production declined by c. 30% in harbor seals inhabiting the Dutch Wadden Sea, and PCB concentrations were significantly higher (by five to seven times) than contiguous populations (Reijnders, 1980). In an experimental setup, seals fed fish from the Wadden Sea showed a decreased reproductive rate (by 50%), and lower levels of oestradiol-17β around the time of implantation, compared to a control group (Reijnders, 1986). Mean PCB levels in the control group ranged from 5 to 11 mg/kg lw, compared to 25–27 mg/kg lw in seals fed contaminated fish (Pierce et al., 2008). Lower levels of estradiol could have impaired endometrial receptivity and prevented successful implantation of the blastocyst, and it was proposed that enhanced OC-induced CYP enzyme activity may have lowered circulating estradiol (Reijnders, 2003). Exposure to PCBs has also been linked to the development of genital cancer in California sea lions (*Zalophus californianus*). A metastatic genital carcinoma was reported in 18% (66 of 370 individuals) of stranded sexually mature individuals between 1979 and 1994 (Gulland et al., 1996; Lipscomb et al., 2000). The cause of urogenital carcinoma in the species is multifactorial, and possible potential causal factors include an otarine herpesvirus-1 (Buckles et al., 2006; King et al., 2002) and infection with β-haemolytic streptococci (Johnson et al., 2006), and it has been proposed that there is a genetic basis as a result of inbreeding depression (Browning et al., 2017). Exposure to pollutants may be an additional factor, as a higher mean concentration of OCs was reported in blubber tissue of females with genital carcinoma relative to those without genital carcinoma (>85% and 30% higher in relation to PCBs and DDT, respectively; Ylitalo et al., 2005). This demonstrates an association between OCs and carcinoma, through possibly affecting the prevalence of carcinomas by acting as immunosuppressive agents or by genotoxic mutation and tumor promotion (Ylitalo et al., 2005). A follow-up study suggested that OCs interact with steroid hormone receptors in the intraepithelial lesions and that alternations in the expression of p53 may also play a role (Browning et al., 2017; Colegrove et al., 2009).

CASE STUDY OF SHORT-BEAKED COMMON DOLPHIN IN THE NORTHEAST ATLANTIC

Within the Northeast Atlantic, a large proportion of necropsied short-beaked common dolphins presented with blubber PCB burdens exceeding Kannan's threshold level for the onset of adverse physiologic health effects (Murphy et al., 2010).

Similar to harbor porpoises in the region, the short-beaked common dolphin Northeast Atlantic population exhibits a lower pregnancy rate and longer calving interval than other conspecific populations, which could be resulting from PCB-mediated effects (Murphy et al., 2009, 2010).

Short-beaked common dolphins are widespread in the Northeast Atlantic, ranging from waters off Norway to Africa. One population has been reported to exist ranging at least from Scotland to Portugal, with a separate population in the Mediterranean Sea (Mirimin et al., 2009a,b; Murphy et al., 2009, 2013; Natoli et al., 2008). Common dolphins have been observed out to the Mid-Atlantic Ridge, but due to a lack of genetic sampling in offshore waters, the range of the Northeast Atlantic population is unknown (Murphy et al., 2013). The most recent abundance estimate from SCANS III suggests at least 467,673 dolphins (CV = 0.26) in continental shelf and adjacent waters (Hammond et al., 2017). Female common dolphins in the population attain sexual maturity at an average length of 188.8 cm and an average age of 8.2 years, with a maximum age of 30 years and a low annual pregnancy rate of 26% reported (Murphy et al., 2009, 2010).

Previous research assessing reproductive effects from exposure to POPs in the region reported that individual feeding history (proxied by blubber fatty acid profiles) was the most important variable explaining individual POP profiles (Pierce et al., 2008). This former EU Fifth Framework funded study called BIOCET analyzed samples collected from stranded and bycaught common dolphins in Irish, Scottish, French, and (Galician) Spanish waters during the period 2001–03. Common dolphins sampled in both French and northwestern Spanish waters had significantly higher PCB concentrations than Irish animals; however, the Galician common dolphin pollutant sample was mainly composed of sexually immature females that had not offloaded their pollutant burdens (>65%). Within the whole mature sample, incidences of pregnancy in individuals was negatively related to blubber PCB and polybrominated diphenyl ether concentrations, which may suggest that higher POP concentrations were inhibiting successful reproduction, though infertility due to other causes may have caused higher POP levels to bioaccumulate (Pierce et al., 2008).

Further analysis of BIOCET data revealed that ovarian corpora number (corpus luteum and albicans) significantly increased with PCB burdens in sexually mature *D. delphis* (Murphy et al., 2010). Approximately 83% of female common dolphins with PCB concentrations exceeding Kannan's threshold level for the onset of adverse physiologic health effects were resting mature individuals with high numbers of ovarian corpora. This suggests that due to high contaminant burdens, females were unable to successfully reproduce (offload) and thus continued ovulating; or some females were not reproducing for other reasons, either physical or social, and therefore accumulated higher levels of PCBs (Murphy et al., 2010). Within the BIOCET sample, 92% of sexually mature (all but two were resting) female common dolphins with PCB burdens above Kannan's threshold level and corresponding high corpora counts (≥15 scars) were obtained from a mass live stranding event of a nursery group at Pleubian, France, in February 2002. Genetic analysis of this nursery group did not reveal evidence for a matriarchal system, as a lack of genetic relatedness among mature individuals was reported (Viricel et al., 2008). The existence of nonreproductive females (based on high contaminant loads and high numbers of ovulations) within this nonmatriarchal female nursery group is remarkable; though, it should be noted that only 52 of the whole mass-stranded group was sampled for genetic analysis, and approximately 50 other individuals were released alive offshore (Viricel et al., 2008).

For animals that stranded as part of the Pleubian mass stranding event, there are no available data pertaining to assessments of previous gravidity or general health status. The possibility of either social or pollutant suppression of reproduction in common dolphins was investigated further using a control group composed of stranded and bycaught individuals collected in English and Welsh waters (Murphy et al., 2010). The control group was composed of "healthy" common dolphins, and in contrast to the BIOCET, common dolphin data results revealed a negative relationship between PCBs and DDT concentrations and increasing corpora number. This suggests that some "healthy" females may go through a large number of (infertile) ovulations prior to a successful pregnancy, birth, and survival of their firstborn offspring during early lactation, when females offload the majority of their OC burden (Murphy et al., 2010).

The current case study aims to further examine the effects of pollutants such as PCBs on reproduction in common dolphins in the Northeast Atlantic using stranded and bycaught animals sampled in English and Welsh waters. Evidence of reproductive failure and reproductive dysfunction, such as abnormalities and disorders, will be assessed, and their association with exposure to PCBs will be investigated.

Study Design

Stranded and bycaught common dolphins were collected and necropsied between 1990 and 2013 by the UK Cetacean Strandings Investigation Programme (CSIP). All animals were in fresh-to-moderate decomposition on necropsy. Cause of death (COD) was determined by specific diagnostic criteria (see Deaville and Jepson, 2011; Jepson et al., 2005), and individuals were categorized into three COD groups: infectious disease, trauma (bycatch, boat/ship strike, bottlenose dolphin attacks, and dystocia), and others (live stranding, starvation, neoplasia, and not established) (after Murphy et al., 2015). Age was determined by counting growth layer groups in the dentine of teeth samples (after Murphy et al., 2014).

Fixed ovarian samples were assessed externally for the presence of ovarian corpora then hand-sectioned into 0.5–2 mm slices and examined internally under a binocular microscope for the presence of additional corpora scars. Total numbers of corpora (number of ovulations) were counted, which was used as an index of reproductive activity (after Murphy et al., 2010). Females were classified as sexually mature if one or more ovarian corpora (corpus luteum or albicans) were present. Pregnancy was established by the presence of an embryo/fetus and an active (nonregressing) corpus luteum, confirmed by histological assessment. Mammary glands were examined for evidence of lactation via gross examination and, in some individuals, histological assessment of mammary gland tissue. Females were classified into five reproductive states: (1) sexually immature, (2) pregnant (fetus present), (3) pregnant and lactating, (4) sexually mature and lactating, and (5) resting mature (not pregnant or lactating) (after Murphy et al., 2010).

The reproductive tracts were assessed for abnormalities and evidence of sterility and infertility, including tumors, uterine stenosis, occlusions and leiomyomas, endometriosis, and vaginal calculi. Ovaries were examined for ovarian cysts and tumors through gross and histopathological assessment. The sample was also assessed for evidence of hermaphroditism and other disorders of genital development. Ovarian lesions and other abnormalities were measured, and shape, color, texture, and position were recorded. Assessment of previous gravidity in mature females and cases of reproductive failure (including fetal death, dystocia and stillborn, recently aborted and aborted, and her calf did not survive) followed classifications outlined in Murphy et al. (2015). Incidences of disease and infection of the reproductive system were confirmed by bacteriology, virology, and histopathology assessments.

Dorsal blubber samples were analyzed by the Centre for Environment, Fisheries, and Aquaculture Science Laboratory for quantification of hexane extractable lipid and wet weight concentrations of 25 individual chlorobiphenyls, and the sum of the concentrations of these chlorobiphenyl congeners (ΣPCB mg/kg lw) was determined (see Jepson et al., 2016; Law et al., 2012; Murphy et al., 2015 for further information). Two toxicity thresholds were applied to the pollutant data, after Jepson et al. (2016). Toxicity threshold concentration was 9 mg/kg lw (as ΣPCB; determined by Jepson et al., 2016) for the onset of physiologic (immunological and reproductive) endpoints in marine mammals (Kannan et al., 2000), and one of the highest PCB toxicity thresholds reported in marine mammal toxicology studies, 41 mg/kg lw (determined for ΣPCB by Jepson et al., 2016 and based on 77 mg/kg for *Clophen 50*), was associated with profound reproductive impairment in Baltic ringed seals (*Pusa hispida*) (Helle et al., 1976). A high prevalence of reproductive tract lesions were observed in Baltic seals, including both ringed and grey seals (*Halichoerus grypus*), during the 1970s and 1980s. These lesions included occlusions and stenosis of the uterine horns (42% of grey seals) and benign uterine leiomyomas (53% of grey seals >4 years) (Bergman, 1999; Bredhult et al., 2008; Helle et al., 1976), which rendered females partially or completely sterile for life (Reijnders, 2003). As some females retained fetal membranes aligned to stenosed sections of the uterine horns, obtrusions may have developed as a result of PCB (or PCB metabolites) induced abortions, or fetal death (Bergman, 2007; Bergman and Olsson,

1985; Helle et al., 1976), and in some seals presenting with obtrusions, purulent endometriosis was also reported (Bergman, 2007). However, unequivocal evidence for a cause-effect relationship was not provided (Reijnders, 2003). Follow-up work using an exposure index indicated that PCBs, more than DDT, were associated with the increased incidence of uterine leiomyomas in Baltic grey seals (Bredhult et al., 2008).

Incidences of Reproductive Abnormalities and Disorders in Common Dolphins

A sample of 107 female common dolphins that stranded along the English and Welsh coastlines were assessed for evidence of reproductive failure and dysfunction. Samples included those from animals that died during a mass live stranding event that occurred in the southwest of the United Kingdom in 2008 ($n=13$ females; Jepson et al. 2013) and a control group of "healthy" female *D. delphis* ($n=43$) collected between 1992 and 2004 along the southwest of the United Kingdom (previously assessed by Murphy et al., 2010). The control group was composed of stranded females whose COD was attributed to incidental capture in fishing gear. Routine general health status assessments revealed that individuals were not suffering from any infectious or noninfectious diseases that might inhibit reproduction (Murphy et al., 2010). The remaining dolphins stranded between 1990 and 2013 ($n=51$) had causes of death that were attributed to trauma ($n=17$), infectious disease ($n=2$), other ($n=26$), or were not established ($n=6$).

Of the 107 animals, 44 were sexually immature and 63 were sexually mature. No incidences of uterine stenosis, occlusions, or leiomyomas were observed within the common dolphin sample (reproductive toxicologic endpoints previously reported in Baltic seals) (Bergman, 1999; Helle et al., 1976). However, 18 proposed cases (16.8% of the whole sample) of reproductive system pathologies were identified. Six females presented within vaginal calculi (5.6%), six females exhibited suspected precocious mammary gland development (5.6%), and there were three possible cases of ovarian tumors (2.8%) (see Table 1.2). The three cases of ovarian neoplasms include a possibly fibroma, a well-differentiated granulosa cell tumor, and a rare case of either a mesothelioma or Schwannoma tumor. In addition to the preceding, females presented with an ovarian cyst, atrophic ovaries, and the first reported case of an ovotestis in a cetacean species. In the majority of cases of reproductive dysfunction, available blubber ΣPCB data were above the toxicity threshold concentration of 9 mg/kg lw (Table 1.2). The only other study assessing evidence of disease and lesions of the reproductive tract in common dolphins was undertaken on the long-beaked common dolphin (*D. capensis*) in the southeast Pacific. Within a sample of 24 bycaught females (14 mature and 10 immature), one case of reproductive dysfunction was observed, which was a case of ovarian cysts (a prevalence of 4.2% of the whole sample; Van Bressem et al., 2006).

TABLE 1.2 Abnormalities and Dysfunction of the Reproductive System in 107 UK Female *D. delphis* Sampled Between 1990 and 2013

Abnormality	Code	COD	Body Length (cm)	Age (year)	Maturity Status	Corpora Scar No.	Previously Gravid	ΣPCB lw
Reproductive Tract								
Vaginal calculi	SW1993/43	BY	195	7	Immature	0	No	27.5
Vaginal calculi	SW1995/4	NE	207	>7	Resting mature	13	NA	NA
Vaginal calculi	SW1996/98	NE	201	>18	Resting mature and lactating	23	Yes	NA
Vaginal calculi	SW2004/131	BY	205	>10	Resting mature	c.34	Yes	41.1
Vaginal calculi	SW2004/336	LS	196	25.5	Resting mature	19	Yes	NA
Vaginal calculi	SW2005/65	LS	196	>10	Resting mature	21	Yes	49.5
Suspected precocious mammary gland development	SW1994/160	BY	177		Immature	0	No	39.2
Suspected precocious mammary gland development	SW2008/94.3	LS	163	3	Immature	0	No	17.6
Suspected precocious mammary gland development	SW2008/94.4	LS	185	5	Immature	0	No	19.4

Precocious mammary gland development	SW2008/94.15	LS	169	7	Immature	0	No	5.72
Precocious mammary gland development	SW2011/562	LS	183	NA	Immature	0	No	14.7
Suspected precocious mammary gland development	SW2013/594	LS?	143	1	Immature	0	No	13.44
Ovary								
Possible granulosa cell tumor	SW1994/24	NE	174	2	Immature	0	No	22.6
Possible fibroma	SW1994/147	BY	192	c. 14	Resting mature	12	Yes	NA
Mesothelioma or Schwannoma tumor	SW1998/148	LS	209	24	Resting mature	26	Yes	48.5
Ovarian cyst	SW2005/2	LS	195	19	Resting mature	27	Yes	26.1
Ovotestis	SW2006/298a	NE	191	6	Immature	0	No	NA
Atrophic ovaries	SW2008/94.11	LS	206	18	Immature	0	No	45.4

BY, bycatch; *COD*, cause of death; *LS*, live stranding; *NA*, not available; *NE*, not established.

Cases of ovarian neoplasms in cetaceans are infrequently reported, with the majority being granulosa cell tumors, the archetypal feminizing sex cord-stromal tumor with cells resembling those of follicular granulosa or luteinized variants (Russell et al., 2009) (see Table 1.1). Other cases of reported cetacean ovarian neoplasms include a dysgerminoma and a mucinous cystadenoma, originating from germ and epithelial cells, respectively. In the current study, a possible ovarian fibroma, benign sex cord-stromal tumor, was observed in SW1994/147. The tumor measured 20 mm × 25 mm and consisted of reddish/black tissue resembling splenic tissue in gross appearance. The ovoid mass was attached to the hilus of the right ovary by a band of mesothelial connective tissue. Fibrous stroma, comprising the core of the mass, was moderately vascular and composed of spindle-shaped or more ovoid and irregular fibroblastic cells frequently orientated at random. These cells were densely packed toward the periphery of the mass and more loosely arranged centrally. Tissue autolysis and poor staining made final diagnosis uncertain. The individual in question was a bycaught 14-year-old resting mature female in moderate nutritional condition.

SW1994/24, a 2-year-old female common dolphin, presented with a well-differentiated granulosa cell tumor in one ovary. The ovarian stroma had many primary follicles and a few secondary follicles, and it contained a vaguely wedge-shaped zone of tissue opposite the hilus that showed irregular tubule-like structures lined with a single layer of polygonal to columnar cells (Murphy et al., 2018b). The tumor was unilateral, as a transverse cross-section of the contralateral ovary revealed only normal follicular development within the ovarian stroma. A third ovarian neoplasm was reported in SW1998/148, a resting mature female that live stranded in moderate nutritional condition. The ethology of the observed tumor on the left ovary remains ambiguous due to few typical histological features and atypical immunostaining patterns, which are common for very poorly differentiated tumors (Murphy et al., 2018b). Results from histological examination and the presence of psammoma bodies indicates either a mesothelioma or Schwannoma tumor, though other tumors types were not ruled out (Murphy et al., 2018b).

The first reported case of an ovotestis in a cetacean species was observed within the sample. A 6-year-old sexually immature female (external phenotype and female reproductive tract) common dolphin (SW2006/298a) that was found stranded on the southwest coast of the United Kingdom was diagnosed as a true hermaphrodite (Murphy et al., 2011). The individual in question had one ovotestis containing both ovarian follicles and testicular tubular elements and a contralateral ovary. Ovarian portions of the ovary appeared normal, demonstrating follicular development, but the testicular tissue tubular elements observed in the medulla (6.8 mm × 5.5 mm) presented with hypoplasia and degeneration (Murphy et al., 2011). COD for this female was not fully established due to scavenger damage, though incidental capture in fishing gear was not ruled out. It is not known if this disorder of genital development was due to abnormalities of genetic or chromosomal origin or inappropriate hormone exposure: blubber samples were not available for retrospective pollutant analysis.

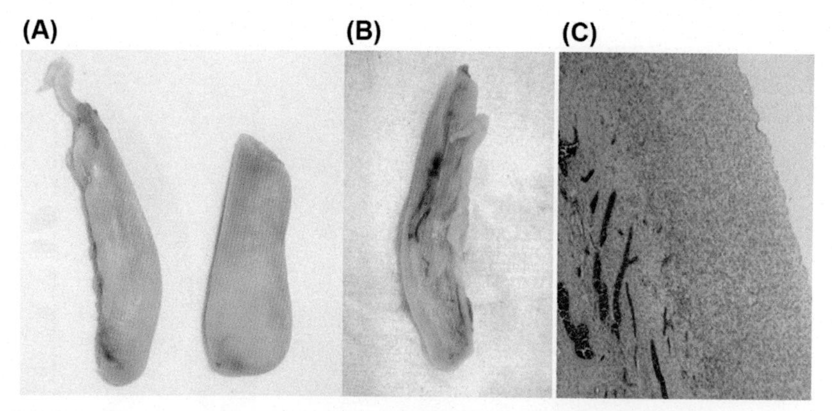

FIGURE 1.2 (A and B) Atrophic ovaries in a 17-year-old common dolphin (SW2008/94.11) that stranded as part of a mass live stranding event in 2008. Note a lack of follicles in the stroma in image (C).

A female common dolphin (SW2008/94.11) from the mass live stranding event sample presented with atrophic ovaries, with only a few follicles present in both ovaries (Fig. 1.2). Although the female was 18 years old, the lack of follicles was not attributed to reproductive senescence, as no ovarian corpora (corpus luteum or albicans) were present. This sexually immature individual was in good nutritional condition and health status, though it presented with a ΣPCB burden of 45.4 mg/kg, which is greater than the threshold for profound reproductive impairment in Baltic ringed seals. Reproductive senescence is rarely observed in cetaceans (Hohn et al., 2007), and not in common dolphins (Murphy et al., 2009). Studies on cetaceans have rarely reported a lack of ovarian follicles or pathologic changes that would result in senescence (Hohn et al., 2007).

A large, thin-walled fluid-filled cyst was noted on the lateral aspect of the left ovary of SW2005/2 (Fig. 1.3). The right ovary exhibited signs of hypostasis with noted congestion, which may be due to the live stranding event. The maximum length of the cyst was c. 47 mm, while the maximum length of the left ovary was only 43 mm, and ovaries presented with a corpora count of 24. This 19-year-old resting mature dolphin was in good nutritional condition, though live stranded in the month of January, i.e., outside the mating period. Nothing abnormal was detected in the uterus or vagina upon gross assessment. The prevalence of ovarian cysts in the mature sample was 1.7%. Ovarian cysts, such as follicular and luteinized cysts (though mainly the former), have been reported in the ovaries of the striped dolphin, long-beaked common dolphin, dusky dolphin (*Lagenorhynchus obscurus*), Pacific white-sided dolphin (*Lagenorhynchus obliquidens*), short-finned pilot whale (*Globicephala macrorhynchus*), and southern minke whale (*Balaenoptera bonaerensis*), among others (Lockyer, 1987; Marsh and Kasuya, 1984; Munson et al., 1998; Robeck et al., 2009; Van Bressem et al., 2000). In cattle, deviation of the preovulatory

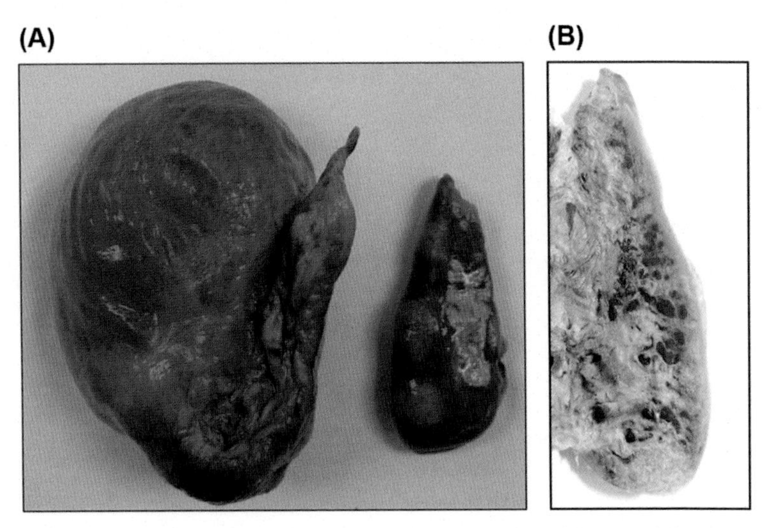

FIGURE 1.3 (A) Lateral fluid-filled ovarian cyst on the left ovary of live stranded common dolphin SW2005/2; cyst measured 47 mm in length; (B) cross-section of congested right ovary.

surge of luteinizing hormone, either the absence or mistiming of the surge, is thought to cause follicular cysts (Kennedy and Miller, 1993; McEntee, 1990; Van Bressem et al., 2006). Though the cyst observed in the current study was not luteinized, luteinized cystic follicles have been known to produce estrogen and progesterone (Robeck et al., 2001).

Of the 107 female common dolphins assessed within the current study, six presented with vaginal calculi within their reproductive tracts (see Table 1.2). These included one sexually immature, four resting mature, and one sexually mature and lactating individual. Corpora count numbers ranged from 13 to 34 in mature individuals presenting with calculi. As common dolphins can possibly ovulate up to five times within one estrus period and have a reproductive lifespan of between 10 and 20 years (Murphy, 2004; Murphy et al., 2009, 2010), this suggests that reproductive dysfunction (large number of ovulations without fertilization) may have occurred in some individuals, conceivably due to the relative size of the calculi (100–490 g within the same) causing an occlusion of the reproductive tract. Some vaginal calculi in cetaceans are believed to represent an incomplete abortion, with retention of part or all of a fetus that subsequently crystallized and coalesced (Benirschke et al., 1984; Woodhouse and Reinne, 1991). Others are possibly produced by a fetal bone that acted as a nidus for calcium phosphate deposition (Benirschke et al., 1984; Woodhouse and Reinne, 1991).

SW2004/131, a bycaught resting mature female who presented with a calculi and evidence of previous gravidity, had the highest corpora count within the whole sample. This bycaught female was in good nutritional condition and health status, though it exhibited adrenal cysts (bilateral). ΣPCB burden in this individual was high at 41.1 mg/kg, suggesting this mature female may never have offloaded her pollutant burden via transplacental and lactational transfer. In SW1996/98, the anterior vagina contained five highly polished faceted calculi that could be assembled into an egg-shaped mass, and analysis revealed they were mainly composed of phosphate salts. Baker (1992) noted a vaginal calculus in a UK common dolphin collected prior to 1992, which was also composed of phosphate salts. Within the current study, a sexually immature female common dolphin (SW1993/43) bore a calcified body within her vagina. This 7-year-old bycaught individual was in good nutritional condition, though it had a high ΣPCB burden (27.3 mg/kg), and its ovaries were heavily congested. This is not the first study to report a vaginal calculus in a sexually immature cetacean. Van Bressem et al. (2000) observed a struvite (not calcium phosphate) calculus in an immature Peruvian dusky dolphin, and suggested an infectious etiology. McFee and Osborne (2004) also reported that a urinary tract infection might have been the underlying cause of a struvite (magnesium-ammoniumphosphate hexahydrate) calculus observed in the vagina of a sexually immature 4-year-old bottlenose dolphin. The presence of struvite calculi (>30) was also reported in a 1-year-old harbor porpoise, which was attributed to mucosal hyperplasia of the distal urogenital tract (Norman et al., 2011).

Six sexually immature females presented with precocious mammary gland development, with individuals ranging in age from 1 to 7 years. Interestingly, three of these dolphins (SW2008/94.3, SW2008/94.4, SW2008/94.15) mass live stranded together in 2008, possibly due to naval activity in the region causing an acoustic disturbance (Jepson et al., 2013). A small quantity of yellow viscous (colostrum-type) milk was present in the mammary glands of all three females, and histological examination revealed mature and minimally lactating mammary gland tissue in the 7-year-old SW2008/94.15, not assessed in the other two dolphins. Premature mammary gland development has been linked to (gestational) pollutant exposure in other animals, including precocious mammary gland development in immature female rats (Fenton, 2009; Moon et al., 2007). Although not all individuals were fully assessed, three cases of mastitis were observed in the sample. SW1996/114, a sexually immature female, presented with mild, multifocal, subacute-chronic mastitis. SW2002/224, another immature individual, presented with enlarged mammary glands and purulent mastitis arising from an infection near the mammary gland tissue. Finally, multifocal chronic inflammation of the lactiferous ducts (mastitis) was observed in a nonlactating mature common dolphin (SW1999/15).

Effects of Contaminants on Reproduction in Common Dolphins

Within the control group of "healthy" bycaught females, all ovarian corpora (lutea and albicantia) were assessed, counted, and where uncertainties existed, verified through histological examination, so the reproductive status of some females has been updated since Murphy et al. (2010). The sample of 43 "healthy" females was composed of 20 immature, 11 resting mature, 1 primiparous pregnant (ovaries contained only one corpus, which was a corpus luteum), 2 pregnant and lactating, 7 mature and lactating, and 2 nonlactating primiparous females that had recently miscarried/aborted. Histological examination of corpora lutea of pregnancy in both females that had recently miscarried/aborted showed clear evidence of regression, with a reduction in luteal cells and an increase in fibrotic connective tissue. Animals had recently miscarried/aborted during what would have been their second trimester based on a mean date of conception of the 19th July and a gestation period of 0.99 years for common dolphins in the Northeast Atlantic (Murphy et al., 2009).

All sexually immature (nulliparous) females (range 10.5–52.9 mg/kg lw), the "primiparous" pregnant female (44.8 mg/kg lw), and the two primiparous females that had recently miscarried/aborted (range 19.1–28.3 mg/kg) had blubber ΣPCB concentrations above the threshold level of 9 mg/kg lw for the onset of adverse health effects (Fig. 1.4A). Other "healthy" females with ΣPCB burdens above this threshold were resting females. Sexually mature females with levels above the 41 mg/kg threshold in the control sample where were either resting ($n=3$) or pregnant ($n=1$), and all three resting females had been previously gravid; and based on the high PCB burden, these females either aborted or their newborn offspring did not survive during early lactation. In common dolphins, an assessment of transfer rates of OC compounds was undertaken on one mother-calf pair, and it was estimated that 41.48% of total PCBs (and 55.41% of total DDT) were offloaded during the first half of lactation, and that complete transfer of the OC burden would possibly occur by the end of lactation (Borrell and Aguilar, 2005). All nine lactating females (i.e., females who successfully reproduced and offloaded) had ΣPCB concentrations ≤10.5 mg/kg, and eight of those females presented with concentrations ≤5.8 mg/kg. A negative relationship was observed between ΣPCB concentration and increasing corpora number in the control group of "healthy" females (see Fig. 1.4A), which suggests that some females may successfully reproduce and offload their pollutant burden after many unsuccessful ovulations/miscarriages and/or early calf mortality.

The noncontrol pollutant sample ($n=19$) comprised nine sexually immature, four resting, five lactating, and one primiparous pregnant female. In contrast to the control sample, a positive relationship was observed between ΣPCB concentration and increasing corpora number (see Fig. 1.4B), which were comparable results to the earlier BIOCET/Pleubian mass stranding pollutant study. What data in the current study suggest, however, is that PCBs may be impacting fetal/newborn survival, similar to that proposed for UK harbor porpoises (Murphy et al., 2015), and high ΣPCB burdens are not due to reproductive suppression of nonbreeding females. All four resting females in the noncontrol sample had a high number of ovarian corpora scars and ΣPCB concentrations ≥26 mg/kg; and of these individuals, three females presented with ΣPCB concentrations ≥41 mg/kg. Pollutant data suggest that these four resting females had not successfully reproduced (offloaded) in the past, though all were previously gravid. Further,

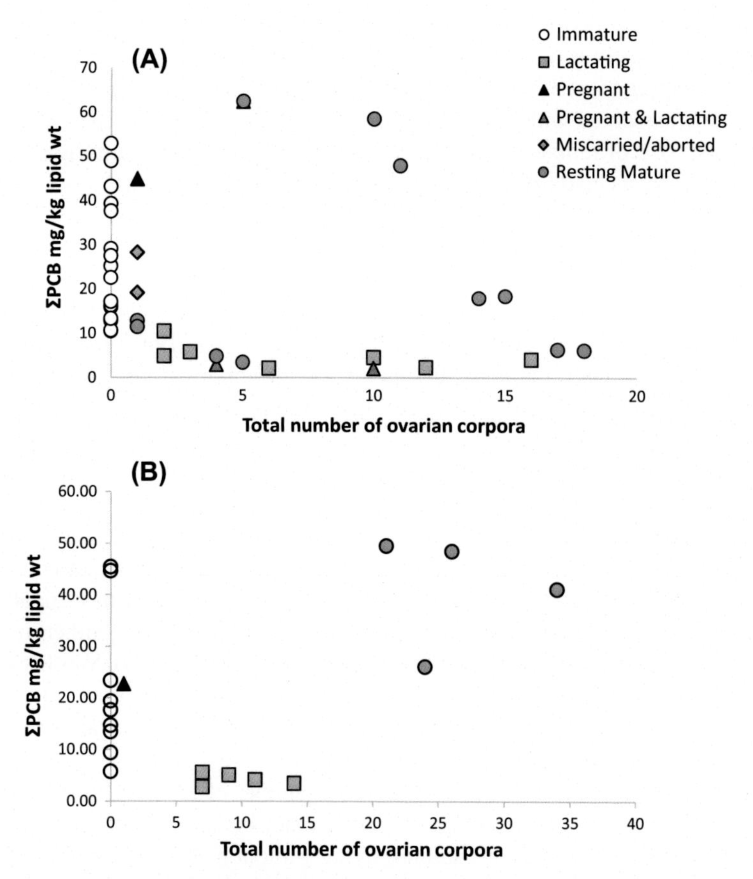

FIGURE 1.4 Blubber Σ25PCBs lipid weight as a function of ovarian corpora number in (A) *D. delphis* control group sample (n=43 bycaught individuals) and (B) *D. delphis* noncontrol sample (n=19).

all four resting females presented with reproductive system pathologies, including an ovarian tumor (n=1; SW1998/148), ovarian cyst (n=1; SW2005/2), and vaginal calculi (n=2; SW2004/131, SW2005/65); the calculi observed in these mature females may have impeded reproduction. The primiparous pregnant female (SW2011/295) was 6 years old and had a ΣPCB pollutant burden of 22.7 mg/kg lw. She presented with a male fetus measuring 88 cm in length in the left uterine horn and a dilated cervix. The female live stranded in the month of July, which may have been associated with an attempt to give birth to a full-term fetus.

Of the 13 females that mass live stranded together in Cornwall in 2008, blubber samples from 10 individuals were processed for pollutant analysis. The sample comprised five sexually immature and five sexually mature females. All five mature females were heavily lactating and presented with low ΣPCB pollutant burdens (<6 mg/kg lw). Three of the immature females presented with suspected precocious mammary gland development and blubber ΣPCB concentrations ranging from 5.7 to 19.4 mg/kg. A fourth immature female was 18 years in age and presented with atrophic ovaries and a high ΣPCB burden of 45.5 mg/kg. Thus, 31% of females assessed from this mass stranding event sample presented with suspected reproductive system pathologies.

PCB and Reproductive Implications for Common Dolphins

16.8% (18 out of 107) of female common dolphins presented with reproductive system pathologies. As 40% of the sample was composed of a control group of "healthy" bycaught animals, the incidences of abnormalities and disorders reported here may be somewhat lower than what is occurring within the wider population. Even so, the current study reported a higher incidence of reproductive system pathologies than observed for other small cetacean species elsewhere. One such similar study assessed for evidence of genital diseases in 264 Peruvian female dusky dolphins. Eleven mature females (4.1% of the whole sample) presented with an ovarian tumor, ovarian cysts, and uterine tumors, and a further three individuals (1.1% of whole sample) presented with vaginal calculi. A vaginal mass was also observed close to the left ovary in another mature female containing the skull of a fetus (Van Bressem et al., 2000). Van Bressem et al. (2000) attributed some of their cases of genital disease in dusky dolphins to reproductive senescence, as animals were large in size: age data were not available for all individuals.

Overall, there have been a relatively low number of reported incidences of neoplasms in cetaceans, resulting from either not being assessed due to autolytic changes or other reasons, animals dying without pathologic investigations, or individuals dying before attaining an older age, namely when most cancers occur (Newman and Smith, 2006). In humans, cancers increase in occurrence with age due, in some cases, to cellular changes and damage (DeGregori, 2012). The age profile of individuals assessed within the current study ranged from neonates to 30 years ($n=77$), though 87% were ≤20 years old. Of the female common dolphins showing evidence of ovarian lesions, individuals ranged from 2–24 years in age. However, many pathologies were reported in sexually immature females, including an ovarian granulosa cell tumor, ovotestis, atrophic ovaries, and precious mammary gland development, which suggests inherited and/or environmental causes.

As noted earlier in the chapter, a recent study on harbor porpoises in UK waters reported that at least 16.5% of females assessed had infections of the reproductive tract or tumors of reproductive tract tissues (Murphy et al., 2015). The number of cases of reproductive system pathologies was only a minimum estimate, as the study did not include occurrences of ovarian tumors and other ovarian disorders that were under investigation. Apart from ovarian abnormalities and lesions, vaginal calculi, precious mammary gland development, and mastitis, no other infections, diseases, or lesions of the reproductive tract were observed in female common dolphins in the current study. Results differing somewhat to harbor porpoises in the same region where cases of endometriosis, vaginitis, lesions of the vagina, uterus, and clitoris, as well as malignant tumors of the reproductive tract (such as squamous cell carcinoma of the cervix, uterine adenocarcinoma, and metastasizing adenocarcinoma) and benign tumors (including clitoral and vaginal papilloma-like lesions, vaginal epithelial plaques, and vaginal wall leiomyoma) were observed (Murphy et al., 2015). In contrast, lesions of the reproductive system were very rare in female harbor porpoises sampled from German waters, with one case of suppurative endometritis reported (Siebert et al., 2001).

High ΣPCB burdens were not inhibiting ovulation, conception, or implantation in common dolphins, as (all except one sexually immature [nulliparous] female and) the four primiparous females in the control and noncontrol samples had ΣPCB levels above the threshold level for the onset of adverse health effects (Murphy et al., 2010; current study). PCBs, however, may be impacting fetal and newborn survival, as some previously gravid resting females had not successfully offloaded their pollutant burdens. After removing the possible effects from stressors such as nutritional and immune issues (93% of individuals were regarded as "healthy" and were overall in good to moderate nutritional condition (Murphy et al. (2010)), evidence of reproductive failure was assessed both directly (e.g., through observations of fetal death/abortion) and indirectly by using individual ΣPCB burdens, after Murphy et al. (2015). Using these data, results suggested that reproductive failure could have occurred in approximately 30% (7 of 23) of mature females sampled in the control study. This was based on that fact that all lactating females (control and noncontrol samples) had ΣPCB burdens ≤10.5 mg/kg. Thus, resting females with pollutant loads >10.5 mg/kg more than likely had not successfully offloaded in the past, and where data were available, these females were previously gravid. Additionally, 8.7% of mature females in the control study showed evidence of recent miscarriage/abortion during their second trimester. The association between higher contaminant burdens in these females and incidence of miscarriage/abortion during the second trimester cannot be discounted.

ΣPCB concentrations in previously gravid resting females ranged from 17.9 to 62.4 mg/kg ($n=9$) in both the control and noncontrol samples, and six of these females had ΣPCB burdens greater than the 41 mg/kg threshold. The presence of vaginal calculi in five resting females may further signify reproductive failure, if they result from an incomplete abortion. Elevated ΣPCB levels may impact uterine and placental health in cetaceans and, subsequently, fetal health and survival (Hohn et al., 2007; Murphy et al., 2010). Further, higher (gestational and lactational) exposure to PCBs in firstborn offspring may increase incidences of mortality in those individuals (Wells et al., 2005). In mink (*Mustela vison*), although ovulation, conception, and implantation occurred, similar to the current study, PCBs (as Clophen A50) increased fetal mortality through causing pathologic changes in the maternal vasculature in the placenta and degenerative changes in the trophoblast and fetal vessels (Bäcklin et al., 1997, 1998). A more recent experimental study focusing on effects from dietary consumption of Aroclor 1268 (dioxin-like PCBs were less prevalent, and it was the most highly chlorinated Aroclor manufactured) on mink reported that mean litter size, kit growth, and kit survival were the main reproductive and growth endpoints that were affected (Folland et al., 2016). As previously documented, female mink continued to reproduce, even at higher exposure concentrations to the chemical mixture. Whereas in harbor seals, experimental studies have shown that the effects from PCB exposure on reproduction occur at the stage of implantation, while the follicular, luteal, and postimplantation phases were not affected (Reijnders, 2003). Since pinnipeds experience delayed implantation/embryonic diapause, they may be more vulnerable than cetaceans at this stage of the reproductive cycle (Murphy et al., 2010).

Common dolphins are not as heavily parasitized as harbor porpoises in the Northeast Atlantic (Jepson, 2005; Pierce et al., 2008), and they did not show a change in nutritional condition during the period 1990–2006 (Murphy et al., 2009). Harbor porpoises are relatively small cetaceans, and individual health status often presents energetic "knife-edge" conditions that can be exacerbated by adding further health complications caused by exposure to anthropogenic pollutants (Murphy et al., 2015). For example, a significant positive association was observed between PCB levels and the number of gastric nematodes in UK harbor porpoises with blubber PCB concentrations >25 mg/kg (ΣPCB) (Bull et al., 2006). As noted earlier, a high incidence of reproductive failure was observed in UK harbor porpoises, with reproductive failure reported to occur in 39% or more of mature females sampled (calculated based on direct evidence such as fetal death and indirectly by using individual PCB burdens). Female health status played an important role in reproductive failure, as 86% of cases of fetal death/spontaneous abortion were observed in females that died from infectious disease or other causes such as starvation and neoplasia. PCBs may be impacting reproduction function indirectly in porpoises through lowering their immunity and increasing susceptibility to diseases (Jepson et al., 2005; Murphy et al., 2015). Terminations during late gestation can incur severe health and reproductive costs, and all reported cases of fetal death/spontaneous abortion in UK harbor porpoises occurred after the first semester (Murphy et al., 2015), similar to the current study. PCBs have been proposed to impact immune function in UK harbor porpoises (Jepson et al., 2005), where the average increase in risk of infectious disease mortality was 2% for each 1 mg/kg increase in blubber PCBs, with a 50% increase in risk occurring around 45 mg/kg lw (Hall et al., 2006a). Female UK harbor porpoises with higher PCB concentrations died due to ill health, as 92% of stranded individuals with ΣPCB concentrations >20 mg/kg died as a result of infectious disease or "other" causes such as starvation (Murphy et al., 2015).

When applying the toxicity thresholds to all available ΣPCB data for common dolphins in the Northeast Atlantic sampled between 1990 and 2013 ($n=183$), 76% of sexually immature males and females had ΣPCB levels above the 9 mg/kg threshold for onset of adverse health effects in marine mammals, and 17% had levels greater than one of the highest toxicity thresholds for marine mammals, 41 mg/kg. ΣPCB ranged from 1.1 to 95.9 mg/kg in sexually immature individuals. Fifty percent of mature males (five out of ten) had blubber ΣPCB concentrations above the 41 mg/kg threshold for profound reproductive effects in female seals (Fig. 1.5). Although the sample size for mature males was small, mean ΣPCB was 45.8 mg/kg, and concentrations ranged from 7.0 to 119.8 mg/kg lw: the highest ΣPCB concentrations were observed in a male stranded in 1992. Males were unable to rid themselves of their lipophilic pollutant burden and accumulated high PCB concentrations, the effect of which is not fully understood in male cetaceans, as very few studies have been undertaken and none on *D. delphis*. One such study reported a negative correlation between testosterone levels and tissue concentrations of DDE in Dall's porpoise (*Phocoenoides dalli*) (Subramanian et al., 1987). It has been suggested that EDCs can cause male disorders such as reduced semen quality, urogenital tract abnormalities (e.g., cryptorchidism, hypospadias, testicular cancer), and altered timing of puberty (reviewed in Diamanti-Kandarakis et al., 2009). Observed effects of OCs on male

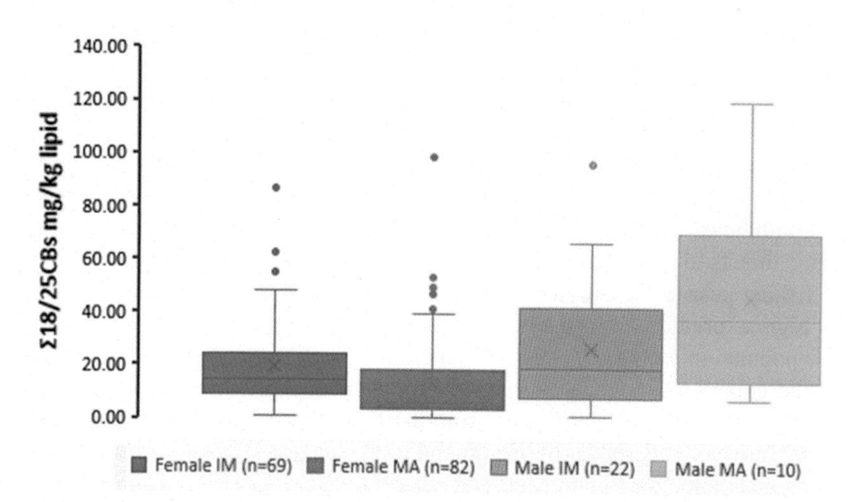

FIGURE 1.5 *Box* plots of male and female common dolphin reproductive status (*IM*, sexually immature; *MA*, sexually mature) and ΣPCB from stranded and bycaught common dolphins (1990–2013, $n=183$). The *dark horizontal line* indicates the median, *x markers* indicate the mean, and outliers are highlighted by *circles*. *(Data obtained from Law (1994), Law et al. (2006), Pierce et al. (2008), Murphy et al. (2010), Jepson et al. (2013);* the current study and Murphy et al. (unpublished data)*)*

rats include decreased spermatogenesis and fertility and delayed puberty (Diamanti-Kandarakis et al., 2009).

CONCLUSIONS

A low reproductive rate of 26% was reported for common dolphins in the Northeast Atlantic for the period 1990–2006 (Murphy et al., 2009). Significantly higher pregnancy rates of 47% and 40.2% were reported in *D. delphis* inhabiting the Eastern Tropical Pacific and *D. capensis* off South Africa, respectively (Danil and Chivers, 2007; Mendolia, 1989; Murphy et al., 2009). In cetaceans, low reproductive rates have been attributed to density-dependent compensatory responses when populations are close to carrying capacity, small population size causes inbreeding depression, interference, and resource competition, or long-term ecosystem change causes a decline in the prey base, leading to nutritional stress, cryptic effects of fishery interactions, such as separation of mother and calves or decreased fecundity due to stress effects, disease, marine biotoxins, and impacts of endocrine disrupting chemicals/anthropogenic pollutants (Cramer et al., 2008; Geraci and Lounsbury, 2009; Geraci et al., 1999; Gerrodette and Forcada, 2005; Murphy et al., 2009, 2015; Reeves et al., 2001). For common dolphins in the Northeast Atlantic, exposure to anthropogenic pollutants may be contributing to the low observed reproductive output (Murphy et al., 2009, 2010).

This study has reported evidence of reproductive failure and reproductive dysfunction in common dolphins inhabiting UK waters, which may be possibly linked to exposure to PCBs. Reproductive failure could have occurred in 30% or more of mature females in the control sample. Where pollutant data were available, all observed cases of reproductive tract pathologies in both control and noncontrol samples were reported in females with ΣPCB burdens >22 mg/kg lw (Table 1.2). However, there could be combined effects from exposure to multiple pollutants, including (low doses of) DDT and other legacy and emerging pollutants, which requires further investigation. One of the main anthropogenic threats to the Northeast common dolphin population is incidental capture in fishing gear (ICES Advice, 2016; Murphy et al., 2013). If bycatch rates are as high as predicted in the region, any suppression of reproduction will limit the population's ability to recover. There has been no evidence of a decline in the common dolphin population in the Northeast Atlantic in recent years (Hammond et al., 2017). Results from the recent SCANS III aerial and shipboard surveys undertaken by Hammond et al. suggest large-scale movements of animals into continental shelf and adjacent waters—possibly from offshore or more southern waters—so more dolphins are now exposed to anthropogenic pollutants in western European waters (Murphy et al., 2018a). Mean levels of ΣPCB in common dolphins are lower than that those observed in bottlenose dolphins, striped dolphins, and killer whales in European waters (Jepson et al., 2016). However, the effects of exposure to lower doses of EDCs may not be of a magnitude less, particularly when exposure occurs during critical periods of development. In humans, observed disorders are more than likely to result from chronic exposure to low amounts of mixtures of chemicals, as these can produce synergistic and additive effects, which may be coupled with generational epigenetic effects (Diamanti-Kandarakis et al., 2009). Consequently, the continued exposure to legacy anthropogenic pollutants such as PCBs and new emerging pollutants raises concerns about the current and future population-level pollutant effects on Northeast Atlantic common dolphins.

ACKNOWLEDGMENTS

We thank Bob Reid, Tony Patterson, Harry Ross, Jason Barley, John Baker, Thijs Kuiken, Simon Northridge, and Christina Lockyer and the staff of the AHVLA at Polwhele (Truro), as well as the volunteers of the Cornwall Wildlife Trust Marine Strandings Network, for their contributions to data collection. This research was supported by a Marie Curie International Outgoing Fellowship within the Seventh European Community Framework Programme (Project Cetacean-stressors, PIOF-GA-2010-276145 to SM and PDJ). Additional funding was provided through the Agreement on the Conservation of Small Cetaceans of the Baltic, North East Atlantic, Irish and North Seas (ASCOBANS) (Grants SSFA/2008 and SSFA/ASCOBANS/2010/5 to SM). Samples examined in this research were collected under the collaborative Cetacean Strandings Investigation Programme (http://ukstrandings.org/), which is funded by the Department for Environment, Food and Rural Affairs (Defra) and the UK's Devolved Administrations in Scotland and Wales (http://sciencesearch.defra.gov.uk/Default.aspx) (grants to PDJ, RD). UK Defra also funded the chemical analysis under a service-level agreement with the Centre for Environment, Fisheries and Aquaculture Science (grants to RJL, JB).

REFERENCES

Aguilar, A., 1985. Compartmentation and reliability of sampling procedures in organochlorine pollution surveys of cetaceans. Residue Reviews 95, 91–114.

Aguilar, A., Borrell, A., 1994. Abnormally high polychlorinated biphenyl levels in striped dolphins (*Stenella coeruleoalba*) affected by the 1990-1992 Mediterranean epizootic. Science of the Total Environment 154, 237–247.

Aguilar, A., Borrell, A., 2005. DDT and PCB reduction in the western Mediterranean from 1987 to 2002, as shown by levels in striped dolphins (*Stenella coeruleoalba*). Marine Environmental Research 59, 391–404.

Aguilar, A., Borrell, A., Pastor, T., 1999. Biological factors affecting variability of persistent pollutant levels in cetaceans. In: Reijnders, P.J.H., et al., (Eds.), Chemical Pollutants and Cetaceans. In: Journal of Cetacean Research and Management, vol. Special Issue 1, pp. 83–116.

Alava, J.J., Salazar, S., Cruz, M., Jiménez-Uzcátegui, G., Villegas-Amtmann, S., Paéz-Rosas, D., Costa, D.P., Ross, P.S., Ikonomou, M.G., Gobas, F.A.P.C., 2011. DDT strikes back: galapagos sea lions face increasing health risks. Ambio 40, 425–430. https://doi.org/10.1007/s13280-011-0136-6.

Bäcklin, B.-M., Madej, A., Forsberg, M., 1997. Histology of ovaries and uteri and levels of plasma progesterone, oestradiol-17β and oestrone sulphate during the implantation period in mated and gonadotrophin-releasing hormone-treated mink (*Mustela vison*) exposed to polychlorinated biphenyls. Journal of Applied Toxicology 17, 297–306.

Bäcklin, B.-M., Persson, E., Jones, C.J.P., Dantzer, V., 1998. Polychlorinated biphenyl (PCB) exposure produces placental vascular and trophoblastic lesions in the mink (*Mustela vison*): a light and electron microscope study. APMIS 106, 785–799.

Baker, J., 1992. Causes of mortality and parasites and incidental lesions in dolphins and whales from British waters. Veterinary Record 130, 569–572.

Béland, P., DeGuise, S., Girard, C., Lagacé, A., Martineau, D., Michaud, R., Muir, D.C.G., Norstrom, R.J., Pelletier, É., Ray, S., Shugart, L.R., 1993. Toxic compounds and health and reproductive effects in St. Lawrence beluga whales. Journal of Great Lakes Research 19, 766–775. https://doi.org/10.1016/s0380-1330(93)71264-2.

Benirschke, K., Marsh, H., 1984. Anatomic and pathologic observations of female reproductive organs in the short-finned pilot whale, *Globicephala macrorhynchus*. International Whaling Commission (Special Issue 6), 451–455.

Benirschke, K., Henderson, J., Sweeney, J., 1984. A vaginal mass, containing bones, in a common dolphin, *Delphinus delphis*. Reports of the International Whaling Commission (Special Issue 6), 457–458.

Bergman, A., 1999. Health condition of the Baltic grey seal (*Halichoerus grypus*) during two decades. Gynaecological health improvement but increased prevalence of colonic ulcers. Acta Pathology Microbiology and Immunology of Scandinavia 107, 270–282.

Bergman, A., 2007. Pathological Changes in Seals in Swedish Waters: The Relation to Environmental Pollution. Tendencies during a 25-year Period. Swedish University of Agricultural Sciences, Uppsala, p. 102.

Bergman, A., Olsson, M., 1985. Pathology of Baltic gray seal and ringed seal females with special reference to adrenocortical hyperplasia: is environmental pollution the cause of a widely distributed disease syndrome? Finnish Game Research 44, 47–62.

Bergman, Å., Heindel, J.J., Jobling, S., Kidd, K.A., Zoeller, R.T., 2013. State of the science of endocrine disrupting chemicals - 2012. In: WHO (World Health Organization)/UNEP (United Nations Environment Programme).

Boon, J.P., Oostingh, I., van der Meer, J., Hillebrand, M.T.J., 1994. A model for the bioaccumulation of chlorobiphenyl congeners in marine mammals. European Journal of Pharmacology: Environmental Toxicology and Pharmacology 270, 237–251. https://doi.org/10.1016/0926-6917(94)90068-X.

Borrell, A., Aguilar, A., 2005. Mother-calf transfer of organochlorine compounds in the common dolphin (*Delphinus delphis*). Bulletin of Environmental Contamination and Toxicology 75, 149–156. https://doi.org/10.1007/s00128-005-0731-y.

Borrell, A., Aguilar, A., 2007. Organochlorine concentrations declined during 1987–2002 in western Mediterranean bottlenose dolphins, a coastal top predator. Chemosphere 66, 347–352. https://doi.org/10.1016/j.chemosphere.2006.04.074.

Borrell, A., Bloch, D., Desportes, G., 1995. Age trends and reproductive transfer of organochlorine compounds in long-finned pilot whales from the Faroe Islands. Environmental Pollution 88, 283–292.

Bredhult, C., Bäcklin, B.-M., Bignert, A., Olovsson, M., 2008. Study of the relation between the incidence of uterine leiomyomas and the concentrations of PCB and DDT in Baltic gray seals. Reproductive Toxicology 25, 247–255. https://doi.org/10.1016/j.reprotox.2007.11.008.

Van Bressem, M.F., Van Waerebeek, K., Siebert, U., Wunschmann, A., Chavez-Lisambart, L., Reyes, J.C., 2000. Genital diseases in the peruvian dusky dolphin (*Lagenorhynchus obscurus*). Journal of Comparative Pathology 122, 266–277.

Van Bressem, M.F., Van Waerebeek, K., Montes, D., Kennedy, S., Reyes, J.C., Garcia-Godos, I.A., Onton-Silva, K., Alfaro-Shigueto, J., 2006. Diseases, lesions and malformations in the long-beaked common dolphin Delphinus capensis from the Southeast Pacific. Diseases of Aquatic Organisms 68, 149–165.

Browning, H.M., Acevedo-Whitehouse, K., Gulland, F.M.D., Hall, A.J., Finlayson, J., Dagleish, M.P., Billington, K.J., Colegrove, K., Hammond, J.A., 2017. Evidence for a genetic basis of urogenital carcinoma in the wild California sea lion. Proceedings of the Royal Society B 281, 20140240.

Buckles, E.L., Lowenstine, L.J., Funke, C., Vittore, R.K., Wong, H.-N., St. Leger, J.A., Greig, D.J., Duerr, R.S., Gulland, F.M.D., Stott, J.L., 2006. Otarine Herpesvirus-1, not papillomavirus, is associated with endemic tumours in California Sea Lions (*Zalophus californianus*). Journal of Comparative Pathology 135, 183–189.

Bull, J.C., Jepson, P.D., Ssuna, R.K., Deaville, R., Allchin, C.R., Law, R.J., Fenton, A., 2006. The relationship between polychlorinated biphenyls in blubber and levels of nematode infestations in harbour porpoises, *Phocoena phocoena*. Parasitology 132, 565–573.

Caserta, D., Maranghi, L., Mantovani, A., Marci, R., Maranghi, F., Moscarini, M., 2008. Impact of endocrine disruptor chemicals in gynaecology. Human Reproduction Update 14, 59–72. https://doi.org/10.1093/humupd/dmm025.

CLEEN, October 2005, 2005. EuroPCB: Inventory PCB Enforcement in Member States Part 1 Final report of the Chemical Legislation European Enforcement Network (CLEEN) http://www.cleen-europe.eu/file/download/71/EuroPCB_part_I_final.pdf.

Cockcroft, V.G., De Kock, A.C., Lord, D.A., Ross, G.J.B., 1989. Organochlorines in bottlenose dolphins *Tursiops truncatus* from the east coast of South Africa. South African Journal of Marine Science 8, 207–217.

Colegrove, K.M., Gulland, F.M., Naydan, D.K., Lowenstine, L.J., 2009. Tumor morphology and immunohistochemical expression of estrogen receptor, progesterone receptor, p53, and Ki67 in urogenital carcinomas of California sea lions (*Zalophus californianus*). Veterinary Pathology 46, 642–655.

Cramer, K.L., Perryman, W.L., Gerrodette, T., 2008. Declines in reproductive output in two dolphin populations depleted by the yellowfin tunaÂ purse-seine fishery. Marine Ecology Progress Series 369, 273–285. https://doi.org/10.3354/meps07606.

Danil, K., Chivers, S.J., 2007. Growth and reproduction of female short-beaked common dolphins, *Delphinus delphis*, in the eastern tropical Pacific. Canadian Journal of Zoology 85, 108–121.

Deaville, R., Jepson, P.D., 2011. CSIP Final Report for the Period 1st January 2005-31st December 2010. UK Cetacean Strandings Investigation Programme: Report to the UK Department for Food and Rural Affairs and the Devolved Administrations. http://randd.defra.gov.uk/Document.aspx?Document=FinalCSIPReport2005-2010_finalversion061211released[1].pdf.

DeGregori, J., 2012. Challenging the axiom: does the occurrence of oncogenic mutations truly limit cancer development with age? Oncogene. https://doi.org/10.1038/onc.2012.281.

Diamanti-Kandarakis, E., Bourguignon, J.-P., Giudice, L.C., Hauser, R., Prins, G.S., Soto, A.M., Zoeller, R.T., Gore, A.C., 2009. Endocrine-disrupting chemicals: an endocrine society scientific statement. Endocrine Reviews 30, 293–342. https://doi.org/10.1210/er.2009-0002.

Diamond, M.L., Melymuk, L., Csiszars, S., Robson, M., 2010. Estimation of PCB stocks, emissions and urban fate: will our policies be effective? Environmental Science and Technology 44, 2777–2783.

EEA Technical Report, 2012. European Environment Agency. The Impacts of Endocrine Disrupters on Wildlife, People and Their Environments. No 2/2012. ISSN: 1725–2237.

Fenton, S.E., 2009. The mammary gland: a tissue sensitive to environmental exposures. Reviews on Environmental Health 24, 319–325.

Folland, W.R., Newsted, J.L., Fitzgerald, S.D., Fuchsman, P.C., Bradley, P.W., Kern, J., Kannan, K., Remington, R.E., Zwiernik, M.J., 2016. Growth and reproductive effects from dietary exposure to Aroclor 1268 in mink (Neovison vison), a surrogate model for marine mammals. Environmental Toxicology and Chemistry 35, 604–618. https://doi.org/10.1002/etc.3201.

Fossi, M.C., Marsili, L., 2003. Effects of endocrine disruptors in aquatic mammals. Pure and Applied Chemistry 75, 2235–2247.

Fukushima, M., Kawai, S., 1981. Variation of organochlorine residue concentration and burden in striped dolphin with growth. In: Fujiyama, T. (Ed.), Studies on the Levels of Organochlorine Compounds and Heavy Metals in the Marine Organisms. University of the Ryukyus, Okinawa, pp. 97–114.

Geraci, J.R., Lounsbury, V.J., 2009. Health. In: William, F.P., et al. (Ed.), Encyclopedia of Marine Mammals, second ed. Academic Press, London, pp. 546–553.

Geraci, J.R., Palmer, N.C., St. Aubin, D.J., 1987. Tumors in cetaceans: analysis and new findings. Canadian Journal of Fisheries and Aquatic Sciences 44, 1289–1300. https://doi.org/10.1139/f87-152.

Geraci, J.R., Harwood, J., Lounsbury, V.J., 1999. Marine mammal die-offs: causes, investigations, and issues. In: Twiss Jr., J.R., Reeves, R.R. (Eds.), Conservation and Management of Marine Mammals. Smithsonian Institution Press, Washington, DC, pp. 367–395.

Gerrodette, T., Forcada, J., 2005. Non-recovery of two spotted and spinner dolphin populations in the eastern tropical Pacific Ocean. Marine Ecology Progress Series 291, 1–21.

De Guise, S., Lagace, A., Beland, P., 1994a. True hermaphroditism in a St. Lawrence beluga whale (*Delphinapterus leucas*). Journal of Wildlife Diseases 30, 287–290.

De Guise, S., Lagacé, A., Béland, P., 1994b. Tumors in St. Lawrence beluga whales (*Delphinapterus leucas*). Veterinary Pathology 31, 444–449.

De Guise, S., Lagacé, A., Béland, P., Girard, C., Higgins, R., 1995. Non-neoplastic lesions in beluga whales (*Delphinapterus leucas*) and other marine mammals from the St lawrence estuary. Journal of Comparative Pathology 112, 257–271. https://doi.org/10.1016/s0021-9975(05)80079-9.

Gulland, F., Trupkiewicz, J., Spraker, T., Lowenstine, L., 1996. Metastatic carcinoma of probable transitional cell origin in 66 free-living California sea lions (*Zalophus californianus*), 1979 to 1994. Journal of Wildlife Diseases 32, 250–258.

Hall, A.J., Hugunin, K., Deaville, R., Law, R.J., Allchin, C.R., Jepson, P.D., 2006a. The risk of infection from polychlorinated biphenyl exposure in the harbor porpoise (*Phocoena phocoena*): a case–control approach. Environmental Health Perspectives 114, 704–711.

Hall, A.J., McConnell, B.J., Rowles, T.K., Aguilar, A., Borrell, A., Schwacke, L., Reijnders, P.J.H., Wells, R.S., 2006b. Individual-based model framework to assess population consequences of polychlorinated biphenyl exposure in bottlenose dolphins. Environmental Health Perspectives 114, 60–64. https://doi.org/10.1289/ehp.8053.

Hammond, P., Lacey, C., Gilles, A., Viquerat, S., Börjesson, P., Herr, H., Macleod, K., Ridoux, V., Santos, M.B., Scheidat, M., Teilmann, J., Vingada, J., Øien, N., 2017. Estimates of Cetacean Abundance in European Atlantic Waters in Summer 2016 from the SCANS-III Aerial and Shipboard Surveys. Sea Mammal Research Unit, University of St Andrews, UK, pp. 40

Helle, E., Olsson, M., Jensen, S., 1976. PCB levels correlated with pathological changes in seal uteri. Ambio 5, 261–263.

Herbst, A.L., Ulfelder, H., Poskanzer, D.C., 1971. Adenocarcinoma of the vagina. New England Journal of Medicine 284, 878–881. https://doi.org/10.1056/NEJM197104222841604.

Hickie, B.E., Ross, P.S., Macdonald, R.W., Ford, J.K.B., 2007. Killer Whales (*Orcinus orca*) face protracted health risks associated with lifetime exposure to PCBs. Environmental Science and Technology 41, 6613–6619. https://doi.org/10.1021/es0702519.

Hohn, A.A., Ewing, R.Y., Zaias, J., 2007. Reproduction in relation to conservation and commercial exploitation. In: Miller, D.L. (Ed.), Reproductive Biology and Phylogeny of Cetacea. Volume 7 of Series: Reproductive Biology and Phylogeny. Science Publishers, Enfield, pp. 371–389.

Houde, M., Hoekstra, P.F., Solomon, K.R., Muir, D.C., 2005. Organohalogen contaminants in delphinoid cetaceans. Reviews of Environmental Contamination and Toxicology 184, 1–57.

ICES Advice, 2016. Bycatch of small cetaceans and other marine animals – review of national reports under Council Regulation (EC) No. 812/2004 and other information. In: ICES Special Request Advice Northeast Atlantic and Adjacent Seas Ecoregions. ICES Advice 2016, Book 1 6 pp. http://www.ices.dk/sites/pub/Publication%20Reports/Advice/2016/2016/Protected_species_bycatch.pdf.

Ingre-Khans, E., Ågerstrand, M., Rudén, C., 2017. Endocrine Disrupting Chemicals in the Marine Environment. ACES Report Number 16. Department of Environmental Science and Analytical Chemistry, Stockholm University.

Jensen, S., 1996. Report of a new chemical hazard. New Scientist 32, 612.

Jensen, S., Johnels, A.G., Olsson, M., Otterlind, G., 1969. DDT and PCB in marine animals from Swedish waters. Nature 224.

Jepson, P.D. (Ed.), 2005. Cetacean Strandings Investigation and Co-ordination in the UK 2000-2004, Final report to the Department for Environment, Food and Rural Affairs. pp. 1–79. http://www.defra.gov.uk/wildlife-countryside/resprog/findings/index.htm.

Jepson, P.D., Law, R.J., 2016. Persistent pollutants, persistent threats. Science 352, 1388–1389. https://doi.org/10.1126/science.aaf9075.

Jepson, P.D., Bennett, P.M., Deaville, R., Allchin, C.R., Baker, J.R., Law, R.J., 2005. Relationships between polychlorinated biphenyls and health status in harbour porpoises (*Phocoena phocoena*) stranded in the United Kingdom. Environmental Toxicology and Chemistry 24, 238–248.

Jepson, P.D., Deaville, R., Acevedo-Whitehouse, K., Barnett, J., Brownlow, A., Brownell Jr., R.L., Clare, F.C., Davison, N., Law, R.J., Loveridge, J., Macgregor, S.K., Morris, S., Murphy, S., Penrose, R., Perkins, M.W., Pinn, E., Seibel, H., Siebert, U., Sierra, E., Simpson, V., Tasker, M.L., Tregenza, N., Cunningham, A.A., Fernández, A., 2013. What caused the UK's largest common dolphin (*Delphinus delphis*) mass stranding event? PLoS One 8, e60953,. https://doi.org/10.1371/journal.pone.0060953.

Jepson, P.D., Deaville, R., Barber, J.L., Aguilar, À., Borrell, A., Murphy, S., Barry, J., Brownlow, A., Barnett, J., Berrow, S., Cunningham, A.A., Davison, N.J., ten Doeschate, M., Esteban, R., Ferreira, M., Foote, A.D., Genov, T., Giménez, J., Loveridge, J., Llavona, Á., Martin, V., Maxwell, D.L., Papachlimitzou, A., Penrose, R., Perkins, M.W., Smith, B., de Stephanis, R., Tregenza, N., Verborgh, P., Fernandez, A., Law, R.J., 2016. PCB pollution continues to impact populations of orcas and other dolphins in European waters. Scientific Reports 6, 18573. https://doi.org/10.1038/srep18573. http://www.nature.com/articles/srep18573#supplementary-information.

Johnson, S., Lowenstine, L., Gulland, F.M.D., Jang, S., Imai, D., Almy, F., Delong, R., Gardner, I., 2006. Aerobic bacterial flora of the vagina and prepuce of California sea lions (*Zalophus californianus*) and investigation of associations with urogenital carcinoma. Veterinary Microbiology 114, 94–103.

Jonsson, B., Gustafsson, O., Axelman, J., Sundberg, H., 2003. Global accounting of PCBs in the continental shelf sediments. Environmental Science and Technology 37, 245–255.

Kannan, K., Blankenship, A., Jones, P., Giesy, J., 2000. Toxicity reference values for the toxic effects of polychlorinated biphenyls to aquatic mammals. Human and Ecological Risk Assessment 6, 181–201.

Kennedy, P.C., Miller, R.B., 1993. The female genital system. In: Jubb, K.V.F., et al. (Ed.), Pathology of Domestic Animals, vol. 3, fourth ed. pp. 349–470.

King, D.P., Hure, M.C., Goldstein, T., Aldridge, B.M., Gulland, F.M.D., Saliki, J.T., Buckles, E.L., Lowenstine, L.J., Stott, J.L., 2002. Otarine herpesvirus-1: a novel gammaherpesvirus associated with urogenital carcinoma in California sea lions (*Zalophus californianus*). Veterinary Microbiology 86, 131–137. https://doi.org/10.1016/s0378-1135(01)00497-7.

Law, R.J., 1994. Collaborative UK Marine Mammal Project: Summary of Data Produced 1988–1992. Fisheries Research Technical. Report 97. Directorate of Fisheries Research, Ministry of Agriculture, Fisheries and Food, Lowestoft, UK.

Law, R.J., Jepson, P.D., 2017. Europe's insufficient pollution remediation. Science 356. https://doi.org/10.1126/science.aam6274.

Law, R., Jepson, P., Deaville, R., Reid, R., Patterson, I., 2006. Collaborative UK marine mammals strandings project: summary of contaminant data for the period 1993-2001. In: Sci. Ser. Tech. Rep., 131. Cefas Lowestoft. 72 pp http://www.cefas.co.uk/publications/techrep/tech131.pdf.

Law, R.J., Barry, J., Barber, J.L., Bersuder, P., Deaville, R., Reid, R.J., Brownlow, A., Penrose, R., Barnett, J., Loveridge, J., Smith, B., Jepson, P.D., 2012. Contaminants in cetaceans from UK waters: status as assessed within the cetacean strandings investigation programme from 1990 to 2008. Marine Pollution Bulletin. 64, 1485–1494. https://doi.org/10.1016/j.marpolbul.2012.05.024.

Letcher, R., Bustnes, J., Dietz, R., Jenssen, B., Jorgensen, E., Sonne, C., Verreault, J., Vijayan, M., Gabrielsen, G., 2010. Exposure and effects assessment of persistent organohalogen contaminants in arctic wildlife and fish. Science of The Total Environment 408, 2995–3043.

Lipscomb, T.P., Scott, D.P., Garber, R.L., Krafft, A.E., Tsai, M.M., Lichy, J.H., Taubenberger, J.K., Schulman, F.Y., Gulland, F.M., 2000. Common metastatic carcinoma of California Sea Lions (*Zalophus californianus*): evidence of genital origin and association with novel gammaherpesvirus. Veterinary Pathology Online 37, 609–617. https://doi.org/10.1354/vp.37-6-609.

Lockyer, C.H., 1987. Observations on the Ovary of the Southern Minke Whale, 38. Scientific Reports of the Whales Research Institute, Tokyo, pp. 75–89.

Lundin, J.I., Ylitalo, G.M., Booth, R.K., Anulacion, B., Hempelmann, J.A., Parsons, K.M., Giles, D.A., Seely, E.A., Hanson, M.B., Emmons, C.K., Wasser, S.K., 2016. Modulation in persistent organic pollutant concentration and profile by prey availability and reproductive status in southern resident killer whale scat samples. Environmental Science and Technology 50, 6506–6516. https://doi.org/10.1021/acs.est.6b00825.

Marsh, H., Kasuya, T., 1984. Changes in the Ovaries of the Short-finned Pilot Whale, *Globicephala Macrorhynchus*, with Age and Reproductive Activity Report of the International Whaling Commission. pp. 331–335.

Martineau, D., 2012. Chapter 17 Contaminants and health of beluga (*D. leucus*) in St Lawrence estuary. In: Levengood, J.M. (Ed.), Ecology and Animal Health. The Baltic University Programme. Uppsala University, Uppsala, Sweden, pp. 139–148.

Martineau, D., Béland, P., Desjardins, C., Lagacé, A., 1987. Levels of organochlorine chemicals in tissues of beluga whales (*Delphinapterus leucas*) from the St. Lawrence Estuary, Quebec Canada. Archives of Environmental Contamination and Toxicology 16, 137–147.

Martineau, D., Lagacé, A., Béland, P., Higgins, R., Armstrong, D., Shugart, L.R., 1988. Pathology of stranded beluga whales (*Delphinapterus leucas*) from the St. Lawrence Estuary, Québec, Canada. Journal of Comparative Pathology 98, 287–310. https://doi.org/10.1016/0021-9975(88)90038-2.

Martineau, D., De Guise, S., Fournier, M., Shugart, L., Girard, C., Lagacé, A., Béland, P., 1994. Pathology and toxicology of beluga whales from the St. Lawrence Estuary, Quebec, Canada. Past, present and future. Science of The Total Environment 154, 201–215. https://doi.org/10.1016/0048-9697(94)90088-4.

Martineau, D., Lemberger, K., Dallaire, A., Labelle, P., Lipscomb, T.P., Michel, P., Mikaelian, I., 2002. Cancer in wildlife, a case study: beluga from the St. Lawrence Estuary, Québec, Canada. Environmental Health Perspectives 110, 1–7.

McAloose, D., Newton, A.L., 2009. Wildlife cancer: a conservation perspective. Nature Reviews Cancer 9, 517–526.

McEntee, K., 1990. Cysts in and around the ovary. In: McEntee, K. (Ed.), Reproductive Pathology of Domestic Mammals. Academic Press, London, pp. 52–68.

McFee, W.E., Osborne, C.A., 2004. Struvite calculus in the vagina of a bottlenose dolphin (*Tursiops truncatus*). Journal of Wildlife Diseases 40, 125–128.

Mendolia, C., 1989. Reproductive Biology of Common Dolphins (*Delphinus delphis* Linnaeus) off the South East Coast of Southern Africa. University of Port Elizabeth, Port Elizabeth, p. 111.

Mirimin, L., Viricel, A., Amaral, A.R., Murphy, S., Ridoux, V., Rogan, E., 2009a. Population Genetic Structure of Common Dolphins in the North-east Atlantic Using Microsatellite Loci and mtDNA Control Region Markers Report to the International Whaling Commission, SC/61/SM27.

Mirimin, L., Westgate, A.J., Rogan, E., Rosel, P., Read, A.J., Coughlan, J., Cross, T., 2009b. Population structure of short-beaked common dolphins (*Delphinus delphis*) in the North Atlantic Ocean as revealed by mitochondrial and nuclear genetic markers. Marine Biology 156, 821–834.

Mongillo, T.M., Ylitalo, G.M., Rhodes, L.D., O'Neill, S.M., Noren, D.P., Hanson, M.B., 2016. Exposure to a Mixture of Toxic Chemicals: Implications for the Health of Endangered Southern Resident Killer Whales. U.S. Dept. Commer., NOAA Tech. Memo. NMFS-NWFSC-135, p. 107. https://doi.org/10.7289/V5/TM-NWFSC-135.

Moon, H.J., Han, S.Y., Shin, J.-H., Kang, I.L.H., Kim, T.S., Hong, J.H., Kim, S.-H., Fenton, S.E., 2007. Gestational exposure to nonylphenol causes precocious mammary gland development in female rat offspring. Journal of Reproduction and Development 53, 333–344. https://doi.org/10.1262/jrd.18055.

Muir, D.C.G., Ford, C.A., Rosenberg, B., Norstrom, R.J., Simon, M., Béland, P., 1996. Persistent organochlorines in beluga whales (*Delphinapterus leucas*) from the St. Lawrence River estuary. I. Concentrations and patterns of specific PCBs, chlorinated pesticides and polychlorinated dibenzo-p-dioxins and dibenzofurans. Environmental Pollution 93, 219–234.

Munson, L., Calzada, N., Kennedy, S., Sorensen, T.B., 1998. Luteinized ovarian cysts in Mediterranean striped dolphins. Journal of Wildlife Diseases 34, 656–660.

Murphy, S., 2004. The Biology and Ecology of the Common Dolphin *Delphinus delphis* in the North-east Atlantic. University College Cork.

Murphy, S., Winship, A., Dabin, W., Jepson, P.D., Deaville, R., Reid, R.J., Spurrier, C., Rogan, E., López, A., González, A.F., Read, F.L., Addink, M., Silva, M., Ridoux, V., Learmonth, J.A., Pierce, G.J., Northridge, S.P., 2009. Importance of biological parameters in assessing the status of *Delphinus delphis*. Marine Ecology Progress Series 388, 273–291. https://doi.org/10.3354/meps08129.

Murphy, S., Pierce, G.J., Law, R.J., Bersuder, P., Jepson, P.D., Learmonth, J.A., Addink, M., Dabin, W., Santos, M.B., Deaville, R., Zegers, B.N., Mets, A., Rogan, E., Ridoux, V., Reid, R.J., Smeenk, C., Jauniaux, T., López, A., Farré, J.M.A., González, A.F., Guerra, A., García-Hartmann, M., Lockyer, C., Boon, J.P., 2010. Assessing the effect of persistent organic pollutants on reproductive activity in common dolphins and harbour porpoises. NAFO/ICES/NAMMCO symposium "The Role of Marine Mammals in the Ecosystem in the 21st Century". Journal of Northwest Atlantic Fishery Science 42, 153–173.

Murphy, S., Deaville, R., Monies, R.J., Davison, N., Jepson, P.D., 2011. True hermaphroditism: first evidence of an ovotestis in a cetacean species. Journal of Comparative Pathology 144, 195–199.

Murphy, S., Pinn, E.H., Jepson, P.D., 2013. The short-beaked common dolphin (*Delphinus delphis*) in the North-eastern Atlantic: distribution, ecology, management and conservation status. In: Hughes, R.N., et al. (Ed.), Oceanography and Marine Biology: An Annual Review, vol. 51. CRC Press, pp. 193–280.

Murphy, S., Perrott, M., McVee, J., Read, F., Stockin, K.A., 2014. Deposition of growth layer groups in dentine tissue of captive common dolphins *Delphinus delphis*. In: NAMMCO Scientific Publication Volume 10: Age Estimation of Marine Mammals with a Focus on Monodontids. https://doi.org/10.7557/3.3017.

Murphy, S., Barber, J.L., Learmonth, J.A., Read, F.L., Deaville, R., Perkins, M.W., Brownlow, A., Davison, N., Penrose, R., Pierce, G.J., Law, R.J., Jepson, P.D., 2015. Reproductive failure in UK harbour porpoises *Phocoena phocoena*: legacy of pollutant exposure? PLoS One 10, e0131085. https://doi.org/10.1371/journal.pone.0131085.

Murphy, S., Evans, P.G.H., Pinn, E., Simmonds, M., Pierce, G.J., 2018a. Conservation and management of common dolphins; lessons learned from the North-east Atlantic. Aquatic Conservation: Marine and Freshwater Ecosystems (in preparation).

Murphy, S., Pocknell, A., Bailey, J., Deaville, R., Perkins, M., Jepson, P.D., 2018b. Rare cetacean ovarian neoplasms. Diseases of Aquatic Organisms (in preparation).

Natoli, A., Cañadas, A., Vaquero, C., Politi, E., Fernandez-Navarro, P., Hoelzel, A., 2008. Conservation genetics of the short-beaked common dolphin (*Delphinus delphis*) in the Mediterranean Sea and in the eastern North Atlantic Ocean. Conservation Genetics 9, 1479–1487.

Newman, S.J., Smith, S.A., 2006. Marine mammal neoplasia: a review. Veterinary Pathology Online 43, 865–880. https://doi.org/10.1354/vp.43-6-865.

Norman, S., Garner, M., Berta, S., Dubpernell, S., Klope, M., 2011. Vaginal calculi in a juvenile harbor porpoise (*Phocoena phocoena*). Journal of Zoo and Wildlife Medicine 42, 335–337.

O'Shea, T.J., Reeves, R.R., Long, A.K., 1999. Marine mammals and persistent ocean contaminants. In: Proceedings of the Marine Mammal Commission Workshop, Keystone, Colorado, 12–15 October 1988, pp. 1–150.

Padmanabhan, V., Sarma, H.N., Savabieasfahani, M., Steckler, T.L., Veiga-Lopez, A., 2010. Developmental reprogramming of reproductive and metabolic dysfunction in sheep: native steroids vs. environmental steroid receptor modulators. International Journal of Andrology 33, 394–404. https://doi.org/10.1111/j.1365-2605.2009.01024.x.

Pierce, G.J., Santos, M.B., Murphy, S., Learmonth, J.A., Zuur, A.F., Rogan, E., Bustamante, P., Caurant, F., Lahaye, V., Ridoux, V., Zegers, B.N., Mets, A., Addink, M., Smeenk, C., Jauniaux, T., Law, R.J., Dabin, W., López, A., A. Farré, J.M., González, A.F., Guerra, A., García-Hartmann, M., Reid, R.J., Moffat, C.F., Lockyer, C., Boon, J.P., 2008. Bioaccumulation of persistent organic pollutants in female common dolphins (*Delphinus delphis*) and harbour porpoises (*Phocoena phocoena*) from western European seas: geographical trends, causal factors and effects on reproduction and mortality. Environmental Pollution 153, 401–415.

Reeves, R.R., Rolland, R., Clapham, P.J., 2001. Causes of reproductive failure in North Atlantic right whales: new avenues of research. Report of a workshop held 26–28 April 2000. In: Northeast Fisheries Science Centre Reference Document 01-16, Falmouth, Massachussetts, p. 46.

Reijnders, P.J.H., 1980. Organochlorine and heavy metal residues in harbour seals from the Wadden sea and their possible effects on reproduction. Netherlands Journal of Sea Research 14, 30–65.

Reijnders, P.J.H., 1986. Reproductive failure in common seals feeding on fish from polluted coastal waters. Nature 324, 456–457.

Reijnders, P., 2003. Reproductive and developmental effects of environmental organochlorines on marine mammals. In: Vos, J.G., et al. (Ed.), Toxicology of Marine Mammals. Taylor and Francis, London, pp. 55–66.

Reijnders, P.J.H., Aguilar, A., Donovan, G.P., 1999. Chemical pollutants and cetaceans. Journal of Cetacean Research and Management (Special Issue 1) (International Whaling Commission, Cambridge).

Rewell, R.E., Willis, R.A., 1950. Some tumours of wild animals. Journal of Pathology and Bacteriology 62, 450–452. https://doi.org/10.1002/path.1700620321.

Robeck, T.R., Atkinson, S.K., Brook, F., 2001. Reproduction. In: Leslie Dierauf and Frances M.D., Gulland (Ed.), CRC Handbook of Marine Mammal Medicine, second ed. 193–236.

Robeck, T.R., Steinman, K.J., Greenwell, M., Ramirez, K., Van Bonn, W., Yoshioka, M., Katsumata, E., Dalton, L., Osborn, S., O'Brien, J.K., 2009. Seasonality, estrous cycle characterization, estrus synchronization, semen cryopreservation, and artificial insemination in the Pacific white-sided dolphin (*Lagenorhynchus obliquidens*). Reproduction 138, 391–405. https://doi.org/10.1530/rep-08-0528.

Russell, P., Robboy, S.J., Prat, J., 2009. Ovarian sex cord-stromal and steroid cell tumors. In: Robboy, S.J., et al. (Ed.), Robboy's Pathology of the Female Reproductive Tract, second ed. Churchill Livingstone Elsevier.

Schwacke, L.H., Voit, E.O., Hansen, L.J., Wells, R.S., Mitchum, G.B., Hohn, A.A., Fair, P.A., 2002. Probabilistic risk assessment of reproductive effects of polychlorinated biphenyls on bottlenose dolphins (*Tursiops truncatus*) from the Southeast United States coast. Environmental Toxicological Chemistry 21, 2752–2764.

Seibel, H., Siebert, U., Schöpper, H., Wohlsein, P., 2012. Granulosa cell tumour in a harbour porpoise (*Phocoena phocoena*) from German waters. Diseases of Aquatic Organisms 99, 79–83. https://doi.org/10.3354/dao02449 PMID: 22585304.

Siebert, U., Wunschmann, A., Weiss, R., Frank, H., Benke, H., Frese, K., 2001. Post-mortem findings in harbour porpoises (*Phocoena phocoena*) from the German North and Baltic Seas. Journal of Comparative Pathology 124, 102–114.

Sinkkonen, S.P.J., 2000. Degradation half-life times for PCDDs, PCDFs and PCBs for environmental fate modelling. Chemosphere 40, 943–949.

Steinberg, R.M., Walker, D.M., Juenger, T.E., Woller, M.J., Gore, A.C., 2008. Effects of perinatal polychlorinated biphenyls on adult female rat reproduction: development, reproductive physiology, and second generational effects. Biology of Reproduction 78, 1091–1101.

Stolk, A., 1950. Tumours in whales. Amsterdam Naturalist 1, 28–33.

Subramanian, A.N., Tanabe, S., Tatsukawa, R., Saito, S.N.M., 1987. Reduction in the testosterone levels by PCBs and DDE in Dall's porpoises of Northwestern North Pacific. Marine Pollution Bulletin 18, 643–646.

Tornero, V., Hanke, G., 2016. Chemical contaminants entering the marine environment from sea-based sources: a review with a focus on European seas. Marine Pollution Bulletin 112, 17–38. https://doi.org/10.1016/j.marpolbul.2016.06.091.

United Nations Environment Programme, 2001. The Stockholm Convention on Persistent Organic Pollutants (UNEP, Nairobi, 2001). http://chm.pops.int/default.aspx.

Viricel, A., Strand, A., Rosel, P., Ridoux, V., Garcia, P., 2008. Insights on common dolphin (*Delphinus delphis*) social organization from genetic analysis of a mass-stranded pod. Behavioral Ecology and Sociobiology 63, 173–185.

Vos, J.G., Bossart, G.D., Fournier, D.A., O'Shea, T.J., 2003. Toxicology of Marine Mammals. New Perspectives: Toxicology and the Environment. Taylor & Francis.

Wells, R.S., Tornero, V., Borrell, A., Aguilar, A., Rowles, T.K., Rhinehart, H.L., Hofmann, S., Jarman, W.M., Hohn, A.A., Sweeney, J.C., 2005. Integrating life-history and reproductive success data to examine potential relationships with organochlorine compounds for bottlenose dolphins (*Tursiops truncatus*) in Sarasota Bay, Florida. Science of the Total Environment 349, 106–119.

Woodhouse, C.D., Reinne, C.J., 1991. Observations of vaginal calculi in dolphins. Journal of Wildlife Diseases 27, 421–427.

Ylitalo, G.M., Stein, J.E., Hom, T., Johnson, L.L., Tilbury, K.L., Hall, A.J., Rowles, T., Greig, D., Lowenstine, L.J., Gulland, F.M.D., 2005. The role of organochlorines in cancer-associated mortality in California sea lions (*Zalophus californianus*). Marine Pollution Bulletin 50, 30–39. https://doi.org/10.1016/j.marpolbul.2004.08.005.

Yordy, J.E., Wells, R.S., Balmer, B.C., Schwacke, L.H., Rowles, T.K., Kucklick, J.R., 2010. Life history as a source of variation for persistent organic pollutant (POP) patterns in a community of common bottlenose dolphins (*Tursiops truncatus*) resident to Sarasota Bay, FL. Science of the Total Environment 408, 2163–2172.

Zoeller, R.T., Brown, T.R., Doan, L.L., Gore, A.C., Skakkebaek, N.E., Soto, A.M., Woodruff, T.J., Vom Saal, F.S., 2012. Endocrine-disrupting chemicals and public health protection: a statement of principles from the endocrine society. Endocrinology 153, 4097–4110.

Zoeller, R.T., Bergman, Å., Becher, G., Bjerregaard, P., Bornman, R., Brandt, I., Iguchi, T., Jobling, S., Kidd, K.A., Kortenkamp, A., Skakkebaek, N.E., Toppari, J., Vandenberg, L.N., 2014. A path forward in the debate over health impacts of endocrine disrupting chemicals. Environmental Health 13, 118. https://doi.org/10.1186/1476-069x-13-118.

Chapter 2

Feeding Ecology Tools to Assess Contaminant Exposure in Coastal Mammals

Elizabeth A. McHuron[1], Sarah H. Peterson[1], Todd M. O'Hara[2]
[1]*University of California Santa Cruz, Santa Cruz, CA, United States;* [2]*University of Alaska Fairbanks, Fairbanks, AK, United States*

CONTAMINANTS AND COASTAL MAMMALS

The exposure of wildlife populations to environmental contamination can be one of many stressors to the health of species and their ecosystems. Understanding the impact of such exposures can inform effective management and conservation of species, with implications for both wildlife and human health. There are numerous and continually emerging contaminants of potential concern to wildlife, but mercury (Hg) and persistent organic pollutants (POPs), a group of contaminants resistant to physical and biological degradation that includes polychlorinated biphenyls (PCBs) and dichlorodiphenyltrichloroethane (DDT), remain the primary focus of most wildlife toxicology studies (Dietz et al., 2013; Fisk et al., 2005; Letcher et al., 2010; Scheuhammer et al., 2007; Wolfe et al., 1998). Monomethyl mercury (MeHg or MMHg), one of the most toxic forms of Hg, adversely impacts neurological and immune functioning (Clarkson and Magos, 2006) and is primarily associated with protein-rich tissues. POPs can impair reproductive, immune, and neurological function and are largely fat soluble and hydrophobic, although some of their metabolites are more water soluble (Jones and de Voogt, 1999; Mackay et al., 2006). Hg concentrations around the globe are significantly elevated above preindustrial levels (Dietz et al., 2011; Lamborg et al., 2014; Sunderland and Mason, 2007), although there is debate as to whether emissions are currently increasing in magnitude (Streets et al., 2009; Zhang et al., 2016). Regardless, models suggest that it will take decades for oceanic concentrations in some ocean basins to plateau (reach steady state) at the top of marine and coastal food webs even if atmospheric Hg concentrations remain constant (Sunderland and Mason, 2007). Environmental concentrations of some POPs have decreased due to international regulation,

Marine Mammal Ecotoxicology. https://doi.org/10.1016/B978-0-12-812144-3.00002-4

resulting in corresponding decreases in wildlife tissue concentrations; however, unregulated and newer POP compounds continue to enter marine ecosystems (Muir and de Wit, 2010; Rigét et al., 2010; Ross et al., 2013).

Exposure to environmental contaminants can occur through a variety of routes and mechanisms, but for mammalian species the vast majority results from dietary intake (Borgå et al., 2004; Gray, 2002; Harley and O'Hara, 2016). Thus, feeding ecology is a prominent component of many ecotoxicological studies and the focus of this chapter. Variation in feeding ecology (also referred to as foraging ecology or foraging behavior throughout this chapter) ranges from broad-scale differences in trophic position at the interspecific level to fine-scale differences in diet and habitat use at the intraspecific level. Because many contaminants of concern magnify as they traverse food webs (Borgå et al., 2004; Lavoie et al., 2013), a process known as biomagnification, herbivorous species typically are exposed to much lower contaminant concentrations than upper trophic level carnivores. There are certain chemicals of concern for primary producers and secondary consumers, but we do not discuss these as part of this chapter. At the intraspecific level, ontogeny, body size (irrespective of ontogeny), and reproductive status influence contaminant exposure because of physiological, morphological, or behavioral differences that ultimately affect feeding ecology and certain elimination processes (e.g., lactation). Although historically ignored as "noise," intraspecific variation in foraging ecology among conspecifics within the same demographic group is now recognized as a common and important ecological process (Bolnick et al., 2011, 2003; Wolf and Weissing, 2012) that likely explains a portion of the wide variation in intraspecific contaminant concentrations detected for many species.

Understanding the pathways and patterns in contaminant exposure and tissue concentrations as they relate to feeding ecology is an important component in identifying the factors that may place individuals, populations, and species at varying risk for adverse effects from contaminants. In this chapter, we explore the feeding ecology tools that are (or can be) used in ecotoxicological studies to explain the contribution of foraging behavior to variation in contaminant concentrations within wildlife populations. We focus on coastal mammals, defined here as (1) any terrestrial species that forages on marine-derived resources ("maritime" mammals) or (2) any marine mammal species (as traditionally defined) that forages in nearshore marine ecosystems. We limit the discussion of pelagic species due to limited information and the need to focus this effort, except where interactions occur at the marine–terrestrial interface. Terrestrial and marine ecosystems and species are typically studied in isolation (independently), but these two ecosystems and their associated mammalian species are inextricably linked at their interface, and use of marine resources is clearly identifiable in the chemical signatures of maritime mammals (Bocharova et al., 2013; Christensen et al., 2005; McGrew et al., 2014; Noël et al., 2014). Thus, we focus on "marine" here with respect to chemical signatures (composition) more than the established approaches for management agencies, particularly as

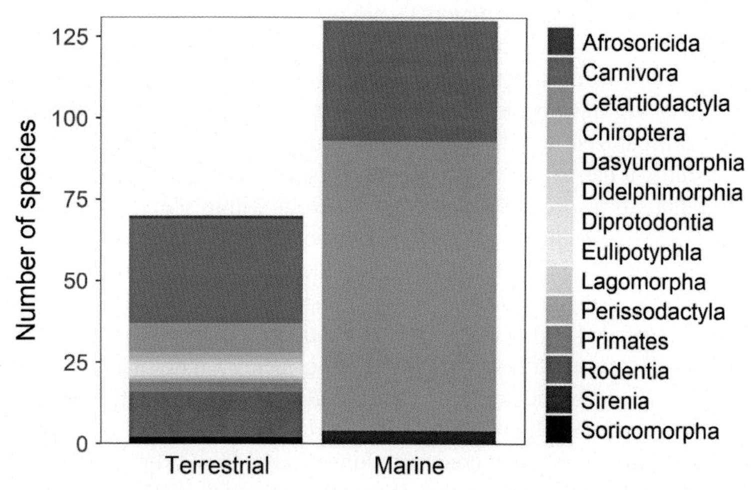

FIGURE 2.1 The number of terrestrial mammal species that have been documented using marine-derived resources (referred to as "maritime" mammals) and the total number of marine mammal species as traditionally defined for comparison. Species are color-coded based on order. Terrestrial mammals were classified as maritime if they have been documented predating or scavenging on any marine-derived resource throughout any part of their range (except in salt marsh and mangrove habitat), and they include freshwater species (e.g., otters). *(Data from Carlton, J.T., Hodder, J., 2003. Maritime mammals: terrestrial mammals as consumers in marine intertidal communities. Marine Ecology Progress Series 256, 271–286. https://doi.org/10.3354/meps256271; Creed, J.C., 2004. Capybara (*Hydrochaeris hydrochaeris *Rodentia: Hydrochaeridae): a mammalian seagrass herbivore. Estuaries 27, 197–200; Moore, P., 2002. Mammals in intertidal and maritime ecosystems: interactions, impacts and implications. Oceanography and Marine Biology: An Annual Review 40, 491–608.)*

dependence on marine resources by maritime mammals appears to be a relatively common yet often overlooked occurrence across a range of mammalian taxonomic groups (Carlton and Hodder, 2003; Fig. 2.1). We start with a brief overview of how these contaminants reach coastal ecosystems and are incorporated into the marine and terrestrial food webs on either side of the land–sea interface. We then discuss the different feeding ecology tools (with a focus on carbon (C), nitrogen (N), and sulfur (S) stable isotope analysis) that can be used to study the foraging ecology of these species, their limitations, and the types of information that can be gleaned from these tools with respect to contaminants.

A NONLINEAR VIEW OF THE FATE AND TRANSPORT OF CONTAMINANTS TO COASTAL ECOSYSTEMS

Hg originates from natural and anthropogenic sources (Driscoll et al., 2013), reaching coastal ecosystems through atmospheric deposition or inputs of point-source pollution where it undergoes transformation from an inorganic (Hg(II) or Hg (0)) to an organic, bioavailable form (MeHg; Driscoll et al., 2013; Fitzgerald et al., 2007). Coastal marine mammals are often assumed to be at an increased risk from Hg due to their trophic position and proximity to inputs of point-source pollution to nearshore ecosystems (Bossart, 2011), and elevated concentrations of Hg have been measured in tissues of marine mammals foraging in polluted urbanized habitats (Brookens et al., 2007; McHuron et al., 2014). More recently, it has been recognized that Hg emitted from diffuse sources can bioaccumulate to surprisingly high concentrations within species that forage in seemingly pristine habitats far from human populations. For example, Hg has been implicated as a potential contributing factor to the continued population decline of Steller sea lions (*Eumetopias jubatus*) in the western Aleutian Islands (Castellini et al., 2012; Rea et al., 2013), and there is increasing concern about the effect of Hg concentrations on the health of many Arctic marine mammals (Dietz et al., 2013). In mesopelagic habitats (200–1000 m in depth), conversion of inorganic to organic Hg is hypothesized to occur via microbial activity (Blum et al., 2013), and the highest concentrations of total Hg (THg) and MeHg are typically associated with the oxygen minimum zone (Gill and Fitzgerald, 1988; Laurier et al., 2004; Sunderland et al., 2009). Consequently, MeHg concentrations are highest in predators that forage within this habitat in the water column (Choy et al., 2009; Polito et al., 2016), or on vertically migrating species that use this habitat (Monteiro et al., 1998). Hg concentrations in open ocean foragers may in some cases rival or exceed concentrations found in coastal marine mammals, as is the case for northern elephant seals (*Mirounga angustirostris*) that forage in mesopelagic habitats thousands of kilometers from the nearest coastline (Peterson et al., 2015a; Fig. 2.2). Although they do not forage in coastal environments, many pelagic species spend a portion of their life cycle

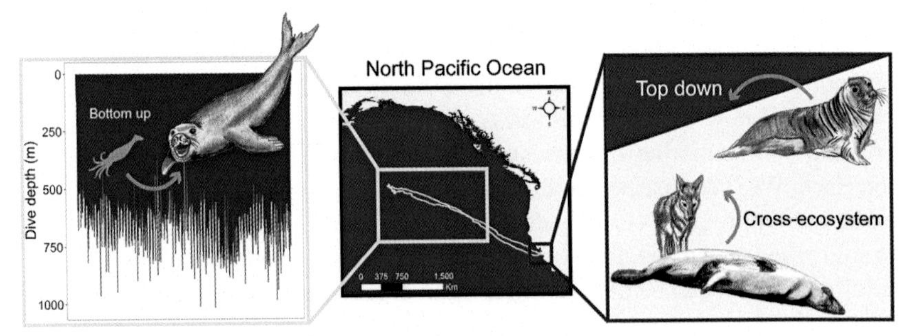

FIGURE 2.2 Visual representation of the long-distance transport of contaminants from pelagic to coastal and terrestrial ecosystems by northern elephant seals, a marine mammal that forages in the mesopelagic zone (depths 200–1000 m) in the North Pacific Ocean. The center panel shows the at-sea locations from a single foraging trip by an adult female, and the left panel shows the dive profile of that same individual over the foraging period. Elephant seals return to land twice throughout the year to either reproduce or molt, where carcasses are scavenged upon by terrestrial species such as coyotes (cross-ecosystem contamination), and molted hair and skin enters the nearshore marine ecosystem (top-down contamination; right panel). *(Illustrations by Chris Law.)*

in nearshore or terrestrial habitats, thereby serving as a vector to transport contaminants between food webs and ecosystems (biotransport).

POPs are a group of organic compounds that are predominately produced as a result of human applications (or unintentional byproducts of these applications), although there are also natural processes that produce a variety of POPs (Fielman et al., 1999; Paul and Pohnert, 2011; Teuten et al., 2005). Detection of POPs in marine ecosystems is geographically widespread across multiple ocean basins (Bossart, 2011; Law, 2014) and may be pervasive across a range of ocean depths, with recent studies detecting POPs in organisms that inhabit the deepest regions of the ocean in seemingly pristine habitats (Jamieson et al., 2017). Similar to Hg, the ocean is increasingly viewed as a long-term sink for legacy and emerging POPs; however, specific properties of individual POP compounds dictate how readily they travel in air or water and become incorporated into and move through marine food webs (Lohmann et al., 2007), resulting in certain POPs traveling further from sources than others (e.g., PCB congeners have varying physical properties that dictate transport outcomes within this group). The majority of POP compounds can be categorized by their chemical transport behavior as swimmers, single-hoppers, and multi-hoppers, based on whether the main mechanism of transport is via water or a combination of air and water (Lohmann et al., 2007). Consequently, POP compounds are unequally distributed, and animal exposure and bioaccumulation potential can vary among ocean habitats (Calambokidis and Barlow, 1991; Jessup et al., 2010; Kucklick et al., 2011; Peterson et al., 2015b).

Transport of contaminants into coastal ecosystems and food webs has classically been viewed as a one-way process, where contaminants originating from terrestrial sources are deposited in the marine environment and subsequently transferred to higher trophic levels in marine food webs through predation. Marine mammals may, however, serve as vectors to transport contaminants from marine into terrestrial food webs, a process that has received recent attention in the ecotoxicological literature (Blais et al., 2007; Elliott et al., 2009; Kurle et al., 2016; McGrew et al., 2014). This process mainly occurs through scavenging of carcasses and less so via targeted predation, and it is particularly prevalent along coastlines with high densities of pinnipeds or whale migratory routes where carcasses are readily available for scavenging by terrestrial species. Human activities may facilitate cross-ecosystem movement of contaminants by providing access to resources (e.g., subsistence hunts, ship strikes), or through direct or indirect interactions that induce behavioral changes in maritime vertebrates. For example, Reid (2014) found that marine-derived resources are a relatively new dietary input for coyotes (*Canis latrans*) from central California despite historical proximity to pinniped rookeries, which may be facilitated by reduced human disturbance of coastal areas or relaxation of interspecific competition from bears and humans that allows for increased access to rookeries. As highlighted earlier, cross-ecosystem transport of contaminants is not limited to inputs from coastal foraging marine mammals because many species that forage in pelagic habitats are highly mobile and migratory (Fig. 2.2). Furthermore, marine mammals are not the sole source of marine-derived resources to terrestrial ecosystems; maritime mammals may scavenge on seabird or turtle eggs, turtle hatchlings, or predate on anadromous fishes returning to rivers to reproduce (Moore, 2002), and seabird guano and decomposition of anadromous fish also serve as point-source pollutants to terrestrial environments (Baker et al., 2009; Blais et al., 2005). Biologically mediated transport of contaminants may be a significant contributor to terrestrial ecosystems even when marine subsidies only occur seasonally (Blais et al., 2007), although to our knowledge there have been no studies to quantify the relative contribution of marine-derived versus terrestrially-derived contaminants to total body contaminant burdens of maritime mammals. We address the issue of biologically mediated cross-ecosystem transport so readers appreciate marine mammals as a source of contaminants for coastal vertebrates, as this process may present a significant risk to some vertebrates not traditionally considered "marine," such as the endangered California condor (*Gymnogyps californianus*; Kurle et al., 2016).

Similar to biologically mediated cross-ecosystem transport, it is important to recognize that mammals are not "dead ends" for contaminants within marine ecosystems because the transfer of energy within food webs does not always occur in a linear fashion. Upper trophic level consumers may serve as a source of relatively high contaminant concentrations to lower trophic levels when the contaminants are recycled back into marine food webs, undergoing further bioaccumulation and biomagnification, a process referred to as "top-down contamination" (Cossaboon et al., 2015). One example of this process is scavenging by seabirds on marine mammal carcasses, such as that observed for the ivory gull (*Pagophila eburnea*) consuming polar bear (*Ursus maritimus*) tissue (Miljeteig et al., 2009). Animals that die at sea can serve as a contaminant source for multiple trophic positions and dimensions in the marine environment, as they may be scavenged upon at the surface before decomposing and releasing stored contaminants into the water column at varying ocean depths (Jamieson et al., 2017). Pinnipeds serve as an input of point-source pollution because they often congregate in large numbers during reproduction and/or molting, where contaminants in feces, fur, and skin return to the marine environment and can enter food webs at lower trophic levels (Cossaboon et al., 2015; Flegal et al., 1981). For example, Cossaboon et al. (2015) estimated that northern elephant seals contribute 10–12 kg of THg to nearshore environments during their annual pelage molt (Fig. 2.2), the majority of which occurs as MeHg (Dietz et al., 2011; Woshner et al., 2008). Because many contaminants of concern are resistant to biological degradation or can undergo reversible transformations (e.g., methylation and demethylation of Hg), iterative recycling of contaminants through food webs can occur many times, progressively increasing contaminant concentrations within marine food webs as entry into circular food webs may exceed exports.

USE OF STABLE ISOTOPES AS A CHEMICAL FEEDING ECOLOGY TOOL IN CONTAMINANT STUDIES

Stable isotope analysis of select elements (e.g., C, N, S) can be broadly used to understand the diet, habitat use, and movements of individuals, populations, and species (Crawford et al., 2008; Hobson, 1999; Kelly, 2000), and it is effectively used to explain how feeding ecology and variation in animal behavior influences contaminant exposure and resultant tissue concentrations. The utility of this approach lies in the fact that some stable isotopes change in a predictable fashion in food webs as a result of physiological and biogeochemical processes (Ben-David and Flaherty, 2012; Peterson and Fry, 1987). These processes result in changes in the ratio of an element's isotopes (e.g., ^{13}C vs. ^{12}C, denoted as $\delta^{13}C$), resulting in isotope values in animal tissues that are reflective of (but not necessarily equivalent to) baseline values of the food web and environments they inhabit. For marine mammals, stable isotopes of C and N are the two most commonly used isotopes, providing information about habitat type/spatial use and trophic level, respectively (Michener and Kaufman, 2007; Montoya, 2007; Newsome et al., 2010b). Sulfur has not been as widely used in the study of coastal mammals, but it has utility for species at the interface of terrestrial, freshwater, and marine environments because it can help distinguish the contribution of different primary producers to food webs, either alone or in combination with C isotopes (Connolly et al., 2004).

Stable isotope analysis can be conducted on a wide range of nonlethally and lethally sampled tissues that, due to differences in turnover or growth rates, integrate information about feeding ecology across different temporal scales. Metabolically active tissues represent behavior across varying timescales, with blood (days to several months depending on the compartment), liver (days to weeks), and muscle (weeks to about a month) representing relatively recent behavior, whereas bone collagen integrates behavior across a period of many months (Caut et al., 2011; Hilderbrand et al., 1996; Hobson and Clark, 1992; Lecomte et al., 2011; Newsome et al., 2006; Tieszen et al., 1983). Teeth and keratinized tissues are metabolically inert and therefore represent a longitudinal record of behavior across the period of tissue growth (Cherel et al., 2009; Hobson and Sease, 1998; McHuron et al., 2016b; McLaren et al., 2015), which may vary from several weeks or months (e.g., hair) to a significant majority if not the entire lifespan of the individual (e.g., teeth, baleen). In some cases, tissues are used for assessing in utero exposure, such as with lanugo (the natal coat) of pinniped pups. Tissues with slow turnover or growth may integrate or record information on feeding ecology at multiple spatial scales depending on animal behavior.

Here, we focus on describing the different applications of stable isotopes in contaminant studies and how their use has contributed to our knowledge of the influence of feeding ecology on inter- and intraspecific differences in contaminant exposure and concentrations in coastal mammals. We also discuss the current limitations of stable isotope analysis and several potential solutions to mitigate these deficiencies. We highlight examples from various systems and species, but we focus on the Arctic as (1) one author has more than 2 decades of experience using this feeding ecology tool in high northern latitudes, (2) this ecosystem represents one of the most comprehensively studied in terms of contaminants, including how feeding ecology influences exposure and concentrations in marine food webs, and (3) there are major ongoing and predicted polar environmental changes that will impact ecological and biological drivers of contaminant exposure in coastal mammals that can affect environmental, animal, and human health (One Health). We start with a brief overview of the statistical approaches typically used in mammalian ecotoxicological studies and divide our discussion of isotopes into three main sections—Trophic Level Interactions, Nontrophic Level Interactions, and Geographic Variation—with the caveat that there is partial overlap among sections, as these processes are not mutually exclusive in driving contaminant exposure.

Statistical Approaches

Generalized linear models are the standard statistical approach used to investigate the relationships between contaminant concentrations and stable isotope values, as they can account for variation associated with other factors that may affect concentrations, including age, sex, season, and tissue lipid content. Multivariate approaches can be useful to visualize complex patterns in contaminant concentrations alone or in conjunction with isotopes (Christensen et al., 2007; Dehn et al., 2006b; Litz et al., 2007), such as those involving comparisons among species, tissue types, and contaminants. Ordination techniques, such as principal components analysis, generate a set of new uncorrelated variables, providing opportunity for additional analyses and overcoming statistical issues associated with correlations between contaminants, contaminant congeners, or stable isotopes (Brown et al., 2014). Principal components analysis in combination with hierarchical cluster analysis is one approach used in feeding ecology studies to identify behavioral clusters (McHuron et al., 2016a; Villegas-Amtmann et al., 2008), which can then be used as a discrete predictor variable of contaminant concentrations in statistical models (Peterson et al., 2015a). For studies on POPs, similar approaches can be used to understand patterns in congener profiles that often reflect geographic variation in environmental contamination (Brown et al., 2014; Litz et al., 2007; Peterson et al., 2015b). Isotopic mixing models using Bayesian approaches provide a quantitative estimate of the proportional contribution of each prey species or prey group to the diet while allowing for the incorporation of variability into input parameters (Newsome et al., 2012; Phillips, 2012), and they are discussed in further detail subsequently. In addition to these methods, researchers should continue to explore existing and new statistical techniques when linking metrics of feeding ecology with contaminant concentrations, including those used for other taxonomic groups, habitats (marine vs. terrestrial), or disciplines (e.g., behavioral or community ecology).

Trophic Level Interactions

Nitrogen stable isotopes are used to investigate the role of trophic variation in contaminant exposure, as $\delta^{15}N$ values are an indicator of relative trophic position, and they are used to calculate the estimated trophic level (TL) of an individual or group of individuals. Estimating TL requires $\delta^{15}N$ values from the predator and the primary herbivore within the food web, as well as a diet-to-tissue trophic discrimination factor (TDF) that describes how isotope values change from prey to predator (Post, 2002). While many studies focus on $\delta^{15}N$ values themselves, they also can provide valuable information about the length (number of levels) of food webs, which is important for studies investigating contaminants that have the potential for biomagnification (St. Louis et al., 2011). Variation in contaminant concentrations driven by strict trophic interactions can also be quantified using biomagnification factors (BMFs) and food web magnification factors (FWMFs) that describe how contaminants magnify between prey and predator or throughout entire food webs, respectively. $\delta^{15}N$-derived estimates of TL are typically a key component of both of these values; BMFs are calculated as the ratio between the contaminant concentrations in predator and prey divided by the ratio of the TL (or in some cases $\delta^{15}N$ values) in predator and prey, whereas FWMFs are calculated from the slope of the regression between contaminant concentrations and TL or $\delta^{15}N$ values (Atwell et al., 1998; Dehn et al., 2006b; Fisk et al., 2001; Hoekstra et al., 2003). A comprehensive review of FWMFs is provided in Borgå et al. (2012), but the use of $\delta^{15}N$-derived TL estimates instead of $\delta^{15}N$ values themselves is a better approach for calculating FWMFs because differences in baseline $\delta^{15}N$ values may vary among food webs and TDFs vary among species (Jardine et al., 2006). Studies that focus on the influence of TL on the patterns of contaminant exposure within or among populations or species typically examine relationships using $\delta^{15}N$ values, whereas calculations of TL, BMFs, and FWMFs are typically employed in food web studies.

In Arctic food webs, there is wide variation in $\delta^{15}N$ values among marine mammals, which is partially attributable to major differences in diet (Dehn et al., 2006a, 2007). Bowhead (*Balaena mysticetus*) and gray whales (*Eschrichtius robustus*) occupy relatively low TLs, foraging on zooplankton and benthic invertebrates, respectively, whereas beluga whales (*Delphinapterus leucas*) have higher $\delta^{15}N$ values and forage on fish and cephalopods (Dehn et al., 2006a). For pinnipeds, spotted seals (*Phoca largha*) tend to have higher $\delta^{15}N$ values than ribbon (*Histriophoca fasciata*), ringed (*Pusa hispida*), or bearded seals (*Erignathus barbatus*) and walrus (*Odobenus rosmarus*) because they are largely piscivorous, whereas other species mainly forage on fish and zooplankton (ringed, ribbon), benthic invertebrates (bearded), and bivalves (walrus) (Dehn et al., 2007). Interspecific comparisons indicate that Hg and some

POPs tend to be greatest in species occupying higher TLs (Dehn et al., 2006a,b; Hoekstra et al., 2003), with concentrations in marine food webs often increasing linearly with $\delta^{15}N$ values (Campbell et al., 2005; Dehn et al., 2006b; Hobson et al., 2002). FWMFs for organochlorines (OCs) and Hg in Arctic food webs range from 0.61 to 18.8 (Hoekstra et al., 2003 and references cited within) and 1.0–1.6 (Atwell et al., 1998; Campbell et al., 2005; Dehn et al., 2006b; Lavoie et al., 2013), respectively, with values <1 for contaminants that undergo active elimination, biotransformation, or regulation. BMFs may vary considerably among species due to physiological or dietary differences that affect metabolism, storage, biotransformation, or contaminant exposure, as has been observed for Hg (Dehn et al., 2006b), cadmium (Cd) (Dehn et al., 2006a,b), and some POPs (Hoekstra et al., 2003).

Trophic interactions also result in intraspecific variation in contaminant concentrations, as evidenced by studies on bottlenose dolphins (*Tursiops truncatus*; Woshner et al., 2008), polar bears (Bentzen et al., 2008), grizzly bears (*U. arctos*; Christensen et al., 2005), Steller sea lions (Rea et al., 2013), and several species of ice-associated seals (Dehn et al., 2005). While there is a continued focus on metabolically active tissues, the use of metabolically inert tissues is increasing because collection is relatively noninvasive, and these tissues often overcome temporal sampling issues (e.g., animals tend to shed and regrow their hair at a similar time irrespective of when sampling occurs) and can provide a longitudinal record of feeding ecology for each individual. Instead of a single isotope value that integrates behavior across the entire period of tissue growth, sectioning metabolically inert tissues results in a longitudinal record that allows for a more comprehensive picture of individual feeding ecology (Hobson et al., 2004; Newsome et al., 2010a). This fine-scale behavior can be particularly useful in understanding why contaminant concentrations of individuals or populations do not follow a hypothesized pattern, such as may occur when isotopes and contaminant concentrations are measured in separate tissues, resulting in incomplete synchrony. Metabolically inactive tissues such as whiskers (vibrissae), baleen, and teeth provide a dual benefit because they also integrate information about foraging behavior over a relatively long timescale compared with most metabolically active tissues, at least for some species. Thus, isotope values in subsections can be linked with contaminant concentrations in tissues that represent short-term accumulation, and isotope values integrated across the entire tissue can be used as a proxy for foraging behavior when compared with contaminant concentrations in tissues that represent lifetime exposure. The ability to obtain a measure of foraging behavior integrated across more than several months is particularly important because many coastal mammals undergo seasonal migrations or dietary shifts. For example, Hobson et al. (2004) found that while $\delta^{13}C$ and $\delta^{15}N$ values were generally similar in muscle and baleen of minke whales (*Balaenoptera acutorostrata*), kidney Hg concentrations were more correlated with isotope values averaged across the entire length of baleen and exhibited stronger relationships than correlations with muscle isotope values.

Stable isotope analysis of metabolically inert tissues presents a unique opportunity to examine the role of maternal foraging behavior on fetal exposure to contaminants, which is of particular concern for Hg because it impacts a critical period of fetal neurodevelopment during late gestation (Davidson et al., 2004; Oken et al., 2008). This approach is only relevant when species retain tissues grown in utero past birth, as is the case with lanugo coat and whiskers of many pinniped species. Whiskers can be serially sampled to produce a longitudinal record of maternal foraging behavior across a portion of gestation and related to contaminant concentrations in the same whisker segments or in lanugo (Rea et al., 2013). Pairing whisker isotope values with lanugo contaminant concentrations is often a more feasible approach because of sample mass limitations (i.e., there may not be enough whisker mass to analyze for both isotopes and contaminants), and because hair is an important excretory tissue for Hg and potentially other metals (Wang et al., 2014). Using this approach, Rea et al. (2013) demonstrated the wide variability in THg concentrations in lanugo of Steller sea lion pups was partially attributable to maternal foraging behavior during late gestation, as concentrations were greater when dams fed at a higher TL. This relationship was strengthened when $\delta^{13}C$ values were included in the analysis, suggesting that there may have been additional differences in maternal foraging behavior not related to TL that also influenced THg concentrations. Because fetal tissues reflect maternal foraging behavior and contaminant exposure, the collection of tissues grown in utero also has utility for studies focused on contaminant exposure to adult females and not solely gestational exposure to the fetus. Pinniped pups are much easier to capture and handle than adult females, which results in increased sample size, hence statistical power, with the added benefit of targeting the cohort of concern for adverse effects and indirectly accessing maternal diet measures during gestation.

Nontrophic Level Interactions

The use of $\delta^{15}N$ values reveals clear trophic relationships for some species and populations, but other sources of dietary variation may be more important in explaining contaminant exposure and tissue concentrations rather than strictly, or predominantly, TL interactions. For example, polar bears often occupy the highest TL within food webs yet do not always have the highest contaminant concentrations (Atwell et al., 1998; Dehn et al., 2006b), with concentrations of Hg and some PCBs in blood and hair more related to $\delta^{13}C$ than $\delta^{15}N$ values (Cardona-Marek et al., 2009; Knott et al., 2011). Dietary variation influences contaminant exposure because individuals or populations use different food webs or simply target different prey species within the same food web due to physiological, morphological, or behavioral differences. Differences in contaminant exposure can also result from selective predation (e.g., if individuals prey on the same species yet do not target the same age class or tissue type), which has been hypothesized as a potential factor driving variation in THg and

POP concentrations in polar bears (Knott et al., 2011). In all of these situations, the use of multiple isotopes ($\delta^{13}C$, $\delta^{15}N$, and potentially $\delta^{34}S$) and methods will likely provide the most comprehensive approach for explaining the role of feeding ecology in driving variation in contaminant concentrations. These methods include different statistical approaches, such as generalized linear models that can account for the influence of other factors (e.g., sex, age) and their interactions, isotopic mixing models, and multivariate techniques, as well as other feeding ecology tools (see Additional Feeding Ecology Tools section).

Because consumption of marine resources by maritime mammals has a strong influence on contaminant exposure (Bocharova et al., 2013; Christensen et al., 2005; McGrew et al., 2014; Noël et al., 2014), identifying populations or individuals using marine-based food webs is of particular interest in understanding patterns of contaminant exposure and potential health risks. $\delta^{13}C$ and $\delta^{15}N$ values of primary producers are influenced by a variety of abiotic and biotic factors, but they generally tend to increase from terrestrial to marine ecosystems (Ben-David and Flaherty, 2012; Kelly, 2000). Thus, $\delta^{13}C$ and $\delta^{15}N$ values can be relatively effective in discriminating the differing dietary inputs of terrestrial- versus marine-derived prey, as has been shown for arctic foxes (*Vulpes lagopus*; Roth, 2002; Tarroux et al., 2012), red foxes (*V. vulpes*; Killengreen et al., 2011), grizzly bears (Christensen et al., 2005), gray wolves (*C. lupus*; Adams et al., 2010; Szepanski et al., 1999), and humans (*Homo sapiens*; Buchardt et al., 2007). In some situations, the range in $\delta^{13}C$ or $\delta^{15}N$ values can be relatively small, often making it challenging to discriminate between these two types of prey resources using these isotopes alone. In these instances, $\delta^{34}S$ values can be used in combination with $\delta^{13}C$ and/or $\delta^{15}N$ values to better discriminate among individuals or populations with varying dependence on terrestrial- versus marine-derived prey, as S exhibits a wide range in values in the transition from terrestrial to freshwater to marine food webs (Connolly et al., 2004; Peterson and Fry, 1987). McGrew et al. (2014) nicely illustrate the utility of this approach, where the authors used all three isotopes ($\delta^{13}C$, $\delta^{15}N$, and $\delta^{34}S$) to examine variation in THg concentrations in liver, kidney, and muscle of Alaskan gray wolves inhabiting coastal and inland areas. There was separation between wolves that had access to coastal areas compared with those that did not using $\delta^{13}C$ and $\delta^{15}N$ (Fig. 2.3A); however, the separation was more distinct with $\delta^{34}S$ because there was large between-group variability in $\delta^{34}S$ values but little within-group variability for coastal wolves (Fig. 2.3B). The dependence on marine resources clearly influenced THg concentrations, with markedly higher muscle THg concentrations in wolves with higher $\delta^{34}S$ values (Fig. 2.3C). Sulfur does require additional laboratory analysis because it cannot be quantified in the same sample as C and N, and many laboratories do not perform this analysis; however, the additional discriminatory power obtained by including a third isotope should be a strong consideration for researchers when undertaking ecotoxicological (and ecological) studies on maritime mammals. We provide this example to show the ecotoxicological overlap of maritime and

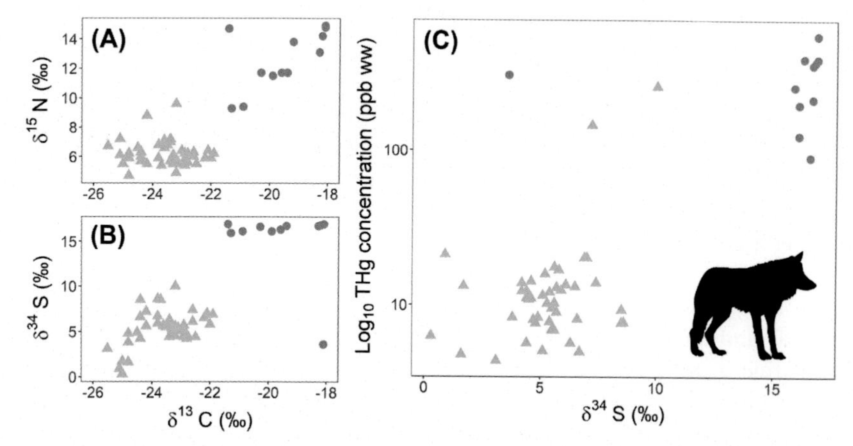

FIGURE 2.3 The relationships between δ^{15}N (A) and δ^{34}S (B) with δ^{13}C values in muscle of interior (*green diamonds*) and coastal (*blue circles*) gray wolves from Alaska. The separation in δ^{15}N, δ^{13}C, and δ^{34}S values between interior and coastal wolves indicates that all three isotopes can differentiate between ecosystem type, but the signal strength is clearer for δ^{34}S due to reduced within-group variability. The relationship between total mercury (THg) concentrations and δ^{34}S values (C) depicts that coastal wolves have much higher muscle THg concentrations than interior wolves, most likely due to differences in exposure from diet. Wolves were classified as interior or coastal based on the presence of a coastline within the Game Management Unit containing the collection site. *(Data were provided by A. McGrew to generate these modified figures from McGrew, A.K., Ballweber, L.R., Moses, S.K., Stricker, C.A., Beckmen, K.B., Salman, M.D., O'Hara, T.M., 2014. Mercury in gray wolves (*Canis lupus*) in Alaska: increased exposure through consumption of marine prey. The Science of the Total Environment 468–469, 609–613. https://doi.org/10.1016/j.scitotenv.2013.08.045.)*

marine mammals, as this technique (use of C, N and S isotopes) can also be used to determine coastal and riparian use by marine mammals and the associated exposure to contaminants, as discussed later.

Differential dependence on food webs can also occur within the marine environment, which influences contaminant exposure due to differences in exposure pathways and food web dynamics. Carbon is the more useful isotope in identifying these differences, as ^{13}C tends to be depleted in pelagic/offshore food webs compared with benthic/nearshore food webs, whereas δ^{15}N values are primarily indicative of TL and not habitat use. This generally results in higher δ^{13}C values in species such as gray whales, bearded seals, and walrus that have benthic-dominated diets compared with more pelagic ice seals, but the magnitude of these differences may be relatively small (<3‰; Dehn et al., 2006b). St. Louis et al. (2011) found that food web length and structure (i.e., benthic vs. pelagic) explained ~67% of the variation in hair THg concentrations in polar bears from the Southern Beaufort Sea and Hudson Bay, an example of how trophic and nontrophic interactions can simultaneously contribute to tissue contaminant concentrations. Southern Beaufort Sea bears had a greater dependence on a longer, pelagic food web than Hudson Bay bears that fed on a shorter, benthic food web (as evidenced by differences in $\delta^{13}C$ values and comparisons of baseline to bear $\delta^{15}N$ values), resulting in higher THg concentrations. Similarly, $\delta^{13}C$ values can be used to examine differential use of inshore versus offshore habitats, which may lead to variation in contaminant exposure due to differences in food web contamination and/or dietary variation (Das et al., 2004; Van De Vijver et al., 2003).

The addition of S should be considered not only for maritime mammals (as noted for wolves earlier), but also for traditionally defined marine mammals that use multiple nearshore marine habitat types, such as harbor seals (*P. vitulina*), harbor porpoise (*Phocoena phocoena*), and bottlenose dolphins, as differences in $\delta^{13}C$ and $\delta^{15}N$ values may be difficult to detect among these habitat types. In marine mammals, $\delta^{34}S$ values have been most widely employed for bottlenose dolphins, and they are effective for differentiating among dolphins using estuarine, inshore, and offshore habitats (Barros et al., 2010). Identification of individuals or populations that depend on estuarine prey is particularly important with respect to contaminant exposure, as they often have high contaminant concentrations due to the proximity of estuaries to human populations, inputs of point-source pollution, and runoff from riverine systems (Fair et al., 2007; Kucklick et al., 2011; McHuron et al., 2014; Tomy et al., 2000). This is apparent in San Francisco Bay, one of the world's largest estuaries with a legacy of Hg contamination that occurred during the Gold Rush in the 1800s. Harbor seals captured in San Francisco Bay have some of the highest THg concentrations measured in pinnipeds (Brookens et al., 2007; McHuron et al., 2014), with the highest concentrations generally found in seals with a greater dependence on estuarine than pelagic species (as inferred from $\delta^{34}S$ values; McHuron et al., 2014).

Dietary mixing models of multiple isotopes represent an alternative and complementary method for examining how inter- and intrapopulation variation in feeding ecology affects contaminant concentrations (Newsome et al., 2012; Phillips, 2012). Isotope values in tissues from the predator and potential prey items are input into mixing models to reconstruct the diet of populations or individuals at multiple dietary levels, ranging from studies that quantify the contribution of individual prey species to those that focus on the contribution of entire functional groups (Phillips et al., 2005). In the Arctic, mixing models using $\delta^{13}C$ and $\delta^{15}N$ have been successfully used to show that the importance of bearded seals and bowhead whales in polar bear diet varies among individuals and years (Bentzen et al., 2007; Rogers et al., 2015), and that the importance of marine versus terrestrial resources to Arctic foxes results from interactions among individual characteristics (sex, breeding status), season, and the spatial distribution of resources (Tarroux et al., 2012). Mixing models have not been widely employed in an ecotoxicological context, but they represent a promising approach to aid in the interpretation of results from generalized linear models, and can overcome the issue that $\delta^{13}C$ and $\delta^{15}N$ values are frequently correlated (interact) with each other, thus complicating their inclusion in the same statistical model. The output can be used not only to corroborate interpretations, but also in further insightful statistical analyses (Peterson et al., 2017). For example, generalized linear models could be used to examine how contaminant concentrations relate to the proportion of a particular prey species or functional group in the diet, or by combining multivariate techniques (e.g., ordination, cluster analysis) to examine the influence of more complex dietary patterns on contaminant patterns (Fig. 2.4). Alternatively, mixing models can be used to refine estimates of predator/prey interactions and trophic models that are used in calculating BMFs and FWMFs (Jones et al., 2014).

Geographic Variation

Species- or population-level differences in geographic feeding locations can also affect contaminant exposure that are irrespective of any differences in dietary variation (Braune et al., 2005; Elfes et al., 2010). We discuss these patterns separately from isotopic gradients resulting from differential use of habitat types/food webs because unlike the former, geographic variation in foraging areas does not necessarily result in dietary differences. In the Arctic, many OCs and several trace elements (e.g., Hg, Cd, Ag) exhibit a general trend of decreasing concentrations from east to west, with higher concentrations measured in tissues of marine mammals from eastern Canada than in the same species from the Beaufort-Chukchi Seas region (Braune et al., 2005; Dehn et al., 2005). These differences can theoretically be detected using stable isotopes, as $\delta^{13}C$ and/or $\delta^{15}N$ values also vary geographically in both marine and terrestrial environments (Graham et al., 2010; McMahon et al., 2013). This approach has been used in several baleen whale and pinniped species to identify migration patterns (Bentaleb et al., 2011; Busquets-Vass et al., 2017; Cherel et al., 2009; Matthews and Ferguson, 2015), such as bowhead whales that migrate between the ^{13}C-depleted Beaufort Sea-Amundsen Gulf and the ^{13}C-enriched Bering Sea (Schell et al., 1989). It can, however, be challenging in practice to disentangle isotopic variation due to geography from the influence of diet and physiological processes on isotope values. $\delta^{15}N$-derived estimates of TL and food web length may be useful in this respect, as they can rule out (or confirm) the contribution of biomagnification to differences in contaminant concentrations among populations from different geographic areas (Hoekstra et al., 2003).

Limitations and Considerations

There are limitations associated with the use of feeding ecology tools that can lead to spurious results or erroneous interpretations and conclusions. For stable isotopes, these limitations primarily influence the interpretation of isotope values and how they relate to patterns in contaminant concentrations and not confidence in the measurement itself, with the exception of $\delta^{15}N$-derived estimates of TL and food web length, and dietary estimates from mixing models. This

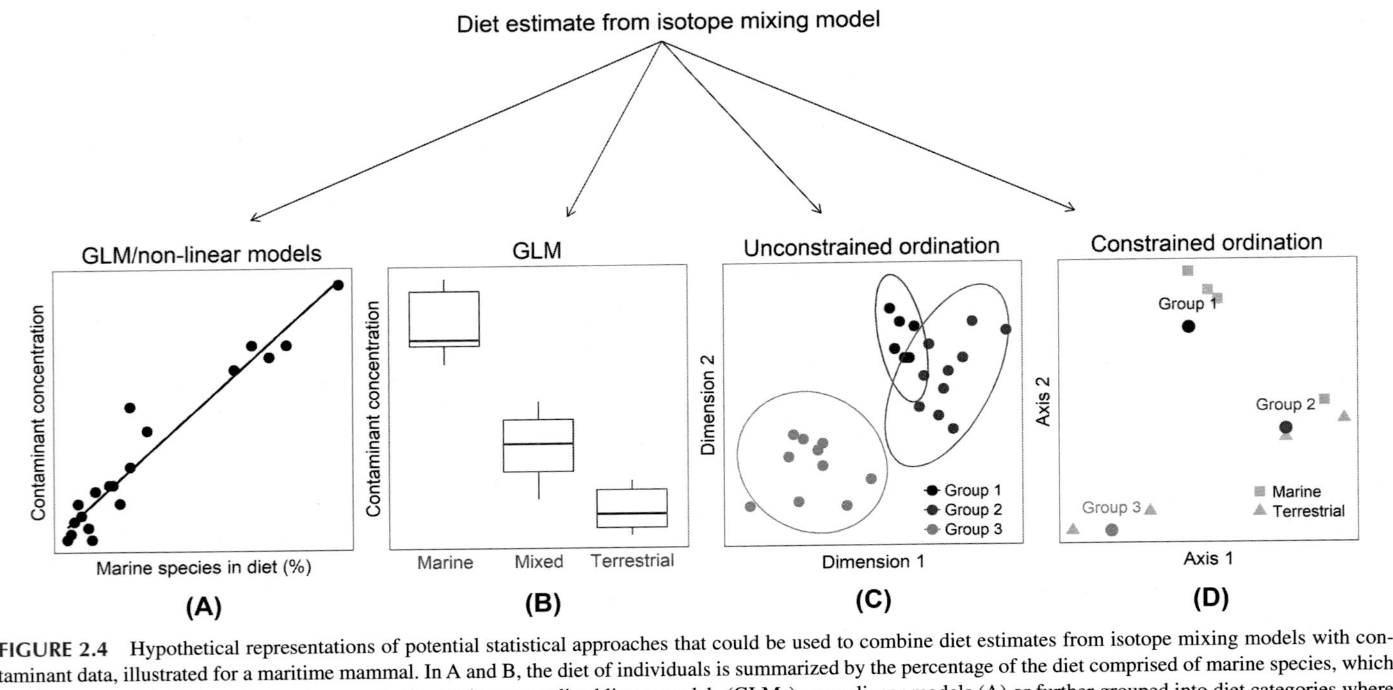

FIGURE 2.4 Hypothetical representations of potential statistical approaches that could be used to combine diet estimates from isotope mixing models with contaminant data, illustrated for a maritime mammal. In A and B, the diet of individuals is summarized by the percentage of the diet comprised of marine species, which can be regressed against contaminant concentrations using generalized linear models (GLMs) or nonlinear models (A) or further grouped into diet categories where differences in contaminant concentrations can be assessed using GLMs (B). In C, an unconstrained ordination (e.g., nonmetric multidimensional scaling; Hussey et al., 2011) can be used to examine patterns in dietary differences among individuals and how this variation relates to contaminant groupings; significant differences among groupings can be identified using an analysis of similarities. Alternatively, constrained ordination (e.g., canonical correspondence analysis; Lundström et al., 2010) using contaminant group as an explanatory factor can be used to identify relationships between diet and contaminants (D). In D, each *gray point* represents a different prey species, with shapes depicting marine or terrestrial species; species close to a contaminant group indicate greater importance in the diet of individuals assigned to that group. In B–D, the identification of unique groupings with respect to diet or contaminants could be done using either a researcher-identified threshold or multivariate techniques, such as cluster analysis.

assumes that appropriate quality assurance and control measures are applied to have confidence in generated values. We discuss these limitations and potential ways to mitigate them, as well as some additional issues that researchers should consider when selecting the target tissue(s) for isotope analysis in ecotoxicological studies.

Physiological processes can have a significant effect on the isotopic composition of tissues (Cherel et al., 2005; Fuller et al., 2005, 2004; Gannes et al., 1998), dependent or independent of feeding ecology variables. The two processes that are most relevant for coastal mammals are reproduction and fasting, the latter of which can be associated with life history events (reproduction, molt, migration) or result from an inability to acquire sufficient prey or nutrients due to environmental conditions, disease, or intrinsic factors (e.g., foraging experience). Captive or other controlled studies can help elucidate the influence of physiological processes on isotope values, as they eliminate or account for other sources of variation, such as diet; however, with the exception of perhaps humans, these types of studies are relatively rare for coastal mammals and are typically limited to inferences made from a limited number of individuals. For example, the influence of pregnancy and lactation on isotope values of pinnipeds is limited to observations made from a single northern fur seal (*Callorhinus ursinus*) that indicated $\delta^{15}N$ values in all blood compartments and $\delta^{13}C$ values in plasma/serum were higher during pregnancy than lactation (Kurle, 2002). A major limitation of captive studies is that it can be challenging to mimic physiological conditions in captive animals that are commonly experienced by their free-ranging counterparts, such as the fact that reproduction and fasting can occur concurrently (e.g., capital breeders, or otariids that fast while on land delivering milk to their pup), thus potentially limiting the inferences that can be made about wild populations from these studies. Targeted studies of free-ranging animals from specific age classes and reproductive groups represent a complementary approach to captive studies, which was used by Polischuk et al. (2001) to show that the direction and magnitude of changes in $\delta^{13}C$ and $\delta^{15}N$ values in adult female polar bears were different when fasting occurred in combination with reproduction rather than in isolation. The simplest way for researchers to largely overcome these issues is to select tissue types that do not integrate information across reproduction or molt (or select tissues where this time period can be excluded), although we recognize that this is not feasible for many studies due to the biology of the target species and/or the toxicological tissue of interest. It may be feasible to apply a correction factor to account for the contribution of physiological processes to isotope values, although additional research is needed to better quantify these processes, and we stress caution in this approach, as individual animals may experience isotopic changes that differ in magnitude from mean values (Polischuk et al., 2001, S. Peterson unpublished data for northern elephant seals).

In addition to physiological influences, isotope values in coastal mammals are also affected by values at the base of the food web (baseline values), which is a more pressing issue for $\delta^{15}N$ because $\delta^{13}C$ and $\delta^{34}S$ values are less influenced by trophic position (McCutchan et al., 2003; Peterson and Fry, 1987). Differences in baseline $\delta^{15}N$ values may obscure actual trends in feeding ecology or give the appearance of a trend when none exists, and without additional data, it is impossible to disentangle these two factors. Isotopic measurements of the primary herbivore within any given food web can greatly aid in distinguishing when differences in $\delta^{15}N$ values among individuals, populations, or species are actually reflective of differences in foraging behavior; however, these values are dynamic because communities and environmental conditions change with time, and temporal overlap between baseline measurements and animal sampling are often lacking due to logistical constraints. Compound-specific stable isotope analysis (CSIA) represents a potential solution to this issue, relying on the measurement of isotope values in certain individual amino acids (C and N) and fatty acids (for C, not N) as compared to a single integrated measurement for the entire tissue (bulk isotope analysis). This approach is based on the finding that isotope values of specific individual amino acids are unaltered by TL and reflect baseline values ("source" amino acids), whereas others change predictably with TL ("trophic" amino acids; Germain et al., 2013; Lorrain et al., 2015; McClelland and Montoya, 2002). As a relatively new technique, CSIA has yet to be widely employed in feeding ecology studies of coastal mammals, and while it faces similar challenges as bulk isotopic measurements with respect to determination of discrimination factors and may be cost prohibitive, it should be considered a complementary method to bulk isotope measurements in ecotoxicological studies that would benefit from more discrete chemical analyses.

TDFs describe the offset in isotope values between a consumer's tissue and their diet, and they are a component of TL (hence BMF and FWBF) calculations as well as an input in mixing models. They are typically determined using studies of captive animals on a known diet (but see Germain et al., 2012, Newsome et al., 2010a), and while the number of studies within this taxonomic group has greatly increased since the early 2000s (Beltran et al., 2016; Browning et al., 2014; Caut et al., 2011; Drago et al., 2015; Lecomte et al., 2011; McLaren et al., 2015; Newsome et al., 2010a; Rode et al., 2016; Stricker et al., 2015; Tyrrell et al., 2013), there has not been a comprehensive assessment of TDFs for coastal mammals. Results from these and other studies on mammalian taxa indicate that TDFs range from −0.6‰ to 6.5‰ ($\delta^{13}C$) and 0.4‰ to 5.5‰ ($\delta^{15}N$) depending on the tissue type, diet, lipid correction, and potentially species, although it is unknown whether apparent species differences represent actual trends or are simply an artifact of small sample sizes or discrepancies among studies due to diet (Beltran et al., 2016). Lecomte et al. (2011) found that TDFs for captive Arctic foxes varied with sex, age, and diet type (in addition to tissue type), with ranges that were similar in magnitude to those observed across all mammalian species. Dietary variables that can influence TDFs in mammals include the protein/fat content and quality of prey and the isotopic value of the diet (Caut et al., 2009; Newsome et al., 2010a;

Robbins et al., 2005; Rode et al., 2016; Wolf et al., 2015; Zhao et al., 2006). For maritime mammals, the use of terrestrial versus marine prey appears to be particularly important in influencing TDFs (Lecomte et al., 2011). Lipids tend to be depleted in ^{13}C relative to proteins and carbohydrates, and the issue of whether to lipid-extract or lipid-correct prey isotope values when calculating TDFs is an important consideration (Newsome et al., 2010a; Post et al., 2007; Wolf et al., 2015). Researchers commonly use a single value for the entire food web (~3.4‰–3.8‰) when calculating TL (Hoekstra et al., 2003; Post, 2002), but this could result in erroneous estimates and affect statistical comparisons with contaminant concentrations given the considerable intra- and interspecific variability in TDFs (Post, 2002). Furthermore, the choice of TDF also influences dietary output from mixing models (Bond and Diamond, 2011; McLaren et al., 2015); current methods allow for the incorporation of prey-specific TDFs as well as variability in these estimates and should be incorporated whenever possible (Newsome et al., 2010a). Further studies that quantify the range and variation of TDFs in coastal mammals are needed, particularly those that focus on diets that are reflective of wild populations.

The purpose of using stable isotopes in ecotoxicological studies is to link values with contaminant concentrations, so tissue selection and an understanding of the timescale represented within each tissue is imperative and should be given careful consideration during study design to achieve temporal congruity. Captive studies have resulted in a general understanding of turnover rates of isotopes in different tissues for a limited number of species (Caut et al., 2011; Hobson and Clark, 1992; Lecomte et al., 2011; Rode et al., 2016; Thomas and Crowther, 2015; Tieszen et al., 1983), although the amount of time represented by some metabolically inert tissues (whiskers, baleen) is less well understood and may vary considerably among species (Beltran et al., 2015; Greaves et al., 2004; Kernaléguen et al., 2015; McHuron et al., 2016b; Rea et al., 2015; Schell et al., 1989; Sumich, 2001; Zhao and Schell, 2004). Temporal mismatches may be less of an issue when contaminants and stable isotopes are measured in the same tissue (e.g., Hg in whole blood); however, they can occur when the target tissue is the primary storage organ for the contaminant of interest. For example, Hg is primarily stored in the liver of most marine mammals, and measurements of THg concentrations represent accumulation processes (very long half-life), yet liver isotope values integrate feeding ecology over a much shorter timescale due to quicker turnover of organic molecules (C, N, and S). For studies of POPs and other lipophilic contaminants, the issue of temporal mismatches is particularly relevant because POPs accumulate in lipid-rich tissues, but stable isotopes (particularly N and S) are best measured in protein-rich tissues. When temporal mismatches are unavoidable and of potential concern, the best approach is to collect multiple tissues for stable isotope analysis that integrate feeding ecology across multiple timescales or use segmental analyses of metabolically inert tissues that represent specific sections (time intervals) from proximal (most recent) to distal (oldest).

ADDITIONAL FEEDING ECOLOGY TOOLS

Due to their widespread use in ecotoxicological studies, the main focus of this chapter with respect to feeding ecology tools is on stable isotopes of C, N, and S. There are, however, a variety of other tools that warrant attention either as stand-alone methods or to be used in combination with stable isotopes, including biologging, photo identification, visual observations, traditional diet metrics, molecular techniques, and fatty acid analysis. We focus on examples that highlight the utility of these approaches, only providing a general explanation of the methods themselves and their limitations; in depth explanations and discussions of each of these methods can be found elsewhere (Bowen and Iverson, 2013; Bromaghin et al., 2016; Budge et al., 2006; Deagle et al., 2009; Hazen et al., 2012; Pompanon et al., 2012).

Biologging refers to the attachment of miniaturized instruments to animals that can collect information about an individual's location, behavior, physiology, and environment, including devices such as satellite tags, time-depth recorders, accelerometers, and video cameras (Bograd et al., 2010; Brown et al., 2013; Rutz and Hays, 2009). This approach is widely used to characterize the habitat use and foraging behavior of marine mammals (McIntyre, 2014), as their aquatic lifestyle and considerable marine range make it difficult to observe animals while they are at sea. The cost of many of these instruments makes it challenging to obtain large sample sizes, but even a limited number of observations can be useful in confirming (or refuting) suspected causes of patterns in isotope values (Busquets-Vass et al., 2017; Rogers et al., 2015). For example, by combining measurements of N isotopes from serially sampled hair sections with telemetry data, Rogers et al. (2015) showed that polar bear dependence on whale carcasses could be detected using $\delta^{15}N$ values, as female bears with depleted ^{15}N-root sections spent the majority of their time within 50 km of the coastline and were located multiple times within 5 km of whale carcasses. The strength of this technique lies in the ability to link contaminant concentrations with high-resolution data on individual behavior (i.e., instruments are attached to sampled individuals); studies by Peterson et al. (2015a,b) on northern elephant seals represent one of the few ecotoxicological applications of animal-borne devices for marine mammals (also see Brown et al., 2014; Lopez et al., 2014). Here, the authors used repetitive sampling at the beginning and end of a foraging trip to show that THg and POP concentrations differed among seals classified into different behavioral clusters identified using satellite tags, time-depth recorders, and stable isotopes of C and N. The underutilization of these devices in ecotoxicological studies is likely due to cost and logistic challenges associated with recapturing animals to obtain tissue samples for contaminant analysis that temporally overlap with biologging data, but they are feasible for many otariid species, sea otters (*Enhydra lutris*), and a limited number of phocid seals where recovery of instruments is already routine practice. Animal-borne devices may still be useful even when it is not possible to obtain temporal overlap in contaminant and foraging data (i.e., tissue samples are collected at the time of instrumentation), as the behavior of individual animals is often consistent through time (Arthur et al., 2015; Baylis et al., 2015; Newsome et al., 2009). Furthermore, animal-borne devices are useful in characterizing the general range of foraging behaviors exhibited by a population, which itself is helpful for understanding contaminant concentrations.

Photo identification (ID) is used to identify habitat use and movements of specific individuals, but this has limited utility for most species because it requires the presence of individually unique markings (natural or applied) that are maintained through time and considerable long-term research (observer) effort. The use of drones may alleviate observer effort and make this approach feasible for species in more remote habitats; however, considerable human effort is still needed to catalog and identify individuals from images. For marine mammals, resightings need to occur while the animal is at sea to provide information about foraging ecology, largely excluding the use of this approach for pinnipeds where observations occur at terrestrial haul-out sites, with the exception that resightings at haul-outs provide information on pup age and identity of sampled mother–pup pairs. Several recent studies on bottlenose dolphins have combined data on fine-scale movements, age, and reproductive histories derived from sighting histories with POP concentrations measured in blubber biopsies to understand geographic and life history variation in POP exposure (Balmer et al., 2011; Litz et al., 2007; Wells et al., 2005). Similar to data collected from satellite tags, resight data can be used to identify home ranges and detect differences in habitat use at small spatial scales (e.g., within an estuary or bay) that may be undetected using C and N isotopic approaches. Due to their long-term nature, an added benefit of photo ID studies is that they also provide information on demographic and life history factors known to affect contaminant concentrations, as well as additional information that could be of future interest in ecotoxicological studies such as social structure and interactions (Titcomb et al., 2017).

Diet can be directly quantified using visual observations of feeding behavior or identification of hard parts in scat, spew, or stomach samples, or inferred from other chemical feeding ecology tools, such as fatty acid analysis. Visual observations are primarily limited to species that exhibit surface foraging behavior (e.g., sea otters, humpback whales [*Megaptera novaeangliae*]) and large, predatory maritime mammals (e.g., wolves) where telemetry can be used to locate kill sites of large prey (Tambling et al., 2012). They can also be indirectly obtained for species foraging underwater by sampling prey patches at feeding aggregations using echosounders or net tows. Scat and stomach samples provide a snapshot of an individual's most recent meal, and hard parts and/or undigested tissues can be used to identify prey to species, size, and age class. DNA barcoding of scat samples can help overcome the issue that certain prey are often overestimated in the diet when using hard parts (Bowen and Iverson, 2013; Tollit et al., 2009), although this technique is not without its own challenges (Thomas et al., 2016, 2014). One of the primary challenges of using scat in ecotoxicological studies is that samples are typically collected without knowledge of the individual, excluding the ability to examine intraspecific variation in contaminant concentrations at the individual level. This could, however, be a useful approach when examining the influence of feeding ecology on geographic, temporal, or species variation in contaminant concentrations, as a general understanding of site- or population-specific diet is important in interpreting patterns in contaminant concentrations. The identification of diet from stomach samples does not suffer from the same limitation, but it typically requires lethal sampling (except when lavage is used), and given the abundance of other methods, it can be avoided unless lethal methods are required for additional reasons (e.g., subsistence hunts) or the animal is deceased for other reasons (e.g., stranded). Similar to stable isotope analysis, fatty acid (FA) analysis is predicated on the concept that chemical signatures in predator tissues reflect prey in a predictable manner. Fatty acid signatures can themselves be used in analyses (McKinney et al., 2011), or much like isotopic mixing models, quantitative FA signature analysis (QFASA) of lipid-rich tissues can be used to predict the input of different prey species to the diet by identifying the combination of prey species that best matches the FA profile of the predator (Budge et al., 2006; Iverson et al., 2004). This requires a prey library to characterize the FA profiles of all potential diet items, which may not be logistically feasible for generalist species that exhibit temporal and/or spatial variation in diet. It also requires an understanding of FA metabolism (but see Bromaghin et al., 2017), which is typically quantified from captive studies and currently limited to a few species (Bromaghin et al., 2016; Iverson et al., 2004; Nordstrom et al., 2008).

CONCLUSIONS

Feeding ecology is an important driver of variation in exposure to contaminants, and advances in the fields of chemistry, physiology, and technology among others have greatly increased our understanding of these relationships in coastal mammals (maritime and marine) within the last several decades. Tools for quantifying foraging behavior have become particularly important in ecotoxicological studies, as it is increasingly recognized that variation in foraging behavior is pervasive even at the intraspecific level, with differences ranging from fine-scale variation in habitat or prey selection to more extreme examples whereby individuals from the same population forage in different food webs within and across ecosystems. It is our intention that this chapter has encouraged readers to look past the more traditional views of contaminant sources and transport of contaminants within coastal ecosystems and adopt a circular or nonlinear view of within and across ecosystem transport of contaminants (primary producers→grazers→multiple levels of predation→scavenging/reuptake by primary producers→repeat). By summarizing the primary feeding ecology tools, their limitations, and the types of information that can be gleaned from these tools with respect to contaminants, we strive to further the use of these tools in ecotoxicological studies and facilitate a deeper understanding of the influence of feeding ecology on contaminant exposure and tissue concentrations of coastal mammals. Although there has traditionally been a unidirectional flow of knowledge and techniques between the fields of ecology and ecotoxicology, this view is changing as more ecologists recognize the utility of contaminants as their own feeding ecology tool that can detect differences in diet, habitat type, geographic foraging regions, and even discriminate among populations (Aguilar, 1987; Fisk et al., 2002; Krahn et al., 2007; Méndez-Fernandez et al., 2017; Peterson et al., 2017). Adopting these cross-disciplinary techniques and employing multiple tools wherever possible to quantify foraging behavior will further the fields of toxicology, ecotoxicology, and ecology and improve our understanding of the cumulative impacts of humans on coastal mammal populations.

REFERENCES

Adams, L.G., Farley, S.D., Stricker, C.A., Demma, D.J., Roffler, G.H., Miller, D.C., Rye, R.O., 2010. Are inland wolf — ungulate systems influenced by marine subsidies of Pacific salmon? Ecological Applications 20, 251–262.

Aguilar, A., 1987. Using organochlorine pollutants to discriminate marine mammal populations: a review and critique of the methods. Marine Mammal Science 3, 242–262. https://doi.org/10.1111/j.1748-7692.1987.tb00166.x.

Arthur, B., Hindell, M., Bester, M., Trathan, P., Jonsen, I., Staniland, I., Oosthuizen, W.C., Wege, M., Lea, M.-A., 2015. Return customers: foraging site fidelity and the effect of environmental variability in wide-ranging Antarctic fur seals. PLoS One 10, e0120888. https://doi.org/10.1371/journal.pone.0120888.

Atwell, L., Hobson, K.A., Welch, H.E., 1998. Biomagnification and bioaccumulation of mercury in an arctic marine food web: insights from stable nitrogen isotope analysis. Canadian Journal of Fisheries and Aquatic Sciences 55, 1114–1121. https://doi.org/10.1139/f98-001.

Baker, M.R., Schindler, D.E., Holtgrieve, G.W., Louis, V.L.S.T., 2009. Bioaccumulation and transport of contaminants: migrating sockeye salmon as vectors of mercury. Environmental Science and Technology 43, 8840–8846. https://doi.org/10.1021/es901798f.

Balmer, B.C., Schwacke, L.H., Wells, R.S., George, R.C., Hoguet, J., Kucklick, J.R., Lane, S.M., Martinez, A., McLellan, W.A., Rosel, P.E., Rowles, T.K., Sparks, K., Speakman, T., Zolman, E.S., Pabst, D.A., 2011. Relationship between persistent organic pollutants (POPs) and ranging patterns in common bottlenose dolphins (*Tursiops truncatus*) from coastal Georgia, USA. The Science of the Total Environment 409, 2094–2101. https://doi.org/10.1016/j.scitotenv.2011.01.052.

Barros, N.B., Ostrom, P.H., Stricker, C.A., Wells, R.S., 2010. Stable isotopes differentiate bottlenose dolphins off west-central Florida. Marine Mammal Science 26, 324–336. https://doi.org/10.1111/j.1748-7692.2009.00315.x.

Baylis, A.M.M., Orben, R.A., Arnould, J.P.Y., Peters, K., Knox, T., Costa, D.P., Staniland, I.J., 2015. Diving deeper into individual foraging specializations of a large marine predator, the southern sea lion. Oecologia 179, 1053–1065. https://doi.org/10.10007/s00442-015-3421-4.

Beltran, R., Connolly Sadou, M., Condit, R., Peterson, S., Reichmuth, C., Costa, D., 2015. Fine-scale whisker growth measurements can reveal temporal foraging patterns from stable isotope signatures. Marine Ecology Progress Series 523, 243–253. https://doi.org/10.3354/meps11176.

Beltran, R.S., Peterson, S.H., McHuron, E.A., Reichmuth, C., Hückstädt, L.A., Costa, D.P., 2016. Seals and sea lions are what they eat, plus what? Determination of trophic discrimination factors for seven pinniped species. Rapid Communications in Mass Spectrometry 30, 1115–1122. https://doi.org/10.1002/rcm.7539.

Ben-David, M., Flaherty, E.A., 2012. Stable isotopes in mammalian research: a beginner's guide. Journal of Mammalogy 93, 312–328. https://doi.org/10.1644/11-MAMM-S-166.1.

Bentaleb, I., Martin, C., Vrac, M., Mate, B., Mayzaud, P., Siret, D., De Stephanis, R., Guinet, C., 2011. Foraging ecology of Mediterranean fin whales in a changing environment elucidated by satellite tracking and baleen plate stable isotopes. Marine Ecology Progress Series 438, 285–302. https://doi.org/10.3354/meps09269.

Bentzen, T.W., Follmann, E.H., Amstrup, S.C., York, G.S., Wooller, M.J., Muir, D.C.G., O'Hara, T.M., 2008. Dietary biomagnification of organochlorine contaminants in Alaskan polar bears. Canadian Journal of Zoology 86, 177–191. https://doi.org/10.1139/Z07-124.

Bentzen, T.W., Follmann, E.H., Amstrup, S.C., York, G.S., Wooller, M.J., O'Hara, T.M., 2007. Variation in winter diet of southern Beaufort Sea polar bears inferred from stable isotope analysis. Canadian Journal of Zoology 85, 596–608. https://doi.org/10.1139/Z07-036.

Blais, J.M., Kimpe, L.E., McMahon, D., Keatley, B.E., Mallory, M.L., Douglas, M.S.V., Smol, J.P., 2005. Arctic seabirds transport marine-derived contaminants. Science 309, 445. https://doi.org/10.1126/science.1112658. (80-.).

Blais, J.M., MacDonald, R.W., Mackay, D., Webster, E., Harvey, C., Smol, J.P., 2007. Biologically mediated transport of contaminants to aquatic systems. Environmental Science and Technology 41, 1075–1084. https://doi.org/10.1021/es061314a.

Blum, J.D., Popp, B.N., Drazen, J.C., Anela Choy, C., Johnson, M.W., 2013. Methylmercury production below the mixed layer in the North Pacific Ocean. Nature Geoscience 6, 879–884. https://doi.org/10.1038/ngeo1918.

Bocharova, N., Treu, G., Czirjak, G.A., Krone, O., Stefanski, V., Wibbelt, G., Unnsteindottir, E.R., Hersteinsson, P., Schares, G., Doronina, L., Goltsman, M., Greenwood, A.D., 2013. Correlates between feeding ecology and mercury levels in historical and modern Arctic foxes (*Vulpes lagopus*). PLoS One 8, 1–11. https://doi.org/10.1371/journal.pone.0060879.

Bograd, S.J., Block, B.A., Costa, D.P., Godley, B.J., 2010. Biologging technologies: new tools for conservation. Endangered Species Research 10, 1–7. https://doi.org/10.3354/esr00269.

Bolnick, D.I., Amarasekare, P., Araújo, M.S., Bürger, R., Levine, J.M., Novak, M., Rudolf, V.H.W., Schreiber, S.J., Urban, M.C., Vasseur, D.A., 2011. Why intraspecific trait variation matters in community ecology. Trends in Ecology and Evolution 26, 183–192. https://doi.org/10.1016/j.tree.2011.01.009.

Bolnick, D.I., Svanbäck, R., Fordyce, J.A., Yang, L.H., Davis, J.M., Hulsey, C.D., Forister, M.L., 2003. The ecology of individuals: incidence and implications of individual specialization. The American Naturalist 161, 1–28. https://doi.org/10.1086/343878.

Bond, A.L., Diamond, A.W., 2011. Recent Bayesian stable-isotope mixing models are highly sensitive to variation in discrimination factors. Ecological Applications 21, 1017–1023.

Borgå, K., Fisk, A.T., Hoekstra, P.F., Muir, D.C.G., 2004. Biological and chemical factors of importance in the bioaccumulation and trophic transfer of persistent organochlorine contaminants in arctic marine food webs. Environmental Toxicology and Chemistry 23, 2367–2385. https://doi.org/10.1897/03-518.

Borgå, K., Kidd, K.A., Muir, D.C.G., Berglund, O., Conder, J.M., Gobas, F.A.P.C., Kucklick, J., Malm, O., Powell, D.E., 2012. Trophic magnification factors: considerations of ecology, ecosystems, and study design. Integrated Environmental Assessment and Management 8, 64–84. https://doi.org/10.1002/ieam.244.

Bossart, G.D., 2011. Marine mammals as sentinel species for oceans and human health. Veterinary Pathology Online 48, 676–690. https://doi.org/10.5670/oceanog.2006.77.

Bowen, W.D., Iverson, S.J., 2013. Methods of estimating marine mammal diets: a review of validation experiments and sources of bias and uncertainty. Marine Mammal Science 29, 719–754. https://doi.org/10.1111/j.1748-7692.2012.00604.x.

Braune, B.M., Outridge, P.M., Fisk, A.T., Muir, D.C.G., Helm, P.A., Hobbs, K., Hoekstra, P.F., Kuzyk, Z.A., Kwan, M., Letcher, R.J., Lockhart, W.L., Norstrom, R.J., Stern, G.A., Stirling, I., 2005. Persistent organic pollutants and mercury in marine biota of the Canadian Arctic: an overview of spatial and temporal trends. The Science of the Total Environment 351–352, 4–56. https://doi.org/10.1016/j.scitotenv.2004.10.034.

Bromaghin, J.F., Budge, S.M., Thiemann, G.W., Rode, K.D., 2017. Simultaneous estimation of diet composition and calibration coefficients with fatty acid signature data. Ecology and Evolution 7, 6103–6113. https://doi.org/10.1002/ece3.3179.

Bromaghin, J.F., Budge, S.M., Thiemann, G.W., Rode, K.D., 2016. Assessing the robustness of quantitative fatty acid signature analysis to assumption violations. Methods in Ecology and Evolution 7, 51–59. https://doi.org/10.1111/2041-210X.12456.

Brookens, T.J., Harvey, J.T., O'Hara, T.M., 2007. Trace element concentrations in the Pacific harbor seal (*Phoca vitulina richardii*) in central and northern California. The Science of the Total Environment 372, 676–692. https://doi.org/10.1016/j.scitotenv.2006.10.006.

Brown, D.D., Kays, R., Wikelski, M., Wilson, R., Klimley, A.P., 2013. Observing the unwatchable through acceleration logging of animal behavior. Animal Biotelemetry 2013, 1–16. https://doi.org/10.1186/2050-3385-1-20.

Brown, T.M., Luque, S., Sjare, B., Fisk, A.T., Helbing, C.C., Reimer, K.J., 2014. Satellite telemetry informs PCB source apportionment in a mobile, high trophic level marine mammal: the ringed seal (*Pusa hispida*). Environmental Science and Technology 48, 13110–13119. https://doi.org/10.1021/es504010q.

Browning, N.E., Dold, C., I-Fan, J., Worthy, G.A.J., 2014. Isotope turnover rates and diet-tissue discrimination in skin of *ex situ* bottlenose dolphins (*Tursiops truncatus*). Journal of Experimental Biology 217, 214–221. https://doi.org/10.1242/jeb.093963.

Buchardt, B., Bunch, V., Helin, P., 2007. Fingernails and diet: stable isotope signatures of a marine hunting community from modern Uummannaq, North Greenland. Chemical Geology 244, 316–329. https://doi.org/10.1016/j.chemgeo.2007.06.022.

Budge, S.M., Iverson, S.J., Koopman, H.N., 2006. Studying trophic ecology in marine ecosystems using fatty acids: a primer on analysis and interpretation. Marine Mammal Science 22, 759–801. https://doi.org/10.1111/j.1748-7692.2006.00079.x.

Busquets-Vass, G., Newsome, S.D., Serra-Valente, G., Jacobsen, J.K., Aguíñiga-García, S., 2017. Estimating blue whale skin isotopic incorporation rates and baleen growth rates: implications for assessing diet and movement patterns in mysticetes. PLoS One 12, e0177880.

Calambokidis, J., Barlow, J., 1991. Chlorinated hydrocarbon concentrations and their use for describing population discreteness in harbor porpoises from Washington, Oregon, and California. In: Mar. mammal strandings United States Proc. Second Mar. Mamm. Stranding Work, vol. 98, pp. 101–110.

Campbell, L.M., Norstrom, R.J., Hobson, K.A., Muir, D.C.G., Backus, S., Fisk, A.T., 2005. Mercury and other trace elements in a pelagic Arctic marine food web (Northwater Polynya, Baffin Bay). The Science of the Total Environment 351–352, 247–263. https://doi.org/10.1016/j.scitotenv.2005.02.043.

Cardona-Marek, T., Knott, K.K., Meyer, B.E., O'Hara, T.M., 2009. Mercury concentrations in southern Beaufort sea polar bears: variation based on stable isotopes of carbon and nitrogen. Environmental Toxicology and Chemistry 28, 1416–1424. https://doi.org/10.1897/08-557.1.

Carlton, J.T., Hodder, J., 2003. Maritime mammals: terrestrial mammals as consumers in marine intertidal communities. Marine Ecology Progress Series 256, 271–286. https://doi.org/10.3354/meps256271.

Castellini, J.M., Rea, L.D., Lieske, C.L., Beckmen, K.B., Fadely, B.S., Maniscalco, J.M., O'Hara, T.M., 2012. Mercury concentrations in hair from neonatal and juvenile Steller sea lions (*Eumetopias jubatus*): implications based on age and region in this northern Pacific marine sentinel piscivore. EcoHealth 9, 267–277. https://doi.org/10.1007/s10393-012-0784-4.

Caut, S., Angulo, E., Courchamp, F., 2009. Variation in discrimination factors ($\Delta^{15}N$ and $\Delta^{13}C$): the effect of diet isotopic values and applications for diet reconstruction. Journal of Applied Ecology 46, 443–453. https://doi.org/10.1111/j.1365-2664.2009.01620.x.

Caut, S., Laran, S., Garcia-Hartmann, E., Das, K., 2011. Stable isotopes of captive cetaceans (killer whales and bottlenose dolphins). Journal of Experimental Biology 214, 538–545. https://doi.org/10.1242/jeb.045104.

Cherel, Y., Hobson, K.A., Bailleul, F., Groscolas, R., 2005. Nutrition, physiology, and stable isotopes: new information from fasting and molting penguins. Ecology 86, 2881–2888.

Cherel, Y., Kernaléguen, L., Richard, P., Guinet, C., 2009. Whisker isotopic signature depicts migration patterns and multi-year intra- and inter-individual foraging strategies in fur seals. Biology Letters 5, 830–832. https://doi.org/10.1098/rsbl.2009.0552.

Choy, C.A., Popp, B.N., Kaneko, J.J., Drazen, J.C., 2009. The influence of depth on mercury levels in pelagic fishes and their prey. Proceedings of the National Academy of Sciences of the United States of America 106, 13865–13869. https://doi.org/10.1073/pnas.0900711106.

Christensen, J.R., MacDuffee, M., MacDonald, R.W., Whiticar, M., Ross, P.S., 2005. Persistent organic pollutants in British Columbia grizzly bears: consequence of divergent diets. Environmental Science and Technology 39, 6952–6960. https://doi.org/10.1021/es050749f.

Christensen, J.R., MacDuffee, M., Yunker, M.B., Ross, P.B., 2007. Hibernation-associated changes in persistent organic pollutant (POP) levels and patterns in British Columbia grizzly bears (*Ursus arctos horribilis*). Environmental Science and Technology 41, 1834–1840.

Clarkson, T.W., Magos, L., 2006. The toxicology of mercury and its chemical compounds. Critical Reviews in Toxicology 36, 609–662. https://doi.org/10.1080/10408440600845619.

Connolly, R.M., Guest, M.S., Melville, A.J., Oakes, J.M., 2004. Sulfur stable isotopes separate producers in marine food-web analysis. Oecologia 138, 161–167. https://doi.org/10.1007/s00442-003-1415-0.

Cossaboon, J.M., Ganguli, P.M., Flegal, A.R., 2015. Mercury offloaded in Northern elephant seal hair affects coastal seawater surrounding rookery. Proceedings of the National Academy of Sciences of the United States of America 112, 12058–12062. https://doi.org/10.1073/pnas.1506520112.

Crawford, K., Mcdonald, R.A., Bearhop, S., 2008. Applications of stable isotope techniques to the ecology of mammals. Mammal Review 38, 87–107.

Creed, J.C., 2004. Capybara (*Hydrochaeris hydrochaeris* Rodentia: Hydrochaeridae): a mammalian seagrass herbivore. Estuaries 27, 197–200.

Das, K., Holsbeek, L., Browning, J., Siebert, U., Birkun, A., Bouquegneau, J.M., 2004. Trace metal and stable isotope measurements ($\delta^{13}C$ and $\delta^{15}N$) in the harbour porpoise *Phocoena phocoena relicta* from the Black Sea. Environmental Pollution 131, 197–204. https://doi.org/10.1016/j.envpol.2004.02.006.

Davidson, P.W., Myers, G.J., Weiss, B., 2004. Mercury exposure and child development outcomes. Pediatrics 113, 1023–1029. https://doi.org/10.1542/peds.113.4.S1.1023.

Deagle, B.E., Kirkwood, R., Jarman, S.N., 2009. Analysis of Australian fur seal diet by pyrosequencing prey DNA in faeces. Molecular Ecology 18, 2022–2038. https://doi.org/10.1111/j.1365-294X.2009.04158.x.

Dehn, L.A., Follmann, E.H., Rosa, C., Duffy, L.K., Thomas, D.L., Bratton, G.R., Taylor, R.J., O'Hara, T.M., 2006a. Stable isotope and trace element status of subsistence-hunted bowhead and beluga whales in Alaska and gray whales in Chukotka. Marine Pollution Bulletin 52, 301–319. https://doi.org/10.1016/j.marpolbul.2005.09.001.

Dehn, L.A., Follmann, E.H., Thomas, D.L., Sheffield, G.G., Rosa, C., Duffy, L.K., O'Hara, T.M., 2006b. Trophic relationships in an Arctic food web and implications for trace metal transfer. The Science of the Total Environment 362, 103–123. https://doi.org/10.1016/j.scitotenv.2005.11.012.

Dehn, L.A., Sheffield, G.G., Follmann, E.H., Duffy, L.K., Thomas, D.L., Bratton, G.R., Taylor, R.J., O'Hara, T.M., 2005. Trace elements in tissues of phocid seals harvested in the Alaskan and Canadian Arctic: influence of age and feeding ecology. Canadian Journal of Zoology 83, 726–746. https://doi.org/10.1139/z05-053.

Dehn, L.A., Sheffield, G.G., Follmann, E.H., Duffy, L.K., Thomas, D.L., O'Hara, T.M., 2007. Feeding ecology of phocid seals and some walrus in the Alaskan and Canadian Arctic as determined by stomach contents and stable isotope analysis. Polar Biology 30, 167–181. https://doi.org/10.1007/s00300-006-0171-0.

Dietz, R., Born, E.W., Rigét, F., Aubail, A., Sonne, C., Drimmie, R., Basu, N., 2011. Temporal trends and future predictions of mercury concentrations in Northwest Greenland polar bear (*Ursus maritimus*) hair. Environmental Science and Technology 45, 1458–1465. https://doi.org/10.1021/es1028734.

Dietz, R., Sonne, C., Basu, N., Braune, B., O'Hara, T., Letcher, R.J., Scheuhammer, T., Andersen, M., Andreasen, C., Andriashek, D., Asmund, G., Aubail, A., Baagøe, H., Born, E.W., Chan, H.M., Derocher, A.E., Grandjean, P., Knott, K., Kirkegaard, M., Krey, A., Lunn, N., Messier, F., Obbard, M., Olsen, M.T., Ostertag, S., Peacock, E., Renzoni, A., Rigét, F.F., Skaare, J.U., Stern, G., Stirling, I., Taylor, M., Wiig, Ø., Wilson, S., Aars, J., 2013. What are the toxicological effects of mercury in Arctic biota? The Science of the Total Environment 443, 775–790. https://doi.org/10.1016/j.scitotenv.2012.11.046.

Drago, M., Franco-Trecu, V., Cardona, L., Inchausti, P., 2015. Diet-to-female and female-to-pup isotopic discrimination in South American sea lions. Rapid Communications in Mass Spectrometry 29, 1513–1520. https://doi.org/10.1002/rcm.7249.

Driscoll, C.T., Mason, R.P., Chan, H.M., Jacob, D.J., Pirrone, N., 2013. Mercury as a global pollutant: sources, pathways, and effects. Environmental Science and Technology 47, 4967–4983. https://doi.org/10.1021/es305071v.

Elfes, C.T., VanBlaricom, G.R., Boyd, D., Calambokidis, J., Clapham, P.J., Pearce, R.W., Robbins, J., Salinas, J.C., Straley, J.M., Wade, P.R., Krahn, M.M., 2010. Geographic variation of persistent organic pollutant levels in humpback whale (*Megaptera novaeangliae*) feeding areas of the North Pacific and North Atlantic. Environmental Toxicology and Chemistry 29, 824–834. https://doi.org/10.1002/etc.110.

Elliott, K.H., Cesh, L.S., Dooley, J.A., Letcher, R.J., Elliott, J.E., 2009. PCBs and DDE, but not PBDEs, increase with trophic level and marine input in nestling bald eagles. The Science of the Total Environment 407, 3867–3875. https://doi.org/10.1016/j.scitotenv.2009.02.027.

Fair, P.A., Mitchum, G., Hulsey, T.C., Adams, J., Zolman, E., McFee, W., Wirth, E., Bossart, G.D., 2007. Polybrominated diphenyl ethers (PBDEs) in blubber of free-ranging bottlenose dolphins (*Tursiops truncatus*) from two southeast Atlantic estuarine areas. Archives of Environmental Contamination and Toxicology 53, 483–494. https://doi.org/10.1007/s00244-006-0244-7.

Fielman, K.T., Woodin, S.A., Walla, M.D., Lincoln, D.E., 1999. Widespread occurrence of natural halogenated organics among temperate marine infauna. Marine Ecology Progress Series 181, 1–12. https://doi.org/10.3354/meps181001.

Fisk, A., Hobson, K., Norstrom, R., 2001. Influence of chemical and biological factors on trophic transfer of persistent organic pollutants in the Northwater Polynya marine food web. Environmental Science and Technology 35, 732–738.

Fisk, A.T., de Wit, C.A., Wayland, M., Kuzyk, Z.Z., Burgess, N., Letcher, R., Braune, B., Norstrom, R., Blum, S.P., Sandau, C., Lie, E., Larsen, H.J.S., Skaare, J.U., Muir, D.C.G., 2005. An assessment of the toxicological significance of anthropogenic contaminants in Canadian arctic wildlife. The Science of the Total Environment 351–352, 57–93. https://doi.org/10.1016/j.scitotenv.2005.01.051.

Fisk, A.T., Tittlemier, S.A., Pranschke, J.L., Norstrom, R.J., 2002. Using anthropogenic contaminants and stable isotopes to assess the feeding ecology of Greenland sharks. Ecology 83, 2162–2172.

Fitzgerald, W.F., Lamborg, C.H., Hammerschmidt, C.R., 2007. Marine biogeochemical cycling of mercury. Chemical Reviews 107, 641–662. https://doi.org/10.1021/cr050353m.

Flegal, A.R., Stephenson, M., Martin, M., Martin, J.H., 1981. Elevated concentrations of mercury in mussels (*Mytilus californianus*) associated with pinniped colonies. Marine Biology 65, 45–48. https://doi.org/10.1007/BF00397066.

Fuller, B.T., Fuller, J.L., Sage, N.E., Harris, D.A., O'Connell, T.C., Hedges, R.E.M., 2005. Nitrogen balance and $\delta^{15}N$: why you're not what you eat during nutritional stress. Rapid Communications in Mass Spectrometry 19, 2497–2506. https://doi.org/10.1002/rcm.2090.

Fuller, B.T., Fuller, J.L., Sage, N.E., Harris, D.A., O'Connell, T.C., Hedges, R.E.M., 2004. Nitrogen balance and $\delta^{15}N$: why you're not what you eat during pregnancy. Rapid Communications in Mass Spectrometry 18, 2889–2896. https://doi.org/10.1002/rcm.1708.

Gannes, L.Z., Martınez del Rio, C., Koch, P., 1998. Natural abundance variations in stable isotopes and their use in animal physiological ecology. Comparative Biochemistry and Physiology Part A: Molecular and Integrative Physiology 119A, 725–737. https://doi.org/10.1016/S1095-6433(98)01016-2.

Germain, L.R., Koch, P.L., Harvey, J., McCarthy, M.D., 2013. Nitrogen isotope fractionation in amino acids from harbor seals: implications for compound-specific trophic position calculations. Marine Ecology Progress Series 482, 265–277. https://doi.org/10.3354/meps10257.

Germain, L.R., McCarthy, M.D., Koch, P.L., Harvey, J.T., 2012. Stable carbon and nitrogen isotopes in multiple tissues of wild and captive harbor seals (*Phoca vitulina*) off the California coast. Marine Mammal Science 28, 542–560. https://doi.org/10.1111/j.1748-7692.2011.00516.x.

Gill, G.A., Fitzgerald, W.F., 1988. Vertical mercury distributions in the oceans. Geochimica et Cosmochimica Acta 52, 1719–1728. https://doi.org/10.1016/0016-7037(88)90240-2.

Graham, B.S., Koch, P.L., Newsome, S.D., McMahon, K., Aurioles, D., 2010. Using isoscapes to trace the movements and foraging behavior of top predators in oceanic ecosystems. In: West, J.B., Bowen, G.J., Dawson, T.E., Tu, K.P. (Eds.), Isoscapes: Understanding Movement, Pattern, and Process on Earth through Isotope Mapping. Springer, pp. 1–487. https://doi.org/10.1007/978-90-481-3354-3.

Gray, J.S., 2002. Biomagnification in marine systems: the perspective of an ecologist. Marine Pollution Bulletin 45, 46–52. https://doi.org/10.1016/S0025-326X(01)00323-X.

Greaves, D.K., Hammill, M.O., Eddington, J.D., Schreer, J.F., 2004. Growth rate and shedding of vibrissae in the gray seal, *Halichoerus grypus*: a cautionary note for stable isotope diet analysis. Marine Mammal Science 20, 296–304.

Harley, J., O'Hara, T.M., 2016. Toxicology and poisons. In: Castellini, M.A., Mellish, J.-A. (Eds.), Marine Mammal Physiology: Requisites for Ocean Living. CRC Press, Boca Raton, FL, pp. 309–336.

Hazen, E.L., Maxwell, S.M., Bailey, H., Bograd, S.J., Hamann, M., Gaspar, P., Godley, B.J., Shillinger, G.L., 2012. Ontogeny in marine tagging and tracking science: technologies and data gaps. Marine Ecology Progress Series 457, 221–240. https://doi.org/10.3354/meps09857.

Hilderbrand, G.V., Farley, S.D., Robbins, C.T., Hanley, T.A., Titus, K., Servheen, C., 1996. Use of stable isotopes to determine diets of living and extinct bears. Canadian Journal of Zoology 74, 2080–2088. https://doi.org/10.1139/z98-165.

Hobson, K.A., 1999. Tracing origins and migration of wildlife using stable isotopes: a review. Oecologia 120, 314–326.

Hobson, K.A., Clark, R.G., 1992. Assessing avian diets using stable isotopes I: turnover of ^{13}C in tissues. The Condor 94, 181–188. https://doi.org/10.2307/1368807.

Hobson, K.A., Fisk, A., Karnovsky, N., Holst, M., Gagnon, J.-M., Fortier, M., 2002. A stable isotope ($\delta^{13}C$, $\delta^{15}N$) model for the North Water food web: implications for evaluating trophodynamics and the flow of energy and contaminants. Deep Sea Research Part II: Topical Studies in Oceanography 49, 5131–5150. https://doi.org/10.1016/S0967-0645(02)00182-0.

Hobson, K.A., Rigét, F., Outridge, P.M., Dietz, R., Born, E., 2004. Baleen as a biomonitor of mercury content and dietary history of North Atlantic Minke whales (*Balaenopetra acutorostrata*): combining elemental and stable isotope approaches. The Science of the Total Environment 331, 69–82. https://doi.org/10.1016/j.scitotenv.2004.03.024.

Hobson, K.A., Sease, J.L., 1998. Stable isotope analyses of tooth annuli reveal temporal dietary records: an example using Steller sea lions. Marine Mammal Science 14, 116–129. https://doi.org/10.1111/j.1748-7692.1998.tb00694.x.

Hoekstra, P.F., O'Hara, T.M., Fisk, A.T., Borgå, K., Solomon, K.R., Muir, D.C.G., 2003. Trophic transfer of persistent organochlorine contaminants (OCs) within an Arctic marine food web from the southern Beaufort-Chukchi Seas. Environmental Pollution 124, 509–522. https://doi.org/10.1016/S0269-7491(02)00482-7.

Hussey, N.E., Dudley, S.F.J., McCarthy, I.D., Cliff, G., Fisk, A.T., 2011. Stable isotope profiles of large marine predators: viable indicators of trophic position, diet, and movement in sharks? Canadian Journal of Fisheries and Aquatic Sciences 68, 2029–2045. https://doi.org/10.1139/F2011-115.

Iverson, S., Field, C., Bowen, W., Blanchard, W., 2004. Quantitative fatty acid signature analysis: a new method of estimating predator diets. Ecological Monographs 74, 211–235. https://doi.org/10.1890/02-4105.

Jamieson, A.J., Malkocs, T., Piertney, S.B., Fujii, T., Zhang, Z., 2017. Bioaccumulation of persistent organic pollutants in the deepest ocean fauna. Nature Ecology and Evolution 1, 51. https://doi.org/10.1038/s41559-016-0051.

Jardine, T.D., Kidd, K.A., Fisk, 2006. Applications, considerations, and sources of uncertainty when using stable isotope analysis in ecotoxicology. Environmental Science and Technology 40, 7501–7511.

Jessup, D.A., Johnson, C.K., Estes, J., Carlson-Bremer, D., Jarman, W.M., Reese, S., Dodd, E., Tinker, M.T., Ziccardi, M.H., 2010. Persistent organic pollutants in the blood of free-ranging sea otters (*Enhydra lutris* spp.) in Alaska and California. Journal of Wildlife Diseases 46, 1214–1233. https://doi.org/10.7589/0090-3558-46.4.1214.

Jones, H.J., Swadling, K.M., Butler, E.C.V., Barry, L.A., Macleod, C.K., 2014. Application of stable isotope mixing models for defining trophic biomagnification pathways of mercury and selenium. Limnology and Oceanography 59, 1181–1192. https://doi.org/10.4319/lo.2014.59.4.1181.

Jones, K.C., de Voogt, P., 1999. Persistent organic pollutants (POPs): state of the science. Environmental Pollution 100, 209–221. https://doi.org/10.1016/S0269-7491(99)00098-6.

Kelly, J.F., 2000. Stable isotopes of carbon and nitrogen in the study of avian and mammalian trophic ecology. Canadian Journal of Zoology 78, 1–27. https://doi.org/10.1139/z99-165.

Kernaléguen, L., Arnould, J.P.Y., Guinet, C., Cherel, Y., 2015. Determinants of individual foraging specialization in large marine vertebrates, the Antarctic and subantarctic fur seals. Journal of Animal Ecology 84, 1081–1091. https://doi.org/10.1111/1365-2656.12347.

Killengreen, S.T., Lecomte, N., Ehrich, D., Schott, T., Yoccoz, N.G., Ims, R.A., 2011. The importance of marine vs. human-induced subsidies in the maintenance of an expanding mesocarnivore in the arctic tundra. Journal of Animal Ecology 80, 1049–1060. https://doi.org/10.1111/j.1365-2656.2011.01840.x.

Knott, K.K., Boyd, D., Ylitalo, G.M., O'Hara, T.M., 2011. Concentrations of mercury and polychlorinated biphenyls in blood of Southern Beaufort Sea polar bears (*Ursus maritimus*) during spring : variations with lipids and stable isotope ($\delta^{15}N$, $\delta^{13}C$) values. Canadian Journal of Zoology 89, 999–1012. https://doi.org/10.1139/Z11-071.

Krahn, M.M., Herman, D.P., Matkin, C.O., Durban, J.W., Barrett-Lennard, L.G., Burrows, D.G., Dahlheim, M.E., Black, N., LeDuc, R.G., Wade, P.R., 2007. Use of chemical tracers in assessing the diet and foraging regions of eastern North Pacific killer whales. Marine Environmental Research 63, 91–114.

Kucklick, J., Schwacke, L., Wells, R., Hohn, A., Guichard, A., Yordy, J., Hansen, L., Zolman, E., Wilson, R., Litz, J., Nowacek, D., Rowles, T., Pugh, R., Balmer, B., Sinclair, C., Rosel, P., 2011. Bottlenose dolphins as indicators of persistent organic pollutants in the western North Atlantic Ocean and northern Gulf of Mexico. Environmental Science and Technology 45, 4270–4277.

Kurle, C.M., 2002. Stable-isotope ratios of blood components from captive northern fur seals (*Callorhinus ursinus*) and their diet: applications for studying the foraging ecology of wild otariids. Canadian Journal of Zoology 80, 902–909. https://doi.org/10.1139/Z02-069.

Kurle, C.M., Bakker, V.J., Copeland, H., Burnett, J., Jones Scherbinski, J., Brandt, J., Finkelstein, M.E., 2016. Terrestrial scavenging of marine mammals: cross-ecosystem contaminant transfer and potential risks to endangered California condors (*Gymnogyps californianus*). Environmental Science and Technology 50, 9114–9123. https://doi.org/10.1021/acs.est.6b01990.

Lamborg, C.H., Hammerschmidt, C.R., Bowman, K.L., Swarr, G.J., Munson, K.M., Ohnemus, D.C., Lam, P.J., Heimbürger, L.-E., Rijkenberg, M.J.A., Saito, M.A., 2014. A global ocean inventory of anthropogenic mercury based on water column measurements. Nature 512, 65–68. https://doi.org/10.1038/nature13563.

Laurier, F.J.G., Mason, R.P., Gill, G.A., Whalin, L., 2004. Mercury distributions in the North Pacific Ocean - 20 years of observations. Marine Chemistry 90, 3–19. https://doi.org/10.1016/j.marchem.2004.02.025.

Lavoie, R.A., Jardine, T.D., Chumchal, M.M., Kidd, K.A., Campbell, L.M., 2013. Biomagnification of mercury in aquatic food webs: a worldwide meta-analysis. Environmental Science and Technology 47, 13385–13394.

Law, R.J., 2014. An overview of time trends in organic contaminant concentrations in marine mammals: going up or down? Marine Pollution Bulletin 82, 7–10. https://doi.org/10.1016/j.marpolbul.2014.03.024.

Lecomte, N., Ahlstrøm, Ø., Ehrich, D., Fuglei, E., Ims, R.A., Yoccoz, N.G., 2011. Intrapopulation variability shaping isotope discrimination and turnover: experimental evidence in arctic foxes. PLoS One 6, e21357. https://doi.org/10.1371/journal.pone.0021357.

Letcher, R.J., Bustnes, J.O., Dietz, R., Jenssen, B.M., Jørgensen, E.H., Sonne, C., Verreault, J., Vijayan, M.M., Gabrielsen, G.W., 2010. Exposure and effects assessment of persistent organohalogen contaminants in arctic wildlife and fish. The Science of the Total Environment 408, 2995–3043. https://doi.org/10.1016/j.scitotenv.2009.10.038.

Litz, J.A., Garrison, L.P., Fieber, L.A., Martinez, A., Contillo, J.P., Kucklick, J.R., 2007. Fine-scale spatial variation of persistent organic pollutants in bottlenose dolphins (*Tursiops truncatus*) in Biscayne Bay, Florida. Environmental Science and Technology 41, 7222–7228. https://doi.org/10.1021/es070440r.

Lohmann, R., Breivik, K., Dachs, J., Muir, D., 2007. Global fate of POPs: current and future research directions. Environmental Pollution 150, 150–165. https://doi.org/10.1016/j.envpol.2007.06.051.

Lopez, J., Hyrenbach, K.D., Littnan, C., Ylitalo, G.M., 2014. Geographic variation of persistent organic pollutants in Hawaiian monk seals *Monachus schauinslandi* in the main Hawaiian Islands. Endangered Species Research 24, 249–262. https://doi.org/10.3354/esr00602.

Lorrain, A., Graham, B.S., Popp, B.N., Allain, V., Olson, R.J., Hunt, B.P.V., Potier, M., Fry, B., Galván-Magaña, F., Menkes, C.E.R., Kaehler, S., Ménard, F., 2015. Nitrogen isotopic baselines and implications for estimating foraging habitat and trophic position of yellowfin tuna in the Indian and Pacific Oceans. Deep Sea Research Part II: Topical Studies in Oceanography 113, 188–198. https://doi.org/10.1016/j.dsr2.2014.02.003.

Lundström, K., Hjerne, O., Lunneryd, S., Karlsson, O., 2010. Understanding the diet composition of marine mammals: grey seals (*Halichoerus grypus*) in the Baltic Sea. ICES Journal of Marine Science 1230–1239.

Mackay, D., Shiu, W.Y., Ma, K.-C., Lee, S.C., 2006. Handbook of Physical-Chemical Properties and Environmental Fate for Organic Chemicals, Second ed. CRC Press, Boca Raton, FL.

Matthews, C.J.D., Ferguson, S.H., 2015. Seasonal foraging behaviour of eastern Canada-West Greenland bowhead whales: an assessment of isotopic cycles along baleen. Marine Ecology Progress Series 522, 269–286. https://doi.org/10.3354/meps11145.

McClelland, J.W., Montoya, J.P., 2002. Trophic relationships and the nitrogen isotopic composition of amino acids in plankton. Ecology 83, 2173–2180. https://doi.org/10.1890/0012-9658(2002)083[2173:TRATNI]2.0.CO;2.

McCutchan, J.H.J., Lewis, W.M., Carol, K., McGrath, C.C., 2003. Variation in trophic shift for stable isotope ratios of carbon, nitrogen, and sulfur. Oikos 102, 378–390.

McGrew, A.K., Ballweber, L.R., Moses, S.K., Stricker, C.A., Beckmen, K.B., Salman, M.D., O'Hara, T.M., 2014. Mercury in gray wolves (*Canis lupus*) in Alaska: increased exposure through consumption of marine prey. The Science of the Total Environment 468–469, 609–613. https://doi.org/10.1016/j.scitotenv.2013.08.045.

McHuron, E.A., Harvey, J.T., Castellini, J.M., Stricker, C.A., O'Hara, T.M., 2014. Selenium and mercury concentrations in harbor seals (*Phoca vitulina*) from central California: health implications in an urbanized estuary. Marine Pollution Bulletin 83, 48–57.

McHuron, E.A., Robinson, P.W., Simmons, S.E., Kuhn, C.E., Fowler, M., Costa, D.P., 2016a. Foraging strategies of a generalist marine predator inhabiting a dynamic environment. Oecologia 182, 995–1005. https://doi.org/10.1007/s00442-016-3732-0.

McHuron, E.A., Walcott, S.M., Zeligs, J., Skrovan, S., Costa, D.P., Reichmuth, C., 2016b. Whisker growth dynamics in two North Pacific pinnipeds: implications for determining foraging ecology from stable isotope analysis. Marine Ecology Progress Series 554, 213–224. https://doi.org/10.3354/meps11793.

McIntyre, T., 2014. Trends in tagging of marine mammals: a review of marine mammal biologging studies. African Journal of Marine Science 36, 409–422. https://doi.org/10.2989/1814232X.2014.976655.

McKinney, M.A., Letcher, R.J., Aars, J., Born, E.W., Branigan, M., Dietz, R., Evans, T.J., Gabrielsen, G.W., Muir, D.C.G., Peacock, E., Sonne, C., 2011. Regional contamination versus regional dietary differences: understanding geographic variation in brominated and chlorinated contaminant levels in polar bears. Environmental Science and Technology 45, 896–902. https://doi.org/10.1021/es102781b.

McLaren, A.A.D., Crawshaw, G.J., Patterson, B.R., 2015. Carbon and nitrogen discrimination factors of wolves and accuracy of diet inferences using stable isotope analysis. Wildlife Society Bulletin. https://doi.org/10.1002/wsb.599.

McMahon, K.W., Hamady, L.L., Thorrold, S.R., 2013. A review of ecogeochemistry approaches to estimating movements of marine animals. Limnology and Oceanography 58, 697–714. https://doi.org/10.4319/lo.2013.58.2.0697.

Méndez-Fernandez, P., Simon-Bouhet, B., Bustamante, P., Chouvelon, T., Ferreira, M., López, A., Moffat, C.F., Pierce, G.J., Russell, M., Santos, M.B., Spitz, J., Vingada, J.V., Webster, L., Read, F.L., González, A.F., Caurant, F., 2017. Inter-species differences in polychlorinated biphenyls patterns from five sympatric species of odontocetes: can PCBs be used as tracers of feeding ecology? Ecological Indicators 74, 98–108. https://doi.org/10.1016/j.ecolind.2016.11.013.

Michener, R., Kaufman, L., 2007. Stable isotope ratios as tracers in marine food webs. In: Michener, R., Lajtha, K. (Eds.), Stable Isotopes in Ecology and Environmental Science. Blackwell Scientific Publications, Boston, pp. 238–282.

Miljeteig, C., Strøm, H., Gavrilo, M.V., Volkov, A., Jenssen, B.M., Gabrielsen, G.W., 2009. High levels of contaminants in ivory gull *Pagophila eburnea* eggs from the Russian and Norwegian arctic. Environmental Science and Technology 43, 5521–5528. https://doi.org/10.1021/es900490n.

Monteiro, L.R., Granadeiro, J.P., Furness, R.W., 1998. Relationship between mercury levels and diet in Azores seabirds. Marine Ecology Progress Series 166, 259–265. https://doi.org/10.3354/meps166259.

Montoya, J., 2007. Natural abundance of ^{15}N in marine planktonic ecosystems. In: Michener, R., Lajtha, K. (Eds.), Stable Isotopes in Ecology and Environmental Science. Blackwell Scientific Publications, Boston, pp. 176–201.

Moore, P., 2002. Mammals in intertidal and maritime ecosystems: interactions, impacts and implications. Oceanography and Marine Biology: An Annual Review 40, 491–608.

Muir, D.C.G., de Wit, C.A., 2010. Trends of legacy and new persistent organic pollutants in the circumpolar arctic: overview, conclusions, and recommendations. The Science of the Total Environment 408, 3044–3051. https://doi.org/10.1016/j.scitotenv.2009.11.032.

Newsome, S.D., Bentall, G.B., Tinker, M.T., Oftedal, O.T., Ralls, K., Estes, J.A., Fogel, M.L., 2010a. Variation in $\Delta^{13}C$ and $\Delta^{15}N$ diet-vibrissae trophic discrimination factors in a wild population of California sea otters. Ecological Applications 20, 1744–1752.

Newsome, S.D., Clementz, M.T., Koch, P.L., 2010b. Using stable isotope biogeochemistry to study marine mammal ecology. Marine Mammal Science 26, 509–572. https://doi.org/10.1111/j.1748-7692.2009.00354.x.

Newsome, S.D., Koch, P.L., Etnier, M.A., Aurioles-Gamboa, D., 2006. Using carbon and nitrogen isotope values to investigate maternal strategies in Northeast Pacific otariids. Marine Mammal Science 22, 556–572. https://doi.org/10.1111/j.1748-7692.2006.00043.x.

Newsome, S.D., Tinker, M.T., Monson, D.H., Oftedal, O.T., Ralls, K., Staedler, M.M., Fogel, M.L., Estes, J.A., 2009. Using stable isotopes to investigate individual diet specialization in California sea otters (*Enhydra lutris nereis*). Ecology 90, 961–974.

Newsome, S.D., Yeakel, J.D., Wheatley, P.V., Tinker, M.T., 2012. Tools for quantifying isotopic niche space and dietary variation at the individual and population level. Journal of Mammalogy 93, 329–341. https://doi.org/10.1644/11-MAMM-S-187.1.

Noël, M., Spence, J., Harris, K.A., Robbins, C.T., Fortin, J.K., Ross, P.S., Christensen, J.R., 2014. Grizzly bear hair reveals toxic exposure to mercury through salmon consumption. Environmental Science and Technology 48, 7560–7567. https://doi.org/10.1021/es500631g.

Nordstrom, C.A., Wilson, L.J., Iverson, S.J., Tollit, D.J., 2008. Evaluating quantitative fatty acid signature analysis (QFASA) using harbour seals *Phoca vitulina richardsi* in captive feeding studies. Marine Ecology Progress Series 360, 245–263. https://doi.org/10.3354/meps07378.

Oken, E., Radesky, J.S., Wright, R.O., Bellinger, D.C., Amarasiriwardena, C.J., Kleinman, K.P., Hu, H., Gillman, M.W., 2008. Maternal fish intake during pregnancy, blood mercury levels, and child cognition at age 3 years in a US cohort. American Journal of Epidemiology 167, 1171–1181. https://doi.org/10.1093/aje/kwn034.

Paul, C., Pohnert, G., 2011. Production and role of volatile halogenated compounds from marine algae. Natural Product Reports 28, 186–195. https://doi.org/10.1039/C0NP00043D.

Peterson, B.J., Fry, B., 1987. Stable isotopes in ecosystem studies. Annual Review of Ecology, Evolution and Systematics 18, 293–320.

Peterson, S.H., Ackerman, J.T., Costa, D.P., 2015a. Marine foraging ecology influences mercury bioaccumulation in deep-diving northern elephant seals. Proceedings of the Royal Society of London B 282, 20150710.

Peterson, S.H., Ackerman, J.T., Eagles-Smith, C.A., 2017. Mercury contamination and stable isotopes reveal variability in foraging ecology of generalist California gulls. Ecological Indicators 74, 205–215. https://doi.org/10.1016/j.ecolind.2016.11.025.

Peterson, S.H., Peterson, M.G., Debier, C., Covaci, A., Dirtu, A.C., Malarvannan, G., Crocker, D.E., Schwarz, L.K., Costa, D.P., 2015b. Deep-ocean foraging northern elephant seals bioaccumulate persistent organic pollutants. The Science of the Total Environment 533, 144–155. https://doi.org/10.1016/j.scitotenv.2015.06.097.

Phillips, D.L., 2012. Converting isotope values to diet composition: the use of mixing models. Journal of Mammalogy 93, 342–352. https://doi.org/10.1644/11-MAMM-S-158.1.

Phillips, D.L., Newsome, S.D., Gregg, J.W., 2005. Combining sources in stable isotope mixing models: alternative methods. Oecologia 144, 520–527. https://doi.org/10.1007/s00442-004-1816-8.

Polischuk, S.C., Hobson, K.A., Ramsay, M.A., 2001. Use of stable-carbon and -nitrogen isotopes to assess weaning and fasting in female polar bears and their cubs. Canadian Journal of Zoology 79, 499–511. https://doi.org/10.1139/z01-007.

Polito, M.J., Brasso, R.L., Trivelpiece, W.Z., Karnovsky, N., Patterson, W.P., Emslie, S.D., 2016. Differing foraging strategies influence mercury (Hg) exposure in an Antarctic penguin community. Environmental Pollution 218, 196–206. https://doi.org/10.1016/j.envpol.2016.04.097.

Pompanon, F., Deagle, B.E., Symondson, W.O.C., Brown, D.S., Jarman, S.N., Taberlet, P., 2012. Who is eating what: diet assessment using next generation sequencing. Molecular Ecology 21, 1931–1950. https://doi.org/10.1111/j.1365-294X.2011.05403.x.

Post, D.M., 2002. Using stable isotopes to estimate trophic position: models, methods, and assumptions. Ecology 83, 703–718.

Post, D.M., Layman, C.A., Arrington, D.A., Takimoto, G., Quattrochi, J., Montaña, C.G., 2007. Getting to the fat of the matter: models, methods and assumptions for dealing with lipids in stable isotope analyses. Oecologia 152, 179–189. https://doi.org/10.1007/s00442-006-0630-x.

Rea, L.D., Castellini, J.M., Correa, L., Fadely, B.S., O'Hara, T.M., 2013. Maternal Steller sea lion diets elevate fetal mercury concentrations in an area of population decline. The Science of the Total Environment 454–455, 277–282. https://doi.org/10.1016/j.scitotenv.2013.02.095.

Rea, L.D., Christ, A., Hayden, A., Stegall, V., Farley, S., Stricker, C., Mellish, J.-A.E., Maniscalco, J.M., Waite, J., Burkanov, V., 2015. Age-specific vibrissae growth rates: a tool for determining the timing of ecologically important events in Steller sea lions. Marine Mammal Science 31, 1213–1233. https://doi.org/10.1111/mms.12221.

Reid, R., 2014. Dietary Ecology of Coastal Coyotes (Canis latrans): Marine-Terrestrial Linkages from the Holocene to Present (Ph.D. dissertation). University of California Santa Cruz.

Rigét, F., Bignert, A., Braune, B., Stow, J., Wilson, S., 2010. Temporal trends of legacy POPs in Arctic biota, an update. The Science of the Total Environment 408, 2874–2884. https://doi.org/10.1016/j.scitotenv.2009.07.036.

Robbins, C.T., Felicetti, L.A., Sponheimer, M., 2005. The effect of dietary protein quality on nitrogen isotope discrimination in mammals and birds. Oecologia 144, 534–540. https://doi.org/10.1007/s00442-005-0021-8.

Rode, K.D., Stricker, C.A., Erlenbach, J., Robbins, C.T., Cherry, S.G., Newsome, S.D., Cutting, A., Jensen, S., Stenhouse, G., Brooks, M., Hash, A., Nicassio, N., 2016. Isotopic incorporation and the effects of fasting and dietary lipid content on isotopic discrimination in large carnivorous mammals. Physiological and Biochemical Zoology 89, 182–197. https://doi.org/10.1086/686490.

Rogers, M.C., Peacock, E., Simac, K., O'Dell, M.B., Welker, J.M., 2015. Diet of female polar bears in the southern Beaufort Sea of Alaska: evidence for an emerging alternative foraging strategy in response to environmental change. Polar Biology 38, 1035–1047. https://doi.org/10.1007/s00300-015-1665-4.

Ross, P.S., Noël, M., Lambourn, D., Dangerfield, N., Calambokidis, J., Jeffries, S., 2013. Declining concentrations of persistent PCBs, PBDEs, PCDEs, and PCNs in harbor seals (*Phoca vitulina*) from the Salish sea. Progress in Oceanography 115, 160–170. https://doi.org/10.1016/j.pocean.2013.05.027.

Roth, J.D., 2002. Temporal variability in arctic fox diet as reflected in stable-carbon isotopes; the importance of sea ice. Oecologia 133, 70–77. https://doi.org/10.1007/s00442-002-1004-7.

Rutz, C., Hays, G.C., 2009. New frontiers in biologging science. Biology Letters 5, 289–292. https://doi.org/10.1098/rsbl.2009.0089.

Schell, D.M., Saupe, S.M., Haubenstock, N., 1989. Bowhead whale (*Balaena mysticetus*) growth and feeding as estimated by $\delta^{13}C$ techniques. Marine Biology 103, 433–443. https://doi.org/10.1007/bf00399575.

Scheuhammer, A.M., Meyer, M.W., Sandheinrich, M.B., Murray, M.W., 2007. Effects of environmental methylmercury on the health of wild birds, mammals, and fish. Ambio 36, 12–18.

St. Louis, V.L., Derocher, A.E., Stirling, I., Graydon, J.A., Lee, C., Jocksch, E., Richardson, E., Ghorpade, S., Kwan, A.K., Kirk, J.L., Lehnherr, I., Swanson, H.K., 2011. Differences in mercury bioaccumulation between polar bears (*Ursus maritimus*) from the Canadian high- and sub-Arctic. Environmental Science and Technology 45, 5922–5928. https://doi.org/10.1021/es2000672.

Streets, D.G., Zhang, Q., Wu, Y., 2009. Projections of global mercury emissions in 2050. Environmental Science and Technology 43, 2983–2988. https://doi.org/10.1021/es802474j.

Stricker, C., Christ, A., Wunder, M., Doll, A., Farley, S., Rea, L., Rosen, D., Scherer, R., Tollit, D., 2015. Stable carbon and nitrogen isotope trophic enrichment factors for Steller sea lion vibrissae relative to milk and fish/invertebrate diets. Marine Ecology Progress Series 523, 255–266. https://doi.org/10.3354/meps11205.

Sumich, J.L., 2001. Growth of baleen of a rehabilitating gray whale calf. Aquatic Mammals 27.3, 234–238.

Sunderland, E.M., Krabbenhoft, D.P., Moreau, J.W., Strode, S.A., Landing, W.M., 2009. Mercury sources, distribution, and bioavailability in the North Pacific Ocean: insights from data and models. Global Biogeochemical Cycles 23, 1–14. https://doi.org/10.1029/2008GB003425.

Sunderland, E.M., Mason, R.P., 2007. Human impacts on open ocean mercury concentrations. Global Biogeochemical Cycles 21. https://doi.org/10.1029/2006GB002876.

Szepanski, M.M., Ben-David, M., Van Ballenberghe, V., 1999. Assessment of anadromous salmon resources in the diet of the Alexander Archipelago wolf using stable isotope analysis. Oecologia 120, 327–335. https://doi.org/10.1007/s004420050866.

Tambling, C.J., Laurence, S.D., Bellan, S.E., Cameron, E.Z., du Toit, J.T., Getz, W.M., 2012. Estimating carnivoran diets using a combination of carcass observations and scats from GPS clusters. Journal of Zoology 286, 102–109. https://doi.org/10.1038/jid.2014.371.

Tarroux, A., Bety, J., Gauthier, G., Berteaux, D., 2012. The marine side of a terrestrial carnivore: intra-population variation in use of allochthonous resources by Arctic foxes. PLoS One 7. https://doi.org/10.1371/journal.pone.0042427.

Teuten, E.L., Xu, L., Reddy, C.M., 2005. Two abundant bioaccumulated halogenated compounds are natural products. Science 307, 917–919 (80-.).

Thomas, A.C., Deagle, B.E., Eveson, J.P., Harsch, C.H., Trites, A.W., 2016. Quantitative DNA metabarcoding: improved estimates of species proportional biomass using correction factors derived from control material. Molecular Ecology Resources 16, 714–726. https://doi.org/10.1111/1755-0998.12490.

Thomas, A.C., Jarman, S.N., Haman, K.H., Trites, A.W., Deagle, B.E., 2014. Improving accuracy of DNA diet estimates using food tissue control materials and an evaluation of proxies for digestion bias. Molecular Ecology 23, 3706–3718. https://doi.org/10.1111/mec.12523.

Thomas, S.M., Crowther, T.W., 2015. Predicting rates of isotopic turnover across the animal kingdom: a synthesis of existing data. Journal of Animal Ecology 84, 861–870. https://doi.org/10.1111/1365-2656.12326.

Tieszen, L.L., Boutton, T.W., Tesdahl, K.G., Slade, N.A., 1983. Fractionation and turnover of stable carbon isotopes in animal tissues: implications for $\delta^{13}C$ analysis of diet. Oecologia 57, 32–37. https://doi.org/10.1007/BF00379558.

Titcomb, E.M., Reif, J.S., Fair, P.A., Stavros, H.-C.W., Mazzoil, M., Bossart, G.D., Schaefer, A.M., 2017. Blood mercury concentrations in common bottlenose dolphins from the Indian River Lagoon, Florida: patterns of social distribution. Marine Mammal Science 1–14. https://doi.org/10.1111/mms.12390.

Tollit, D.J., Schulze, A.D., Trites, A.W., Olesiuk, P.F., Crockford, S.J., Gelatt, T.S., Ream, R.R., Miller, K.M., 2009. Development and application of DNA techniques for validating and improving pinniped diet estimates. Ecological Applications 19, 889–905. https://doi.org/10.1890/07-1701.1.

Tomy, G.T., Muir, D.C.G., Stern, G.A., Westmore, J.B., 2000. Levels of C_{10} - C_{13} polychloro-n-alkanes in marine mammals from the Arctic and the St. Lawrence River Estuary. Environmental Science and Technology 34, 1615–1619.

Tyrrell, L.P., Newsome, S.D., Fogel, M.L., Viens, M., Bowden, R., Murray, M.J., 2013. Vibrissae growth rates and trophic discrimination factors in captive southern sea otters (*Enhydra lutris nereis*). Journal of Mammalogy 94, 331–338. https://doi.org/10.1644/12-MAMM-A-035.1.

Van De Vijver, K.I., Hoff, P.T., Das, K., Van Dongen, W., Esmans, E.L., Jauniaux, T., Bouquegneau, J.M., Blust, R., De Coen, W., 2003. Perfluorinated chemicals infiltrate ocean waters: link between exposure levels and stable isotope ratios in marine mammals. Environmental Science and Technology 37, 5545–5550. https://doi.org/10.1021/es0345975.

Villegas-Amtmann, S., Costa, D., Tremblay, Y., Salazar, S., Aurioles-Gamboa, D., 2008. Multiple foraging strategies in a marine apex predator, the Galapagos sea lion *Zalophus wollebaeki*. Marine Ecology Progress Series 363, 299–309. https://doi.org/10.3354/meps07457.

Wang, W., Evans, R.D., Hickie, B.E., Rouvinen-Watt, K., Evans, H.E., 2014. Methylmercury accumulation and elimination in mink (*Neovison vison*) hair and blood: results of a controlled feeding experiment using stable isotope tracers. Environmental Toxicology and Chemistry 33, 2873–2880. https://doi.org/10.1002/etc.2762.

Wells, R.S., Tornero, V., Borrell, A., Aguilar, A., Rowles, T.K., Rhinehart, H.L., Hofmann, S., Jarman, W.M., Hohn, A.A., Sweeney, J.C., 2005. Integrating life-history and reproductive success data to examine potential relationships with organochlorine compounds for bottlenose dolphins (*Tursiops truncatus*) in Sarasota Bay, Florida. The Science of the Total Environment 349, 106–119. https://doi.org/10.1016/j.scitotenv.2005.01.010.

Wolf, M., Weissing, F.J., 2012. Animal personalities: consequences for ecology and evolution. Trends in Ecology and Evolution 27, 452–461. https://doi.org/10.1016/j.tree.2012.05.001.

Wolf, N., Newsome, S.D., Peters, J., Fogel, M.L., 2015. Variability in the routing of dietary proteins and lipids to consumer tissues influences tissue-specific isotopic discrimination. Rapid Communications in Mass Spectrometry 29, 1448–1456. https://doi.org/10.1002/rcm.7239.

Wolfe, M.F., Schwarzbach, S., Sulaiman, R.A., 1998. Effects of mercury on wildlife: a comprehensive review. Environmental Toxicology and Chemistry 17, 146–160. https://doi.org/10.1002/etc.5620170203.

Woshner, V., Knott, K., Wells, R., Willetto, C., Swor, R., O'Hara, T., 2008. Mercury and selenium in blood and epidermis of bottlenose dolphins (*Tursiops truncatus*) from Sarasota Bay, FL: interaction and relevance to life history and hematologic parameters. EcoHealth 5, 360–370. https://doi.org/10.1007/s10393-008-0164-2.

Zhang, Y., Jacob, D.J., Horowitz, H.M., Chen, L., Amos, H.M., Krabbenhoft, D.P., Slemr, F., St. Louis, V.L., Sunderland, E.M., 2016. Observed decrease in atmospheric mercury explained by global decline in anthropogenic emissions. Proceedings of the National Academy of Sciences of the United States of America 113, 526–531. https://doi.org/10.1073/pnas.1516312113.

Zhao, L., Schell, D., 2004. Stable isotope ratios in harbor seal *Phoca vitulina* vibrissae: effects of growth patterns on ecological records. Marine Ecology Progress Series 281, 267–273. https://doi.org/10.3354/meps281267.

Zhao, L., Schell, D.M., Castellini, M.A., 2006. Dietary macronutrients influence ^{13}C and ^{15}N signatures of pinnipeds: captive feeding studies with harbor seals (*Phoca vitulina*). Comparative Biochemistry and Physiology Part A: Molecular and Integrative Physiology 143, 469–478. https://doi.org/10.1016/j.cbpa.2005.12.032.

Chapter 3

The Effects of Oil Exposure on Cetaceans

Céline A.J. Godard-Codding[1], Tracy K. Collier[2]
[1]The Institute of Environmental and Human Health Texas Tech University, Lubbock, TX, United States; [2]Ocean Associates, Inc., Bainbridge Island, WA, United States

INTRODUCTION

Properties of Oil

The extraction and transportation of fossil fuels and their various refined products inevitably leads to accidental releases of these substances into receiving waters, and these releases are collectively termed "oil spills." The use of a single term to encompass a wide range of accidental releases of oil, generally petroleum products, can imply a relatively straightforward relationship between oil spills and their effects in aquatic systems, but this is not the case. Oil spills exhibit a wide range of initial compositions, depending on the source of the oil. Crude oils contain thousands of different organic and inorganic compounds (Scholz et al., 1999) and can vary widely depending on the geographic areas, depths, and methods used to extract them (Neff, 1990). Once extracted, crude oils can be highly manipulated or distilled to be reformulated into different end products (Barber et al., 1996), making the understanding of the associated toxicity and the determination of the source of releases difficult. Spilled oil is also subject to complex physical and chemical processes that markedly change the initial composition of the oil, and these processes are collectively termed weathering. Crude (i.e., unrefined) oils are complex mixtures comprised of thousands of compounds, all with varying degrees of toxicity. An excellent review of the characteristics of oils was recently published by the Royal Society of Canada (Lee et al., 2015). Briefly, crude oils are comprised of four major classes of compounds, designated as saturates, aromatics, resins, and asphaltenes.

Of these four classes of compounds, aromatics are the most studied and generally of most concern regarding their toxicity to biota. Aromatic compounds contain one or more rings of six carbons, each connected by alternating carbon-carbon double bonds, and they are generally considered the most harmful compounds in oil (Neff, 1979), with smaller single-ring aromatics, such as benzene,

Marine Mammal Ecotoxicology. https://doi.org/10.1016/B978-0-12-812144-3.00003-6

strongly associated with carcinogenicity, organ damage, and even death at high exposure levels in vertebrates (ATSDR, 1995a). These single-ring compounds are also readily available to biologic systems due to their relatively high water solubility and volatility in air; however, they are often not found in large concentrations except immediately following a spill. Conversely, compounds containing two or more aromatic rings (also called polycyclic aromatic hydrocarbons, or PAHs) can be carcinogens and cause reproductive failure, developmental abnormalities, organ damage, genotoxicity, and immunotoxicity in laboratory and field settings (Conney, 1982; ATSDR, 1995b; White, 2002; Incardona et al., 2004; Collier et al., 2014) and are less volatile. Thus, while PAHs may be found at comparatively low concentrations in oil products and are fairly insoluble in water, they are of concern due to their persistence and potential to cause a wide range of adverse effects. The PAHs present in petroleum products generally are highly alkylated.

Weathering refers to the range of physical, chemical, and biologic processes that change the composition and properties of spilled oils. Spilled oil products begin to weather immediately after they are released into the environment, and weathering occurs over spans of a few hours to a few years. There are many reviews and discussions of weathering of oil, and again one recent and highly informative treatment of this complex subject is found in the recent publication by the Royal Society of Canada (Lee et al., 2015). Studies of weathering are generally compartmentalized into considerations of weathering at the water's surface, in the water column, on shorelines, and in sediments underlying the water column.

Exposure of Cetaceans to Oil

Cetaceans can be exposed to oil and oil-derived compounds via several routes of exposure. These include direct contact, ingestion of oil or oil-contaminated prey, inhalation of volatilized or aerosolized oil and/or oil components, and aspiration of oil directly into lungs. However, most components of oil, including PAHs, are rapidly metabolized by vertebrate species, including fish and marine mammals. Thus, detection of oil exposure is complicated by the highly complex chemical nature of oils and weathered oils, as described before, and by the rapid degradation of oil-derived compounds by the animals themselves. Because cetaceans of necessity must breathe at the air/water interface, where most oil spills also are obviously found, determining the cooccurrence of cetaceans and the footprint of spilled oil is the most robust method for establishing exposure. This was very successfully done following the Deepwater Horizon (DWH) oil spill, and as described later, our understanding of the effects of oil on cetaceans has increased exponentially as a result of that tragic oil spill (DWH NRDA, 2016). Because the metabolites of PAHs are primarily excreted via the bile, urine, or in the feces of exposed animals, it can also be possible to determine exposure by analyzing these matrices for PAH metabolites. A review of the utility of measuring PAH metabolites in bile was recently published (Beyer et al., 2010), but obviously there are ethical and logistic issues associated with the collection of bile from cetaceans. The DWH oil spill pointed to the need to develop more sophisticated and noninvasive methods for assessing oil exposure in cetaceans.

EFFECTS OF OIL ON CETACEANS

Range of Organismal Effects Following Oil Exposure

The effects of oil in cetaceans have been investigated to some extent following oil spills and in a small number of past experimental exposure studies, which are now prohibited. The wide range of reported effects include poor body condition, calcium imbalance, inflammation, reproductive failure, organ damage (lung disease and failure, adrenal gland disease), altered hepatobiliary function, immune changes and increased susceptibility to infections, impaired stress response, and death (Geraci and St. Aubin, 1982; Engelhardt, 1983; Engelhardt, 1987; Matkin et al., 2008; Geraci, 1990; Schwacke et al., 2014; Lane et al., 2015; Venn-Watson et al., 2013, 2015a; Colegrove et al., 2016; DWH NRDA Trustees, 2016; De Guise et al., 2017; Kellar et al., 2017; Smith et al., 2017; Jessup and Leighton, 1996; Johnson and Ziccardi, 2006; Helm et al., 2015).

To date, the greatest and most detailed source of information on this topic is related to the 2010 DWH oil spill that released an estimated 794.9 million liters of crude oil in the northern Gulf of Mexico (nGoM) following a drilling platform explosion (DWH NRDA Trustees, 2016). Fig. 3.1 depicts the range of effects shown across a wide range of taxa following the DWH oil spill. A total of 21 cetacean species inhabiting the nGoM had ranges overlapping with the DHW surface oil footprint, and the habitats of two specific bottlenose dolphin (*Tursiops truncatus*) stocks, the Barataria Bay (BB) and Mississippi Sound (MS) stocks, were heavily oiled (Waring et al., 2013; Dias et al., 2017). The BB dolphin stock has been characterized as exhibiting long-term and year-round residency, suggesting prolonged exposure to oil (Wells et al., 2017). The increased mortality and adverse health effects reported in coastal bottlenose dolphins from the nGoM were deemed consistent with, and likely caused by, exposure to oil from the DWH spill (DWH NRDA Trustees, 2016; Takeshita et al., 2017; Wallace et al., 2017). These adverse health effects included poor body condition, pulmonary disease (pneumonia, lung abscesses, and lung infections), chronic adrenal gland disease, reproductive failure, serum hepatobiliary abnormalities, and a compromised stress response identified by abnormally low measures of serum cortisol and aldosterone suggestive of hypoadrenocorticism (Schwacke et al., 2014; Lane et al., 2015; Venn-Watson et al., 2015a; Colegrove et al., 2016; Kellar et al., 2017). Immune changes compatible with an increase in both intra- and extracellular bacterial infections and suggestive of oil exposure were also observed in BB dolphins for 3 years following the spill (De Guise et al., 2017). Investigations of a perinatal dolphin stranding cluster

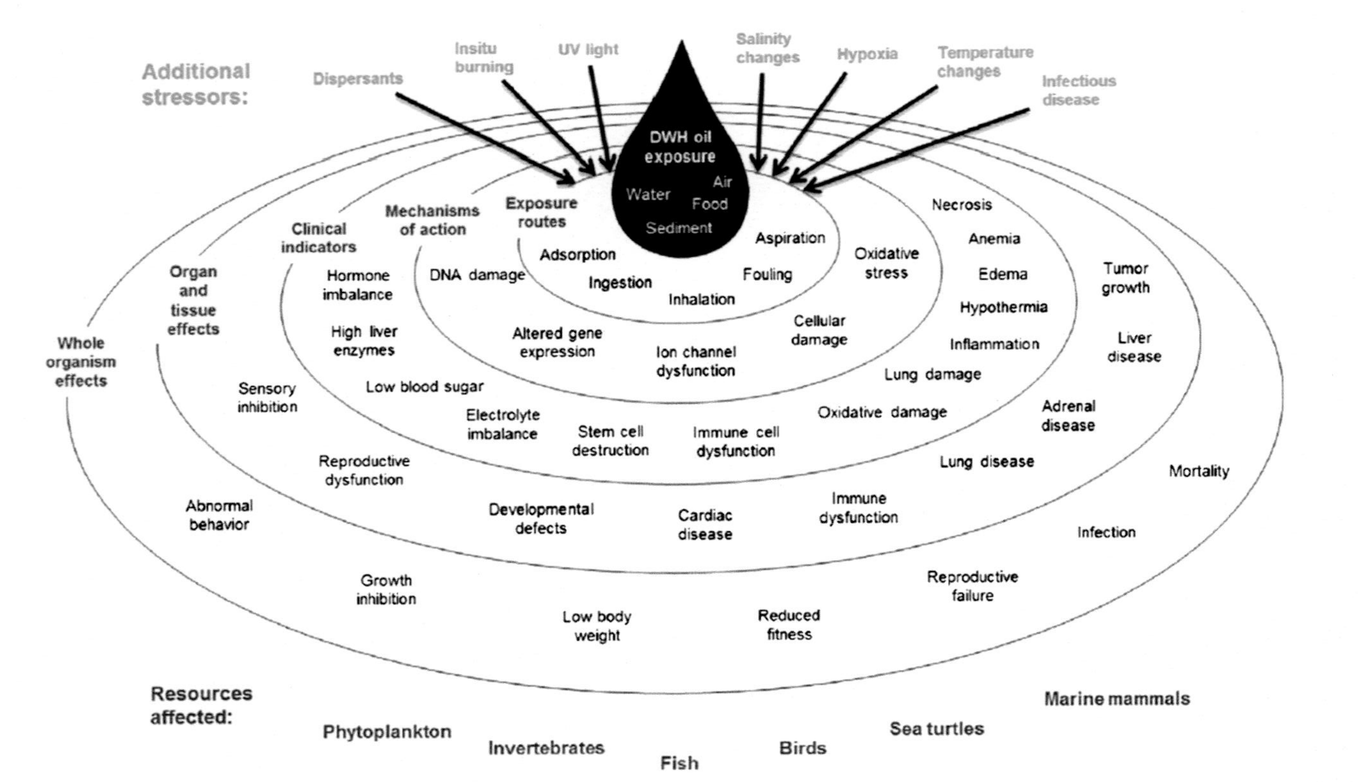

FIGURE 3.1 Biologic effects of the DWH oil spill. This conceptual figure illustrates the relationships between oil exposure and toxicologic effects in organisms affected by the spill. All exposure and effects elements shown here are supported by findings in the DWH oil spill literature. *(From Beyer, J., Trannum, H.C., Bakke, T., Hodson, P.V., Collier, T.K., 2016. Environmental effects of the deepwater horizon oil spill: a review. Marine Pollution Bulletin 110, 28–51. https://doi.org/10.1016/j.marpolbul.2016.06.027.)*

observed during the DWH-associated unusual mortality event highlighted an increased susceptibility in nGoM populations to late-term pregnancy failures and in utero infections (Colegrove et al., 2016). Close to 4 years after the spill, only 20% of pregnant dolphins that had been part of a 2011 live-capture health assessment produced viable calves, while a reference population presented an 83% pregnancy success rate (Lane et al., 2015). The reproductive success rates estimated for the BB and MS stocks, during and after the spill, were less than a third of those previously observed in nonimpacted areas (Kellar et al., 2017). Severe lung disease (including alveolar interstitial syndrome, pulmonary masses, and lung consolidation) was estimated to be five times more likely in BB dolphins compared to dolphins from a reference site (Schwacke et al., 2014) and persisted to some degree for at least 4 years post spill (Smith et al., 2017). Histologic evaluation of 46 nonperinatal nGoM dolphins that stranded in 2010–12 showed a higher prevalence of primary bacterial pneumonia and rare adrenal gland disease characterized by thin cortices, compared to reference dolphins stranded outside the unusual mortality event associated with the DWH spill (Venn-Watson et al., 2015a). The primary bacterial pneumonia caused or significantly contributed to death in 70% of these dolphins, while their adrenal cortical atrophy was consistent with the low blood cortisol and aldosterone and evidence of hypoadrenocorticism observed in live-captured BB dolphins in 2011 (Venn-Watson et al., 2015a). Close to half of 29 BB dolphins evaluated during a 2011 capture-release health assessment were given a guarded or worse prognosis, while 17% were not expected to survive (Schwacke et al., 2014). Overall, poor health, high stranding rates, and low survival estimates pointed to increased mortality in BB dolphins for at least 3 years following the spill (Lane et al., 2015; McDonald et al., 2017). The proportion of guarded or worse prognoses was still higher in BB dolphins than in reference dolphins 4 years after the spill but had decreased over time (Smith et al., 2017). Overall, the link between oil exposure and observed injuries was established by the combined evidence resulting from the stranding data, necropsy investigations, and live health evaluations that were part of the thorough natural resource damage assessment following the DWH spill and by the ruling out of alternative causes (Litz et al., 2014; Venn-Watson et al., 2015a,c; Fauquier et al., 2017; Takeshita et al., 2017).

Organismal Effects Related to Specific Routes of Exposure

The three main routes of oil exposure in cetaceans are respiratory, dermal, and oral, as in most animals. Assigning the effects reported in cetaceans to a specific route is challenging when in vivo experiments are prohibited. A few in vitro PAH exposure studies have provided valuable information on the resulting cellular response, cytotoxicity, and immunotoxicity in cetaceans but are not reviewed further here (Carvan et al., 1995; Godard et al., 2006; Fossi et al., 2014; White et al., 2017). Published reports including historical in vivo studies relating to effects linked to a specific route are limited in number and discussed subsequently.

Inhalation and Aspiration

The unique respiratory adaptations of cetaceans likely play a role in the frequency or severity of effects resulting from exposure via inhalation. Cetaceans breathe directly above the surface and are thus susceptible to inhaling petroleum compounds in both volatile as well as aerosolized forms. Cetaceans have four to nine times deeper lung air exchange than humans, an extensive blood supply to the lungs, and deep inhalations followed by breath holding but no nasal turbinates or cilia to filter air (Ridgway et al., 1969; Green, 1972; Ridgway, 1972). These adaptations can lead to increased coating of the lung surface, prolonged exchange with blood, and a more rapid and effective distribution of inhaled substances into the systemic circulation. Moreover, when petroleum compounds are inhaled and absorbed into the circulatory system at the lungs, that blood is then circulated to the rest of the body by way of the heart, thus bypassing the liver, a major site for detoxification (Fig. 3.2). The special adaptations of cetacean lungs are also relevant in the case of oil aspiration, a route of lung exposure that could potentially occur in cetaceans by direct aspiration of oil at the air/water interface, or by regurgitation of vomited oil following oral exposure. Hydrocarbon vapor concentrations following an oil spill have been shown to reach levels injurious to cetacean health (Geraci and St Aubin, 1982; Geraci, 1990) and may lead to mucous membrane inflammation, lung congestion, pneumonia, liver disorders, and neurological issues (Geraci, 1990; Neff, 1990). The respiratory distress and ensuing death of seven dolphins exposed to oil by

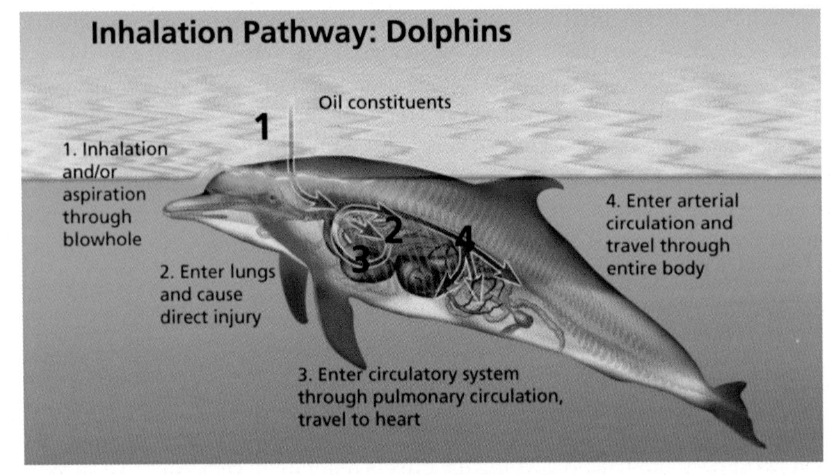

FIGURE 3.2 After inhalation of petroleum components (or aspiration of oil) into a cetacean's respiratory system, substances can be absorbed into the blood from the lungs, and are then distributed throughout the animal's body by the heart, initially bypassing the liver. Illustration by Kate Sweeney. *(From DWH NRDA (Deepwater Horizon Natural Resource Damage assessment) Trustees, 2016. Deepwater Horizon Oil Spill Programmatic Damage Assessment and Restoration Plan and Programmatic Environmental Impact Statement. DWH NRDA. www.gulfspillrestoration.noaa.gov/restoration-planning/gulf-plan/ and Takeshita et al. (2017).)*

inhalation after a spill in the Arabian Sea was previously reported (Griffiths et al., 1987 as reported in Matkin et al., 2008). Captive US Navy dolphins exposed to high levels of PAHs released into the air during two separate firestorms in California exhibited mild physiologic effects, including alterations in pulmonary function, immune response, and calcium homeostasis (Venn-Watson et al., 2013). However, the PAHs resulting from combustion processes (e.g., pyrogenic PAHs) are higher in molecular weight and considerably less alkylated, compared to petroleum-derived, or petrogenic, PAHs (Collier et al., 2014).

Dermal Exposure

Effects of oil on skin are determined in part by whether oil can adhere to the skin and subsequently penetrate it. There are many reports of dolphins and whales swimming and feeding through oil slicks (Evans, 1982; Geraci, 1990; Harvey and Dahlheim, 1994; Gubbay and Earll, 2000; Matkin et al., 2008; Schwacke et al., 2014; Takeshita et al., 2017), and photos collected during and after the DWH spill show that oil can adhere and persist on cetacean skin (Dias et al., 2017). The viscosity and other characteristics of the oil and the texture of the exposed surface are likely to influence the extent of adherence (Engelhardt, 1983). The thick epidermis (10–20 times thicker than in humans) and tight intercellular bridges characteristic of cetacean skin are thought to be likely to limit oil penetration (Geraci, 1990; O'Hara and O'Shea, 2001). In vivo experiments in which crude oil or gasoline was applied for up to 75 min to restricted areas of the skin of three species of odontocetes (bottlenose dolphins, Risso's dolphin *Grampus griseus*, and sperm whale *Physeter catodon*) did not elicit the irritation or inflammation responses typically observed in rodents (Nessel et al., 1999) but resulted in transient damage in epidermal cells (Geraci and St. Aubin, 1982).

Oral Exposure

Cetaceans can be exposed to oil orally through direct exposure while eating or whenever the mouth is open and indirectly through contaminated prey (Goodale et al., 1981; Engelhardt, 1983; Matkin et al., 2008). Exposure to viscous oil may interfere with filtering capability in mysticetes due to coating of baleen plates (Geraci and St. Aubin, 1982; Braithwaite, 1983; Engelhardt, 1983). No clinical or hematologic changes were observed after small quantities (5 mL/day) of machine oil were fed 5 days a week for 10 weeks to one captive bottlenose dolphin (Caldwell and Caldwell, 1982). Oiled harbor seals were reported as significantly more approachable due to lethargy following Exxon Valdez spill (EVOS) and would likely have been easy prey for killer whales in the area (Lowry et al., 1994; Matkin et al., 2008). Contaminated prey may thus be an important source of oral exposure to oil, especially in case of prolonged environmental contamination, and it has been suggested as a possible cause of sublethal effects that could eventually have led to decreased fitness and death in the case of EVOS killer whales (Matkin et al., 2008). Numerous dolphins were seen feeding in areas of surface, subsurface, and sediment oiling during and after the DWH spill (Schwacke et al., 2014). When petroleum compounds are absorbed from the gastrointestinal tract after ingestion, they are transported first to the liver via the hepatic portal vein, where significant detoxification is thought to occur (Fig. 3.3). Also, ingested oil may lead to aspiration of oily vomitus, which has been reported in humans and cattle (Takeshita et al., 2017).

Effects Observed in Other Relevant Species

Additional effects of oil exposure reported in seals, sea otters, and surrogate laboratory species (mink) as well as humans and terrestrial mammals are presumably relevant to some extent to cetaceans, even though species-specific responses have been reported (Engelhardt, 1983). Because deliberate in vivo contaminant exposures of cetaceans are no longer acceptable, every effort should be made to collect data relevant to elucidating these effects during oil spills and other accidental exposure scenarios and to refine our capacity to extrapolate between

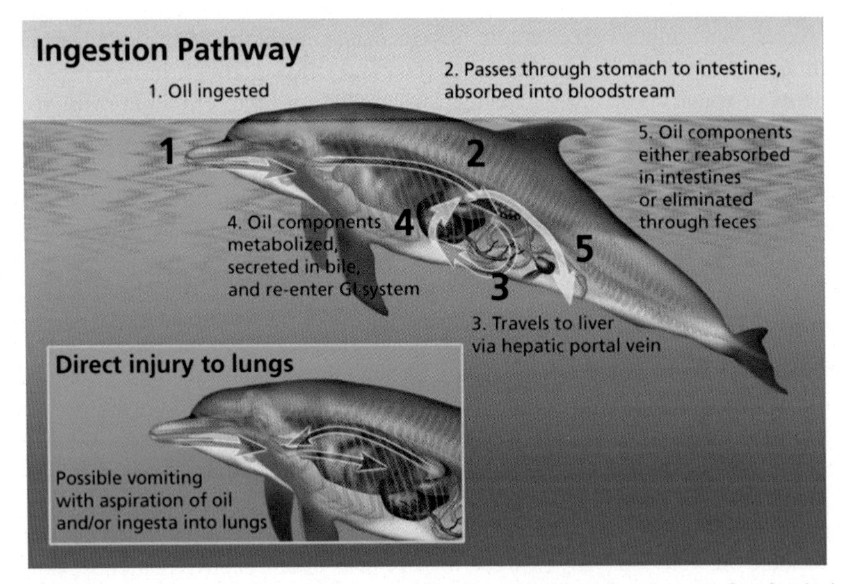

FIGURE 3.3 After ingestion of oil or oil-derived compounds, substances can be absorbed into the blood from the gastrointestinal tract, and are then transported via the hepatic portal vein to the liver, before circulation to the rest of the animal's body. Animals that ingest oil may become nauseous and vomit oil and ingesta, which may then be aspirated into the lungs Illustration by Kate Sweeney. *(From From DWH NRDA (Deepwater Horizon Natural Resource Damage assessment) Trustees, 2016. DWH Oil Spill Programmatic Damage Assessment and Restoration Plan and Programmatic Environmental Impact Statement. DWH NRDA. www.gulfspillrestoration.noaa.gov/restoration-planning/gulf-plan/ and Takeshita et al. (2017).)*

species. Related publications are too numerous to summarize, but a selected number of relevant studies are reviewed next.

Aquatic Species

Seals exposed to oil following the EVOS in Prince William Sound, Alaska, in 1989 were significantly affected by intramyelinic edema in the midbrain, neuronal swelling and necrosis, axonal swelling and degeneration, and hyperkeratosis of the epidermis compared to nonoiled seals (Spraker et al., 1994). The neuronal lesions, when severe, can cause behavior alterations, preventing normal swimming and feeding (Spraker et al., 1994). Skin and neuronal injuries along with eye irritation including conjunctivitis and corneal abrasions have been reported in other instances where seals have been exposed to oil in their environment or experimentally (Geraci and Smith, 1976; St. Aubin, 1990; O'Hara and O'Shea, 2001). In sea otters, reproductive failure, interstitial pulmonary emphysema, centrilobular hepatic necrosis, lipidosis in the liver and kidney, gut lesions and hemorrhage, and multiple alterations in blood chemistry, hematology, and liver enzymes were recorded following exposure to oil from the EVOS (Lipscomb et al., 1993; Lipscomb et al., 1994; Rebar et al., 1995; Tuomi and Williams, 1995; O'Hara and O'Shea, 2001). Oil ingestion is known to cause irritation of the gut mucosa, vomiting, and regurgitation in cattle (Rowe et al., 1973 and Edwards, 1989) and has been linked to issues in sea otter (bodkin 12 MEPS). Experimental chronic crude or fuel oil ingestion in mink (*Mustela vison*) resulted in alteration of hepatic metabolism, adrenal physiology, and blood parameters such as leukocyte numbers and erythrocyte homeostasis (Mazet et al., 2001; Schwartz et al., 2004; Mohr et al., 2008, 2010). The adrenal issues included hypoadrenocorticism, similar to what was reported in live-captured dolphin after the DWH spill (Schwacke et al., 2014), and cortical hypertrophy with vacuolation of corticocytes (Mohr et al., 2008, 2010). The reproductive output and kit survival rates of mink sows fed crude oil and bunker C oil was significantly reduced by twofold to fivefold, respectively, compared to control unexposed minks, even though they exhibited no signs of toxicosis (Mazet et al., 2001).

Humans and Other Terrestrial Mammalian Species

Inflammatory response, lung diseases (atelectasis, hemorrhage, edema, pneumonia, fibrosis, lower respiratory tract symptoms, pneumonary dysfunction), central nervous system effects (mydriasis, tremors, prostration), and death have been reported in humans and other terrestrial mammals upon exposure to oil or selected petroleum hydrocarbons via inhalation (Carpenter et al., 1975, 1976; Coppock et al., 1995, 1996; Eade et al., 1974; Zock et al. 2007, 2012; Sim et al., 2010; Tormoehlen et al., 2014), as well as aspiration of regurgitate following ingestion (Eade et al., 1974; Scharf et al., 1981; Tormoehlen et al., 2014; Makrygianni et al., 2016). Respiratory symptoms in oil spill clean-up workers have been shown to persist for years after exposure (Zock et al., 2007, 2012). A review of acute and chronic effects observed in humans (including clean-up workers and fishermen) after both inhalation and dermal exposure to oil spills also included additional eye and skin symptoms, reproductive and endocrine abnormalities, tiredness, fatigue, and psychological disorders (Levy and Nassetta, 2011). A substantial number of studies suggest exposure to oil and gas activities has a negative impact on human reproduction with evidence for an increase in spontaneous abortion rate, poorer semen quality, and disruption of the estrogen, androgen, and progesterone receptors (as reviewed in Mehri, 2010 and Balise et al., 2016). Additionally, maternal lung and adrenal cortex issues similar to those reported in humans and dolphins following oil exposure can cause prenatal and neonatal complications and mortalities (Hobel and Culhane, 2003; Goodnight and Soper, 2005; Beehner et al., 2006; Schwacke et al., 2014; Smith et al., 2017). Alterations in blood profiles and liver enzymes reported in bottlenose dolphins exposed to DWH oil (Schwacke et al., 2014) were consistent with those observed in DWH clean-up workers (D'Andrea and Reddy, 2013), while their reproductive failure (Kellar et al., 2017) was consistent with effects suggested in humans (Mehri, 2010; Levy and Nassetta, 2011; Balise et al., 2016).

Population Effects Following Oil Exposure

Effects observed following oil exposure in cetaceans go beyond the organismal level. Population effects are discussed here with a focus on mortality events, reproductive success, and growth rate.

Unusual Mortality Events (UME)

A UME is defined under the Marine Mammal Protection Act of 1972 as "a stranding that is unexpected; involves a significant die-off of any marine mammal population; and demands immediate response." Oil exposure following a spill can lead to adverse effects at a population level known to contribute to marine mammal UMEs (Litz et al., 2014; Venn-Watson et al., 2015b; Colegrove et al., 2016; Wallace et al., 2017). In the case of the DWH oil spill, these adverse effects were found to cause the largest and longest (from March 2010 until July 2014) marine mammal UME tracked by NOAA, with over 1140 recorded cetacean strandings (Litz et al., 2014; Venn-Watson et al., 2015b; Colegrove et al., 2016; NOAA Fisheries, 2016).

Population Effects Beyond UME

Population or stock abundance can be greatly affected by oil exposure, as observed following the DWH spill. The reproductive success rates estimated for the BB and MS stocks, during and after the DWH spill, were less than a third of those previously observed in nonimpacted areas (Lane et al., 2015; Kellar et al., 2017). Ultimately, reductions of up to 51%–62% of stock abundance were estimated for BB and MS bottlenose dolphins (DWH NRDA Trustees, 2016; Schwacke et al., 2017; Wallace et al., 2017). The DWH NRDA Trustees (2016) assessed that recovery from these drastic reductions in population abundance would likely take several decades and dedicated restoration actions. Schwacke et al. (2017) estimated a sharp decline in population 9 years post spill, over 30,000 lost cetacean years, and 39 years to recover to 95% of baseline for the BB bottlenose dolphin stock. Similar recovery estimates were projected for other nGoM dolphin stocks (DWH NRDA Trustees, 2016; Schwacke et al., 2017). Similarly, two populations of killer whales (*Orcinus orca*) in Prince William Sound, Alaska, were severely affected following EVOS (Loughlin, 1994; Matkin et al., 2008). One year after the spill, one resident and one transient populations had suffered losses of 33% and 41%, respectively (Matkin et al., 2008). The resident population had not recovered to prespill numbers 16 years post spill and exhibited a rate of yearly increase half that of reference resident populations (Dahlheim and Matkin, 1994; Matkin et al., 2008). The transient population's continuous decline included the significant loss of juvenile and reproductive-age females and unexpected mortalities in years after the spill (Matkin et al., 2008). It resulted in the listing of this population as depleted under the Marine Mammal Protection Act (Matkin et al., 2008). A clear link was suggested between these effects and exposure to the oil spill, but the disruption of social groups, sharp decline in prey, and exposure to persistent pollutants may have also been contributing factors (Ylitalo et al., 2001; Ver Hoef and Frost, 2003; Matkin et al., 2008; Wade et al., 2012).

FUTURE RESEARCH NEEDS AND RECOMMENDATIONS

The thorough natural resource damage assessment following the DWH spill included a population model parameterized specifically for six nGoM bottlenose dolphin stocks including the BB population (DWH NRDA Trustees, 2016; Takeshita et al., 2017; Schwacke et al., 2017). The model featured a density-dependent fecundity function and compared population trajectories between the observed oil exposure scenario and a no oil exposure scenario. The link between exposure and effects and the resulting lost dolphin years and time to recovery were determined by a weight of evidence approach that included density, stranding, and habitat data as well as contaminant and pathogen analyses from health assessments of live-captured animals. This model provided a powerful paradigm to holistically quantify long-term population impacts based on available information. It is recommended that a careful and sustained observation of the affected stocks be continued to allow for refinement of the model as new data becomes available. The design of similar population models would benefit our understanding of the long-term effects of oil exposure and is recommended for future oil spill scenarios. Further research on how best to evaluate reduced reproductive success and survival when no data is available, and on the limitations of assuming no effects in offspring of affected individuals in such population models, is needed.

The scale and scope of the DWH NRDA required extensive logistics, financial resources, and interdisciplinary collaborations (Wilkin et al., 2017). The ensuing benefit to the field of cetacean toxicology, cetacean conservation and management, and ocean health warrants similar assessments to be conducted in future oil spill scenarios whenever possible. It is recommended that information on at least some of the parameters critical for successful damage assessment, such as density, stranding, and habitat data, be collected on a regular basis in areas of oil production to provide baseline information in preparedness for future spills. On a human logistical note, the 2010 DWH spill highlighted the need for marine mammal stranding network personnel with Hazardous Waste Operations and Emergency Response (HAZWOPER) certification (Wilkin et al., 2017). It is recommended that such personnel, along with state and federal agents and researchers likely to be involved with spill response and damage assessment, should obtain and maintain current HAZWOPER certification.

Literature on the effect of oils on other marine mammals and on surrogate species has provided important information for understanding possible effects in cetaceans (Engelhardt 1983; Mazet et al. 2000; Mazet et al. 2001; Schwartz et al. 2004; Mohr et al. 2008; Mohr et al. 2010). Due to the currently limited body of knowledge pertaining to cetaceans specifically, further research characterizing the validity and limitations of extrapolating effects seen in related taxa to cetaceans should prove valuable.

Finally, cetaceans along with other marine mammals have been identified as focal species for ecosystem health and function assessment (Knap, 2000; Wells et al., 2004; Moore, 2008; Bossart, 2011; Godard-Codding et al., 2011). This is due largely to their long life spans, large bodies, substantial fatty reserves, and wide habitat use, all predisposing them to bioaccumulation and biomagnification of persistent lipophilic pollutants.

Overall, the historical literature, combined with recent findings from the DWH oil spill, suggest that cetaceans are at high risk of adverse effects from oil exposures, and these effects have importance at both the individual and population levels. Determining the usefulness and limitations of cetaceans as focus species in assessing ecosystem health in the specific case of oil spills is an area of future research made relevant by the substantial literature related to the DWH oil spill and the inevitability of future spills.

The potential relationship between oil exposure and the following specific endpoints remain to be further characterized:

1. effects of in utero exposure on disease susceptibility, reproduction, and survival
2. low maternal body mass and its impact on reproductive success
3. increased susceptibility to pathogens and their impact on reproductive success and survival
4. high white blood cell counts and their impact on reproductive success
5. massive tooth loss in dolphins and its impact on health and survival
6. effects of in vitro exposure on cellular dysfunction and toxicity

Similarly, general research questions regarding long-term oil toxicity in cetaceans remain unanswered and are listed subsequently. Some of these are directly related to several key research themes of global importance recently identified for cetacean conservation (Parsons et al., 2015).

1. What are the long-term impacts of oil exposure on individual survival and population density?
2. Which monitoring and experimental methods are most appropriate to assess oil toxicity in cetaceans?
3. What is the potential role of oil toxicity in chronic and progressive lung diseases?
4. What is the potential role of immune changes related to oil exposure in loss of tolerance to fetus and fetal loss?
5. What are the population recovery processes for cetaceans exposed to oil spills?
6. How do ecosystem, social, and behavioral factors influence models of population growth in cetaceans exposed to an oil spill?
7. How may differences in reproductive biology and ecology between odontocete and mysticete populations influence their resilience following a large oil spill?
8. How can we best extrapolate the effects observed in vitro in cetaceans to in vivo scenarios?
9. How can we best extrapolate the effects observed in other species to cetaceans?
10. What wildlife and laboratory species may be best suited as model or surrogate species to infer the effects of oil in cetaceans?
11. What are the usefulness and limitations of cetaceans as focus species in assessing ecosystem health after oil spills?

REFERENCES

ATSDR, 1995a. Toxicological Profile for Benzene (Update). U.S. Dept. of Health and Human Services, Public Health Service, Agency for Toxic Substances and Disease Registry, Atlanta.

ATSDR, 1995b. Toxicological Profile for Polycyclic Aromatic Hydrocarbons. U.S. Dept. of Health and Human Services, Public Health Service, Agency for Toxic Substances and Disease Registry, Atlanta, p. 458.

Balise, V.D., et al., 2016. Systematic review of the association between oil and natural gas extraction processes and human reproduction. Fertility and Sterility 106 (4), 795–819.

Barber, R., et al., 1996. Motor Gasolines Technical Review. Chevron Products Company, San Francisco. 67 p.

Beehner, J.C., et al., 2006. The endocrinology of pregnancy and fetal loss in wild baboons. Hormones and Behavior 49 (5), 688–699.

Beyer, J.G., et al., 2010. Analytical methods for determining metabolites of polycyclic aromatic hydrocarbon (PAH) pollutants in fish bile: a review. Environmental Toxicology and Pharmacology 30, 224–244.

Beyer, J., Trannum, H.C., Bakke, T., Hodson, P.V., Collier, T.K., 2016. Environmental effects of the deepwater horizon oil spill: a review. Marine Pollution Bulletin. 110, 28–51. https://doi.org/10.1016/j.marpolbul.2016.06.027.

Bossart, G.D., 2011. Marine mammals as sentinel species for oceans and human health. Veterinary Pathology 48 (3), 676–690.

Braithwaite, L.F., 1983. The Effects of Oil on the Feeding Mechanism of the Bowhead Whale Final Report, Contract N0 AA851-CTO-55. U.S. Dept. of the Interior, Washington, D.C.

Caldwell, M.C., Caldwell, D.K., 1982. A study of the effects of oil ingestion on a bottlenose dolphin, *Tursiops truncatus*. Final Report, Contract No AA551-CT9-29. In: Geraci, St Aubin (Eds.), Study of the Effects of Oil on Cetaceans. U.S. Dept. of the Interior, Washington, D.C.

Carpenter, C.P., et al., 1975. Petroleum hydrocarbon toxicity studies: 5. Animal and human response to vapors of mixed xylenes. Toxicology and Applied Pharmacology 33 (3), 543–558.

Carpenter, C.P., et al., 1976. Petroleum hydrocarbon toxicity studies: 13. Animal and human response to vapors of toluene concentrate. Toxicology and Applied Pharmacology 36 (3), 473–490.

Carvan, M.J., et al., 1995. Effects of benzo(a)pyrene and tetrachlorodibenzo(p)dioxin on fetal dolphin kidney cells: inhibition of proliferation and initiation of DNA damage. Chemosphere 30 (1), 187–198.

Colegrove, K.M., et al., 2016. Fetal distress and in utero pneumonia in perinatal dolphins during the Northern Gulf of Mexico unusual mortality event. Diseases of Aquatic Organisms 119 (1), 1–16.

Collier, T.K., et al., 2014. Effects on fish of polycyclic aromatic hydrocarbon (PAH) and naphthenic acid exposures. In: Tierney, K.B. (Ed.), Fish Physiology Volume 33: Organic Chemical Toxicology of Fishes. Elsevier, pp. 195–255.

Conney, A.H., 1982. Induction of microsomal enzymes by foreign chemicals and carcinogenesis by polycyclic aromatic hydrocarbons: GHA Clowes Memorial Lecture. Cancer Research 42, 4875–4917.

Coppock, R.W., et al., 1995. Toxicology of oil-field pollutants in cattle: A review. Veterinary and Human Toxicology 37 (6), 569–576.

Coppock, R.W., et al., 1996. Toxicopathology of oilfield poisoning in cattle: A review. Veterinary and Human Toxicology 38 (1), 36–42.

D'Andrea, M.A., Reddy, G.K., 2013. Health consequences among subjects involved in Gulf oil spill clean-up activities. The American Journal of Medicine 126 (11), 966–974.

Dahlheim, M.E., Matkin, C.O., 1994. Assessment of injuries to Prince William Sound killer whales. In: Loughlin, T.R. (Ed.), Marine Mammals and the 'Exxon Valdez'. Academic Press, San Diego, CA, pp. 163–172.

De Guise, S., et al., 2017. Changes in immune functions in bottlenose dolphins in the northern Gulf of Mexico associated with the Deepwater Horizon oil spill. Endangered Species Research 33, 290–303.

Dias, L., Litz, J., Garrison, L., Martinez, A., Barry, K., Speakman, T., 2017. Exposure of cetaceans to petroleum products following the deepwater horizon oil spill in the Gulf of Mexico. Endangered Species Research 33, 119–125.

DWH NRDA (Deepwater Horizon Natural Resource Damage assessment) Trustees, 2016. Deepwater Horizon Oil Spill Programmatic Damage Assessment and Restoration Plan and Programmatic Environmental Impact Statement. DWH NRDA. www.gulfspillrestoration.noaa.gov/restoration-planning/gulf-plan/.

Eade, N.R., Taussig, L.M., Marks, M.I., 1974. Hydrocarbon pneumonitis. Pediatrics 54 (3), 351–357.

Edwards, W.C., 1989. Toxicology of oil field wastes. Hazards to livestock associated with the petroleum industry. Veterinary Clinics: Food Animal Practice 5, 363–374.

Engelhardt, F.R., 1983. Petroleum effects on marine mammals. Aquatic Toxicology 4, 199–217.

Engelhardt, F.R., 1897. Assessment of the vulnerability of marine mammals to oil pollution. In: Kiuper, J., Van Den Brink, W.J. (Eds.), Fate and Effects of Oil in Marine Ecosystems. Martinus Nijhoff Publishing, Boston, MA, pp. 101–115.

Evans, W., 1982. A study to determine if gray whales detect oil. In: Geraci, J.R., St. Aubin, D.J. (Eds.), Study on the Effects of Oil on Cetaceans, pp. 47–61 Contract AA 551-CT9-22. Final report to U.S. Dept. of Interior, BLM, Washington, DC.

Fauquier, D.A., et al., 2017. Evaluation of morbillivirus exposure in cetaceans from the northern Gulf of Mexico 2010–2014. Endangered Species Research 33, 211–220.

Fossi, M.C., et al., 2014. An "ex vivo" model to evaluate toxicological responses to mixtures of contaminants in cetaceans: integumentum biopsy slices. Environmental Toxicology 29 (10), 1107–1121.

Geraci, J.R., Smith, T.G., 1976. Direct and indirect effects of oil on ringed seals (Phoca-Hispida) of Beaufort sea. Journal of the Fisheries Research Board of Canada 33 (9), 1976–1984.

Geraci, J.R., St. Aubin, D.J., 1982. Study of the Effects of Oil on Cetaceans. Contract 14–12–0001–29169. Final Report. U.S. Dept. of Interior, Bur. Land Manage, Washington, D.C.

Geraci, J.R., 1990. Physiologic and toxic effects on cetaceans. In: Geraci, J.R., St. Aubin, D.J. (Eds.), Sea Mammals and Oil: Confronting the Risks. Academic Press, San Diego, CA.

Godard, C.A.J., et al., 2006. Benzo[a]pyrene cytotoxicity in right whale (*Eubalaena glacialis*) skin, testis and lung cell lines. Marine Environmental Research 62, S20–S24.

Godard-Codding, C.A.J., et al., 2011. Pacific ocean-wide profile of CYP1A1 expression, stable carbon and nitrogen isotope ratios, and organic contaminant burden in sperm whale skin biopsies. Environmental Health Perspectives 119 (3), 337–343.

Goodale, D.R., Hyman, M.A.M., Winn, H.E., 1981. Cetacean responses in association with Regal Sword oil spill, ch. XI. Cetacean and Turtle Assessment Program, Annual Report 1979. In: Edel, R.K., Hyman, M.A., Tyrell, M.F. (Eds.), A Characterization of Marine Mammals and Turtles in the Mid- and North Atlantic Areas of the U.S. Outer Continental Shelf. University of Rhode Island, pp. XI-I–XI-15.

Goodnight, W.H., Soper, D.E., 2005. Pneumonia in pregnancy. Critical Care Medicine 33, S390–S397.

Green, R.F., 1972. Observations of the anatomy of some cetaceans and pinnipeds. In: Ridgway, S.H. (Ed.), Mammals of the Sea: Biology and Medicine. Charles C. Thomas Publishers, Springfield, IL, pp. 247–297.

Griffiths, D.J., Oritsland, N.A., Oritsland, T., 1987. Marine Mammals and Petroleum Activities in Norwegian Waters, p. 179 Fisken og Havet 1.

Gubbay, S., Earll, R., 2000. Review of Literature on the Effects of Oil Spills on Cetaceans. Scottish Natural Heritage, United Kingdom.

Harvey, J.T., Dahlheim, M.E., 1994. Cetaceans in oil. In: Loughlin, T.R. (Ed.), Marine mammals and the 'Exxon Valdez'. Academic Press, San Diego, CA, pp. 257–264.

Helm, R.C., Coasta, D.P., DeBruyn, T.D., O'Shea, T.J., Wells, R.S., Williams, T.M., 2015. Overview of effects of oil spills on marine mammals. In: Fingas, M. (Ed.), Handbook of Oil Spill Science and Technology. John Wiley & Sons, Hoboken, NJ, pp. 455–475.

Hobel, C., Culhane, J., 2003. Role of psychosocial and nutritional stress on poor pregnancy outcome. Journal of Nutrition 133 (5), 1709s–1717s.

Incardona, J.P., et al., 2004. Defects in cardiac function precede morphological abnormalities in fish embryos exposed to polycyclic aromatic hydrocarbons. Toxicology and Applied Pharmacology 196 (2), 191–205.

Jessup, D.A., Leighton, F.A., 1996. Oil pollution and petroleum toxicity to wildlife. In: Fairbrother, A., Lock, L.N., Hoff, G.L. (Eds.), Noninfectious Diseases of Wildlife. Iowa State University Press, Ames, IA, pp. 141–156.

Johnson, S., Ziccardi, M., 2006. Marine mammal Oil Spill Response Guidelines. NOAA Tech Memo www.nmfs.noaa.gov/pr/pdfs/health/eis_appendixl.pdf.

Kellar, N.M., et al., 2017. Low reproductive success rates of common bottlenose dolphins *Tursiops truncatus* in the northern Gulf of Mexico following the Deepwater Horizon disaster (2010–2015). Endangered Species Research 33, 143–158.

Knap, A., 2000. NIEHS/UNESCO meeting on indicators of ocean and human health. Marine Pollution Bulletin 40 (5), 461–462.

Lane, S.M., et al., 2015. Reproductive outcome and survival of common bottlenose dolphins sampled in Barataria Bay, Louisiana, USA, following the Deepwater Horizon oil spill. In: Proceedings of the Royal Society B-biological Sciences 282(1818).

Lee, K. (chair) et al., 2015. Expert Panel Report on the Behaviour and Environmental Impacts of Crude Oil Released into Aqueous Environments. Royal Society of Canada, Ottawa, ON. ISBN: 978-1-928140-02-3.

Levy, B.S., Nassetta, W.J., 2011. The adverse health effects of oil spills: a review of the literature and a framework for medically evaluating exposed individuals. International Journal of Occupational and Environmental Health 17 (2), 161–167.

Lipscomb, T.P., et al., 1994. In: Loughlin, T.R. (Ed.), Pathology of Sea Otters in Marine Mammals and the Exxon Valdez. Academic Press, San Diego, CA, pp. 265–280.

Lipscomb, T.P., et al., 1993. Histopathologic lesions in sea otters exposed to crude-oil. Veterinary Pathology 30 (1), 1–11.

Litz, J.A., et al., 2014. Review of historical unusual mortality events (UMEs) in the Gulf of Mexico (1990-2009): providing context for the multi-year northern Gulf of Mexico cetacean UME declared in 2010. Diseases of Aquatic Organisms 112 (2), 161–175.

Loughlin, T.R. (Ed.), 1994. Marine Mammals and the 'Exxon Valdez'. Academic Press, San Diego, CA.

Lowry, L.F., Frost, K.J., Pitcher, K.W., 1994. Observations of oiling of harbor seals in Prince Williams Sound. In: Loughlin, T.R. (Ed.), Marine Mammals and the Exxon Valdez. Academic Press, San Diego, CA, pp. 209–225.

Makrygianni, E.A., et al., 2016. Respiratory complications following hydrocarbon aspiration in children. Pediatric Pulmonology 51 (6), 560–569.

Matkin, C.O., et al., 2008. Ongoing population-level impacts on killer whales *Orcinus orca* following the 'Exxon Valdez' oil spill in Prince William Sound, Alaska. Marine Ecology Progress Series 356, 269–281.

Mazet, J.K., et al., 2000. Evaluation of changes in hematologic and clinical biochemical values after exposure to petroleum products in mink (Mustela vison) as a model for assessment of sea otters (Enhydra lutris). American Journal of Veterinary Research 61 (10), 1197–1203.

Mazet, J.A.K., et al., 2001. Effects of petroleum on mink applied as a model for reproductive success in sea otters. Journal of Wildlife Diseases 37 (4), 686–692.

McDonald, T.L., et al., 2017. Survival, density, and abundance of common bottlenose dolphins in Barataria Bay (USA) following the deepwater horizon oil spill. Endangered Species Research 33, 193–209.

Mehri, Z., 2010. Gulf coast oil disaster: impact on human reproduction. Fertility and Sterility 94 (5), 1575–1577.

Mohr, F.C., et al., 2008. Chronic oral exposure to bunker C fuel oil causes adrenal insufficiency in ranch mink (*Mustela vison*). Archives of Environmental Contamination and Toxicology 54 (2), 337–347.

Mohr, F.C., et al., 2010. Fuel oil-induced adrenal hypertrophy in ranch mink (*Mustela vison*): effects of sex, fuel oil weathering, and response to Adrenocorticotropic Hormone. Journal of Wildlife Diseases 46 (1), 103–110.

Moore, S.E., 2008. Marine mammals as ecosystem sentinels. Journal of Mammalogy 89 (3), 534–540.

Neff, J.M., 1979. Polycyclic Aromatic Hydrocarbons in the Aquatic Environment: Sources, Fates, and Biological Effects. Applied Science Publishers, London. 262 p.

Neff, J.M., 1990. Composition and fate of petroleum and spill-treating agents in the marine environment. In: Geraci, J.R., St. Aubin, D.J. (Eds.), Sea Mammals and Oil: Confronting the Risks. Academic Press, San Diego, CA, pp. 35–54.

Nessel, C.S., et al., 1999. The role of dermal irritation in the skin tumor promoting activity of petroleum middle distillates. Toxicological Sciences 49 (1), 48–55.

NOAA Fisheries, 2016. Cetacean Unusual Mortality Event in Northern Gulf of Mexico (2010–2014). http://www.nmfs.noaa.gov/pr/health/mmume/cetacean_gulfofmexico.htm.

O'Hara, T.M., O'Shea, T.J., 2001. Toxicology. In: Dierauf, L.A., Gulland, F.M.D. (Eds.), CRC Handbook of Marine Mammal Medicine, second ed. CRC Press, Boca Raton, FL, pp. 471–520.

Parsons, E.C.M., et al., 2015. Key research questions of global importance for cetacean conservation. Endangered Species Research 27 (2), 113–118.

Rebar, A.H., et al., 1995. Clinical and clinical laboratory correlates in sea otters dying unexpectedly in rehabilitation centers following the Exxon-Valdez oil-spill. Veterinary Pathology 32 (4), 346–350.

Ridgway, S.H., Scronce, B.L., Kanwisher, J., 1969. Respiration and deep diving in the bottlenose porpoise. Science 166, 1651–1654.

Ridgway, S.H. (Ed.), 1972. Mammals of the Sea: Respiration System. Charles C. Thomas, Springfield, Illinois, pp. 260–264.

Rowe, L.D., Dollahite, J.W., Camp, B.J., 1973. Toxicity of two crude oils and kerosine to cattle. Journal of the American Veterinary Medical Association 162, 61–66.

Scharf, S.M., et al., 1981. Pathologic and physiologic effects of aspiration of hydrocarbons in the rat. American Review of Respiratory Disease 124 (5), 625–629.

Scholz, D.K., et al., 1999. Fate of Spilled Oil in Marine Waters. American Petroleum Institute, Cape Charles. 43 p.

Schwartz, J., et al., 2004. Chronic fuel oil toxicity in American mink (*Mustela vison*): systemic and hematological effects of ingestion of a low-concentration of bunker C fuel oil. Toxicology and Applied Pharmacology 200, 146–158.

Schwacke, L.H., et al., 2014. Health of common bottlenose dolphins (*Tursiops truncatus*) in Barataria Bay, Louisiana, following the deepwater horizon oil spill. Environmental Science and Technology 48 (1), 93–103.

Schwacke, L.H., et al., 2017. Quantifying injury to common bottlenose dolphins from the Deepwater Horizon oil spill using an age-, sex- and class-structured population model. Endangered Species Research 33, 265–279.

Sim, M.S., et al., 2010. Acute health problems related to the operation mounted to clean the Hebei Spirit oil spill in Taean, Korea. Marine Pollution Bulletin 60 (1), 51–57.

Smith, C.R., et al., 2017. Slow recovery of Barataria Bay dolphin health following the Deepwater Horizon oil spill (2013-2014), with evidence of persistent lung disease and impaired stress response. Endangered Species Research 33, 127–142.

Spraker, T.R., Lowry, L.F., Frost, K.J., 1994. Gross necropsy and histopathologic lesions found in harbor seals. In: Loughlin, T.R. (Ed.), Marine Mammals and the Exxon Valdez. Academic Press, San Diego, CA, pp. 281–312.

St Aubin, D.J., 1990. Physiologic and toxic effects of oil on pinnipeds. In: Geraci, J.R., St. Aubin, D.J. (Eds.), Sea Mammals and Oil: Confronting the Risks. Academic Press, San Diego, CA, pp. 103–127.

Takeshita, R., et al., 2017. The Deepwater Horizon oil spill marine mammal injury assessment. Endangered Species Research 33, 95–106.

Tormoehlen, L.M., et al., 2014. Hydrocarbon toxicity: a review. Clinical Toxicology 52 (5), 479–489.

Tuomi, P., Williams, T., 1995. Rehabilitation of pregnant sea otters and females with newborn pups. In: Williams, T.M., Davis, D.W. (Eds.), Emergency Care and Rehabilitation of Oiled Sea Otters: A Guide for Oil Spills Involving Fur-Bearing Marine Mammals. University of Alaska Press, Fairbanks, AK, pp. 121–132.

Venn-Watson, S., et al., 2013. Assessing the potential health impacts of the 2003 and 2007 firestorms on bottlenose dolphins (*Tursiops trucatus*) in San Diego Bay. Inhalation Toxicology 25 (9), 481–491.

Venn-Watson, S., et al., 2015a. Adrenal gland and lung lesions in Gulf of Mexico common bottlenose dolphins (*Tursiops truncatus*) found dead following the deepwater horizon oil spill. PLoS One 10 (5).

Venn-Watson, S., et al., 2015b. Demographic clusters identified within the northern Gulf of Mexico common bottlenose dolphin (Tursiops truncates) unusual mortality event: January 2010–June 2013. PLoS One 10 (2).

Venn-Watson, S., Colegrove, K.M., Litz, J., Kinsel, M., et al., 2015c. Morbidity and Mortality in Bottlenose Dolphins: Summary of Alternative Hypotheses. (DWH Marine Mammal NRDA Technical Working Group Report).

Ver Hoef, J.M., Frost, K.J., 2003. A Bayesian hierarchical model for monitoring harbor seal changes in Prince William Sound, Alaska. Environmental and Ecological Statistics 10, 201–219.

Wade, P.R., et al., 2012. Social and behavioural factors in cetacean responses to overexploitation: are odontocetes less "resilient" than mysticetes? Journal of Marine Biology:567276 15 pages.

Wallace, B.P., et al., 2017. Effects of the Deepwater Horizon oil spill on protected marine species. Endangered Species Research 33, 1–7.

Waring, G.T., Josephson, E., Maze-Foley, K., Rosel, P.E. (Eds.), 2013. US Atlantic and Gulf of Mexico Marine Mammal Stock Assessments – 2012. NOAA Tech Memo NMFSNE-223.

Wells, R., et al., 2004. Bottlenose dolphins as marine ecosystem sentinels: developing a health monitoring system. EcoHealth 1, 246–254.

Wells, R.S., et al., 2017. Ranging patterns of common bottlenose dolphins *Tursiops truncatus* in Barataria Bay, Louisiana, following the deepwater horizon oil spill. Endangered Species Research 33, 159–180.

White, N.D., et al., 2017. Immunotoxic effects of in vitro exposure of dolphin lymphocytes to Louisiana sweet crude oil and Corexit. Journal of Applied Toxicology 37, 676–682.

White, P.A., 2002. The genotoxicity of priority polycyclic aromatic hydrocarbons in complex mixtures. Mutation Research: Genetic Toxicology and Environmental Mutagenesis 515 (1–2), 85–98.

Wilkin, S.M., et al., 2017. Marine mammal response operations during the deepwater horizon oil spill. Endangered Species Research 33, 107–118.

Ylitalo, G.M., et al., 2001. Influence of life-history parameters on organochlorine concentrations in free-ranging killer whales (*Orcinus orca*) from Prince William Sound, AK. The Science of the Total Environment 281 (1–3), 183–203.

Zock, J.P., et al., 2007. Prolonged respiratory symptoms in clean-up workers of the Prestige oil spill. American Journal of Respiratory and Critical Care Medicine 176 (6), 610–616.

Zock, J.P., Rodríguez-Trigo, G., Rodríguez-Rodríguez, E., Espinosa, A., et al., 2012. Persistent respiratory symptoms in clean-up workers 5 years after the *Prestige* oil spill. Occupational and Environmental Medicine 69, 508–513.

FURTHER READING

Oritsland, N.A., Engelhardt, F.R., Juck, F.A., Hurst, R.J., Watts, P.D., 1981. Effects of Crude Oil on Polar Bears. Environmental Studies No 24. Department of Indian and Northern Affairs Canada, Northern Affairs Program, Ottawa, Canada. 268 p.

Warner, R.E., 1969. Environmental effects of oil pollution in Canada. An evaluation of problems and research needs. In: Can. Wildl. Serv. MS Rep., pp. 16–17.

Chapter 4

Legacy Contamination in Estuarine Dolphin Species From the South American Coast

Salvatore Siciliano[1], Jailson F. Moura[2], Davi C. Tavares[3,4], Helena A. Kehrig[3,4], Rachel A. Hauser-Davis[5], Isabel Moreira[6], Ricardo Lavandier[6], Leila S. Lemos[4,5], Renata Emin-Lima[7], Natalia S. Quinete[8]

[1]Instituto Oswaldo Cruz/Fiocruz, Laboratório de Enterobactérias (LABENT), Pavilhão Rocha Lima, Rio de Janeiro, Brazil; [2]Leibniz Center for Tropical Marine Ecology – ZMT, Systems Ecology Group, Bremen, Germany; [3]Universidade Estadual do Norte Fluminense, Campos dos Goytacazes, Brazil; [4]Grupo de Estudos de Mamíferos Marinhos da Região dos Lagos (GEMM-Lagos), Rio de Janeiro, Brazil; [5]Escola Nacional de Saúde Pública (ENSP), Centro de Estudos em Saúde do Trabalhador e Ecologia Humana (CESTEH), Fiocruz, Rio de Janeiro, Brazil; [6]Pontifícia Universidade Católica do Rio de Janeiro, Rio de Janeiro, Brazil; [7]Museu Paraense Emílio Goeldi, Coordenação de Zoologia, Setor de Mastozoologia, Grupo de Estudos de Mamíferos Aquáticos da Amazônia (GEMAM) and Programa de Capacitação Institucional, Belém, Brazil; [8]Southeast Environmental Research Center (SERC), Florida International University, Miami, FL, United States

INTRODUCTION

Marine mammals are considered excellent marine environmental health sentinels, since most are top predators that can accumulate contaminants in their tissues through bioaccumulation and biomagnification processes, in amounts that are proportional to pollutant concentrations in the environment (water, sediment), but mainly to the amounts present in ingested food.

Two small cetaceans, Guiana (*Sotalia guianensis*) and Franciscana dolphins (*Pontoporia blainvillei*) have been used as valuable model species regarding the assessment of marine pollution along the South American coast. Both species are sympatric along the southwestern Atlantic Ocean, and they coexist throughout the Brazilian coast with minimal feeding overlap. They exploit mainly shallow waters (up to around 30 m, or a little further), preying upon pelagic and demersal neritic species, with fish being the most representative items in the diet of both species. Teleost fish (up to 10 cm), mainly sciaenid fish, as well as cephalopods and shrimp, are also important in *P. blainvillei* diets.

Marine Mammal Ecotoxicology. https://doi.org/10.1016/B978-0-12-812144-3.00004-8

These species represent examples of species significantly threatened by pollutant effects, due to the following reasons:

- small home range and site fidelity (Crespo et al., 2010b; Oshima et al., 2010)
- long life span: ~30 years for *S. guianensis* and ~13 years for *P. blainvillei* (Pinedo and Hohn, 2000; Di Beneditto and Ramos, 2004)
- the fact that they are apex food chain predators
- the presence of multianthropogenic stressors in the coastal environment (Moura et al., 2014);

The combination of the aforementioned descriptions makes both coastal species very sensitive to pollutant effects, and specimen carcasses are a valuable source for the assessment of contaminant levels in their tissues. However, this approach can be affected by certain sources of bias since (1) the carcasses can drift from different areas, and (2) mortality causes are normally unknown when necropsy procedures are not conducted, and they may not be representative of any specific condition on a population level. One way of reducing the bias caused by drifting patterns is to restrict the analyses to fresh specimens according to their decomposition stage according to Geraci and Lounsbury (2005), since corpses farther than 30 km from the coastline are unlikely to be recovered at beaches due to predation, sinking to the ocean floor, or quickly decomposing, especially in warm tropical waters (Moura et al., 2016; Tavares et al., 2016). However, information on the cause of death and body condition is crucial to assess the contribution of contaminant levels to mortality patterns (Fort et al., 2015). As both species present a coastal distribution, and their deaths are mainly related to by-catch events through entrapment in passive gillnets species (Ott et al., 2002; Crespo et al., 2010a,b), the potential for bias is, thus, reduced, making them adequate sentinel species regarding environmental contamination. Among important environmental contaminants detected in these species, metals and certain organic pollutants, such as polychlorinated biphenyls (PCBs) and polybrominated diphenyl ethers (PBDEs) are noteworthy, and studies on their presence and levels in these organisms have been increasingly conducted in the last years.

ESSENTIAL AND NONESSENTIAL ELEMENTAL CONTAMINATION IN GUIANA AND FRANCISCANA DOLPHINS

Although metals occur naturally (e.g., soil erosion, volcanic activity), human activities have substantially mobilized great amounts of these elements and released them into the environment (Walsh, 1990; Booth and Zeller, 2005; Moura et al., 2011). Since the mid-20th century, the effect of chemical contaminants discharged into the aquatic environment as a result of human activities has become a serious issue. Consequently, chemical contamination loaded with substantial amounts of different contaminants, including metals, may affect ecosystems, causing changes in the functions of living organisms or modifying the physical properties of the environment. The relationship among the contaminants, environment, and organisms may, under certain conditions, result in the degradation of toxic compounds through their modification, inducing changes in the environment and producing negative effects on living organisms.

Metal absorbance and accumulation depends on the biologic, ecologic, and physiologic mechanisms of each species, and it is also a function of the exposure, absorption, and elimination of the metals present in lower trophic level organisms (Driscoll et al., 2013; Newman, 2015). Certain metals can bioaccumulate over time to reach sublethal, or even lethal, levels in organisms, unless they are excreted or detoxified. They are capable of interacting with nuclear proteins and DNA, causing oxidative deterioration of biologic macromolecules. Individuals chronically exposed to high concentrations of harmful nonessential elements, such as Al, As, Cd, Hg, and Pb, are the most vulnerable. Additionally, long-lived species that occupy apex trophic levels are more affected by pollutants, frequently presenting high levels of persistent elements, such as methylmercury (Kehrig et al., 2004; Driscoll et al., 2013).

In this context, certain coastal dolphins have been thoroughly applied as important biologic models to assess metal concentrations in ecosystems and their population effects, such as Guiana (*S. guianensis*) and Franciscana dolphins (*P. blainvillei*).

Several studies assessing elemental levels in Guiana and Franciscana dolphins have been conducted in the last decades (Gerpe et al., 2002; Carvalho et al., 2008; Seixas et al., 2008, 2009; Panebianco et al., 2011; Botte et al., 2013; Lemos et al., 2013; Romero et al., 2016). However, an extensive assessment of contaminant levels is still required to acquire a better understanding of elemental accumulation patterns in these species and possible consequent population impacts. In general, elemental concentrations (e.g., Hg) normally detected in biologic samples from the coastal Guiana and Franciscana dolphins may be considered low when compared to other species that inhabit areas such as the Mediterranean Sea or Asia (Moura et al., 2014). Nevertheless, alarming levels have still been detected in both species. For example, liver Hg concentrations have reached up to 211 μg/g (dry weight) in Guiana dolphins sampled along the Rio de Janeiro state (Lemos et al., 2013), although *P. blainvillei* from the same region present very low Hg levels. Mercury differences between the species have been mainly attributed to their trophic level dissimilarities. Both species seem to share some of same resources (e.g., largehead hairtail: *Trichiurus lepturus*); *S. guianensis* normally select larger organisms, which contributes to the uptake of a higher load of chemicals. This observation is reinforced by the study performed by Carvalho et al. (2008) that investigated Cd, Hg, and Zn concentrations in liver and muscle samples from Franciscana and Guiana dolphins off the coast of Rio de Janeiro, Brazil, between 1998 and 2000. In that study, liver samples exhibited higher Hg and Zn concentrations in *S. guianensis* compared to *P. blainvillei*, and the authors identified the prey size preference between both species as the probable cause for this discrepancy.

On the other hand, environmental availability of the element may also be an important factor. For example, mercury concentrations in Guiana dolphin muscle samples from Rio de Janeiro were higher (means of 3.82 μg/g; converted to dry weight (cdw) according to Yang and Miyazaki, 2003) (Moura et al., 2012a) when compared to muscle samples collected in Amapá, Brazil, from the same species (mean of 1.36 μg/g cdw) (Moura et al., 2012b). The authors concluded that higher values in Rio de Janeiro were associated to a higher Hg discharge along the coast by the existing anthropogenic activities in the area. Thus, both environmental availability and different prey species are probably the main factors driving the differences in the assimilation of persistent metals by coastal cetaceans (Baptista et al., 2016).

Guiana and Franciscana dolphins are thought to have developed special metabolic mechanisms to cope with harmful nonessential metals. Kehrig et al. (2008) examined the demethylation process and the mercury differentiation in liver of Guiana dolphins ($n=29$) from the northern coast of Rio de Janeiro, Brazil, and they observed a lower methylmercury (MeHg) fraction (9%) compared to mercury selenide (HgSe), which was found in a higher proportion (53%). The authors suggested that this difference is related to the different MeHg detoxification strategies in liver. Indeed, HgSe has been observed in the liver of both species, identified as amorphous crystals in Kupffer cells (macrophages) by ultrastructural analyses (Lailson-Brito et al., 2012; Romero et al., 2016). Selenium plays an important role in mercury detoxification processes, as commonly observed in several different species worldwide (Savery et al., 2013). Data on Hg and Se in liver usually show a clear linear and positive correlation, indicating that Se provides a neutralization action against the toxic effects of Hg. However, this detoxification mechanism seems to be activated after a still undefined Hg threshold is reached, demanding essential Se levels to interact with MeHg molecules, offsetting their effects in the organism (Savery et al., 2013).

Although the main Hg uptake route is via ingestion, its accumulation can begin before organisms develop self-foraging capabilities. For example, organic mercury can be transferred from mothers to their fetuses via placenta, thereby establishing the first contact of the fetus with these toxic elements. In this regard, Romero et al. (2016), analyzing Hg levels in hepatic tissues of individuals belonging to different age classes, including fetuses, calves, juveniles and adults, observed that Hg concentrations were transferred via placenta from mothers to their fetuses. In fact, fetuses presented even higher concentrations than calves, while adults and juveniles presented the highest Hg levels, as expected.

Several studies have also investigated the accumulation of other essential and nonessential elements, and they have determined interesting relationships. For example, Seixas et al. (2009) examined six different trace elements (Ag, As, Cd, Cu, Pb, and Se) in liver samples from Guiana and Franciscana dolphins from the coast of Rio de Janeiro. Ag, As, and Cu levels were similar in both species, while variations in Cd and Pb levels were observed. The study concluded that environmental conditions and biologic parameters, such as total body length and diet, significantly contributed to the accumulation of the elements in these species. Another study analyzed five elemental concentrations (Ag, Cd, Cu, Hg, and Se) in liver samples of Guiana dolphins from the northern coast of Rio de Janeiro (Seixas et al., 2009). Higher Hg levels (27.77 μg/g dry weight) were observed, followed by Cu, Se, Ag, and Cd, respectively. A positive correlation was also identified between Hg and Se, as well as between Ag and Se, indicating detoxification processes of the nonessential elements by selenium. In addition, age was shown to influence Ag, Cd, Hg, and Se levels.

Lemos et al. (2013) analyzed both essential (Cu, Mn, Se, and Zn) and nonessential (Cd and Hg) elements in Guiana and Franciscana dolphin liver samples from Rio de Janeiro, Brazil. Element concentrations were similar to other studies in general, especially regarding Franciscana dolphins (Cd: 0.53 μg/g; Cu: 5.80 μg/g; Hg: 1.10 μg/g; Mn: 1.73 μg/g; Se: 3.03 μg/g; and Zn: 100.13 μg/g; dw). The authors related this to the species coastal habits, diet, and age of the analyzed specimens.

Regarding the dietary route of elemental exposure, some cephalopods naturally concentrate high amounts of Cd, and they are considered potential Cd vectors to predators that exhibit a dietary preference for this taxonomic group (Bustamante et al., 1998; Dorneles et al., 2007). However, Franciscana diets are composed of squids from the Loliginidae family (Dorneles et al., 2007) that presents low Cd concentration, justifying the low levels of this metal in dolphins sampled from southeastern Brazil. Franciscana prey specimens are usually smaller than 10 cm, therefore also limiting metal transference (e.g., Hg; Siciliano et al., 2006). Regarding Guiana dolphins, the authors reported concerning Cd levels (maximum value of 2.670 μg/g dry weight) and considerably high levels of Zn (maximum value of 530.82 μg/g). However, similar Cd levels (2.190 μg/g dry weight) were reported for *S. guianensis* from São Paulo and Paraná, Brazil, by Kunito et al. (2004), thus seemingly indicating a trend for high levels of these elements in this species off the Brazilian coastline.

Kunito et al. (2004) also reported higher concentrations of Hg, Mn, and Se (380 μg/g; 15.1 μg/g; 170 μg/g, respectively) for Guiana dolphins along the coast of São Paulo, as well as high Cu and Zn values (446 μg/g; 522 μg/g, respectively), while Moreira et al. described high Se values (54.3 μg/g) in samples from Rio Grande do Sul, Brazil. Cu concentrations determined in samples from the state of Rio de Janeiro, also in Brazil, presented the highest values in the same species (83.8 μg/g). Regarding Franciscana dolphins, Lailson-Brito et al. (2002) described high Cd, Hg, and Mn levels (5.67 μg/g; 156 μg/g; 27.3 μg/g, respectively), converted to dry weight according to Yang and Miyazaki (2003), in Rio de Janeiro, Brazil.

Fig. 4.1 presents a representation of the spatial variability of mean levels of some essential and nonessential elements for the two assessed coastal dolphin species along the Southwest Atlantic coast. The map indicates that most studies

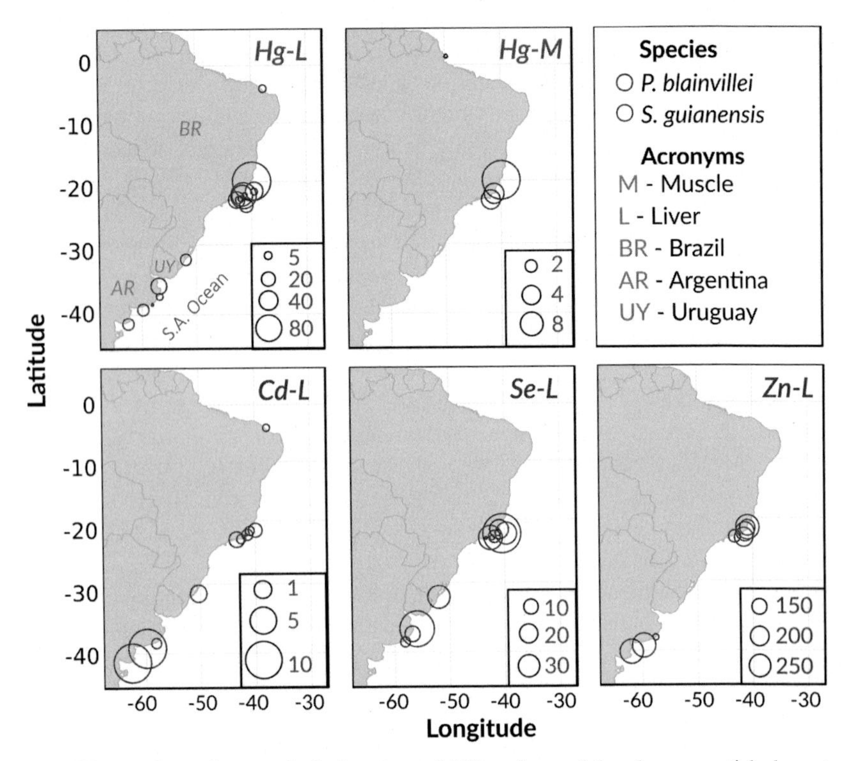

FIGURE 4.1 Synoptic map displaying the variability of essential and nonessential elements between *S. guianensis* and *P. blainvillei* along the Southwest Atlantic coast. The circle sizes are proportional to the average concentrations of the elements (in μg/g dry weight) in muscle (*M*) and liver (*L*) tissues. Some values presented in wet weight form were converted to dry weight according to Yang and Miyazaki (2003). The elemental data used to prepare this map were compiled from the following publications: Marcovecchio et al. (1990, 1994), Lailson-Brito et al. (2002), Monteiro-Neto et al. (2003), Kunito et al. (2004), Dorneles et al. (2007), Carvalho et al. (2008), Moreira et al. (2009), Seixas et al. (2009), Moura et al. (2012a,b), Panebianco et al. (2012), Lemos et al. (2013), Baptista et al. (2016), Romero et al. (2016).

have been carried out on specimens sampled along the coasts of southeastern Brazil and northeastern Argentina. Further studies are, thus, encouraged, to identify population vulnerabilities linked with environmental contamination by metals. The figure also indicates that *S. guianensis* generally presents higher concentrations of the most pervasive and persistent metal, Hg. The highest concentrations of this metal for this species were observed along the coast of southeastern Brazil, while, in contrast, studies carried out in northern Brazil in liver and muscle samples have indicated very low levels of this metal. Conversely, *P. blainvillei* seems to accumulate comparatively higher levels of Cd, especially in organisms sampled in Argentinean waters. Regarding Zn and Se, levels are usually not highly different either between the species or spatially along the Southwest Atlantic coast.

However, despite the considerable knowledge gained with the development of studies regarding metal contamination in South American coastal dolphins, further studies are required to acquire a better understanding of the possible toxic effects of metals on different levels of organization, from cellular to population segments. Additionally, it is extremely relevant to comprehend how such possible impacts affect ecosystem functions.

THE SPECIFIC CASE OF MERCURY ACCUMULATION AND TOXICITY IN MARINE MAMMALS

Hg is an exogenous and harmful metal, known to naturally occur in high concentrations in several regions, which can affect the productivity, reproduction, and survival of marine mammals. Environmental exposure to Hg in its more toxic organic form (methylmercury) in marine mammals, mainly via the food chain, is significantly higher than in other organisms, since MeHg presents high toxicity and the ability to undergo biomagnification along marine trophic chains.

Hg is a metal of environmental interest that occurs in different chemical and physical forms. Its most abundant forms are elemental Hg (Hg°), divalent mercury (Hg^{+2}), methylmercury (MeHg, CH_3Hg^+), dimethylmercury (DMHg, CH_3HgCH_3), and ethylmercury (EtHg, $CH_3CH_2Hg^+$), each of which is unique with regard to exposure pattern, metabolism, and toxic effects. Hg is known to be toxic in all forms, but some forms are more toxic than others, depending on the chemistry of the Hg-containing molecule, which determines its absorption, distribution, and excretion pattern in the body. Hg toxicity depends not only on its chemical form and entry route, but also on bioavailable amounts. The inorganic and organic forms of Hg are preferably accumulated in the protein-rich tissues of liver, kidneys, hair, and blood, while significantly lower concentrations are found in muscle, urine, feces, and brain.

In general, Hg concentrations increase with length or age in long-lived marine mammals due to this metal accumulation in tissues throughout their lives. This is probably related to the continuous accumulation of Hg through their diet combined with a much slower turnover rate in the body, as well as the rather long biologic half-life of this metal in these animals, since Hg shows a strong affinity with the sulfhydryl groups (SH) present in cysteine.

Hg can be transferred via placental or lactational transfer from mother to fetus and suckling calves, affecting them during their most sensitive development periods (Gerpe et al., 2002). It is noteworthy that Hg, in the form of MeHg, passes through the placental barrier and accumulates in dolphin fetal tissues, whereas the placenta is an important barrier against inorganic Hg (Neathery and Miller, 1975). However, knowledge about the maternal transfer of metals to offspring is very limited in marine mammals. Hg also has the potential to cause neurotoxicity in marine mammals, leading to major concerns, especially for females of childbearing age and their developing fetuses (Weihe and Joensen, 2012).

However, other effects have also been reported for Hg in marine mammals, including immune suppression and endocrine disruption. Thus, exposure to Hg is thought to be a severe problem for marine mammals because this contaminant is highly similar to hormones and is a toxic cellular substance that negatively affects the neuroendocrine, reproductive, developmental, and immune functions in mammals. Therefore, other parameters, such as the osteoid tissue, internal organs, endocrine glands, liver enzymes, and hormone and vitamin homeostasis are also negatively influenced by Hg, which adds multiple stresses toward the mammalian organism.

POLYCHLORINATED BIPHENYL AND POLYBROMINATED DIPHENYL ETHER CONTAMINATION IN GUIANA AND FRANCISCANA DOLPHINS

PCBs and PBDEs are synthetic organic compounds classified as persistent organic pollutants (POPs) due to their persistence in the environment and their capacity to bioaccumulate in fatty tissues and to biomagnify throughout the food chain (Tanabe et al., 2000; Magalhães et al., 2012). When these compounds are released in the environment, they can be transported over long distances by the atmosphere and ocean currents, even reaching remote polar regions. They are, therefore, spread throughout the globe and can trigger harmful effects in living organisms (Hakk and Letcher, 2003; Watanabe and Sakai, 2003; de Wit et al., 2006). There are indications that PCBs, PBDEs, and their respective metabolites can act as endocrine disruptors, as their chemical structures bear similarities to certain hormones (Streets et al., 2006; Richardson et al., 2008).

PCBs are organochlorine compounds commercially produced by the catalytic chlorination of biphenyls, yielding a complex mixture of multiple isomers with different degrees of chlorination, sold under the trade names of Aroclor in the United States, Clophen in Germany, Phenoclor in France, Fenclor in Italy, and Kanechlor in Japan. Theoretically, 209 PCB compounds (congeners) are possible (UNEP, 1999; Storelli et al., 2003). PCBs have high thermal and chemical stability, a high dielectric constant, are not flammable, and present low acute toxicity. In virtue of these properties, they were widely used in industry as insulating fluids in capacitors and transformers, in cooling systems, in plastics, paints, and lubricants, and as flame retardants (Wu et al., 2008). They were first produced commercially in 1929, but it was only in 1966 that they were accidentally detected in environmental samples, when researchers conducted analyses on the pesticide DDT (Jensen, 1966).

The Chemical Treaty on Persistent Organic Pollutants cited PCBs as a priority on its list of chemicals that should be eliminated by 2025 (Borja et al., 2005; Davis et al., 2007). The use and production of PCBs was banned in Japan in 1972, in the United States in 1979, in the United Kingdom in 1981, in Italy in 1984, and in Argentina in 2002 (Arias et al., 2016). Some countries have never produced PCBs and relied on imports from other countries, as is the case of Brazil.

However, many transformers still in use contain PCBs imported from elsewhere, despite restrictions on the use of organochlorine compounds in Brazil as of the early 1970s (de Souza et al., 2008) and their subsequent prohibition in the country in the early 1980s (Penteado and Vaz, 2001; Azevedo de Silva et al., 2007), thus leading to possible environmental PCB sources.

The oceans constitute a sink for PCBs, where they are quickly adsorbed to particulate matter and taken up by plankton, thereby entering the food chain. Marine mammals, like dolphins, often present high concentrations of these contaminants because of the effects of biomagnification. Since 1968, several adverse effects in aquatic mammals due to these compounds have been reported, such as immune impairment, reproductive failure, tumors, population decline, and mass mortality (Colborn and Smolen, 1996; Kajiwara et al., 2004).

PBDEs are produced by bromination of diphenyl ethers in the presence of a catalyst in a solvent such as dibromomethane. Also known as brominated compounds, they present 209 different congeners. They are structurally similar to PCBs, but they present an ether bond between the two phenyl rings, and the chlorine atoms are replaced by bromine atoms (Alaee et al., 2003). They are used in industry in three technical formulations: penta-BDE, octa-BDE, and deca-BDE. Traditionally, PBDEs have been used as flame retardants in electronic equipment, upholstered furniture, polyurethane foams, etc. (Alaee et al., 2003; Richardson et al., 2008). Some PBDEs are still produced in large amounts, even though they are listed as POPs under the Stockholm Convention (UNEP, 2009) and pose a risk to human beings and wildlife, leading to neurobehavioral development problems, altered thyroid hormone levels, and fetal toxicity and teratogenicity (Darnerud, 2003). In many cases the highest levels of these chemicals are found in marine mammals, fatty fish species, and fish that inhabit highly industrialized coastal regions (Darnerud, 2003). Table 4.1 displays published results on PCB and PBDE concentrations (ng/g lipid weight) in *S. guianensis* and *P. blainvillei* from Brazil and Argentina.

These pollutants have different residence times in each biologic tissue, so each type of tissue provides different information. For example, blubber gives information on bioaccumulation over time, while liver, kidney, and muscle give information about concentrations over time. Most of the published studies investigate blubber samples due to the lipophilic nature of PCBs and PBDEs (Kajiwara et al., 2004; Lailson-Brito et al., 2010; Yogui et al., 2010; Leonel et al., 2014; Santos-Neto et al., 2014). A few investigate liver, kidney, and muscle tissue (Dorneles et al., 2010; Quinete et al., 2011; Alonso et al., 2012; Cardoso et al., 2014; Moura et al., 2015; Lavandier et al., 2016a,b).

The PCB and PBDE concentrations reported in Table 4.1 in Brazil are a cause for concern. Some authors have found similar results in Brazil to those described in the literature for cetaceans in industrialized regions of the northern hemisphere (Kajiwara et al., 2004; Dorneles et al., 2010; Lailson-Brito et al., 2010), indicating the high exposure of these organisms to these contaminants in some coastal parts of the country, such as the coastal waters of the state of

TABLE 4.1 Published Results on PCB and PBDE Concentrations (ng/g Lipid Weight) in Blubber, Liver, Kidney, and Muscle of *S. guianensis* and *P. blainvillei* From Brazil and Argentina

Species	References	Location	Year	N	Tissue	∑PCBs (n. Congeners)	∑PBDEs (n. Congeners)
Sotalia guianensis	Yogui et al. (2003)	Brazil (São Paulo)	1996–2001	9	Blubber	200–9220 (27)	
	Kajiwara et al. (2004)	Brazil (São Paulo and Paraná)	1997 and 1999	26	Blubber	1300–79000 (29)	
	Lailson-Brito et al. (2010)	Brazil (Rio de Janeiro, Guanabara Bay)	1995–2005	12	Blubber	6663–99175 (27)	
		Brazil (Rio de Janeiro, Sepetiba Bay)	1995–2005	5	Blubber	1745–25482 (27)	
		Brazil (Paraná, Paranaguá Bay)	1995–2005	15	Blubber	765–14333 (27)	
	Yogui et al. (2010)	Brazil (São Paulo)	1997–2003	1	Blubber	1950–9710 (27)	
	Alonso et al. (2010)	Brazil (São Paulo, Baixada Santista)	2004–2005	6	Blubber	27860–61340	
		Brazil (São Paulo, Ubatuba)	2004–2005	6	Blubber	25870–66030	
	Dorneles et al. (2010)	Brazil (Rio de Janeiro)	1994–2006	13 M	Liver		260–1620 (9)
			1994–2006	6 F	Liver		13–450 (9)
	Yogui et al. (2011)	Brazil (São Paulo)	1996–2003	9	Blubber		65.6 ± 59.3 (30)
	Quinete et al. (2011)	Brazil (Rio de Janeiro)		10	Liver	24312 (28)	1104 (9)
				2	Kidney		142 (9)
	Santos-Neto et al. (2014)	Brazil (Ceará)	2005–2011	25	Blubber	20–17300 (25)	
	Lavandier et al. (2015)	Brazil (Rio de Janeiro)	2003–2012	8	Muscle	6340 ± 2860 (24)	74 ± 33 (9)
			2003–2012	8	Liver	9410 ± 7020 (24)	150 ± 76 (9)

Pontoporia blainvillei	Kajiwara et al. (2004)	Brazil (São Paulo and Paraná)	1997 and 1999	26	Blubber	320–12000 (29)	
	Leonel et al. (2010)	Brazil (Rio Grande do Sul)	1994–2004	73	Blubber	1135–10555 (209)	
	Yogui et al. (2010)	Brazil (São Paulo)	1997–2003	8	Blubber	484–9710 (27)	
	Lailson-Brito et al. (2011)	Brazil (São Paulo and Paraná)		10	Blubber	909–5849 (27)	
	Yogui et al. (2011)	Brazil (São Paulo)	1996–2003	8	Blubber		60.3 ± 74.8 (39)
	Alonso et al. (2012)	Brazil (From Espírito Santo to Rio Grande do Sul)	1994–2009	53	Liver		6–1797 (8)
	Leonel et al. (2014)	Brazil (Rio Grande do Sul)	1994–2004	73	Blubber		7.9–65 (39)
		Brazil (São Paulo)	2002–2005	41	Blubber		67.8–763.7 (39)
	Lavandier et al. (2016a,b)	Brazil (Rio de Janeiro)	2011-2012	9	Liver	6107–26199 (28)	24.6–123 (9)
	Castello et al. (2000)	Argentina (Buenos Aires)	1991–1996	8	Blubber	3350 ± 1950	
	Arias et al. (2016)	Argentina (Necochea)	2004–2011	16	Blubber	440 ± 950 (28)	
		Argentina (Claromecó)	2006–2009	10	Blubber	170 ± 240 (28)	
		Argentina (Monte Hermoso)	2004–2011	12	Blubber	300 ± 870 (28)	
		Argentina (Bahia Blanca)	2007–2009	2	Blubber	5 ± 6 (28)	

The number of congeners analyzed for PCBs and PBDEs in each study is shown in parentheses.

São Paulo, near the Cubatão industrial hub, the largest in Brazil (Alonso et al., 2010). Lailson-Brito et al. (2010) observed that cetaceans from southeastern and southern Brazilian region waters are exposed to high levels of organochlorine compounds, including PCBs. These relatively high levels represent a risk not only to health, but also to the survival of *S. guianensis* and *P. blainvillei*, both endangered species, due to the endocrine-disruptive action of these compounds (Lailson-Brito et al., 2011), and the fact that placental transfer of organobromine compounds has been confirmed in *S. guianensis* (Dorneles et al., 2010). In addition, several studies show a predominance of certain specific PCB and PBDE congeners, which suggest the use of Aroclor 1254, 1260, and penta-BDE mixtures and sources of these contaminants along the Brazilian coast (Lailson-Brito et al., 2010; Lavandier et al., 2015; Lavandier et al., 2016a,b).

The results presented in Table 4.1 indicate that the PCB concentrations are higher than PBDEs in both coastal dolphin species. This becomes clearer when both compounds are analyzed in the same dolphin samples (Quinete et al., 2011; Lavandier et al., 2015, 2016a,b), where, in some cases, the differences between PCB and PBDE concentrations exceeded one order of magnitude. This difference, of higher PCB levels compared to PBDE concentrations in dolphins, has been observed in several aquatic ecosystem studies in Brazil (Magalhães et al., 2012), and it may be due to the fact that PCBs started to be produced and sold a few decades before PBDEs in Brazil, and that, although PCBs were banned in Brazil in 1981, some electrical equipment containing PCBs, such as dielectric fluid, is still in use in the country. In addition, Brazilian legislation requires that PBDE-containing computers and computer accessories and components (either imported or produced locally) must present PBDE concentrations lower than 0.1% (Brasil, 2009), further accounting for these differences.

Comparing both species, higher concentrations of PCBs have been observed in *S. guianensis* compared to *P. blainvillei*, sometimes exceeding one order of magnitude. This may be explained by certain differences between the two species, namely that *S. guianensis* has a longer life span, of about 30 years, while that of *P. blainvillei* is usually much shorter, rarely exceeding 15 years, and that *S. guianensis* feeds mainly on Atlantic cutlassfish, which is a high-level carnivorous fish in the food chain, while *P. blainvillei* prefers small estuarine fish. As PCBs show the capacity to biomagnify throughout the food chain, the preference for larger, high-level carnivore fish species exposes *S. guianensis* to higher levels of these contaminants.

Regarding PBDEs, only one study has been conducted on both species using the same type of biologic tissue, blubber (Yogui et al., 2011), where concentrations were of 65.6 ± 59.3 ng/g and 60.3 ± 74.8 ng/g for *S. guianensis* and *P. blainvillei*, respectively. In the case of liver tissue, four studies on *S. guianensis* and four on *P. blainvillei* are available, but none were conducted with both species together. Minimum concentrations in *S. guianensis* were relatively higher than those in *P. blainvillei*, while maximum concentrations were of 1620 ng/g for *S. guianensis* (Dorneles et al., 2010) and 1797 ng/g for *P. blainvillei* (Alonso et al., 2012).

In view of the limited number of results on these compounds, it is not possible to infer which of the species shows higher PBDE contamination levels or whether statistically significant differences are present. Thus, further studies are necessary to reach some conclusion on this matter.

Regarding other regions of Latin America other than Brazil, the only other study on PCBs and PBDEs in either coastal dolphin species is a recent investigation on the contamination of *P. blainvillei* by PCBs in Argentina (Arias et al., 2016). The general PCB total average concentration in that study was four to seven times lower than the values reported around 20 years earlier in the same species from Argentinean coastal wasters (Castello et al., 2000). According to the authors, this indicates that greater care is being taken concerning the disposal of these contaminants in the environment and that the international program in place to eliminate these compounds from the environment is indeed working. Unfortunately, no such decrease has been recorded in Brazil.

Thus, monitoring programs must continue to better understand the distribution, transportation, and final deposits of PCBs and PBDEs in the southern hemisphere. The results should be used by the competent authorities to introduce measures to improve environmental conditions and reduce the risk to which wildlife and humans themselves are exposed.

APPLICATION OF BIOMARKERS OF RESPONSE TO ENVIRONMENTAL POLLUTANTS IN DOLPHINS

Biomarkers are defined as measurements in body fluids, cells, or tissues representative of biochemical or cellular modifications due to the presence of pollutants or of the host's response to them (NRC, 1987). In an environmental context, biomarkers of response indicate the effects of toxicants on the organism, and they can be broadly classified into those that indicate pathologic damage and those that detect biochemical changes or responses (Timbrell, 1998). These indicators are extremely useful since they demonstrate that toxicants have indeed entered exposed organisms, have suffered tissue distribution, and are eliciting toxic effects on critical targets (McCarthey and Shugart, 1990). The use of biomarkers, in fact, often increases the possibility of identifying the underlying causes behind toxic effects and provides information about pollutant bioavailability, early biologic responses, and potential ecologic damage (Albertsson et al., 2007).

Some biomarkers of response have been applied in the evaluation of the health status of marine mammals, although studies are still scarce in this regard due to specific peculiarities in marine mammal investigations. For example, it is important to note that although enzymatic activities of several biotransformation enzymes, such as glutathione-S-transferase and glutathione peroxidase, as well as endpoints usually linked to certain contaminants, such as acetylcholinesterase inhibition after pesticide exposure, are adequate biomarkers of response regarding environmental contamination, they are not very useful with regard to marine mammals, since most marine mammal samplings are conducted on carcasses, and enzymatic activities degrade after death.

However, other biomarkers of response used to indicate environmental contamination can still be applied in this scenario. Metallothioneins, for example, are heat-stable proteins that are somewhat resistant to proteolysis degradation and can, thus, be applied as useful biomarkers regarding, mainly, metal contamination. This metalloprotein has been investigated in both *S. guianensis* and *P. blainvillei*. The synthesis of these cysteine-rich proteins that bind to metals entirely through metal-thiolate bonds is significantly induced in the presence of excess metals, both toxic and essential, and they show high affinity to Cd, Zn, Cu, and inorganic mercury (Hg_{inorg}), among others (Kagi and Schaffer, 1988; Viarengo et al., 1999). As MT concentrations have been shown to correlate well with biota metal exposure, they have been increasingly applied as a biomarker of response with regard to metal contamination in the environment (Hauser-Davis et al., 2014; Polizzi et al., 2014; Viarengo et al., 1999). However, discrepancies in the data regarding this biomarker in marine mammals are observed. For example, Kehrig et al. (2016) evaluated metal and trace element loads and MT concentrations in both liver and muscle samples, using MT as a biomarker for trace element exposure in *P. blainvillei* and *S. guianensis* specimens sampled after entanglement in gill-net fisheries along the northern Rio de Janeiro state. The authors observed that MT concentrations in the livers of *P. blainvillei* and *S. guianensis* appear to play an important role in the detoxification process of total Hg and Hg_{inorg}, although not MeHg. This indicates that MT in both species could be considered an adequate biomarker of response to Hg exposure. However, another study, conducted by Romero et al. (2016), evaluated Hg and Se loads and MT levels in fetuses, calves, juveniles, and adult *P. blainvillei* specimens from Argentina, and it indicated that no significant relationship between Hg and MT concentrations was found for liver, kidney, muscle, or brain tissue. Following this trend of noncorrelations between MT levels and metals, the study performed by Polizzi et al. (2014) on *P. blainvillei* specimens from the Argentinean coast also observed no correlations between environmental metal levels and MT levels, and it stated that this metalloprotein seems to be closely related to the homeostasis processes for the regulation of Cu and Zn levels according to ontogenetic changes in this coastal dolphin species (Polizzi et al., 2014). Thus, further investigations in this regard should be conducted to validate the application of this biomarker in marine mammals regarding metal contamination in the environment.

Another biomarker of response, reduced glutathione (GSH), has also been applied to investigate metal exposure in marine mammals, although reports in this regard are still limited. This is a tripeptide rich in thiol groups, possessing significant antioxidant capacity, and acting as a first line of defense against metal cytotoxicity (Canesi et al., 1999), including Hg, Cd, and Pb. Kehrig et al. (2016) observed significant correlations between Hg and GSH in *P. blainvillei* and *S. guianensis* fresh muscle samples, which may be associated to the role played by Hg as an oxidative stress inducer. This, however, is the only study available to date in these species regarding the use of GSH as a biomarker of response to metal contamination.

Exposure to certain environmental pollutants, notably PCBs, organochlorine pesticides, and metals, has also been linked to alterations in both the innate and adaptive immune systems, including cellular and humoral immunity (Desforges et al., 2016). Several categories of immunological biomarkers of response to these compounds have been reported in marine mammals, including immune tissue histopathology, hematology/circulating immune cell populations, functional immune assays (lymphocyte proliferation, phagocytosis, respiratory burst, and natural killer cell activity), immunoglobulin production, and cytokine gene expression (Desforges et al., 2016). However, although hematological alterations associated with exposure to environmental contaminants have been evaluated in some dolphin species, such as bottlenose dolphins (*Tursiops truncatus*) (Schaefer et al., 2011; Fair and Romano, 2013), no studies are available for either *P. blainvillei* or *S. guianensis*, and further progress in this regard is paramount to evaluate chronic and sublethal environmental contamination effects in these species using these biomarkers of response.

FINAL REMARKS

This chapter has discussed several important marine environment contaminants and their presence and distribution in two coastal dolphin species, *S. guianensis* and *P. blainvillei*. Although the available literature has increased substantially in the last decades with regard to studies on metal and organic pollutant contamination in these species, the peculiarities of marine mammal sampling make this type of study difficult. Therefore, considerations on carcass degradation and individual age, sex, and diet should always be taken into account to produce reliable results. Historical data series on metal, PCB, and PBDE concentrations in these organisms are extremely useful in the evaluation of environmental contamination trends, since they are useful for decision-making processes regarding their use and release into the environment and allow for information of whether certain contaminants have decreased over time. Finally, the application of biomarkers of response in these threatened species is a very useful approach in determining possible investigating pollutant effects in these animals, since the possibility of identifying the underlying causes behind toxic effects is increased, allowing for inferences on pollutant bioavailability, early biologic responses, and potential ecologic damage that may arise from environmental contamination exposure.

REFERENCES

Alaee, M., Arias, P., Sjödin, A., Bergman, Å., 2003. An overview of commercially used brominated flame retardants, their applications, their use patterns in different countries/regions and possible modes of release. Environment International 29, 683–689. https://doi.org/10.1016/S0160-4120(03)00121-1.

Albertsson, E., Kling, P., Gunnarsson, L., Larsson, D.G.J., Fo, L., 2007. Proteomic analyses indicate induction of hepatic carbonyl reductase/20 b-hydroxysteroid dehydrogenase B in rainbow trout exposed to sewage effluent 68, 33–39. https://doi.org/10.1016/j.ecoenv.2007.02.003.

Alonso, M.B., Eljarrat, E., Gorga, M., Secchi, E.R., Bassoi, M., Barbosa, L., Bertozzi, C.P., Marigo, J., Cremer, M., Domit, C., Azevedo, A.F., Dorneles, P.R., Torres, J.P.M., Lailson-Brito, J., Malm, O., Barceló, D., 2012. Natural and anthropogenically-produced brominated compounds in endemic dolphins from Western South Atlantic: another risk to a vulnerable species. Environmental Pollution 170, 152–160. https://doi.org/10.1016/j.envpol.2012.06.001.

Alonso, M.B., Marigo, J., Bertozzi, C., Santos, M.C.O., Taniguchi, S., Montone, R.C., 2010. Occurrence of chlorinated pesticides and polychlorinated biphenyls (PCBs) in Guiana dolphins (*Sotalia guianensis*) from Ubatuba and Baixada Santista, São Paulo, Brazil. Latin American Journal of Aquatic Mammals 8, 123–130. https://doi.org/10.5597/lajam00161.

Arias, A.H., Panebianco, M.V., Net, S., Dumoulin, D., Souissi, S., Ouddane, B., Cappozzo, H.L., Marcovecchio, J.E., 2016. Franciscana dolphins as PCBs marine biomonitors in Argentina, south-west Atlantic Ocean. Journal of the Marine Biological Association of the United Kingdom 96, 979–992. https://doi.org/10.1017/S0025315415000776.

Azevedo de Silva, C.E., Torres, J.P.M., Malm, O., 2007. Toxicologia das bifenilas policloradas. Oecologia Brasiliensis 11, 179–187. https://doi.org/10.4257/oeco.2007.1102.02.

Baptista, G., Kehrig, H.A., Di Beneditto, A.P.M., Hauser-Davis, R.A., Almeida, M.G., Rezende, C.E., Siciliano, S., Moura, J.F., Moreira, I., 2016. Mercury, selenium and stable isotopes in four small cetaceans from the Southeastern Brazilian coast: influence of feeding strategy. Environmental Pollution 218, 1298–1307. https://doi.org/10.1016/j.envpol.2016.08.088.

Booth, S., Zeller, D., 2005. Mercury, food webs, and marine mammals: implications of diet and climate change for human health. Environmental Health Perspectives 113, 521–526. https://doi.org/10.1289/ehp.7603.

Borja, J., Taleon, D.M., Auresenia, J., Gallardo, S., 2005. Polychlorinated biphenyls and their biodegradation. Process Biochemistry 40, 1999–2013. https://doi.org/10.1016/j.procbio.2004.08.006.

Botte, S.E., Marcovecchio, J.E., Cappozzo, H.L., 2013. Essential and non-essential heavy metals in skin and muscle tissues of franciscana dolphins (*Pontoporia blainvillei*) from the southern Argentina coast. Chemistry and Ecology 29, 511–518. https://doi.org/10.1080/02757540.2013.810727.

Brasil, 2009. Parecer 173 de 2009. Comissão de Meio Ambiente. Câmara dos Senadores, Brasília.

Bustamante, P., Caurant, F., Fowler, S.W., Miramand, P., 1998. Cephalopods as a vector for the transfer of cadmium to top marine predators in the north-east Atlantic Ocean. The Science of the Total Environment 220, 71–80. https://doi.org/10.1016/S0048-9697(98)00250-2.

Canesi, L., Viarengo, A., Leonzio, C., Filippelli, M., Gallo, G., 1999. Heavy metals and glutathione metabolism in mussel tissues. Aquatic Toxicology 46, 67–76.

Cardoso, M.D., Moura, J.F., Tavares, D.C., Gonçalves, R.A., Colabuono, F.I., Roges, E.M., de Souza, R.L., Rodrigues, D. dos P., Montone, R.C., Siciliano, S., 2014. The Manx shearwater (*Puffinus puffinus*) as a candidate sentinel of Atlantic Ocean health. Aquatic Biosystems 10, 1–10. https://doi.org/10.1186/2046-9063-10-6.

Carvalho, C.E.V., Di Beneditto, A.P.M., Souza, C.M.M., Ramos, R.M., Rezende, C.E., 2008. Heavy metal distribution in two cetacean species from Rio de Janeiro State, south-eastern Brazil. Journal of the Marine Biological Association of the United Kingdom 88, 1117–1120. https://doi.org/10.1017/S0025315408000325.

Castello, H.P., Junin, M., Rotman, F., Sarti, G.C., 2000. Analisis de contaminantes organoclorados y metals pesados en franciscana, *Pontoporia blainvillei*, de Argentina y Brasil.

Colborn, T., Smolen, M.J., 1996. Epidemiological analysis of persistent organochlorine contaminants in cetaceans. In: Ware, G.W. (Ed.), Reviews of Environmental Contamination and Toxicology: Continuation of Residue Reviews. Springer, New York, NY, pp. 91–172. https://doi.org/10.1007/978-1-4613-8478-6_4.

Crespo, E.A., Alarcon, D., Alonso, M., Bazzalo, M., Borobia, M., Cremer, M., Filla, G., Lodi, L., Magalhães, F.A., Marigo, J., Queiróz, H.L., Reynolds, J.E., Schaeffer, Y., Dorneles, P.R., Lailson-Brito, J., Wetzel, D.L., 2010a. Report of the working group on major threats and conservation. Latin American Journal of Aquatic Mammals 8, 47–56.

Crespo, E.A., Pedraza, S.N., Grandi, M.F., Dans, S.L., Garaffo, G.V., 2010b. Abundance and distribution of endangered Franciscana dolphins in Argentine waters and conservation implications. Marine Mammal Science 26, 17–35. https://doi.org/10.1111/j.1748-7692.2009.00313.x.

Darnerud, P.O., 2003. Toxic effects of brominated flame retardants in man and in wildlife. Environment International 29, 841–853. https://doi.org/10.1016/S0160-4120(03)00107-7.

Davis, J.A., Hetzel, F., Oram, J.J., McKee, L.J., 2007. Polychlorinated biphenyls (PCBs) in San Francisco Bay. Environmental Research 105, 67–86. https://doi.org/10.1016/j.envres.2007.01.013.

de Souza, A.S., Torres, J.P.M., Meire, R.O., Neves, R.C., Couri, M.S., Serejo, C.S., 2008. Organochlorine pesticides (OCs) and polychlorinated biphenyls (PCBs) in sediments and crabs (*Chasmagnathus granulata*, Dana, 1851) from mangroves of Guanabara Bay, Rio de Janeiro State, Brazil. Chemosphere 73, 186–192. https://doi.org/10.1016/j.chemosphere.2007.04.093.

de Wit, C.A., Alaee, M., Muir, D.C.G., 2006. Levels and trends of brominated flame retardants in the Arctic. Chemosphere 64, 209–233. https://doi.org/10.1016/j.chemosphere.2005.12.029.

Desforges, J.-P.W., Sonne, C., Levin, M., Siebert, U., De Guise, S., Dietz, R., 2016. Immunotoxic effects of environmental pollutants in marine mammals. Environment International 86, 126–139. https://doi.org/10.1016/j.envint.2015.10.007.

Di Beneditto, A.P.M., Ramos, R.M.A., 2004. Biology of the marine tucuxi dolphin (*Sotalia fluviatilis*) in south-eastern Brazil. Journal of the Marine Biological Association of the United Kingdom 84, 1245–1250. https://doi.org/10.1017/S0025315404010744h.

Dorneles, P.R., Lailson-Brito, J., Dirtu, A.C., Weijs, L., Azevedo, A.F., Torres, J.P.M., Malm, O., Neels, H., Blust, R., Das, K., Covaci, A., 2010. Anthropogenic and naturally-produced organobrominated compounds in marine mammals from Brazil. Environment International 36, 60–67. https://doi.org/10.1016/j.envint.2009.10.001.

Dorneles, P.R., Lailson-Brito, J., Secchi, E.R., Bassoi, M., Lozinsky, C.P.C., Torres, J.P.M., Malm, O., 2007. Cadmium concentrations in franciscana dolphin (*Pontoporia blainvillei*). Brazilian Journal of Oceanography 55, 179–186. https://doi.org/10.1590/S1679-87592007000300002.

Driscoll, C.T., Mason, R.P., Chan, H.M., Jacob, D.J., Pirrone, N., 2013. Mercury as a global pollutant: sources, pathways, and effects. Environmental Science and Technology 47, 4967–4983. https://doi.org/10.1021/es305071v.

Fair, P.A., Romano, T., 2013. Associations between perfluoroalkyl compounds and immune and clinical chemistry parameters in highly exposed bottlenose dolphins (*Tursiops truncatus*). Environmental Toxicology and Chemistry 32, 736–746. https://doi.org/10.1002/etc.2122.

Fort, J., Lacoue-Labarthe, T., Nguyen, H.L., Boué, A., Spitz, J., Bustamante, P., 2015. Mercury in wintering seabirds, an aggravating factor to winter wrecks? The Science of the Total Environment 527, 448–454.

Geraci, J.R., Lounsbury, V.J., 2005. Marine Mammals Ashore: A Field Guide for Strandings. National Aquarium, Baltimore.

Gerpe, M.S., Rodríguez, D.H., Moreno, V.J., Bastida, R.O., Moreno, J.A.E., 2002. Accumulation of heavy metals in the franciscana (*Pontoporia blainvillei*) from Buenos Aires Province, Argentina. Latin American Journal of Aquatic Mammals 1, 95–106. https://doi.org/10.5597/lajam00013.

Hakk, H., Letcher, R.J., 2003. Metabolism in the toxicokinetics and fate of brominated flame retardants - a review. Environment International 29, 801–828. https://doi.org/10.1016/S0160-4120(03)00109-0.

Hauser-Davis, R.A., Bastos, F.F., Tuton, B., Chávez, R., Saint, T., Ziolli, R.L., Arruda, M.A.Z., 2014. Bile and liver metallothionein behavior in copper-exposed fish. Journal of Trace Elements in Medicine and Biology 28, 70–74. https://doi.org/10.1016/j.jtemb.2013.09.003.

Jensen, S., 1966. Report of a new chemical hazard. New Scientist 32, 612.

Kagi, J.H.R., Schaffer, A., 1988. Biochemistry of metallothionein. Biochemistry 27, 8509–8515.

Kajiwara, N., Matsuoka, S., Iwata, H., Tanabe, S., Rosas, F.C.W., Fillmann, G., Readman, J.W., 2004. Contamination by persistent organochlorines in cetaceans incidentally caught along Brazilian coastal waters. Archives of Environmental Contamination and Toxicology 46, 124–134. https://doi.org/10.1007/s00244-003-2239-y.

Kehrig, H.A., Hauser-Davis, R.A., Seixas, T.G., Pinheiro, A.B., Di Beneditto, A.P.M., 2016. Mercury species, selenium, metallothioneins and glutathione in two dolphins from the southeastern Brazilian coast: mercury detoxification and physiological differences in diving capacity. Environmental Pollution 213, 785–792. https://doi.org/10.1016/j.envpol.2016.03.041.

Kehrig, H.A., Seixas, T.G., Baeta, A.P., Brito, J.L., Moreira, I., Malm, O., 2004. Total mercury, methylmercury and selenium in the livers and muscle of different fishes and a marine mammal from a tropical estuary-Brazil. In: 7th International Conference on Mercury as Global Pollutant. RMZ-M{&}G, vol. 51, pp. 1111–1114.

Kehrig, H.A., Seixas, T.G., Palermo, E.A., Di Beneditto, A.P.M., Souza, C.M.M., Malm, O., 2008. Different species of mercury in the livers of tropical dolphins. Analytical Letters 41, 1691–1699. https://doi.org/10.1080/00032710802122677.

Kunito, T., Nakamura, S., Ikemoto, T., Anan, Y., Kubota, R., Tanabe, S., Rosas, F.C.W., Fillmann, G., Readman, J.W., 2004. Concentration and subcellular distribution of trace elements in liver of small cetaceans incidentally caught along the Brazilian coast. Marine Pollution Bulletin 49, 574–587. https://doi.org/10.1016/j.marpolbul.2004.03.009.

Lailson-Brito, J., Azeredo, M.A.A., Malm, O., Ramos, R.R., Di Beneditto, A.P.M., Saldanha, M.F.C., 2002. Trace metals in liver and kidney of the franciscana (*Pontoporia blainvillei*) from the northern coast of Rio de Janeiro State, Brazil. Latin American Journal of Aquatic Mammals 1, 107–114. https://doi.org/10.5597/lajam00014.

Lailson-Brito, J., Dorneles, P.R., Azevedo-Silva, C.E., Azevedo, A., de Freitas Azevedo, A., Vidal, L.G., Marigo, J., Bertozzi, C., Zanelatto, R.C., Bisi, T.L., Malm, O., Torres, J.P.M., 2011. Organochlorine concentrations in franciscana dolphins, *Pontoporia blainvillei*, from Brazilian waters. Chemosphere 84, 882–887. https://doi.org/10.1016/j.chemosphere.2011.06.018.

Lailson-Brito, J., Dorneles, P.R., Azevedo-Silva, C.E., Azevedo, A.F., Vidal, L.G., Zanelatto, R.C., Lozinski, C.P.C., Azeredo, A., Fragoso, A.B.L., Cunha, H.A., Torres, J.P.M., Malm, O., 2010. High organochlorine accumulation in blubber of Guiana dolphin, *Sotalia guianensis*, from Brazilian coast and its use to establish geographical differences among populations. Environmental Pollution 158, 1800–1808. https://doi.org/10.1016/j.envpol.2009.11.002.

Lailson-Brito, J., Cruz, R., Dorneles, P.R., Andrade, L., Azevedo, A.F., Fragoso, A.B., Vidal, L.G., Costa, M.B., Bisi, T.L., Almeida, R., Carvalho, D.P., Bastos, W.R., Malm, O., 2012. Mercury-selenium relationships in liver of Guiana dolphin: the possible role of Kupffer cells in the detoxification process by tiemannite formation. PLoS One 7, 42162. https://doi.org/10.1371/journal.pone.0042162.

Lavandier, R., Arêas, J., Dias, P.S., Taniguchi, S., Montone, R., Moura, J.F., Quinete, N., Siciliano, S., Moreira, I., 2015. An assessment of PCB and PBDE contamination in two tropical dolphin species from the Southeastern Brazilian coast. Marine Pollution Bulletin 99, 1–7. https://doi.org/10.1016/j.marpolbul.2015.10.039.

Lavandier, R., Arêas, J., Quinete, N., Moura, J.F., Taniguchi, S., Montone, R., Siciliano, S., Moreira, I., 2016a. PCB and PBDE levels in a highly threatened dolphin species from the Southeastern Brazilian coast. Environmental Pollution 207, 442–449. https://doi.org/10.1016/j.envpol.2015.10.013.

Lavandier, R., Arêas, J., Quinete, N., Moura, J.F., Taniguchi, S., Montone, R.C., Siciliano, S., Moreira, I., 2016b. PCB and PBDE levels in a highly threatened dolphin species from the Southeastern Brazilian coast. Environmental Pollution 208, 442–449.

Lemos, L., Moura, J.F., Hauser-Davis, R.A., Campos, R.C., Siciliano, S., 2013. Small cetaceans found stranded or accidentally captured in southeastern Brazil: bioindicators of essential and non-essential trace elements in the environment. Ecotoxicology and Environmental Safety 97, 166–175. https://doi.org/10.1016/j.ecoenv.2013.07.025.

Leonel, J., Sericano, J.L., Fillmann, G., Secchi, E., Montone, R.C., 2010. Long-term trends of polychlorinated biphenyls and chlorinated pesticides in franciscana dolphin *(Pontoporia blainvillei)* from Southern Brazil. Marine Pollution Bulletin 60, 412–418.

Leonel, J., Sericano, J.L., Secchi, E.R., Bertozzi, C., Fillmann, G., Montone, R.C., 2014. PBDE levels in franciscana dolphin (*Pontoporia blainvillei*): temporal trend and geographical comparison. The Science of the Total Environment 493, 405–410. https://doi.org/10.1016/j.scitotenv.2014.06.003.

Magalhães, C.A., Taniguchi, S., Cascaes, M.J., Montone, R.C., 2012. PCBs, PBDEs and organochlorine pesticides in crabs *Hepatus pudibundus* and *Callinectes danae* from Santos Bay, state of São Paulo, Brazil. Marine Pollution Bulletin. https://doi.org/10.1016/j.marpolbul.2011.12.020.

Marcovecchio, J.E., Gerpe, M.S., Bastida, R.O., Rodríguez, D.H., Morón, S.G., 1994. Environmental contamination and marine mammals in coastal waters from Argentina: an overview. The Science of the Total Environment 154, 141–151. https://doi.org/10.1016/0048-9697(94)90084-1.

Marcovecchio, J.E., Moreno, V.J., Bastida, R.O., Gerpe, M.S., Rodríguez, D.H., 1990. Tissue distribution of heavy metals in small cetaceans from the Southwestern Ocean. Marine Pollution Bulletin 21, 299–304. https://doi.org/10.1016/0025-326X(90)90595-Y.

McCarthey, J.F., Shugart, L.S., 1990. Biomarkers of Environmental Contamination. Lewis Publishing, Boca Raton, FL.

Monteiro-Neto, C., Itavo, R.V., Moraes, L.E.S., 2003. Concentrations of heavy metals in *Sotalia fluviatilis* (Cetacea: Delphinidae) off the coast of Ceará, northeast Brazil. Environmental Pollution 123, 319–324. https://doi.org/10.1016/S0269-7491(02)00371-8.

Moreira, I., Seixas, T., Kehrig, H., Fillmann, G., Di Beneditto, A., Souza, C., Malm, O., 2009. Selenium and mercury (total and organic) in tissues of a coastal small cetacean, *Pontoporia blainvillei*. Journal of Coastal Research (SI56), 866–870.

Moura, J.F., Emin-Lima, R., Hacon, S.S., Vega, C.M., Campos, R.C., Siciliano, S., 2012a. Mercury status of the Amazon Continental Shelf: Guiana dolphins (*Sotalia guianensis*, Van Benédén 1864) as a bioindicator. Bulletin of Environmental Contamination and Toxicology 89, 412–418. https://doi.org/10.1007/s00128-012-0663-2.

Moura, J.F., Hacon, S.S., Vega, C.M., Hauser-Davis, R.A., Campos, R.C., Siciliano, S., 2012b. Guiana dolphins (*Sotalia guianensis*, Van Benédén 1864) as indicators of the bioaccumulation of total mercury along the coast of Rio de Janeiro state, Southeastern Brazil. Bulletin of Environmental Contamination and Toxicology 88, 54–59. https://doi.org/10.1007/s00128-011-0448-z.

Moura, J.F., Hauser-Davis, R.A., Lemos, L., Emin-lima, R., Siciliano, S., 2014. Guiana dolphins (*Sotalia guianensis*) as marine ecosystem sentinels: ecotoxicology and emerging diseases. Reviews of Environmental Contamination and Toxicology 228, 1–29. https://doi.org/10.1007/978-3-319-01619-1.

Moura, J.F., Soares, M., Belo, P., Hacon, S., Siciliano, S., 2011. The interface of public healthcare with the health of the oceans: proliferation of disease, socio-economic impacts and beneficial relationships. Ciência and Saúde Coletiva 16, 3469–3480.

Moura, J.F., Merico, A., Montone, R.C., Silva, J., Seixas, T.G., Godoy, J.M.O., Saint'Pierre, T.D., Hauser-Davis, R.A., Di Beneditto, A.P.M., Reis, E.C., Tavares, D.C., Lemos, L.S., Siciliano, S., 2015. Assessment of trace elements, POPs, ^{210}Po and stable isotopes (^{15}N and ^{13}C) in a rare filter-feeding shark: the megamouth. Marine Pollution Bulletin 95, 402–406. https://doi.org/10.1016/j.marpolbul.2015.03.038.

Moura, J.F., Acevedo-Trejos, E., Tavares, D.C., Meirelles, A.C.O., Silva, C.P.N., Oliveira, L.R., Santos, R.A., Wickert, J.C., Machado, R., Siciliano, S., Merico, A., 2016. Stranding events of *Kogia* whales along the Brazilian coast. PLoS One 11, e0146108. https://doi.org/10.1371/journal.pone.0146108.

Neathery, M.W., Miller, W.J., 1975. Metabolism and toxicity of cadmium, mercury, and lead in animals: a review. Journal of Dairy Science 58, 1767–1781. https://doi.org/10.3168/jds.S0022-0302(75)84785-0.

Newman, M.C., 2015. Fundamentals of Ecotoxicology: The Science of Pollution, Fourth ed. CRC Press, Boca Raton.

NRC, 1987. Biological markers in environmental health research. Environmental Health Perspectives 74, 7–9.

Oshima, G.E.F., Santos, M.C.O., Bazzalo, M., Flores, P.A.C., Pupim, F.N., 2010. Home ranges of Guiana dolphins (*Sotalia guianensis*) (Cetacea: Delphinidae) in the Cananeia estuary, Brazil. Journal of the Marine Biological Association of the United Kingdom 90, 1641–1647. https://doi.org/10.1017/S0025315410001311.

Ott, P.E., Secchi, E.R., Moreno, I.B., Danilewicz, D., Crespo, E.A., Bordino, P., Ramos, R., Di Beneditto, A.P., Bertozzi, C., Bastida, R., Zanelatto, R., Perez, J.E., Kinas, P.G., 2002. Report of the working group on fishery interactions. Latin American Journal of Aquatic Mammals 1, 55–64. https://doi.org/10.5597/lajam00008.

Panebianco, M.V., Botte, S.E., Negri, M.F., Marcovecchio, J.E., Cappozzo, H.L., 2012. Heavy metals in liver of the franciscana dolphin, *Pontoporia blainvillei*, from the Southern coast of Buenos Aires, Argentina. Journal of the Brazilian Society of Ecotoxicology 7, 33–41. https://doi.org/10.5132/jbse.2012.01.006.

Panebianco, M.V., Negri, M.F., Botté, S.E., Marcovecchio, J.E., 2011. Metales pesados en el riñón del delfín franciscana, *Pontoporia blainvillei* (Cetacea: Pontoporiidae) y su relación con parámetros biológicos. Latin American Journal of Aquatic Research 39, 526–533. https://doi.org/10.3856/vol39-issue3-fulltext-12.

Penteado, J.C.P., Vaz, J.M., 2001. O legado das bifenilas policloradas (PCBs). Quimica Nova 24, 390–398. https://doi.org/10.1590/S0100-40422001000300016.

Pinedo, M.C., Hohn, A.A., 2000. Growth layer patterns in teeth from the franciscana, *Pontoporia blainvillei*: developing a model for precision in age estimation on. Marine Mammal Science 16, 1–27.

Polizzi, P.S., Romero, M.B., Chiodi Boudet, L.N., Das, K., Denuncio, P.E., Rodríguez, D.H., Gerpe, M.S., 2014. Metallothioneins pattern during ontogeny of coastal dolphin, *Pontoporia blainvillei*, from Argentina. Marine Pollution Bulletin 80, 275–281. https://doi.org/10.1016/j.marpolbul.2013.10.037.

Quinete, N., Lavandier, R., Dias, P., Taniguchi, S., Montone, R., Moreira, I., 2011. Specific profiles of polybrominated diphenylethers (PBDEs) and polychlorinated biphenyls (PCBs) in fish and tucuxi dolphins from the estuary of Paraíba do Sul River, Southeastern Brazil. Marine Pollution Bulletin. https://doi.org/10.1016/j.marpolbul.2010.11.021.

Richardson, V.M., Staskal, D.F., Ross, D.G., Diliberto, J.J., DeVito, M.J., Birnbaum, L.S., 2008. Possible mechanisms of thyroid hormone disruption in mice by BDE 47, a major polybrominated diphenyl ether congener. Toxicology and Applied Pharmacology 226, 244–250. https://doi.org/10.1016/j.taap.2007.09.015.

Romero, M.B., Polizzi, P., Chiodi, L., Das, K., Gerpe, M., 2016. The role of metallothioneins, selenium and transfer to offspring in mercury detoxification in franciscana dolphins (*Pontoporia blainvillei*). Marine Pollution Bulletin 109, 650–654. https://doi.org/10.1016/j.marpolbul.2016.05.012.

Santos-Neto, E.B., Azevedo-Silva, C.E., Bisi, T.L., Santos, J., Meirelles, A.C.O., Carvalho, V.L., Azevedo, A.F., Guimarães, J.E., Lailson-Brito, J., 2014. Organochlorine concentrations (PCBs, DDTs, HCHs, HCB and MIREX) in delphinids stranded at the northeastern Brazil. The Science of the Total Environment 472, 194–203. https://doi.org/10.1016/j.scitotenv.2013.10.117.

Savery, L.C., Evers, D.C., Wise, S.S., Falank, C., Wise, J., Gianios, C., Kerr, I., Payne, R., Thompson, W.D., Perkins, C., Zheng, T., Zhu, C., Benedict, L., Wise, J.P., 2013. Global mercury and selenium concentrations in skin from free-ranging sperm whales (*Physeter macrocephalus*). The Science of the Total Environment 450–451, 59–71. https://doi.org/10.1016/j.scitotenv.2013.01.070.

Schaefer, A.M., Stavros, H.-C.W., Bossart, G.D., Fair, P.A., Goldstein, J.D., Reif, J.S., 2011. Associations between mercury and hepatic, renal, endocrine, and hematological parameters in Atlantic bottlenose dolphins (*Tursiops truncatus*) along the eastern coast of Florida and South Carolina. Archives of Environmental Contamination and Toxicology 61, 688–695. https://doi.org/10.1007/s00244-011-9651-5.

Seixas, T.G., Kehrig, H.A., Paula, A., Di Beneditto, M., Souza, C.M.M., Malm, O., Moreira, I., 2009. Trace elements in different species of cetacean from Rio de Janeiro coast. Journal of the Brazilian Chemical Society 20, 243–251. https://doi.org/10.1590/S0103-50532009000200008.

Seixas, T.G., Kehrig, H.D.A., Costa, M., Fillmann, G., Di Beneditto, A.P.M., Secchi, E.R., Souza, C.M.M., Malm, O., Moreira, I., 2008. Total mercury, organic mercury and selenium in liver and kidney of a South American coastal dolphin. Environmental Pollution 154, 98–106. https://doi.org/10.1016/j.envpol.2008.01.030.

Siciliano, S., Moreno, I.B., Silva, E.D.E., Alves, V.C., 2006. Baleias, botos e golfinhos na Bacia de Campos, first ed. ENSP/FIOCRUZ, Rio de Janeiro.

Storelli, M.M., Storelli, A., Marcotrigiano, G.O., 2003. Coplanar polychlorinated biphenyl congeners in the liver of *Galeus melastomus* from different areas of the Mediterranean Sea. Bulletin of Environmental Contamination and Toxicology 71, 276–282. https://doi.org/10.1007/s00128-003-0161-7.

Streets, S.S., Henderson, S.A., Stoner, A.D., Carlson, D.L., Simcik, M.F., Swackhamer, D.L., 2006. Partitioning and bioaccumulation of PBDEs and PCBs in Lake Michigan. Environmental Science and Technology 40, 7263–7269. https://doi.org/10.1021/es061337p.

Tanabe, S., Prudente, M.S., Kan-atireklap, S., Subramanian, A., 2000. Mussel watch: marine pollution monitoring of butyltins and organochlorines in coastal waters of Thailand, Philippines and India. Ocean and Coastal Management 43, 819–839. https://doi.org/10.1016/S0964-5691(00)00060-0.

Tavares, D.C., Moura, J.F., Siciliano, S., 2016. Environmental predictors of seabird wrecks in a tropical coastal area. PLoS One 11, e0168717.

Timbrell, J.A., 1998. Biomarkers in toxicology. Toxicology 129, 1–12. https://doi.org/10.1016/S0300-483X(98)00058-4.

UNEP, 2009. The 9 new POPs under the Stockholm convention. Stockholm Convention on Persistent Organic Pollutants. United Nations Environment Programme.

UNEP, 1999. Chemicals Guidelines for the Identification of PCBs and Materials Containing PCBs.

Viarengo, A., Burlando, B., Dondero, F., Marro, A., Fabbri, F., 1999. Metallothionein as a tool in biomonitoring programmes. Biomarkers 4, 455–466. https://doi.org/10.1080/135475099230615.

Walsh, P.M., 1990. Heavy Metals in the Marine Environment. CRC Press, Boca Raton, FL.

Watanabe, I., Sakai, S., 2003. Environmental release and behavior of brominated flame retardants. Environment International 29, 665–682. https://doi.org/10.1016/S0160-4120(03)00123-5.

Weihe, P., Joensen, H.D., 2012. Dietary recommendations regarding pilot whale meat and blubber in the Faroe Islands. International Journal of Circumpolar Health 71, 18594. https://doi.org/10.3402/ijch.v71i0.18594.

Wu, J.-P., Luo, X.-J., Zhang, Y., Luo, Y., Chen, S.-J., Mai, B.-X., Yang, Z.-Y., 2008. Bioaccumulation of polybrominated diphenyl ethers (PBDEs) and polychlorinated biphenyls (PCBs) in wild aquatic species from an electronic waste (e-waste) recycling site in South China. Environment International 34, 1109–1113. https://doi.org/10.1016/j.envint.2008.04.001.

Yang, J., Miyazaki, N., 2003. Moisture content in Dall's porpoise (*Phocoenoides dalli*) tissues: a reference base for conversion factors between dry and wet weight trace element concentrations in cetaceans. Environmental Pollution 121, 345–347. https://doi.org/10.1016/S0269-7491(02)00239-7.

Yogui, G.T., de Oliveira Santos, M.C., Montone, R.C., 2003. Chlorinated pesticides and polychlorinated biphenyls in marine tucuxi dolphins *(Sotalia fluviatilis)* from the Cananeia estuary, southeastern Brazil. Environmental Pollution 312, 67.

Yogui, G.T., Santos, M.C.O., Bertozzi, C.P., Montone, R.C., 2010. Levels of persistent organic pollutants and residual pattern of DDTs in small cetaceans from the coast of São Paulo, Brazil. Marine Pollution Bulletin. https://doi.org/10.1016/j.marpolbul.2010.07.022.

Yogui, G.T., Santos, M.C.O., Bertozzi, C.P., Sericano, J.L., Montone, R.C., 2011. PBDEs in the blubber of marine mammals from coastal areas of São Paulo, Brazil, southwestern Atlantic. Marine Pollution Bulletin 62, 2666–2670. https://doi.org/10.1016/j.marpolbul.2011.09.024.

Chapter 5

Poly- and Perfluoroalkyl Substances in Marine Mammals

Patricia A. Fair[1], Magali Houde[2]
[1]Medical University of South Carolina, Charleston, SC, United States; [2]Aquatic Contaminants Research Division, Environment and Climate Change Canada, Montreal, QC, Canada

INTRODUCTION

Poly- and perfluoroalkyl substances (PFASs) represent one of the more contemporary chemicals of environmental concern for marine mammals. These chemicals have over 200 industrial and consumer applications and are often used for water-, soil-, and stain-resistant coatings for textiles, carpets, and food packaging, as well as other products such as floor polishes, insecticides, and aqueous firefighting foam (AFFF) (Renner, 2006). PFASs are highly stable compounds because of the high energy of the carbon–fluorine covalent bond. Buck et al. (2011) published a comprehensive review on the terminology and classification of these chemicals. Two well-studied groups include the perfluoroalkyl sulfonates (PFSAs) and perfluoroalkyl carboxylates (PFCAs). Although PFAS production began in the 1950s and has scaled up steadily (Renner, 2006), the first detection of these chemicals in wildlife was reported 50 years later (Giesy and Kannan, 2001). Since then, studies have documented PFASs in species throughout the world. Reviews have focused attention on the presence of these chemicals in the environment, wildlife, and humans (Beach et al., 2006; Butt et al., 2014; D'Hollander et al., 2010; Giesy et al., 2010; Kovarova and Svobodova, 2008; Lau et al., 2004, 2007; Lau, 2015; Lindstrom et al., 2011; Suja et al., 2009). Limited reviews have, however, been directed specifically toward marine mammals (Houde et al., 2011), although some reviews on aquatic environment and organisms have included marine mammals (Ahrens and Bundschuh, 2014; Butt et al., 2010; Houde et al., 2006a; Letcher et al., 2010). The objectives of the present chapter are to summarize the recent literature on marine mammal exposure to PFASs in relation to their spatial distribution, temporal trends, and toxicity. In addition, a specific case study and future perspectives are discussed.

Polyfluoroalkyl chemicals are emitted either directly into the environment as manufacturing by-products (Paul et al., 2009) or indirectly through degradation

Marine Mammal Ecotoxicology. https://doi.org/10.1016/B978-0-12-812144-3.00005-X

of precursors and intermediates leading to the production of persistent metabolites such as perfluorooctane sulfonate (PFOS) (Wang et al., 2013). As such, PFASs have been distributed globally, even occurring in remote locations of the Arctic and Antarctic (Armitage et al., 2009; Yamashita et al., 2008; Young and Mabury, 2010). The widespread global environmental contamination and toxicologic concerns on long-chain PFSAs and PFCAs have brought about the phase-out of PFOS and perfluorooctanoic acid (PFOA), two predominant PFASs, by the major global manufacturers in 2000–02 (3M Company, 2000; USEPA, 2000) and in 2013 when DuPont ceased production of PFOA. In 2006, the PFOA Stewardship Agreement formed between the US Environmental Protection Agency (USEPA) and eight leading global companies reduced emissions and product content of PFOA and related chemicals by 95% and the goal of elimination by 2015 was met (USEPA, 2015). Although PFOA and PFOS are no longer manufactured in the United States, they are still produced in other locations around the globe, and may continue to be imported into the United States and used in consumer goods such as carpets, leather, apparel, textiles, paper, packaging, coatings, rubber, and plastics.

In May 2009, PFOS and its related compounds were added to the Stockholm Convention on Persistent Organic Pollutants as an Annex B substance (i.e., restricted in its use); however, many "acceptable purposes" remain, such as firefighting foams and use in insecticides, as well as exemptions in production of stain-repellent surface coatings (UNEP, 2009). Other regions have also enacted provisions to limit the use of PFASs. In Canada, a similar agreement to that promulgated by the USEPA was developed between the Canadian environmental and health authorities and five companies to restrict PFCAs in products (Environment Canada, 2010). The European Union Marketing and Use Directive restricted the use of "perfluorooctane sulfonates" in the European Union (European Parliament, 2006).

Despite the international phase-out initiatives, production and use of PFOS and related substances continues in several countries. Although the production of PFOS was significantly reduced in 2001, China continues to produce PFOS with scaled-up production from 30 tons/year in 2001 to an average of 250–300 tons/year since 2006 (Lim et al., 2011; Wang et al., 2013, 2014). Large increases in emission of PFOA and its perfluorotelomer alcohol precursors were predicted during 2005–08 as production of these chemicals continued (Schenker et al., 2008). Brazil is among the top three current producers and consumers of PFOS-related substances globally, having an exemption from the Stockholm Convention for manufacturing the pesticide Sulfluramid to control leaf-cutting ants, an agricultural pest in neotropical regions (Carloni, 2009; Löfstedt Gilljam et al., 2016). During 2003–13 the production of Sulfluramid in Brazil increased from 30 to 60 tons/year for *N*-ethyl perfluorooctane sulfonamide (EtFOSA) used in manufacturing (Löfstedt Gilljam et al., 2016), with most products exported within South America. With the increased production and use of Sulfluramid, the active ingredient EtFOSA, which degrades to PFOS and perfluorooctane sulfonamide (FOSA), may be an important contamination source for these congeners. Information on PFOS contamination in this region is lacking, although high levels have been reported in surface water (Gonzalez-Gaya et al., 2014) and tucuxi dolphins (*Sotalia guianensis*) (Dorneles et al., 2008).

Due to the recycling of these substances and stockpiles of PFAS-containing products such as AFFF, this class of chemicals continues to be used and, thus, continues to contribute to environmental contamination (Buck et al., 2011). After manufacturers ceased production of PFOS and PFOA a host of replacement compounds have emerged, which are typically polyfluorinated rather than perfluorinated or have a carbon backbone with ether linkages (Wang et al., 2013). Alternatives for the eight-carbon chemicals have consisted of shorter chained PFASs such as perfluorobutanesulfonic acid (PFBS) and perfluorobutyric acid (PFBA) (Ritter, 2010). However, short-chain PFCA and PFSA homologues, such as perfluorohexanoic acid (PFHxA) and PFBS, may be as persistent in the environment as the long-chain homologues, and their increased production could lead to environmental exposures that will be long-lasting (Wang et al., 2013). The concern over potential environmental and human health impacts of PFASs has led to the scientists' Helsingør Statement on poly- and perfluoroalkyl substances, outlining their concerns over continued development of these chemicals (Scheringer et al., 2014).

GLOBAL DISTRIBUTION OF PFASs IN MARINE MAMMALS

Kannan et al. (2001) first reported PFOS in tissues of 15 marine mammal species collected worldwide. As previously mentioned, PFAS concentrations have since been reviewed in biota including marine mammals. Other studies have reported PFSA levels in tissues of these top-of-the-food-chain animals in recent years. A summary of this scientific data (i.e., articles published between 2011 and 2017) can be found in the following section. An update of the PFOS spatial distribution presented by Houde et al. (2011) is shown in Fig. 5.1 (results from temporal trend studies were not included in the figure).

The most recent published information for PFSA contamination in cetaceans comes from a 2002–14 study in stranded Pacific humpback dolphins (*Sousa chinensis*) and finless porpoise (*Neophocaena phocaenoides*) from the South China Sea in which high PFOS levels (mean for 2013–14 samples: 2600 ng/g dry weight) were reported in liver samples of dolphins (Lam et al., 2016). PFOS levels were also high (>500 ng/g ww) in livers of East Greenland polar bear (*Ursus maritimus*) (mean > 3000 ng/g ww) (Greaves et al., 2012; Greaves and Letcher, 2013), harbor porpoises (*Phocoena phocoena*) from the Baltic and North Sea (up to 2425 ng/g ww) (Huber et al., 2012), harbor seals (*Phoca vitulina*) from the North Sea (maximum: 1284 ng/g ww) (Galatius et al., 2013), and plasma of bottlenose dolphins (*Tursiops truncatus*) from the US East Coast (up to 6260 ng/g ww) (Fair et al., 2012) collected during the 2000s and harbor seals from San Francisco Bay, California (up to 796 ng/g ww). Long-chain PFCAs

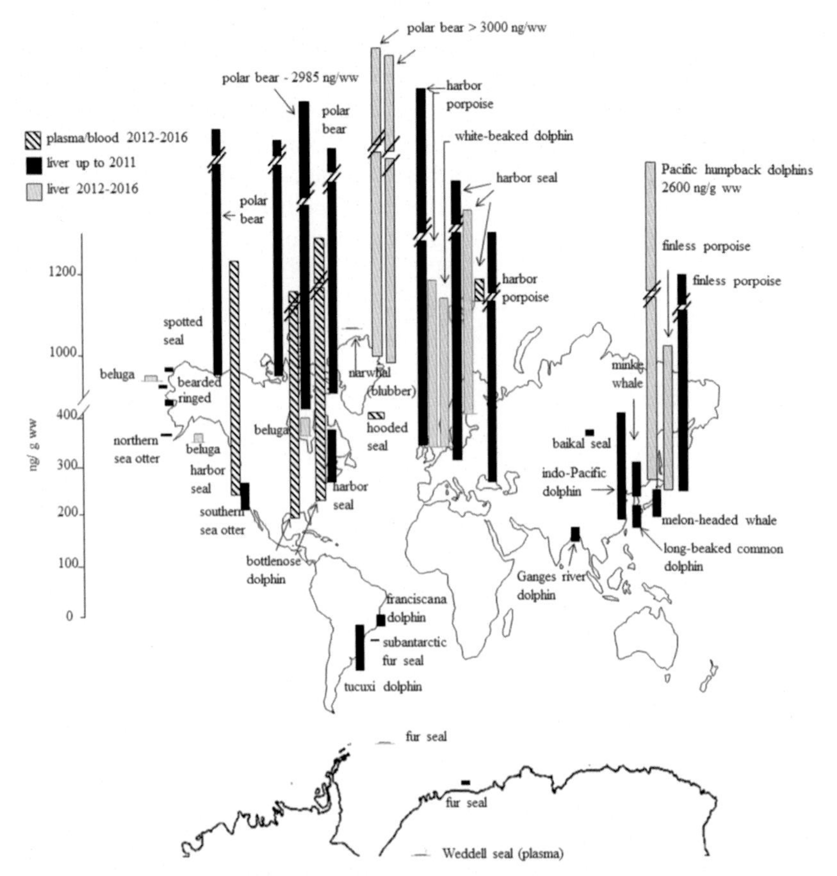

FIGURE 5.1 PFOS concentrations (ng/g ww) reported in marine mammals worldwide. *Black bars* were presented in Houde et al. (2011) for liver concentrations reported between 2006 and 2010. *Grey bars* represent liver levels, and *striped bars* are plasma concentrations reported between 2011 and 2016 (Bengtson Nash et al., 2010; Carlsson et al., 2013; Dietz et al., 2012; Fair et al., 2012; Galatius et al., 2013; Greaves and Letcher, 2013; Greaves et al., 2013; Grønnestad et al., 2016; Lam et al., 2016; NCP, 2015; Reiner et al., 2011; Routti et al., 2014, 2015).

(typically C_9–C_{13}) were still found to be predominant in marine mammals from southern and northern locations (Routti et al., 2014, 2016; Bytingsvik et al., 2012; Dietz et al., 2012; Reiner et al., 2011), including recent quantification of perfluoroundecanoate (PFUnA) in plasma of Weddell seals from Antarctica (Routti et al., 2015).

PFASs may be transferred to offspring as previously reported in mother–fetus pairs in melon-headed whales (*Peponocephala electra*) (Hart et al., 2008a), harbor seal pups (Shaw et al., 2009), and bottlenose dolphin mother–calf pairs from Sarasota Bay, Florida (Houde et al., 2006b). More recent publications support the transfer of PFAS via milk and placenta in hooded seals and a killer whale (*Orcinus orca*) pair from East Greenland (Gebbink et al., 2016; Grønnestad et al., 2016) as well as mother–cub polar bear pairs from Svalbard, Norway (Bytingsvik et al., 2012). Significant differences were also found between age groups of harbor porpoises, suggesting that this could be due to maternal transfer or metabolic differences (Galatius et al., 2011).

Species differences in metabolic capacity to transform and eliminate PFAS have been described in multiple studies. Analysis of liver samples from cetaceans and pinnipeds from the North Sea, collected between 1999 and 2002, indicated a very low contribution of FOSA in harbor seals (0.1% of all PFSA) and harbor porpoises (8.3%) compared to white-beaked dolphins (*Lagenorhynchus albirostris*; 26%), suggestive of species differences in the capacity to transform FOSA (Galatius et al., 2013). Another notable difference among the species was that the two smaller species (seals and porpoises) with supposedly higher metabolic rates had lower concentrations of the PFCAs, which are generally more readily excreted than perfluorinated sulfonamides. Analyses of PFOS:FOSA ratios indeed indicated lower capacities to degrade FOSA in cetaceans compared to pinnipeds and polar bears (Gebbink et al., 2016; Greaves et al., 2012). *In vitro* assays using liver microsomes supported these observations by indicating that polar bears and ringed seals (*Phoca hispida*) can rapidly transform N-EtFOSA to FOSA and that beluga whales (*Delphinapterus leucas*) seemed to lack the enzymatic ability or have a rate too slow to deplete levels of this precursor (Letcher et al., 2014). Therefore, species-specific characteristics should be recognized when studying PFAS contamination in marine mammals.

The distribution of these substances in marine mammals inhabiting several remote locations of the globe has been reported. Since the first identification and quantification of PFSA in Arctic marine mammals (Kannan et al., 2001; Martin et al., 2004; De Silva and Mabury, 2004), several studies have described the distribution and accumulation of these compounds at the poles. PFOS and long-chained PFCAs (C_9–C_{12}) were predominantly detected in Canadian Arctic and Greenland ringed seals (Bossi et al., 2005; Butt et al., 2007; Reiner et al., 2011). FOSA was also found as an important congener in Alaskan beluga whales collected between 1989 and 2006 (Reiner et al., 2011). Polar bears, the apex predator in Arctic regions, have been found to be among the most contaminated wildlife (Houde et al., 2011). The specific distribution of PFSAs has been studied in brains from East Greenland polar bears (Greaves et al., 2013; Pedersen et al., 2015), showing that these chemicals cross the blood-brain barrier of bears and that inner regions of the brain contained higher concentrations of PFSAs (Greaves et al., 2013). Long-chain PFCAs (C_{11-15}) and PFOS were also found to be predominant in brain of polar bears.

PFASs have also been detected in wildlife inhabiting the South Pole (Giesy and Kannan, 2001; Schiavone et al., 2009; Tao et al., 2006). More recently, Bengtson Nash et al. (2010) have reported PFOS in liver of fur seals collected in 2004 from the sub-Antarctic region. Results from that study indicate occasional occurrence of PFSA in migratory marine vertebrates with foraging ranges extending into or north of the Antarctic Circumpolar Current and suggest the hydrospheric transport of these substances to the remote region. Routti et al. (2015) reported the detection of PFASs in plasma of lactating Weddell seals (*Leptonychotes weddellii*). PFUnDA was detected in all samples (0.08–0.23 ng/mL) and PFOS, PFHxA, and PFTriDA were sporadically detected. The authors suggest that these top Antarctic predators may have been exposed to PFCAs by long-range atmospheric transport of fluorotelomer alcohol precursors and/or from a local source. Overall, recent results from studies at the South Pole suggest that PFCAs are more predominant in species feeding south of the Antarctic Circumpolar Current and PFOS in organisms feeding north of the current (Routti et al., 2015).

Tissue-specific isomer patterns have been reported in these recent studies. In polar bears, branched PFOS isomers were quantified in blood (<15% of total PFOS) and liver (<7%) but not in brain, muscle, or fat (Greaves and Letcher, 2013). Branched isomers (10%–15%) were also quantified in livers of harbor porpoises from the Baltic and North Seas (Huber et al., 2012). Linear PFOS was detected in plasma of Antarctic Weddell seals (Routti et al., 2015); no branched PFOS isomers were found. A range of 83%–91% of linear PFOS in relation to total PFOS was also found in livers of ringed seals from the Faroe Islands and NW Greenland and 93%–94% in pilot whales (*Globicephala melas*) (Rotander et al., 2012). Moreover, similar proportions of linear PFOS have been reported in ringed seals (92%), polar bears (88%), and killer whales (90%) (Gebbink et al., 2016). Mean linear FOSA proportions also ranged from 94% to 99%. Linear PFOS and FOSA dominated the composition profile in liver of grey seals (*Halichoerus grypus*) from the Baltic Sea where no branched PFCAs were observed (Kratzer et al., 2011). The predominance of industrial production of linear PFOS (70% of electrochemical fluorination production), longer half-lives (Benskin et al., 2010), and greater potential for bioaccumulation (Greaves and Letcher, 2013; Powley et al., 2008) could explain these results in marine mammals.

Isomeric patterns have additionally been studied for PFCAs in marine mammals. PFOA branched isomers (5%) have been reported in Greenland polar bears, while only linear PFOA was found in Canadian polar bear samples (De Silva and Mabury, 2004). Branched isomers of perfluorodecanoate (PFDA), PFUnA, and perfluorododecanoate (PFDoA) have also been found in polar bears (<4%); linear perfluorononanoate (PFNA) and perfluorotridecanoate (PFTriA) were found to predominate (>99%) in all animals (De Silva and Mabury, 2004). Multiple branched PFNA isomers were also reported in plasma of bottlenose dolphins from the Gulf of Mexico and Atlantic Ocean as well as in ringed seal livers from the Canadian Arctic (De Silva et al., 2009). The presence of branched isomers in marine mammals may be indicative of electrochemical fluorination sources of contamination, as structural isomers are produced by this technology, compared to linear isomers typically produced by telomerization.

The distribution of the common PFSA and PFCA in wildlife has been relatively well studied. However, knowledge on the presence and distribution of new or emerging PFSA, many used as replacement products to regulated congeners, is still scarce. Other subcategories of PFSA, such as perfluoroalkyl phosphinic acids (PFPIAs) also used as surfactants and leveling/wetting agents, have been consistently detected in plasma samples of bottlenose dolphins from the United States (De Silva et al., 2016). Concentrations found in dolphins ranged between 0.27 and 1.5 ng/g ww and were lower than PFSA and PFCA levels with 6:6 and 6:8 congeners mainly found. PFBS and 6:2 chlorinated polyfluorinated ether sulfonate (6:2 Cl-PFAES with trade name F-53B) were also reported in liver of ringed seals, polar bears, and killer whales from East Greenland (Gebbink et al., 2016). Ratios of PFBS:PFOS indicated a shift from PFOS to PFBS in South China Sea dolphins and finless porpoises (Lam et al., 2016). No PFBSs were detected in Western and Southern Hudson Bay polar bears, but N-EtFOSA and FOSA as well as low (ppb) levels of perfluorobutane sulfonamide (PFBA; C_4) and perfluoro-4-ethylcyclohexane sulfonic acid (PFECHS) were quantified in liver (NCP, 2016). This represented the first report of PFECHS and PFBA in Arctic biota. The detection of shorter-chain PFAS in top predators from different locations suggests the transport and the accumulation of these congeners, the use of which has increased since the regulations of longer-chain compounds were introduced.

TEMPORAL TRENDS IN THE BIOACCUMULATION OF PFASs IN MARINE MAMMALS

Assessments of temporal trends for PFAS have been published in marine mammals around the globe; the trends for major congeners are summarized in Table 5.1. Law (2014) has written a short review of temporal trends of PFAS in marine mammals. Additionally, Butt et al. (2010) and Sturm and Ahrens (2010) have reviewed trends of PFAS in Arctic marine mammals. Some patterns can be observed through time by geographic regions. Concentrations in seals and whales collected in Asia have predominantly been increasing for PFSAs and PFCAs (Hart et al., 2008a; Ishibashi et al., 2008a). These trends continue as shown by data from Pacific humpback dolphins and finless porpoises for 2002–14 (Lam et al., 2016). The shift in PFAS production to countries in Asia where environmental regulations are less robust (Wang et al., 2014) may explain the continuing increase of these compounds in marine top predators residing in this area.

Similar increasing levels of PFCAs have been reported in studies conducted in northern Canada and Alaska, United States, for beluga whales, otters, seals, and polar bears (Butt et al., 2007; Hart et al., 2008b; Reiner et al., 2011; Shaw et al., 2009; Smithwick et al., 2006). Samples for these studies were collected before the 2006 PFOA Stewardship Agreement. Trends for PFCAs were also found to be increasing in East Greenland ringed seals (2006–11) (Rigét et al., 2013) and polar bear (1984–2006) (Dietz et al., 2008), mother–cub polar bear pairs from Norway (1998–2008) (Bytingsvik et al., 2012), harbor porpoises

TABLE 5.1 Temporal Trends of Major PFAS Congeners in Marine Mammals Worldwide Classified According to Continents and Date of Publication of the Study

Locations	Species	Years	PFOS	FOSA	PFOA	PFNA	PFDA	PFUnA	References
Asia									
South China Sea	Pacific humpback dolphins	2002–2014	↑		↑*	↑	↑		Lam et al. (2016)
	Finless porpoise	2002–2014	↑		↑*		↑		
Lake Baikal, Russia	Baikal seal (*Pusa sibirica)*	1992 and 2005	↑*			↑	↑*		Ishibashi et al. (2008a)
Japanese coast	Melon-headed whale (*Pepono-cephala electra*)	1982–2001/02	↑*	↑*		↑*	↑	↑*	Hart et al. (2008a)
		2001/02–2006	↑	↓*		↑*	↑*		
Europe									
Faroe Islands	North Atlantic pilot whale	1986–2013	↑*	↑* (up to 2006)		↑*	↑*	↑*	Dassuncao et al. (2017)
Kongsfjorden, Norway	Ringed seal	1990–2004	↔			↑*	↑*	↑*	Routti et al. (2016)
		2004–2010	↓			↓	↓	↓	
East Greenland	Polar bear	1984–2006	↑*	↑	↑	↑*	↑*	↑*	Rigét et al. (2013)
		2006–2011	↓	↓*	↓	↓	↓	↓	
East Greenland	Ringed seal	1986–2006	↑			↑	↑	↑	
		2006–2010	↓			↑	↑	↑	

West Greenland	Ringed seal	1982–2006	↑			↑	↑	↑*	
		2006–2010	↓			↓	↓	↓	
Svalbard, Norway	Polar bear (mother–cub pairs)	1998 and 2008	↓*		↓*	↑	↑	↑	Bytingsvik et al. (2012)
Baltic and North Sea	Harbor porpoise	1991–2008	↓	↓		↑*	↑*	↑*	Huber et al. (2012)
North Atlantic	White-sided dolphin	2001–2006				↑	↑*	↑*	Rotander et al. (2012)
North West Greenland	Ringed seal	1984–2006	↔			↑*	↑*	↑*	
	Pilot whale	1986–2006	↑* (up to 2002)			↑*	↑*	↑*	
	Harbor porpoise	1992–1997				↑*	↔	↔	
	Hooded seal	1990–2007	↓*			↑*	↔	↔	
	Fin whale	1986–2009						↑*	
Danish North Sea	Harbor porpoise[a]	1980–2005	↔	↑			↑	↑*	Galatius et al. (2011)
Baltic Sea	Grey seal	1974–1997	↑*	↑	↑*	↑*	↑*	↑*	Kratzer et al. (2011)
			(1974–1987)						
		1997–2008	↓	↓*	↓*	↓*	↔	↔	
			(1987–2008)						

Continued

TABLE 5.1 Temporal Trends of Major PFAS Congeners in Marine Mammals Worldwide Classified According to Continents and Date of Publication of the Study—cont'd

Locations	Species	Years	PFOS	FOSA	PFOA	PFNA	PFDA	PFUnA	References
German Bight	Harbor seal	1999–2008	↓	↑*	↑*	↔	↔	↔	Ahrens et al. (2009)
East Greenland	Polar bear	1984–2006	↑*	↔	↑*	↑*	↑*	↑*	Dietz et al. (2008)
East Greenland	Ringed seal	1986–2003	↑*				↑*	↑*	Bossi et al. (2005)
West Greenland	Ringed seal	1982–2003	↑*				↑*	↑*	
North America									
San Francisco, CA	Harbor seal								Sedlak et al. (2017)
Central Bay		2004–2014	↓*		↔	↔	↔	↔	
South Bay		2004–2014	↔ (decline in 2014)			↔	↔	↔	
Hudson Bay, NU, Canada	Polar bear	2007–2015	↔			↔	↔	↔	NCP (2016)
Sanikiluaq, NU	Beluga	2005–2013	↓	↓		↓	↓	↓	NCP (2015)
Alaska, U.S.A.	Beluga	1989–2006	↑*	↑*		↑*	↑*	↑*	Reiner et al. (2011)
South Central Alaska	Northern sea otter (*Enhydra lutris kenyoni*)	1998–2007	↓*	↓*		↑*			Hart et al. (2008b)

Northwest Atlantic	Harbor seal	2000–2007	↑			↑	↑	↑	Shaw et al. (2009)
Arviat, NU	Ringed seal	1992–2005				↑*	↑*	↑*	Butt et al. (2007)
		1992–1998	↑	↑					
		1998–2005	↓	↓					
Resolute Bay, NU	Ringed seal	1993–2005				↑*	↑*	↑*	
		1992–2000	↑	↑					
		2000–2005	↓	↓					
Sachs Harbour, NT	Ringed seal	2010–2014	↓						NCP (2016)
West Hudson Bay		2010–2014	↑						
Resolute Bay		2010–2014	↑						
California	Southern sea otter (*Enhydra lutris nereis*)	1992–2002	↑* (up to 1998)		↑*				Kannan et al. (2006)
Baffin Island, NU	Polar bear	1972–2002	↑	↓*	↑*	↑	↑	↑	Smithwick et al. (2006)
Barrow, Alaska	Polar bear	1972–2002	↑*	↓		↑*	↑*	↑*	

*Statistically significant trend. Refer to the list of abbreviations at the end of the chapter for acronyms. Case study on bottlenose dolphins from Charleston, South Carolina, U.S.A.
[a]*Neonates, sucklings, and lactating females excluded from analyses.*

from the Baltic and North Sea (Huber et al., 2012), and harbor seals from the Danish North Sea (1980–2005) (Galatius et al., 2011), as well as white-sided dolphins (2001–06; *Lagenorhynchus acutus*), ringed seals (1984–2006), and pilot whales from Northwest Greenland (1986–2006; Rotander et al., 2012) and the Faroe Islands (1986–2013; Dassuncao et al., 2017). In contrast, PFCA levels have been recently found to be decreasing in ringed seals from Norway (2004–10) (Routti et al., 2016) and West Greenland (2006–10), polar bears from East Greenland (2006–11) (Rigét et al., 2013), and grey seals from the Baltic Sea (1997–2008) (Kratzer et al., 2011).

Except for belugas from Alaska and harbor seals from the Northwest Atlantic, PFOS and FOSA tissue concentrations have been decreasing in North American marine mammals when early 2000s years are included in trends. Recent data published in a Synopsis of research report from the Indigenous and Northern Affairs Canada indicated stable PFAS levels in polar bears (2007–14) from Hudson Bay and decreasing concentrations in beluga whales harvested in Sanikiluaq (NCP, 2015). Trends in liver of ringed seals varied depending on the locations with decreasing PFOS concentrations observed in seals collected from Sachs Harbour, Northwest Territories, and increasing trends reported in seals from the Western Hudson Bay (NCP, 2016). Information from other marine mammal species from southern locations would be useful to obtain a clearer understanding of time trends. Rigét et al. (2013) recently summarized the temporal trends of PFAS in Greenland ringed seals and polar bears. The authors report that peaks for PFAS in Greenland ringed seals and polar bears occurred later (2006 for sulfonates) than in Canadian seals and bears and from other southern species of seals (Rigét et al., 2013). They suggest that distance from sources and long-range transport to this Arctic site may explain this later peaking in levels. Trends for PFOS in European marine mammals have generally declined when samples collected after 2005–06 were included in the analyses (Ahrens et al., 2009; Kratzer et al., 2011; Bytingsvik et al., 2012; Rigét et al., 2013; Routti et al., 2016; Rotander et al., 2012), except for East Greenland polar bears (1984–2006), where PFOS levels had significantly increased (Dietz et al., 2008). Sulfonate trends seem to respond to the change in the phase-out of perfluorooctanesulfonylfluoride (POSF)-based compounds, the production and use of different homologues, as well as shifts in manufacturing locations. Oceanic transport of PFAS congeners and atmospheric transport of precursors such as fluorotelomer alcohols (FTOH), sulphonamido ethanols (FOSE) and FOSA may be different from site to site and explain differences in PFCA and PFSA trends. Species also have different habitats, diets, metabolism, and migration patterns that can influence the accumulation of PFASs. A recent study by Tartu et al. (2017) indeed reported that diet and metabolic rate were the main factors predicting PFAS concentrations in plasma of female polar bears from Svalbard with the highest contaminated individuals feeding on high trophic level sea ice–associated prey, fasting, and with small cubs.

Continued global monitoring of PFASs is of importance. It is also crucial to assess the presence and contamination levels of replacement congeners such as the shorter-chain PFSA as well as precursors. The first report of PFBS temporal trends in marine mammals was published by Lam et al. (2016), which showed increases of this short-chain PFSA in liver of finless porpoise and significant increase in Pacific humpback dolphins from the South China Sea from 2002 to 2014. Shifts in the contamination patterns may be occurring following the regulation of PFOS and its salts and the use of alternatives substances by the industry (Lam et al., 2016). Recent partitioning and bioaccumulation modeling results for pilot whales from the Faroe Islands also indicated that atmospheric inputs of FOSA were important exposure sources from the marine food webs (Dassuncao et al., 2017), warranting the monitoring of precursor compounds in the marine ecosystems.

BOTTLENOSE DOLPHINS FROM CHARLESTON, SOUTH CAROLINA, U.S. - A CASE STUDY

The Atlantic Bottlenose Dolphin Health and Environmental Risk Assessment (HERA) Project was initiated as a multidisciplinary, integrated collaborative effort in 2003 to assess individual and population health in two southeast estuarine coastal regions of the United States: Charleston, South Carolina, and the Indian River Lagoon, Florida (Fair et al., 2006). As reported by Houde et al. (2006a), the concentrations of PFASs found in plasma of dolphins from Charleston were among the highest detected in wildlife. Plasma PFAS concentrations, as well as a suite of persistent organic contaminants, measured in Charleston dolphins were significantly higher than in the Indian River Lagoon dolphins and PFAS levels similar to those reported for occupationally exposed humans (Fair et al., 2013). PFOS was the most prominent of the 13 PFASs determined in dolphin samples collected in 2003–05, with concentrations in plasma ranging from 574 to 8670 ng/g ww in dolphins from Charleston and 112 to 4470 ng/g ww in dolphins from the Indian River Lagoon. Age-related differences in PFAS bioaccumulation were found in both populations with decreasing concentrations with age. The distribution of PFASs in blubber, feces, and urine was also examined (Fair et al., 2007). As with plasma, PFOS was the most prominent PFAS in blubber, ranging in concentration from 104 to 1534 ng/g ww at Charleston and from 71 to 790 ng/g ww at the Indian River Lagoon. In general, the pattern of PFASs in all tissues was dominated by PFOS, but detectable levels of PFCAs, particularly, PFNA, PFDA, and PFUnA were found. PFAS concentrations in dolphin blubber was lower compared to plasma; however, since blubber tissue comprises a large proportion of the total body weight the remobilization of PFASs and other contaminants from blubber during periods of fasting, starvation, lactation, or other physiologic demands may be considerable.

The identification of high PFAS levels in Charleston dolphins led to an investigation of the dolphin food web and the environment (Houde et al., 2006c).

PFOS, C_8, and C_{10} PFCAs predominated most matrices, including waste water treatment effluents, whole fish, and dolphin tissues. In Charleston, biomagnification factors (BMFs) ranged from <1 to 30, and trophic magnification factors (TMFs) for PFOS and C_8–C_{11} PFCAs indicated biomagnification in this marine food web. It was also determined that plasma and liver PFAS concentrations as surrogate to whole body burden in a top marine predator overestimates the BMFs and TMFs. The PFAS concentrations detected in environmental samples and fish from the Charleston Harbor area and the BMFs calculated in this study may explain why PFASs in dolphins from Charleston are among the highest reported in marine mammals.

Further, higher body burdens of specific PFASs were found in Charleston dolphins inhabiting areas with greater developed land use (Adams et al., 2008). Specifically, plasma PFOS, PFDA, and PFUnA levels in dolphins were positively associated with industrialized urban areas compared to residential and less developed areas. This study showed that the impact of urbanization on the distribution of PFASs in the environment can manifest itself on a much more local scale than previously observed. The bottlenose dolphin is a valuable sentinel species for monitoring the health of the environment that can provide important information on regional environmental and public health issues (Bossart, 2011).

Sediment levels were additionally measured in the Charleston area. The concentrations were found to be higher than any other urban US area, with over half of the sites exceeding the median global PFOS sediment concentration (White et al., 2015). The high PFAS concentrations in the Charleston area and dolphins from this region alerted researchers to investigate the potential exposure to these chemicals in humans. One study focused on the Gullah African-Americans in the Charleston area, as this community resides along the coastal sea island and is a unique study population to define environmental factors for autoimmune disease due to their genetic, environmental-geographic homogeneity, and high prevalence of disease, such as systemic lupus erythematosus (Kamen et al., 2012). Since fish is an important dietary route of PFAS exposure in humans (Haug et al., 2010; Holzer et al., 2011; Yamaguchi et al., 2013) and because locally obtained seafood is a dietary staple, trends of PFAS in the Gullah African American population of the Charleston area were investigated. While temporal trends in this population during 2003–13 showed decreases in several PFASs, substantial heterogeneity exists (Gribble et al., 2015). Recent findings for PFPIA, a newly detected PFAS in Charleston dolphins (De Silva et al., 2016), emphasize future concerns and implications for wildlife and human health from exposure to replacement and alternative PFAS.

The effects of environmental exposure of bottlenose dolphins to these compounds remained unclear. Detection of high concentrations of PFASs in Charleston dolphins led to the first investigation of associations between these compounds and health effects in dolphins. Significant relationships between exposure to total PFASs and/or individual analytes and markers of immune, hematologic, hepatic, and renal function were found (Fair et al., 2013). Chronic exposures to high levels of PFASs in Charleston dolphins were associated with stimulation of both innate and adaptive immunity (both humoral and cell-mediated), in addition to modulation of various clinicopathologic parameters including liver enzymes. Findings from this study reveal that the dolphins' chronic exposure to these contaminants appears to be producing immune perturbations and tissue toxicity, which are further discussed in the Health-Related Effects of PFASs section of this chapter. Identification of effects of exposure to chemical contaminants on free-ranging dolphins led to laboratory investigations using rodent models designed to corroborate *in situ* results, assess the sensitivities between the mouse model and dolphins, and to evaluate risk (Wirth et al., 2014, 2015). Assessment of immunotoxic effects of PFSAs in a murine model showed that immunological functions and gene expression were sensitive to exposure to these chemicals at concentrations found in the Charleston dolphins (Fair et al., 2011; Mollenhauer et al., 2009, 2011). Compared to other toxicological endpoints, the immune system appears to be particularly sensitive to exposure by PFASs. Using dolphin lymphocytes, T cell–specific responses were examined *ex vivo* and *in vitro* to exogenous PFOS exposure, and results indicated proinflammatory and proliferative responses (Soloff et al., 2018), which are further discussed in the next section.

HEALTH-RELATED EFFECTS OF PFASs IN MARINE MAMMALS

The hazard assessment published by the Organization for Economic Cooperation and Development concluded that PFOS is persistent, bioaccumulative, and toxic to mammalian species (OECD, 2002). Most studies to date have focused primarily on PFOS, although studies indicate related properties for the entire category of PFAS chemical substances. Because of these properties, the potential toxic effects of these chemicals are of great concern.

While a large body of literature has been published on the presence and trends of PFASs in the aquatic environment and marine mammals, few studies exist on the biological effects of these chemicals on marine mammals since this research is difficult to conduct under regulations and impeded by challenging and expensive logistics. Effects related to PFAS exposure have been observed in laboratory animals with a wide range of alterations reported, including reproductive toxicity, hepatotoxicity, neurotoxicity, immunotoxicity, and endocrine and metabolic effects. For a more detailed summary, a number of comprehensive toxicological reviews have been published (Ahrens and Bundschuh, 2014; Corsoni et al., 2014; DeWitt et al., 2012, DeWitt, 2015; Kennedy et al., 2004, Lau et al., 2004, Lau, 2012, Mariussen, 2012; Olsen et al., 2009; Stahl et al., 2011; White et al., 2011). In laboratory animal studies, liver hypertrophy and peroxisome proliferation are typically found with significant body burden of PFASs concurrent with activation of genes related to the peroxisome proliferator-activated receptor alpha (PPARα). (Lau, 2015). PFASs have been determined to cause endocrine-disrupting effects, and PFOA-induced effects have been reported in several animal and epidemiological studies (Jensen and Leffers, 2008; White et al., 2011).

PFASs are primarily retained in blood plasma and internal organs (e.g., liver and kidney) due to their physicochemical properties. The potential effects of PFASs are of concern in marine mammals since high levels have been reported in multiple species. One of the highest PFAS plasma levels were found in bottlenose dolphins, as described in the case study previously presented. The importance of PFAS as a contaminant of concern in the Arctic environment is also an area of active research. Polar bears, an apex predator, bioaccumulate a range of contaminants with PFAS liver concentrations comparable to polychlorinated biphenyls (PCBs, >3000 ng/g ww) (Greaves et al., 2012). Sonne (2010) reviewed the health impacts from the long-range transported contaminants, including PFASs, in polar bears. Limited data is available from wildlife field studies, particularly marine mammals, which demonstrate a link between exposure levels and biologic effects. However, in recent years, toxico-pathologic studies conducted with marine mammals have reported correlations between blood and tissue PFASs concentrations and markers of biological effects. Table 5.2 summarizes several studies that examined the relationship of concentrations of PFASs related to health in Baikal seals (Ishibashi et al., 2008b), sea otters (Kannan et al., 2006), bottlenose dolphins (Fair et al., 2013), and polar bears (Pedersen et al., 2015, 2016; Sonne et al., 2009; Bourgeon et al., 2017).

A study by Kannan et al. (2006) investigated the association between PFOS and PFOA exposures and potential effects, comparing concentrations among adult female otters that died from infectious diseases, noninfectious causes, and apparent emaciation. Concentrations of both PFOA and PFOS were significantly higher in sea otters in the infectious disease category than in the noninfectious. Residue concentrations of PFOS and PFOA in sea otter livers were two to three orders of magnitude lower than concentrations that demonstrate an effect in laboratory animals, although the authors noted these studies have focused more on reproductive and developmental endpoints than immunotoxic effects of these chemicals.

Significant relationships have also been found between PFSA and multiple neurochemical markers in specific regions of the brain in polar bears from East Greenland (Pedersen et al., 2015). Specifically, monamine oxidase enzyme activity and receptor density of gamma-butyric acid type A were positively associated with PFAS levels (Pedersen et al., 2015). Negative associations were also found between concentrations of PFASs and glutamine synthetase receptor densities and dopamine-2 levels. The authors indicated that these findings support the hypothesis that PFAS concentrations in polar bears from East Greenland exceed threshold limits for neurochemical alterations. The fundamental role of these neurotransmitter systems in behavior and reproduction underscores the need to better understand effects of PFAS exposure on marine mammals. Furthermore, Pedersen et al. (2016) found positive correlations between PFASs and steroid hormones (i.e., testosterone, progesterone, pregnenolone, androstendione) in

TABLE 5.2 Health-Related Effects Associated With PFASs in Marine Mammals. Refer to the List of Abbreviations for all Acronyms

Species	Region	Year	Tissue/ Number	PFAS Concentration	Health Effect	References
Southern sea otter (*Enhydra lutris nereis*)	California, USA	1992–2002	Liver (nondiseased *n*=26; emaciated *n*=27; diseased *n*=27)	Mean and ranges (ng/g) PFOS nondiseased 31(1–102) PFOS emaciated 39 (4–169) PFOS diseased 95 (5–884) PFOA nondiseased 49 (5–98) PFOA emaciated 62 (5–147) PFOA diseased 69 (5–139)	Significant association of elevated PFOS and PFOA with "diseased" status	Kannan et al. (2006)
Bottlenose dolphins (*Tursiops truncatus*)	Charleston, South Carolina USA	2003–2005	Blood plasma (*n*=79)	∑PFAS mean (range) 1970 (574–8670) ng/ml	Significant association of elevated PFASs with multiple immunological and hematological parameters	Fair et al. (2013)
Baikal seal (*Pusa sibirica*)	Lake Baikal, Russia	1992	Liver-cloned cDNA PPARα; in vitro PPARα receptor assay	PFNA, PFDA, PFUnA concentrations had LOEC values 125, 125 and 62.5 μM, respectively Liver PFNA levels 3.3–72 ng/g PFDA <0.56–35 ng/g	Expression of hepatic CYP4A-like protein significantly correlated with hepatic PFNA and PFDA levels	Ishibashi et al. (2008b)

Continued

TABLE 5.2 Health-Related Effects Associated With PFASs in Marine Mammals. Refer to the List of Abbreviations for all Acronyms—cont'd

Species	Region	Year	Tissue/ Number	PFAS Concentration	Health Effect	References
Polar bear (*Ursus maritimus*)	East Greenland	1990–2006	Liver (52 subadults, 23 adult females, 31 adult males)	PFOS mean (range) 3108 (1332–6160) ng/g	Risk Quotient >1 for reproduction based on critical body residue and PBPK model	Sonne et al. (2009)
Polar bear	Scoresby Sound, East Greenland	2011–2012	Brain ($n = 9$)	Mean brain ∑PFSA 25 ng/g (91% PFOS) ∑PFCA 88 ng/g (79% comprised of PFUnA, PFDoA, and PFTriDA)	Significant positive correlations between PFAS in brain regions and MAO activity; negative correlation to dopamine-2	Pedersen et al. (2015)
Polar bear	Scoresby Sound, East Greenland	2011–2012	Brain ($n = 10$; 5 males, 5 females)	PFAS data from Pedersen et al., (2015)	Positive associations between PFASs and two hormones, 17α-hydropregnenolone and testosterone, in brain regions	Pedersen et al. (2016)
Polar bear	Barents Sea, Svalbard, Norway	2012–2013 (April and September)	Blood plasma ($n = 78$ females with several recaptures)	Plasma ∑PFAS mean 352.64 ± 15.99 ng/g; ∑PFSA mean 264.35 ± 12.45 ng/g; ∑PFCA mean 88.25 ± 3.86 ng/g	Alterations of thyroid hormone homeostasis; TT3 and FT3 negatively associated with ∑PFAS; FT3 negatively associated with ∑PFAS in spring but not autumn	Bourgeon et al. (2017)

LOEC, lowest observed effect concentration; *TT3*, total tri-iodothyronine; *FT3*, free tri-iodothyronine.

brains from polar bears. Given the importance of steroids for functions such as brain plasticity, postnatal development, and reproductive function, a better understanding on how PFASs may affect the endocrine system is needed. Additionally, Bourgeon et al. (2017) reported on alterations of thyroid homeostasis in polar bears associated with blood plasma PFAS concentrations. Moreover, pharmacokinetic modeling work for PFOS in polar bears has suggested increased risk for adverse reproductive effects (Sonne et al., 2009). Using this approach, PFOS was found to be second, after PCBs, for effects on the reproductive and immune systems, as well as carcinogenicity/genotoxicity (Dietz et al., 2015).

Immunotoxicity is one of the more sensitive and notable effects of PFASs affecting both cell-mediated and humoral immunity (Corsoni et al., 2014; DeWitt et al., 2012). Immunotoxic effects may involve activation PPARα receptors; however, some immune-related effects may also be independent of this receptor (Keil, 2015). Increasing evidence suggests that immune effects in laboratory models occur below and within the range of reported levels from highly exposed wildlife and humans (Takacs and Abbott, 2007). In dolphins, significant positive associations with one or more PFASs and immunological parameters include absolute numbers of CD2+T cells, CD4+helper T cells, CD19+immature B cells, CD21+mature B cells, CD2/CD21 ratio, MHCII+ cells, B cell proliferation, serum IgG1, granulocytic, and monocytic phagocytosis. Additionally, several PFAS analyte groups were positively associated with serum alanine aminotransferase, gamma-glutamyl-transferase, creatinine, and negatively associated with cholesterol levels, creatinine phosphokinase, eosinophils, and monocytes. These findings by Fair et al. (2013) suggest that chronic PFAS exposure in Charleston bottlenose dolphins may have effects on immune, hematopoietic, kidney, and liver functions. Further, *in vitro* examination of T cell–specific responses of dolphin lymphocytes to environmentally relevant concentrations of PFOS promoted pro-inflammatory interferon-gamma (IFNγ), but not immunoregulatory interleukin-4 (IL-4) production, in T cells. These responses may establish a state of chronic immune activation known to be associated with susceptibility to disease (Soloff et al., 2018). Taken together, these findings suggest PFOS dysregulates the dolphin cellular immune system and has implications for health (Soloff et al., 2018). In contrast, an *in vitro* exposure study with PFOS and PFOA (1000 ng/g) found no alterations in seal lymphocyte proliferation (Levin et al., 2016). Differences in species, experimental methods, and dose may potentially explain the variability in the results of these *in vitro* studies.

Laboratory investigations have identified the receptor involved with hepatotoxic action of PFASs as PPARα, a nuclear receptor that is highly expressed in liver (Takacs and Abbott, 2007). Results by Ishibashi et al. (2008b) suggest PFASs, particularly long-chain PFCAs such as PFNA and PFDA, may exert biologic effects by disruption of the PPARα-CYP4A signaling pathways in the livers of wild Baikal seals. PPARα cloned cDNA in Baikal seals showed remarkable similarity with other species such as dog, human, rat, and mouse. The *in vitro* reporter gene assay developed using Baikal seal PPARα-transfected cells supports the observations of significant correlations between hepatic CYP4A-like protein and hepatic concentrations of PFAS in wild seals. Thus, the PPARα-mediated response may be a useful biomarker to evaluate biologic effects of PFASs in wildlife. Laboratory-based studies have also found that effects of PFASs vary depending upon perfluoroalkyl chain lengths and functional groups, animal species, and sex. Factors contributing to these differences include pharmacokinetic differences between species as well as the PFAS homologues. Serum half-lives can vary from hours to years with differential rates of renal reabsorption between species contributing to differences in pharmacokinetic characteristics between species (Andersen et al., 2008; Han et al., 2012). Both pharmacokinetics and serum half-lives of PFAS vary among species, making extrapolation between studies and animal models difficult.

Studies demonstrating correlations between PFAS exposures and effects in wild populations can add to weight of evidence and assist in developing a framework for future investigations. Multiple PFASs are found in the environment as well as in humans and wildlife, therefore, studies need to consider cumulative risks and potential interactions. An inability to control for confounding factors is one of the limitations of correlative analysis in wildlife studies; associations may in fact be due to other factors or to exposures to chemicals other than the one studied. Moreover, contaminants in the environment exist in complex mixtures, and interactions can occur between the contaminants as well as other factors such as stress, nutrition, and disease.

Further research is needed to determine the role of PFASs on adverse health outcomes. As such, future studies should focus on modes of action found in animal and human studies to identify similar adverse outcome pathways. Weight of evidence approaches are useful and will assist in supporting cases for health impacts of PFASs in marine mammals. The toxicologic properties of PFASs are wide-ranging, and the high levels found in marine mammals raise concerns about their potential developmental, reproductive, immune, and systemic toxicity, especially during critical developmental stages. Studies on marine mammals have reported effects of PFASs on neuro-, immune-, and endocrine systems, as well as reduced survival after infection in animals with high PFAS levels, as summarized herein. Given the importance of these systems, coupled with laboratory animal studies characterizing toxicity at environmentally relevant concentrations, and results from correlative and *in vivo* and *in vitro* marine mammal studies, there is an urgent need to better understand the effects of chronic PFAS exposure on marine mammals.

FUTURE PERSPECTIVES

It is clear from the reported results that additional exposure data from more recent years are needed in all regions of the world to evaluate the impacts of PFAS regulations on marine wildlife. The identification of specific areas of concern where local sources or long-range transport specifically affect marine wildlife PFAS contamination should be targeted in investigations. Time trend information for PFAS as described in this paper remains limited across the globe. Trends in PFASs levels in marine mammals continue to be variable and diverse, but generally concentrations of regulated compounds such as PFOS have declined in many populations, and those of PFCAs have increased. Additionally, short-chain PFAS and novel compounds, such as PFPIA and chlorinated polyfluorinated ether sulfonate, are being detected in marine mammal tissues with unknown potential impacts. Data for spatial and temporal trends, not only in the Arctic regions where many of the studies have been focused, but also in industrialized regions where higher concentrations are typically found are needed. Modeling regional and global fate and exposure of PFASs in marine mammals and their environment as well as further biomonitoring is necessary to be able to draw conclusions and to assess the overall threat to marine mammals worldwide. These chemicals also need to be considered in the context of the mixture of chemicals to which marine mammals are exposed, so measuring multiple contaminants is essential. Climate-mediated ecologic changes may alter the ecology and the diet of marine organisms as well as the transport, the degradation and the bioaccumulation of chemical substances, therefore changing the exposure of marine mammals to contaminants including PFAS. It is essential to gather baseline information and long-term follow-up on the fate and distribution of PFAS in marine food webs to be able to understand changes occurring in their environment. Evaluation of climate change consequences on population level and health-related effects caused by a combination of contaminant exposures and environmental changes is needed, especially for marine mammals living in ecosystems highly impacted such as the Arctic. There is also a need for continued characterization of PFASs in marine ecosystems, as well as identifying novel precursors. Food web researchers would benefit from pharmacokinetic studies on a wide variety of relevant PFAS compounds such as perfluorooctanesulfonate precursors (PreFOS) and FOSA, allowing determination of the extent to which environmental concentrations and body burdens of PFOS in marine mammals are due to degradation of precursors. Better understanding the metabolic capacities of marine mammals for precursor compounds is also essential. Such studies would allow more accurate predictions of future risks of exposed marine mammals and other wildlife. To date, few studies have examined health outcomes and cause-effect relationships for PFAS exposure. Such research are difficult to establish in marine mammals. Existing studies suggest adverse effects and increased risks for populations with elevated PFAS levels. Additional studies are needed to assess health-related outcomes and population level parameters such as survival and reproduction. Relevant biomarkers of immune and endocrine disruption in marine mammals would assist investigations aimed at establishing health impacts. Determining exposure and threshold levels of PFASs that are deleterious to marine mammals is critical so that risks can be better assessed in these populations. Identifying risks to marine mammals will also constitute a part of the rationale for continued examination and refinement of existing regulations to protect wildlife and human health.

REFERENCES

3M Company, 2000. Voluntary Use and Exposure Information Profile for Perfluorooctanoic Acid and Salts. Document EPA-HQ-OPPT-2002-0051-0009. Available from: http://www.regulations.gov.

Adams, J., Houde, M., Muir, D., Speakman, T., Bossart, G., Fair, P., 2008. Land use and the spatial distribution of perfluoroalkyl compounds as measured in the plasma of bottlenose dolphins (*Tursiops truncatus*). Marine Environmental Research 66, 430–437.

Ahrens, L., Bundschuh, M., 2014. Fate and effects of poly- and perfluoroalkyl substances in the aquatic environment: a review. Environmental Toxicology and Chemistry 33, 1921–1929.

Ahrens, L., Siebert, U., Ebinghaus, R., 2009. Temporal trends of polyfluoroalkyl compounds in harbor seals (*Phoca vitulina*) from the German Bight, 1999–2008. Chemosphere 76, 151–158.

Andersen, M., Butenhoff, J., Chang, S., Farrar, D., Kennedy, G., Lwu, C., Olsen, G., Seed, J., Wallace, K., 2008. Perfluoroalkyl acids and related chemistries – toxicokinetics and modes of action. Toxicological Sciences 102, 3–14.

Armitage, J., Schenker, U., Scheringer, M., Martin, J., Macleod, M., Cousins, I., 2009. Modeling the global fate and transport of perfluorooctane sulfonate (PFOS) and precursor compounds in relation to temporal trends in wildlife exposure. Environmental Science and Technology 43, 9274–9280.

Beach, S., Newsted, J., Coady, K., Giesy, J., 2006. Ecotoxicology evaluation of perfluorooctane-sulfonate (PFOS). Reviews of Environmental Contamination and Toxicology 186, 133–174.

Bengtson Nash, S., Rintoul, S.R., Kawaguchi, S., Staniland, I., Hoff, J.V.D., Tierney, M., Bossi, R., 2010. Perfluorinated compounds in the Antarctic region: ocean circulation provides prolonged protection from distant sources. Environmental Pollution 158, 2985–2991.

Benskin, J., De Silva, A.O., Martin, J.W., 2010. Isomer profiling of perfluorinated substances as a tool for source tracking: a review of early findings and future applications. Reviews of Environmental Contamination and Toxicology 208, 111–160.

Bossart, G., 2011. Marine mammals as sentinel species for oceans and human health. Veterinary Pathology 48, 676–690.

Bossi, R., Riget, F.F., Dietz, R., Sonne, C., Fauser, P., Dam, M., Vorkamp, K., 2005. Preliminary screening of perfluorooctane sulfonate (PFOS) and other fluorochemicals in fish, birds and marine mammals from Greenland and the Faroe Islands. Environmental Pollution 136, 323–329.

Bourgeon, S., Riemer, A.K., Tartu, S., Aars, J., Polder, A., Jenssen, B.M., Routti, H., 2017. Potentiation of ecological factors on the disruption of thyroid hormones by organo-halogenated contaminants in female polar bears (*Ursus maritimus*) from the Barents Sea. Environmental Research 158, 94–104.

Buck, R., Franklin, J., Conder, J., Cousins, I., Devoogt, P., Jensen, A., Kannan, K., Mabury, S., Leeuwen, S., 2011. Perfluorolkyl and polyfluoroalkyl substances in the environment: Terminology, classification, and origins. Integrated Environmental Assessment and Management 7, 513–541.

Butenhoff, J., Olsen, G., Pfahles-Hutchens, A., 2006. The applicability of biomonitoring data for perfluoroctanesulfonate to the environmental public health continuum. Environmental Health Perspectives 114, 1783–1789.

Butt, C.M., Muir, D.C.G., Stirling, I., Kwan, M., Mabury, S.A., 2007. Rapid responses of Arctic ringed seals to changes in perfluoroalkyl production. Environmental Science and Technology 41, 42–49.

Butt, C.M., Berger, U., Bossi, R., Tomy, G.T., 2010. Levels and trends of poly- and perfluorinated compounds in the arctic environment. The Science of the Total Environment 408, 2936–2965.

Butt, C., Muir, D., Mabury, S., 2014. Biotransformation pathways of fluorotelomer-based compounds in the Arctic environment. The Science of the Total Environment 408, 2936–2965.

Bytingsvik, J., Van Leeuwen, S.P.J., Hamers, T., Swart, K., Aars, J., Lie, E., Nilsen, E.M.E., Wiig, Ø., Derocher, A.E., Jenssen, B.M., 2012. Perfluoroalkyl substances in polar bear mother–cub pairs: a comparative study based on plasma levels from 1998 and 2008. Environment International 49, 92–99.

Carloni, D., 2009. Perfluorooctane Sulfonate (PFOS) Production and Use: Past and Current Evidence. United Nations Industrial Development Organization (UNIDO). 38 pages.

Carlsson, P., Herzke, D., Kallenborn, R., 2013. Polychlorinated biphenyls (PCBs), polybrominated diphenyl ethers (PBDEs) and perfluorinated alkylated substances (PFASs) in traditional seafood items from western Greenland. Environmental Science and Pollution Research 21, 4741–4750.

Corsoni, E., Luebke, R., Germolec, D., Dewitt, J., 2014. Perfluorinated compounds: emerging POPs with potential immunotoxicity. Toxicology Letters 230, 263–270.

Dassuncao, C., Hu, X.C., Zhang, X., Bossi, R., Dam, M., Mikkelsen, B., Sunderland, E.M., 2017. Temporal shifts in poly- and perfluoroalkyl substances (PFASs) in North Atlantic pilot whales indicate large contribution of atmospheric precursors. Environmental Science and Technology 51, 4512–4521.

De Silva, A.O., Mabury, S.A., 2004. Isolating isomers of perfluorocarboxylates in polar bears (*Ursus maritimus*) from two geographical locations. Environmental Science and Technology 38, 6538–6545.

De Silva, A.O., Muir, D.C.G., Mabury, S.A., 2009. Distribution of perfluorocarboxylate isomers in select samples from the North American environment. Environmental Toxicology and Chemistry 28, 1801–1814.

De Silva, A.O., Spencer, C., Ho, K.C.D., Al Tarhuni, M., Go, C., Houde, M., De Solla, S.R., Lavoie, R.A., King, L.E., Muir, D.C.G., Fair, P.A., Wells, R.S., Bossart, G.D., 2016. Perfluoroalkylphosphinic acids in northern pike (*Esox lucius*), double-crested cormorants (*Phalacrocorax auritus*), and bottlenose dolphins (*Tursiops truncatus*) in relation to other perfluoroalkyl acids. Environmental Science and Technology 50, 10903–10913.

DeWitt, J., 2015. Toxicological effects of perfluoroalkyl and polyfluoroalkyl substances. In: DeWitt, J. (Ed.), Molecular and Integrative Toxicology Series. Springer International Publishing, New York. 495 pages.

DeWitt, J.C., Peden-Adams, M.M., Keller, J.M., Germolec, D.R., 2012. Immunotoxicity of perfluorinated compounds: recent developments. Toxicologic Pathology 40, 300–311.

Dietz, R., Bossi, R., Rigét, F., Sonne, C., Born, E.W., 2008. Increasing perfluoroalkyl contaminants in East Greenland polar bears (*Ursus maritimus*): a new toxic threat to the Arctic bears. Environmental Science and Technology 42, 2701–2707.

Dietz, R., Rigét, F.F., Galatius, A., Sonne, C., Teilmann, J., Bossi, R., 2012. Spatial trends of perfluorochemicals in harbor seals (*Phoca vitulina*) from Danish waters. The Science of the Total Environment 414, 732–737.

Dietz, R., Gustavson, K., Sonne, C., Desforges, J.-P., Rigét, F.F., Pavlova, V., Mckinney, M.A., Letcher, R.J., 2015. Physiologically-based pharmacokinetic modelling of immune, reproductive and carcinogenic effects from contaminant exposure in polar bears (*Ursus maritimus*) across the Arctic. Environmental Research 140, 45–55.

Dorneles, P., Lailson-Brito, J., Azevedo, A., Meyer, J., Vida, L., Fragoso, A., Torres, J., Malm, O., Blust, R., Meyer, J., Vidal, L., 2008. High accumulation of perfluorooctane sulfonate (PFOS) in marine tucuxi dolphins (*Sotalia guianensis*) from the Brazilian coast. Environmental Science and Technology 42, 5368–5373.

DuPont, 2013. DuPont Position Statement on PFOA. Available from: http://www.dupont.com/corporate-functions/our-company/insights/articles/position-statements/articles/pfoa.html.

D'Hollander, W., De Voogt, P., De Coen, W., Bervoets, L., 2010. Perfluorinated substances in human food and other sources of human exposure. Reviews of Environmental Contamination and Toxicology 208, 179–215.

Environment Canada, 2010. Environmental Performance Agreement Respecting Perfluorinated Carboxylic Acids (PFCAs) and Their Precursors in Perfluorochemical Products Sold in Canada. Available from: http://ec.gc.ca/epe-epa/default.asp?lang=En&n=81AE80CE-1.

European Parliament, 2006. Directive 2006/122/EC of the European Parliament and of the Council of 12 December 2006. Official Journal of the European Union 3 pages.

Fair, P.A., Hulsey, T., Varela, R.A., Goldstein, J., Adams, J.A., Zolman, E., Bossart, G.D., 2006. Hematology, serum chemistry and cytology results from Atlantic bottlenose dolphins (*Tursiops truncatus*) inhabiting the coastal waters of Charleston, South Carolina. Aquatic Mammals 32, 182–195.

Fair, P.A., Small, J., Sturman, S., Adams, J., Houde, M., Bossart, G.D., Muir, D.C.G., 2007. Tissue distribution of perfluoroalkyl compounds in bottlenose dolphins (*Tursiops truncatus*) from southeast coastal USA. Organohalogen Compounds 69, 849–852.

Fair, P.A., Driscoll, E., Mollenhauer, M.A.M., Bradshaw, S.G., Hun Yun, S., Kannan, K., Bossart, G.D., Keil, D.E., Peden-Adams, M.M., 2011. Effects of environmentally relevant levels of perfluorooctane sulfonate on clinical parameters and immunological functions in B6C3F1 mice. Journal of Immunotoxicology 8, 17–29.

Fair, P.A., Houde, M., Hulsey, T.C., Bossart, G.D., Adams, J., Balthis, L., Muir, D.C.G., 2012. Assessment of perfluorinated compounds (PFCs) in plasma of bottlenose dolphins from two southeast US estuarine areas: relationship with age, sex and geographic locations. Marine Pollution Bulletin 64, 66–74.

Fair, P.A., Romano, T., Schaefer, A.M., Reif, J.S., Bossart, G.D., Houde, M., Muir, D., Adams, J., Rice, C., Hulsey, T.C., Peden-Adams, M., 2013. Associations between perfluoroalkyl compounds and immune and clinical chemistry parameters in highly exposed bottlenose dolphins (*Tursiops truncatus*). Environmental Toxicology and Chemistry 32, 736–746.

Galatius, A., Dietz, R., Rigét, F.F., Sonne, C., Kinze, C.C., Lockyer, C., Bossi, R., 2011. Temporal and life history related trends of perfluorochemicals in harbor porpoises from the Danish North Sea. Marine Pollution Bulletin 62, 1476–1483.

Galatius, A., Bossi, R., Sonne, C., Rigét, F.F., Kinze, C.C., Lockyer, C., Teilmann, J., Dietz, R., 2013. PFAS profiles in three North Sea top predators: metabolic differences among species? Environmental Science and Pollution Research 20, 8013–8020.

Gebbink, W.A., Bossi, R., Rigét, F.F., Rosing-Asvid, A., Sonne, C., Dietz, R., 2016. Observation of emerging per- and polyfluoroalkyl substances (PFASs) in Greenland marine mammals. Chemosphere 144, 2384–2391.

Giesy, J.P., Kannan, K., 2001. Global distribution of perfluorooctane sulfonate in wildlife. Environmental Science and Technology 35, 1339–1342.

Giesy, J., Naile, J., Khim, J., Jones, P., Newsted, J., 2010. Aquatic toxicology of perfluorinated chemicals. Reviews of Environmental Contamination and Toxicology 1–52.

Gonzalez-Gaya, B., Dachs, J., Roscales, J., Cabellero, G., Jimenz, B., 2014. Perfluoroalkylated substances in the global tropical and subtropical surface oceans. Environmental Science and Technology 48, 13076–13084.

Greaves, A.K., Letcher, R.J., 2013. Linear and branched perfluorooctane sulfonate (PFOS) isomer patterns differ among several tissues and blood of polar bears. Chemosphere 93, 574–580.

Greaves, A.K., Letcher, R.J., Sonne, C., Dietz, R., Born, E.W., 2012. Tissue-specific concentrations and patterns of perfluoroalkyl carboxylates and sulfonates in East Greenland polar bears. Environmental Science and Technology 46, 11575–11583.

Greaves, A.K., Letcher, R.J., Sonne, C., Dietz, R., 2013. Brain region distribution and patterns of bioaccumulative perfluoroalkyl carboxylates and sulfonates in East Greenland polar bears (*Ursus maritimus*). Environmental Toxicology and Chemistry 32, 713–722.

Gribble, M.O., Scott, M., Bartell, S.M., Kannan, K., Fair, P.A., Kamen, D.L., 2015. Longitudinal measures of perfluoroalkyl substances (PFAS) in serum of Gullah-ethnicity African-Americans in Charleston, South Carolina: 2003-2013. Environmental Research 143 (Part B), 82–88.

Grønnestad, R., Villanger, G.D., Polder, A., Kovacs, K.M., Lydersen, C., Jenssen, B.M., Borgå, K., 2016. Maternal transfer of perfluoroalkyl substances in hooded seals. Environmental Toxicology and Chemistry 36, 763–770.

Han, X., Nabb, D., Russell, M., Kennedy, G., Rickard, R., 2012. Renal elimination of perfluorocarboxylates (PFCAs). Chemical Research in Toxicology 25, 35–46.

Hart, K., Kannan, K., Isobe, T., Takahashi, S., Yamada, T.K., Miyazaki, N., Tanabe, S., 2008a. Time trends and transplacental transfer of perfluorinated compounds in melon-headed whales stranded along the Japanese coast in 1982, 2001/2002, and 2006. Environmental Science and Technology 42, 7132–7137.

Hart, K., Gill, V.A., Kannan, K., 2008b. Temporal Trends (1992–2007) of perfluorinated chemicals in Northern sea otters (*Enhydra lutris kenyoni*) from South-Central Alaska. Archives of Environmental Contamination and Toxicology 56, 607–614.

Haug, L., Thomsen, C., Brantsaeter, A., Kvalem, H.E., Haugen, M., Becher, G., Alexander, J., Meltzer, H.M., Knutsen, H.K., 2010. Diet and particularly seafood are major sources of perfluorinated compounds in humans. Environmental Science and Technology 36, 772–778.

Holzer, J., Goen, T., Just, P., Reupert, R., Rauchfuss, K., Kraft, M., Müller, J., Wilhelm, M., 2011. Perfluorinated compounds in fish and blood of anglers at Lake Mohne, Sauerland area, Germany. Environmental Science and Technology 45, 8046–8052.

Houde, M., Martin, J.W., Letcher, R.J., Solomon, K.R., Muir, D.C.G., 2006a. Biological monitoring of polyfluoroalkyl substances: a review. Environmental Science and Technology 40, 3463–3473.

Houde, M., Balmer, B.C., Brandsma, S., Wells, R.S., Rowles, T.K., Solomon, K.R., Muir, D.C.G., 2006b. Perfluoroalkyl compounds in relation to life-history and reproductive parameters in bottlenose dolphins (*Tursiops truncatus*) from the Sarasota Bay, Florida, USA. Environmental Toxicology and Chemistry 25, 2405–2412.

Houde, M., Bujas, T.A.D., Small, J., Wells, R.S., Fair, P., Bossart, G.D., Solomon, K.R., Muir, D.C.G., 2006c. Biomagnification of perfluoroalkyl compounds in the bottlenose dolphin (*Tursiops truncatus*) food web. Environmental Science and Technology 40, 4138–4144.

Houde, M., De Silva, A.O., Muir, D.C.G., Letcher, R.J., 2011. Monitoring of perfluorinated compounds in aquatic biota: an updated review. Environmental Science and Technology 45, 7962–7973.

Huber, S., Ahrens, L., Bårdsen, B.-J., Siebert, U., Bustnes, J.O., Víkingsson, G.A., Ebinghaus, R., Herzke, D., 2012. Temporal trends and spatial differences of perfluoroalkylated substances in livers of harbor porpoise (*Phocoena phocoena*) populations from Northern Europe, 1991-2008. The Science of the Total Environment 419, 216–224.

Ishibashi, H., Iwata, H., Kim, E.-Y., Tao, L., Kannan, K., Amano, M., Miyazaki, N., Tanabe, S., Batoec, V.B., Petrov, E.A., 2008a. Contamination and effects of perfluorochemicals in Baikal seal (*Pusa sibirica*). 1. Residue level, tissue distribution, and temporal trend. Environmental Science and Technology 42, 2295–2301.

Ishibashi, H., Iwata, H., Kim, E.-Y., Tao, L., Kannan, K., Tanabe, S., Batoec, V.B., Petrov, E.A., 2008b. Contamination and effects of perfluorochemicals in Baikal seal (*Pusa sibirica*). 2. Molecular characterization, expression level, and transcriptional activation of peroxisome proliferator-activated receptor α. Environmental Science and Technology 42, 2302–2308.

Jensen, A.H., Leffers, H., 2008. Emerging endocrine disrupters: perfluoroalkylated substances. International Journal of Andrology 31, 161–169.

Kamen, D., Peden-Adams, M., Vena, J., Gilkeson, G., Hulsey, T., Moultrie, L., Stevens, B., 2012. Seafood consumption and persistent organic pollutants as triggers of autoimmunity among Gullah African Americans. Arthritis Research and Therapy 14 (Suppl. 3), A19–A23.

Kannan, K., Koistinen, J., Beckmen, K., Evans, T., Gorzelany, J.F., Hansen, K.J., Jones, P.D., Helle, E., Hyman, M., Giesy, J.P., 2001. Accumulation of perfluorooctane sulfonate in marine mammals. Environmental Science and Technology 35, 1593–1598.

Kannan, K., Perrotta, E., Thomas, N.J., 2006. Association between perfluorinated compounds and pathological conditions in southern sea otters. Environmental Science and Technology 40, 4943–4948.

Keil, D.E., 2015. Immunotoxicity of perfluoroalkylated compounds. In: Dewitt, J.C. (Ed.), Toxicological Effects of Perfluoroalkyl and Polyfluoroalkyl Substances. Molecular and Integrative Toxicology Series. Springer International Publishing, New York, pp. 1–21.

Kennedy, G., Butenhoff, J., Olsen, G., O'connor, J., Seacat, A., Perkins, R., Biegel, L., Murphy, S., Farrar, D., 2004. The toxicology of perfluorooctanoate. Critical Reviews in Toxicology 34, 351–384.

Kovarova, J., Svobodova, Z., 2008. Perfluorinated compounds: occurrence and risk profile. Neuro Endocrinology Letters 29, 599–608.

Kratzer, J., Ahrens, L., Roos, A., Bäcklin, B.-M., Ebinghaus, R., 2011. Temporal trends of polyfluoroalkyl compounds (PFCs) in liver tissue of grey seals (*Halichoerus grypus*) from the Baltic Sea, 1974–2008. Chemosphere 84, 1592–1600.

Lam, J.C.W., Lyu, J., Kwok, K.Y., Lam, P.K.S., 2016. Perfluoroalkyl substances (PFASs) in marine mammals from the South China sea and their temporal changes 2002-2014: concern for alternatives of PFOS? Environmental Science and Technology 50, 6728–6736.

Lau, C., 2012. Perfluorinated compounds. In: Luch, A. (Ed.), Molecular Clinical and Environmental Toxicology. Birkhauser-Verlag, Basel, pp. 47–86.

Lau, C., 2015. Perfluorinated compounds: an overview. In: Dewitt, J.C. (Ed.), Toxicological Effects of Perfluoroalkyl and Polyfluoroalkyl Substances. Molecular and Integrative Toxicology Series. Springer International Publishing, New York, pp. 1–21.

Lau, C., Butenhoff, J., Rogers, J., 2004. The developmental toxicity of perfluoroalkyl acids and their derivatives. Toxicology and Applied Pharmacology 15, 231–241.

Lau, C., Anitole, K., Hodes, C., Lai, D., Pfahles-Hutchens, A., Seed, J., 2007. Perfluoroalkyl acids: a review of monitoring and toxicological findings. Toxicological Sciences 99, 66–94.

Law, R.J., 2014. An overview of time trends in organic contaminant concentrations in marine mammals: going up or down? Marine Pollution Bulletin 82, 7–10.

Letcher, R.J., Bustnes, J.O., Dietz, R., Jenssen, B.M., Jørgensen, E.H., Sonne, C., Verreault, J., Vijayan, M.M., Gabrielsen, G.W., 2010. Exposure and effects assessment of persistent organohalogen contaminants in Arctic wildlife and fish. The Science of the Total Environment 408, 2995–3043.

Letcher, R.J., Chu, S., Mckinney, M.A., Tomy, G.T., Sonne, C., Dietz, R., 2014. Comparative hepatic *in vitro* depletion and metabolite formation of major perfluorooctane sulfonate precursors in arctic polar bear, beluga whale, and ringed seal. Chemosphere 112, 225–231.

Levin, M., Gebhard, E., Jasperse, L., Desforges, J.-P., Dietz, R., Sonne, C., Eulaers, I., Covaci, A., Bossi, R., De Guise, S., 2016. Immunomodulatory effects of exposure to polychlorinated biphenyls and perfluoroalkyl acids in East Greenland ringed seals (*Pusa hispida*). Environmental Research 151, 244–250.

Lim, T., Wang, B., Huang, J., Deng, S., Yu, G., 2011. Emission inventory for PFOS in China: a review of past methodologies and suggestions. The Scientific World Journal 11, 1963–1980.

Lindstrom, A., Strynar, M., Libelo, E., 2011. Polyfluorinated compounds: past, present, and future. Environmental Science and Technology 45, 7954–7961.

Löfstedt Gilljam, J., Leonel, J., Cousins, I.T., Benskin, J.P., 2016. Is ongoing Sulfluramid use in South America a significant source of perfluorooctanesulfonate (PFOS)? Production inventories, environmental fate, and local occurrence. Environmental Science and Technology 50, 653–659.

Mariussen, E., 2012. Neurotoxic effects of perfluoroalkylated compounds: mechanisms of action and environmental relevance. Archives of Toxicology 86, 1349.

Martin, J.W., Smithwick, M., Braune, B.M., Hoekstra, P.F., Muir, D.C.G., Mabury, S.A., 2004. Identification of long-chain perfluorinated acids in biota from the Canadian Arctic. Environmental Science and Technology 38, 373–380.

Mollenhauer, M.A.M., Carter, B.J., Peden-Adams, M.M., Bossart, G.D., Fair, P.A., 2009. Gene expression changes in bottlenose dolphin, *Tursiops truncatus*, skin cells following exposure to methylmercury (MeHg) or perfluorooctane sulfonate (PFOS). Aquatic Toxicology 91, 10–18.

Mollenhauer, M.A.M., Bradshaw, S.G., Fair, P.A., McGuinn, W.D., Peden-Adams, M., 2011. Effects of perfluorooctane sulfonate (PFOS) exposure on markers of inflammation in female B6C3F1 mice. Journal of Environmental Science and Health Part A Toxic/Hazardous Substances and Environmental Engineering 46, 97–108.

NCP, 2015. Synopsis of Research Conducted under the 2014-2015 Northern Contaminants Program. Indigenous and Northern Affairs Canada, Ottawa, Canada. 479 p.

NCP, 2016. Synopsis of Research Conducted under the 2015-2016 Northern Contaminants Program. Indigenous and Northern Affairs Canada, Ottawa, Canada. 390 p.

OECD, 2002. Hazard assessment of perfluorooctane sulfonate (PFOS) and its salts. In: Organization for Economic Cooperation and Development (Ed.), Joint Meeting of the Chemical Committee and the Working Party of Chemicals, Pesticides and Biotechnology. ENV/JM/RD, pp. 1–362.

Olsen, G., Butenhoff, J., Zobel, L., 2009. Perfluoroalkyl chemicals and human fetal development: an epidemiologic review with clinical and toxicological perspectives. Reproductive Toxicology 27, 212–230.

Paul, A., Jones, K., Seetman, A., 2009. A first global production, emission, and environmental inventory for perfluoroctane sulfonate. Environmental Science and Technology 43, 386–392.

Pedersen, K.E., Basu, N., Letcher, R., Greaves, A.K., Sonne, C., Dietz, R., Styrishave, B., 2015. Brain region-specific perfluoroalkylated sulfonate (PFSA) and carboxylic acid (PFCA) accumulation and neurochemical biomarker responses in east Greenland polar bears (*Ursus maritimus*). Environmental Research 138, 22–31.

Pedersen, K., Letcher, R., Sonne, C., Dietz, R., Styrishave, B., 2016. Per- and polyfluoroalkyl substances (PFASs) – new endocrine disruptors in polar bears (*Ursus maritimus*)? Environment International 96, 180–189.

Powley, C.R., George, S.W., Russell, M.H., Hoke, R.A., Buck, R.C., 2008. Polyfluorinated chemicals in a spatially and temporally integrated food web in the Western Arctic. Chemosphere 70, 664–672.

Reiner, J.L., O'Connell, S.G., Moors, A.J., Kucklick, J.R., Becker, P.R., Keller, J.M., 2011. Spatial and temporal trends of perfluorinated compounds in beluga whales (*Delphinapterus leucas*) from Alaska. Environmental Science and Technology 45, 8129–8136.

Renner, R., 2006. The long and the short of perfluorinated replacements. Environmental Science and Technology 40, 12–13.

Rigét, F., Bossi, R., Sonne, C., Vorkamp, K., Dietz, R., 2013. Trends of perfluorochemicals in Greenland ringed seals and polar bears: indications of shifts to decreasing trends. Chemosphere 93, 1607–1614.

Ritter, S., 2010. Fluorochemicals go short. Chemical and Engineering News 88, 12–17.

Rotander, A., Kärrman, A., Bavel, B.V., Polder, A., Rigét, F., Auðunsson, G.A., Víkingsson, G., Gabrielsen, G.W., Bloch, D., Dam, M., 2012. Increasing levels of long-chain perfluorocarboxylic acids (PFCAs) in Arctic and North Atlantic marine mammals, 1984–2009. Chemosphere 86, 278–285.

Routti, H., Lydersen, C., Hanssen, L., Kovacs, K.M., 2014. Contaminant levels in the world's northernmost harbor seals (*Phoca vitulina*). Marine Pollution Bulletin 87, 140–146.

Routti, H., Krafft, B.A., Herzke, D., Eisert, R., Oftedal, O., 2015. Perfluoroalkyl substances detected in the world's southernmost marine mammal, the Weddell seal (*Leptonychotes weddellii*). Environmental Pollution 197, 62–67.

Routti, H., Gabrielsen, G.W., Herzke, D., Kovacs, K.M., Lydersen, C., 2016. Spatial and temporal trends in perfluoroalkyl substances (PFASs) in ringed seals (*Pusa hispida*) from Svalbard. Environmental Pollution 214, 230–238.

Schenker, U., Scheringer, M., Macleod, M., Martin, J., Cousins, I., Hungerbühler, K., 2008. Contribution of volatile precursor substances to the flux of perfluorooctanoate to the Arctic. Environmental Science and Technology 42, 3710–3716.

Scheringer, M., Trier, W., Cousins, I., De Voogt, P., Fletcher, T., Wang, Z., Webster, T., 2014. Helsingør statement on poly- and perfluorinated alkyl substances (PFASs). Chemosphere 114, 337–339.

Schiavone, A., Corsolini, S., Kannan, K., Tao, L., Trivelpiece, W., Torres, D., Focardi, S., 2009. Perfluorinated contaminants in fur seal pups and penguin eggs from South Shetland, Antarctica. The Science of the Total Environment 407, 3899–3904.

Sedlak, M.D., Benskin, J.P., Wong, A., Grace, R., Greig, D.J., 2017. Per- and polyfluoroalkyl substances (PFASs) in San Francisco Bay wildlife: temporal trends, exposure pathways, and notable presence of precusor compounds. Chemosphere 185, 1217–1226.

Shaw, S., Berger, M.L., Brenner, D., Tao, L., Wu, Q., Kannan, K., 2009. Specific accumulation of perfluorochemicals in harbor seals (*Phoca vitulina concolor*) from the northwest Atlantic. Chemosphere 74, 1037–1043.

Smithwick, M., Norstrom, R.J., Mabury, S.A., Solomon, K.R., Evans, T.J., Stirling, I., Taylor, M.K., Muir, D.C.G., 2006. Temporal trends of perfluoroalkyl contaminants in polar bears (*Ursus maritimus*) from two locations in the North American Arctic, 1972-2002. Environmental Science and Technology 40, 1139–1143.

Soloff, A.C., Wolf, B.J., White, N.D., Muir, D.C.G., Courtney, S., Hardiman, G., Bossart, G.D., Fair, P.A., 2018. Environmental perfluorooctane sulfonate (PFOS) exposure drives T-cell activation in bottlenose dolphins. Journal of Applied Toxicology 37, 1108–1116.

Sonne, C., 2010. Health effects from long-range transported contaminants in Arctic top predators: an integrated review based on studies of polar bears and relevant model species. Environment International 36, 461–491.

Sonne, C., Gustavson, K., Rigét, F.F., Dietz, R., Birkved, M., Letcher, R.J., Bossi, R., Vorkamp, K., Born, E.W., Petersen, G., 2009. Reproductive performance in East Greenland polar bears (*Ursus maritimus*) may be affected by organohalogen contaminants as shown by physiologically-based pharmacokinetic (PBPK) modelling. Chemosphere 77, 1558–1568.

Stahl, T., Matern, D., Brunn, H., 2011. Toxicity of perfluorinated compounds. Environmental Sciences Europe 23, 38–90.

Sturm, R., Ahrens, L., 2010. Trends of polyfluoroalkyl compounds in marine biota and in humans. Environmental Chemistry 7, 457.

Suja, F., Pramanik, B.K., Zain, S.M., 2009. Contamination, bioaccumulation and toxic effects of perfluorinated chemicals (PFCs) in the water environment: a review paper. Water Science and Technology 60, 1533.

Takacs, M., Abbott, B., 2007. Activation of mouse and human peroxisome proliferator–activated receptors (α, β/δ, γ) by perfluorooctanoic acid and perfluorooctane sulfonate. Toxicological Sciences 95, 108–117.

Tao, L., Kannan, K., Kajiwara, N., Costa, M.M., Fillman, G., Takahashi, S., Tanabe, S., 2006. Perfluorooctanesulfonate and related fluorochemicals in albatrosses, elephant seals, penguins, and polar skuas from the Southern Ocean. Environmental Science and Technology 40, 7642–7648.

Tartu, S., Bourgeon, S., Aars, J., Andersen, M., Lone, K., Jenssen, B.M., Polder, A., Thiemann, G.W., Torget, V., Welker, J.M., Routti, H., 2017. Diet and metabolic state are the main factors determining concentrations of perfluoroalkyl substances in female polar bears from Svalbard. Environmental Pollution 229, 146–158.

UNEP, 2009. Governments Unite to Step-Up Reduction on Global DDT Reliance and Add Nine New Chemicals under International Treaty. Available from: http://chm.pops.int/Convention/Pressrelease/COP4Geneva8May2009/tabid/542/language/en-US/Default.aspx.

USEPA, 2000. EPA and 3M Announce Phase-Out of PFOS. Available from: http://yosemite.epa.gov/opa/admpress.nsf/0/33aa946e6cb11f35852568e1005246b4.

USEPA, 2015. Fact Sheet: 2010/2015 PFOA Stewardship Program. Available from: https://www.epa.gov/assessing-and-managing-chemicals-under-tsca/fact-sheet-20102015-pfoa-stewardship-program.

Wang, Z., Cousins, I., Scheringer, M., Hungerbuhler, K., 2013. Fluorinated alternatives to long-chain perfluoroalkyl carboxylic acids (PFCAs), perfluoroalkane sulfonic acids (PFSAs) and their potential precursors. Environment International 60, 242–248.

Wang, Z., Cousins, I., Scheringer, M., Buck, R., Hungerbu¨hler, K., 2014. Global emission inventories for C_4–C_{14} perfluoroalkyl carboxylic acid (PFCA) homologues from 1951 to 2030 Part I: Production and emissions from quantifiable sources. Environment International 69, 166–176.

White, S., Fenton, S., Hines, E., 2011. Endocrine disrupting properties of perfluorooctanoic acid. The Journal of Steroid Biochemistry and Molecular Biology 127, 16–26.

White, N.D., Balthis, L., Kannan, K., De Silva, A.O., Wu, Q., French, K.M., Daugomah, J., Spencer, C., Fair, P.A., 2015. Elevated levels of perfluoroalkyl substances in estuarine sediments of Charleston, SC. The Science of the Total Environment 521–522, 79–89.

Wirth, J.R., Peden-Adams, M.M., White, N., Bossart, G.D., Fair, P.A., 2014. *In vitro* PFOS exposure on immune endpoints in bottlenose dolphins (*Tursiops truncatus*) and mice. Journal of Applied Toxicology 34, 658–666.

Wirth, J.R., Peden-Adams, M.M., White, N., Bossart, G.D., Fair, P.A., 2015. *In vitro* exposure of BDE-71, a penta-PBDE mixture, on immune endpoints in bottlenose dolphins (*Tursiops truncatus*) and mice. Journal of Applied Toxicology 35, 191–198.

Yamaguchi, M., Arisawa, K., Uemura, H., Katsuura-Kamano, S., Takami, H., Sawachika, F., Nakamoto, M., Juta, T., Toda, E., Mori, K., Hasegawa, M., Tanto, M., Shima, M., Sumiyoshi, Y., Morinaga, K., Kodama, K., Suzuki, T., Nagai, M., Satoh, H., 2013. Consumption of seafood, serum liver enzymes, and blood levels of PFOS and PFOA in the Japanese population. Journal of Occupational Health 55, 184–194.

Yamashita, N., Taniyasu, S., Petrick, G., Wei, S., Gamo, T., Lam, P., Kannan, K., 2008. Perfluorinated acids as novel chemical tracers of global circulation of ocean waters. Chemosphere 70, 1247–1255.

Young, C., Mabury, S., 2010. Atmospheric perfluorinated acid precursors: chemistry, occurrence, and impacts. Reviews of Environmental Contamination and Toxicology 208, 1–109.

Chapter 6

Impacts of Marine Litter on Cetaceans: A Focus on Plastic Pollution

Maria Cristina Fossi[1], Matteo Baini[1], Cristina Panti[1], Sarah Baulch[2]
[1]University of Siena, Siena, Italy; [2]Environmental Investigation Agency (EIA), London, United Kingdom

IMPACT OF MARINE LITTER ON CETACEANS

Although the impacts of marine litter on marine fauna first began to be documented in the 1960s, it is as plastic production has increased exponentially in recent decades that quantities of marine plastic pollution have significantly increased and the severity of its impacts on marine ecosystems and species has intensified. It is now deemed a major threat to marine biodiversity (Secretariat of the Convention on Biological Diversity, 2016). The issue has gained international prominence and has seen a resurgence in scientific research (Galgani et al., 2015). Plastic pollution is now abundant in all the world's oceans and can be found throughout the water column, floating on the sea surface, accumulating in deep sea trenches and sediments and sequestered in polar sea ice (Obbard et al., 2014; Woodall et al., 2014).

Plastics, of various types—from food packaging to fishing nets—form the vast majority of marine litter (Derraik, 2002). Over the last 50 years, plastic use has increased 20-fold and is expected to double again in the next 20 years, with the largest application being single-use plastic packaging (World Economic Forum et al., 2016). It is estimated that 32% of plastic packaging is escaping collection systems (World Economic Forum et al., 2016), with 4.6–12.7 million tons of plastic waste currently entering oceans annually from land, the equivalent of one rubbish truck every minute (Jambeck et al., 2015). Sources at sea (such as shipping, fisheries, offshore platforms, and recreational vessels) contribute an additional 1.75 million tons per year (Eunomia Research and Consulting Ltd., 2016).

Impacts on marine fauna occur throughout the food chain, with adverse impacts documented on over 800 species (Secretariat of the Convention on

Marine Mammal Ecotoxicology. https://doi.org/10.1016/B978-0-12-812144-3.00006-1

Biological Diversity, 2016). For marine mammals, impacts can be divided into those arising from entanglement, which can result in injury, drowning, or strangulation, and those from ingestion, with pathology ranging from no discernible impact to blockage of the digestive tract, suffocation, and starvation (Laist, 1997). Sublethal impacts may include injury, compromised feeding and digestion and associated malnutrition, disease, and reduced reproduction, growth, and longevity (Allen et al., 2012; Katsanevakis, 2008; McCauley and Bjomdal, 1999; Moore et al., 2013a). While individual strandings provide indications of the range of pathology that can occur, evaluation of the frequency and severity of impacts of marine litter on cetaceans is complicated by low sample sizes due to the low rate of detection (with as few as 0%–6.2% of carcasses recovered from cetacean deaths at sea) and the compounding effects of a low necropsy and publication rate.

Building on previous reviews (Cornish et al., 2011; Katsanevakis, 2008; Laist, 1997; Simmonds, 2011, 2012; Walker and Coe, 1990), Baulch and Perry (2014) undertook an updated review, consulting published literature and stranding networks to compile data on the prevalence of marine litter impacts on cetaceans, with a focus on macroplastic ingestion. The review collated over 500 records of such interactions from the published literature and responses from stranding networks in 11 countries.

Data show an increase in the number of cases being reported over the last five decades (see Fig. 6.1), in particular related to the ingestion of marine litter. Necropsy and reporting rates are also likely to have risen over this period,

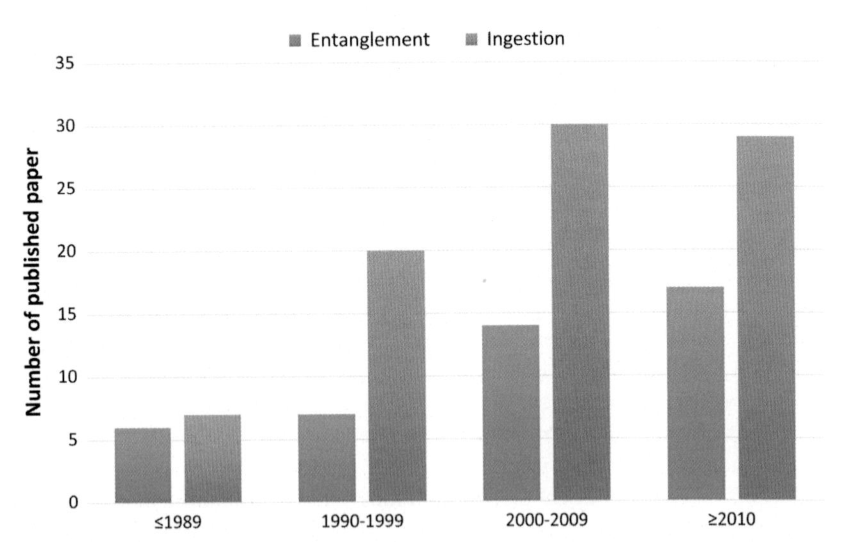

FIGURE 6.1 Temporal trends of scientific papers published on impact of marine litter on cetaceans. Total published documents = 131; total documents on entanglement = 45; total documents on ingestion = 86 (last update January 2018).

contributing to this rise, though the gradient is such that it suggests a concomitant rise in interaction rates, as would be expected with the increasing densities of marine plastic pollution.

Among the 14 families of cetaceans (Committee on Taxonomy, 2017), interactions with marine litter has been reported for 11 families (Table 6.1); data are missing for the river dolphins belonging to the families Platanistidae, Iniidae, and Lipotidae. The number of publications is unlikely to represent the extent of impact on these families: rather what has been observed that will be impacted by the extent of study of different species and factors such as differential rates of stranding and necropsy.

Ingestion

Ingestion of plastic litter has been documented in over 60% of all cetacean species, in species employing a variety of feeding techniques at different levels of the feeding column (Baulch and Perry, 2014; Kühn et al., 2015). Items ingested are most commonly plastic and range in size from small fragments to large plastic sheeting and netting. Pathology can range from no discernible impact to complete obstruction of the digestive tract. Analyzing the species that have been reported to have ingested marine litter, 50 out of 86 species (58.1%) (Table 6.2) have at least one documented occurrence of ingestion. This data is relative to the number of species and not to the number of individuals analyzed in each paper/report.

Macrolitter Ingestion

In the 2014 Baulch and Perry review, relatively few stranding networks were found to currently collect data on rates of marine litter ingestion. However, based on available data where the sample size of animals necropsied exceeded 10 animals, ingestion rates varied from 0% to 31% of animals necropsied, with high geographic and intra- and interspecific variation in rates. The highest rates were documented in Franciscana, with 17%–31% of bycaught animals necropsied found to have ingested marine litter, across three separate studies (Bassoi, 1997; Bastida et al., 2000; Denuncio et al., 2011). Prevalence was also relatively high in Guiana dolphins (*Sotalia guianensis*) in Brazil (12%), sperm whales (*Physeter macrocephalus*) in Southern Australia (11%), and northern bottlenose whales (*Hyperoodon ampullatus*) in the United Kingdom (16.7%) (Baulch and Perry, 2014; Evans and Hindell, 2004). A recent comparison of debris ingestion by Franciscana (*Pontoporia blainvillei*) and Guiana dolphins found that prevalence of debris ingestion was higher in Franciscana (15.7% compared to 1.3% in *Sotalia guianensis*), a trend attributed to the feeding activity of Franciscana, which mainly feeds near the sea bed, the main zone of accumulation of marine litter in the study area (Di Beneditto and Awabdi, 2014; Di Beneditto and Ramos, 2014). In the German North and Baltic Sea, ingested marine litter items were found in 0.7% of all necropsied harbor porpoises (*Phocoena phocoena*)

TABLE 6.1 Details on Cetacean Species Affected (Entanglement and Ingestion) by Plastic Litter With Relative Reference. Search Closed January 2018

Family	Common Name	Scientific Name	Entanglement	Ingestion
Baleen Whales (Mysticeti)				
Balaenidae	Southern right whale	*Eubalaena australis*	Cawthorn (1985) and Ceccarelli (2009)	–
Balaenidae	North Atlantic right whale	*Eubalaena glacialis*	Cassoff et al. (2011), Johnson et al. (2005), Knowlton et al. (2012), Kraus (1990), Marine Mammal Commission (1993, 2006), Moore et al. (2009, 2013a), van der Hoop et al. (2014) and Waring et al. (2004)	GEF Council Meeting (2012) and Waring et al. (2004)
Balaenidae	Bowhead whale	*Balaena mysticetus*	Citta et al. (2014), Heyning and Lewis (1990), Jackson (2011), Philo et al. (1992) and Sternfeld (2004, 2005)	Finley (2001) and Lowy (1993)
Neobalaenidae	Pygmy right whale	*Caperea marginata*	Ceccarelli (2009)	(Australian Antarctic Division in: Baulch and Perry, 2014)
Eschrichtiidae	Gray whale	*Eschrichtius robustus*	Carretta et al. (2013), Hare and Mead (1987), Heyning and Lewis (1990), Jackson (2010, 2012, 2013), Mate (1985) and Sternfeld (2005)	–
Balaenopteridae	Common minke whale	*Balaenoptera acutorostrata*	Bogomolni et al. (2010), Cassoff et al. (2011), Cawthorn (1985), Ceccarelli (2009), Deaville et al. (2010), Gill et al. (2000), Hare and Mead (1987), Lusher et al. (2018), Mate (1985) and Waring et al. (2004)	De Pierrepont et al. (2005), Hare and Mead (1987), Mauger et al. (2002) and Tarpley and Marwitz (1993)

Balaenopteridae	Bryde's whale	*Balaenoptera edeni*	Cassoff et al. (2011)	Ceccarelli (2009) and Haynes and Limpus (2000)
Balaenopteridae	Blue whale	*Balaenoptera musculus*	–	Baulch and Perry (2014) and Baxter (2009)
Balaenopteridae	Fin whale	*Balaenoptera physalus*	Arbelo et al. (2013), Jackson (2013), Lusher et al. (2018) and Waring et al. (2004)	Fossi et al. (2012), Sadove and Morreale (1990) and Smiddy et al. (2002)
Balaenopteridae	Humpback whale	*Megaptera novaeangliae*	Bogomolni et al. (2010), Cassoff et al. (2011), Ceccarelli (2009), Humpback Whale Recovery Team (1991), Jackson (2010, 2011, 2012, 2013, 2014), Johnson et al. (2005), Mate (1985), Mattila and Lyman (2006), Sternfeld (2004, 2005, 2006) and Waring et al. (2004)	Berrow et al. (2007) and Besseling et al. (2015)
Toothed Whales (Odontoceti)				
Physeteridae	Sperm whale	*Physeter macrocephalus*	Arbelo et al. (2013), Jackson (2013), Laist (1997), Lusher et al. (2018), Moore et al. (2009), Sadove and Morreale (1990) and Waring et al. (2004)	Carretta et al. (2013), de Stephanis et al. (2013), Evans and Hindell (2004), Fernández et al. (2009), Jacobsen et al. (2010), Katsanevakis (2008), Martin and Clarke (1986), Mate (1985), Mazzariol et al. (2011), Roberts (2003), Sadove and Morreale (1990), Spence (1994), Unger et al. (2016), Vitale et al. (1992) and Walker and Coe (1990)

Continued

TABLE 6.1 Details on Cetacean Species Affected (Entanglement and Ingestion) by Plastic Litter With Relative Reference. Search Closed January 2018—cont'd

Family	Common Name	Scientific Name	Entanglement	Ingestion
Kogiidae	Pygmy sperm whale	*Kogia breviceps*	Coleman and Wehle (1984) and Scott (2001)	Barros et al. (1990), Fernández et al. (2009), Sadove and Morreale (1990), Schofield et al. (1994), Stamper et al. (2006), Tarpley (1990), Tarpley and Marwitz (1993) and Walker and Coe (1990)
Kogiidae	Dwarf sperm whale	*Kogia sima*	–	Barros et al. (1990) and Walker and Coe (1990)
Ziphiidae	Shepherd's beaked whale	*Tasmacetus shepherdi*	–	Baulch and Perry (2014)
Ziphiidae	Baird's beaked whale	*Berardius bairdii*	–	Walker and Coe (1990)
Ziphiidae	Longman's beaked whale	*Mesoplodon pacificus*	–	Yamada et al. (2012a)
Ziphiidae	Sowerby's beaked whale	*Mesoplodon bidens*	–	Berrow et al. (2010), Deaville et al. (2010) and Kühn et al. (2015)
Ziphiidae	Blainville's beaked whale	*Mesoplodon densirostris*	–	Laist (1997), Secchi and Zarzur (1999) and Walker and Coe (1990)
Ziphiidae	Gervais' beaked whale	*Mesoplodon europaeus*	–	Fernández et al. (2009), Walker and Coe (1990) and Waring et al. (2004)

Ziphiidae	Gray's beaked whale	*Mesoplodon grayi*	–	Baulch and Perry (2014)
Ziphiidae	Stejneger's beaked whale	*Mesoplodon stejnegeri*	–	Walker and Hanson (1999) and Yamada et al. (2012b)
Ziphiidae	True's beaked whale	*Mesoplodon mirus*	–	Gassner et al. (2005), Lusher et al. (2015) and Souza et al. (2005)
Ziphiidae	Ginkgo-toothed beaked whale	*Mesoplodon ginkgodens*	–	Baulch and Perry (2014)
Ziphiidae	Hubbs' beaked whale	*Mesoplodon carlhubbsi*	–	Yamada et al. (2012b)
Ziphiidae	Deraniyagala's beaked whale	*Mesoplodon hotaula*	–	Cuenca et al. (2016)
Ziphiidae	Cuvier's beaked whale	*Ziphius cavirostris*	Waring et al. (2004)	Arbelo et al. (2013), Coleman and Wehle (1984), Lusher et al. (2018), MacLeod (2007), Poncelet et al. (2000), Santos et al. (2007), Simmonds (2012) and Walker and Coe (1990)
Ziphiidae	Northern bottlenose whale	*Hyperoodon ampullatus*	–	Baird and Hooker (2000) and Deaville et al. (2010)
Pontoporiidae	Franciscana dolphin	*Pontoporia blainvillei*	–	Denuncio et al. (2011), Di Beneditto and Ramos (2014), Laist (1997), Pinedo (1982)
Monodontidae	Narwhal	*Monodon monoceros*	–	Haelters et al. (2017)

Continued

TABLE 6.1 Details on Cetacean Species Affected (Entanglement and Ingestion) by Plastic Litter With Relative Reference. Search Closed January 2018—cont'd

Family	Common Name	Scientific Name	Entanglement	Ingestion
Monodontidae	Beluga	*Delphinapterus leucas*	Jackson (2011, 2012, 2013, 2014) and Sternfeld (2005)	–
Phocoenidae	Harbor porpoise	*Phocoena phocoena*	Bogomolni et al. (2010), Hare and Mead (1987), Jackson (2010, 2011, 2013, 2014), Jackson and Sternfeld (2009), Lusher et al. (2018), Sternfeld (2005) and Unger et al. (2017)	Baird and Hooker (2000), Bogomolni et al. (2010), Carretta et al. (2013), Deaville et al. (2010), Hare and Mead (1987), Kasteleine and Lavaleye (1992), Lusher et al. (2018), Tonay et al. (2007), Unger et al. (2017), van Franeker et al. (2018) and Walker and Coe (1990)
Phocoenidae	Burmeister's porpoise	*Phocoena spinipinnis*	–	Baulch and Perry (2014)
Phocoenidae	Finless porpoise	*Neophocaena phocaenoides*	–	Baird and Hooker (2000)
Phocoenidae	Dall's porpoise	*Phocoenoides dalli*	Degange and Newby (1980) Jackson (2010), Jackson and Sternfeld (2009), Jones and Ferrero (1985) and Sternfeld (2004)	Walker and Coe (1990)
Delphinidae	Rough-toothed dolphin	*Steno bredanensis*	Coleman and Wehle (1984)	Oliveira de Meirelles and Duarte do Rego Barros (2007) and Walker and Coe (1990)

Delphinidae	Indo-Pacific humpback dolphin	*Sousa chinensis*	Ceccarelli (2009), Kiessling (2003) and Slooten et al. (2013)	–
Delphinidae	Tucuxi	*Sotalia fluviatilis*	Laist (1997)	Geise and Gomes (1992) and Laist (1997)
Delphinidae	Guiana river dolphin	*Sotalia guianensis*	–	Di Beneditto and Awabdi (2014) and Geise and Gomes (1992)
Delphinidae	White-beaked dolphin	*Lagenorhynchus albirostris*	–	Baird and Hooker (2000)
Delphinidae	Atlantic white-sided dolphin	*Lagenorhynchus acutus*	Lusher et al. (2018)	–
Delphinidae	Pacific white-sided dolphin	*Lagenorhynchus obliquidens*	–	Caldwell et al. (1965) and Walker and Hanson (1999)
Delphinidae	Risso's dolphin	*Grampus griseus*	Frantzis (2007) and Lusher et al. (2018)	Henderson (1988), Laist (1997), Lusher et al. (2018), Shoham-Frider et al. (2002) and Walker and Coe (1990)
Delphinidae	Fraser's dolphin	*Lagenodelphis hosei*	–	Fernández et al. (2009)
Delphinidae	Common bottlenose dolphin	*Tursiops truncatus*	Lusher et al. (2018)	Adimey et al. (2014), Barros et al. (1990), Ceccarelli (2009), Deaville et al. (2010), Gomerčić et al. (2009), Gorzelany (1998), Levy et al. (2009), Lusher et al. (2018), Mann et al. (1995), O'Brien and Berrow (2006), Schwartz et al. (1992) and Walker and Coe (1990)

Continued

TABLE 6.1 Details on Cetacean Species Affected (Entanglement and Ingestion) by Plastic Litter With Relative Reference. Search Closed January 2018—cont'd

Family	Common Name	Scientific Name	Entanglement	Ingestion
Delphinidae	Indo-Pacific bottlenose dolphin	*Tursiops aduncus*	Bossley (n.d.), Ceccarelli (2009) and Chatto and Warneke (2000)	–
Delphinidae	Atlantic spotted dolphin	*Stenella frontalis*	–	Arbelo et al. (2013)
Delphinidae	Pantropical spotted dolphin	*Stenella attenuata*	–	Baird and Hooker (2000)
Delphinidae	Striped dolphin	*Stenella coeruleoalba*	Frantzis (2007) and Lusher et al. (2018)	Arbelo et al. (2013), Fernández et al. (2009), Hernandez-Milian (2014), Lusher et al. (2018) and Walker and Coe (1990)
Delphinidae	Common dolphin	*Delphinus delphis*	Ceccarelli (2009)	Curran et al. (2014), Deaville et al. (2010), Hernandez-Milian (2014) and Walker and Coe (1990)
Delphinidae	Long-beaked common dolphin	*Delphinus capensis*	–	Carretta et al. (2013)

Delphinidae	Northern right whale dolphin	*Lissodelphis borealis*	–	Walker and Coe (1990)
Delphinidae	False killer whale	*Pseudorca crassidens*	–	Barros et al. (1990)
Delphinidae	Killer whale	*Orcinus orca*	Cawthorn (1985)	Baird and Hooker (2000), Hernandez-Milian (2014) and Lusher et al. (2018)
Delphinidae	Long-finned pilot whale	*Globicephala melas*	Lusher et al. (2018)	Laist (1997)
Delphinidae	Short-finned pilot whale	*Globicephala macrorhynchus*	–	Barros et al. (1997) and Walker and Coe (1990)
Delphinidae	Irrawaddy dolphin	*Orcaella brevirostris*	–	Baulch and Perry (2014)
Delphinidae	Australian snubfin dolphin	*Orcaella heinsohni*	Ceccarelli (2009)	–

TABLE 6.2 Number of Cetacean Species With Documented Records of Ingestion of Marine Litter

	Family	Species Total (*n*)	Ingestion	
			(*n*)	%
Baleen whales (*Mysticeti*)	Balaenidae	4	2	50
	Neobalaenidae	1	1	100
	Eschrichtiidae	1	0	0
	Balaenopteridae	8	5	62.5
Toothed whales (*Odontoceti*)	Physeteridae	1	1	100
	Kogiidae	2	2	100
	Ziphiidae	22	14	63.6
	Pontoporiidae	1	1	100
	Monodontidae	2	1	50
	Phocoenidae	7	4	57.1
	Delphinidae	37	19	51.4
	Total	86	50	58.1

(Unger et al., 2017), while 7% of harbor porpoises analyzed from 2003 to 2013 in The Netherlands presented plastic litter in their stomach (van Franeker et al., 2018). Although, van Franeker et al. underline that the results are biased by the method and protocol used, following a standard protocol a higher percentage of litter was detected. This demonstrates the difficulties in comparing data obtained with different protocols.

With regard to mortality due to ingestion, in 14 of 21 species (where necropsy sample sizes exceeded 10 animals), mortality rates were 0%. In the other seven species, mortality rates due to marine litter ingestion ranged between 1.3% and 16.7% of animals necropsied. The highest rates recorded were in Gervais beaked whales (*Mesoplodon europaeus*) necropsied by the Canary Islands stranding network, with the cause of death of 2 of the 12 animals necropsied attributed to ingestion of marine litter, though based on a small sample size. Although in many of the regions and species analyzed ingestion rarely caused mortality, the limited data available indicated that in specific populations it could be a significant cause of mortality, as well as having health and welfare impacts on individuals; though given the low sample sizes, the accuracy of such interpretations must be used with caution.

Severe cases abound, including a sperm whale that ingested 134 different net types of up to 16 m^2; a Cuvier's beaked whale (*Ziphius cavirostris*) that ingested 378 items weighing 33 kg; another Cuvier's that stranded in Norway in 2017 and was found to have consumed 30 bags and other plastic packaging waste; and a Longman's beaked whale (*Mesoplodon pacificus*) in India thought to have died due to ingestion of four thick plastic bags (Jacobsen et al., 2010; Kaladharan et al., 2014; Poncelet et al., 2000).

In the global data collation exercise conducted by Baulch and Perry (2014), the authors highlighted that the paucity and heterogeneity of data available prevented a robust identification of whether, at a species level, there are certain cetacean species particularly prone to ingesting litter. However, there are some indications from other studies that species such as sperm whales and beaked whales may be more likely to ingest marine litter due to the depth and way in which they feed, their geographic overlap with high densities of marine litter, and the physiology of the beaked whale's digestive tract (de Stephanis et al. 2013; IWC, 2013).

Microplastics Ingestion

The study of microplastic ingestion by cetaceans is a challenging task, due to the high difficulties in obtaining accurate samples during necropsies and analysis of large volumes of gut contents, in particular for large cetaceans.

Few studies have directly identified microplastics in the digestive tracts of stranded cetaceans. Applying standard protocols for the detection and identification of microplastics in the digestive tract, microplastics were found throughout the stomach/intestine of seven odontocetes species: *Ziphius cavirostris*, *Delphinus delphis*, *Stenella coeruleaolba*, *Phocoena phocoena*, *Orcinus orca*, and *Tursiops truncatus* (Lusher et al., 2018; van Franeker et al., 2018). Only one study on mysticetes, a stranded humpback whale (*Megaptera novaeangliae*), has so far recorded the presence of microplastic in its intestines, including fragments and threads (Besseling et al., 2015).

There are multiple possible routes of microplastic uptake, including ingestion from the water column while feeding, inhalation at the air–water interface, or via trophic transfer from prey items (IWC, 2013). Uptake of microplastics has been demonstrated in zooplankton (Frias et al., 2014; Setälä et al., 2014) including copepods and euphausiids, which are some of the main prey species of baleen whales (Desforges et al., 2015), shellfish, benthic organisms (e.g., Pellini et al., 2018), planktivorous fish (Collard et al., 2015), and pelagic and demersal fish (Lusher et al., 2013; Murphy et al., 2017) and may thus be secondarily transferred to cetaceans.

For the Mediterranean fin whale, it has been estimated that animals could consume more than 3000 microplastic particles per day, along with associated persistent, bioaccumulative, and toxic (PBT) chemicals (Fossi et al., 2014). Using phthalates (a common plastic additive that leaches from plastic debris) as a tracer of microplastic uptake, Fossi et al. (2014) found that concentrations of the phthalate metabolite and organochlorines were markedly higher in the fin whale compared to the basking shark (*Cetorhinus maximus*). They attributed this to a difference in the total plankton consumed daily and excretory activity, in particular the potential excretion of such contaminants through the gills in fish versus bioaccumulation in adipose tissue in cetaceans. Particularly high levels of microplastics have been documented in the Ligurian Sea, the summer feeding ground of Mediterranean fin whale. These are in the same order of magnitude as the North Pacific Gyre.

Entanglement

Entanglement in marine litter has been documented less frequently and in fewer cetacean species than ingestion, with 27 species and a total of 78 incidences documented worldwide (Baulch and Perry, 2014; Kühn et al., 2015).

Twenty-seven out of 86 species (31.4%) (Table 6.3) have at least one documented event of entanglement. This data is relative to the number of species and not to the number of individuals analyzed in each paper/report.

Almost all entanglements were in "ghost" fishing gear (also known as abandoned, lost, or otherwise discarded fishing gear). Cetaceans tend be entangled

TABLE 6.3 Number of Cetacean Species With Documented Records of Entanglement in Marine Litter

			Entanglement	
	Family	Spp. Total (*n*)	(*n*)	%
Baleen whales (*Mysticeti*)	Balaenidae	4	3	75
	Neobalaenidae	1	1	100
	Eschrichtiidae	1	1	100
	Balaenopteridae	8	4	50
Toothed whales (*Odontoceti*)	Physeteridae	1	1	100
	Kogiidae	2	1	50
	Ziphiidae	22	1	4.5
	Pontoporiidae	1	0	0
	Monodontidae	2	1	50
	Phocoenidae	7	2	28.6
	Delphinidae	37	12	32.4
	Total	86	27	31.4

around the head, flippers, and flukes, often simultaneously in several types of fishing gear (Moore et al., 2013a; van der Hoop et al., 2014). However, the number of records is not necessarily indicative of the scale of the issue, due to the aforementioned low detection rate and, moreover, the additional difficulty in determining whether fishing gear was operational or ghost gear at the time of entanglement; many records were omitted from the review because this was unclear. A recent case study where ghost gear is clearly responsible is the entanglement of bowhead whales in gear from crab fisheries in the Bering Sea (Citta et al., 2014). Whales do not enter the fishing area until after the seasonal fisheries end, so it is clear that lost gear is the most likely source of entanglement. In this region, pack ice can rapidly advance, covering large numbers of pots and lines and causing them to be lost; for example, one severe incident resulted in the loss of 800 crab pots and lines, 6% of the total number set. In a review of humpback whale entanglements off the coast of Hawaii, it was also suggested that ghost gear may account for a number of the documented entanglements (Mattila and Lyman, 2006).

To better identify hotspots of gear loss and develop effective mitigation measures, it is necessary to, wherever possible, distinguish entanglements in abandoned, lost, or discarded fishing gear (ALDFG) from those in active fishing gear, named as commercial and other active fishing gears (COAFG). In addition, it is generally agreed that the numbers of both types of entanglements are widely and severely underreported.

Development of methods to distinguish entanglements in ALDFG from those in COAFG is at an early stage. Spatiotemporal overlap is one indicator that can signal whether ghost gear is responsible for an entanglement. Other potential indicators include the number of different gear types per entanglement or a ghost gear risk metric, based on modeling of spatiotemporal distribution of fisheries and cetaceans and rates of gear loss. While existing gear marking schemes often provide only limited clues as to the origin of gear, efforts are underway within the FAO to improve gear marking. Such measures are vital to correctly identify when entanglements are due to ALDFG as opposed to interaction with COAFG, and thus allow appropriate mitigation.

CURRENT EXISTING PROTOCOLS AND APPROACH FOR THE ANALYSIS OF MARINE LITTER

Stranded Organisms

Protocols for the analysis of marine litter in stranded organisms were developed at an International Whaling Commission (IWC) workshop (IWC, 2013) and recently reviewed according to the existing protocols for other marine taxa (Lusher et al., 2017, 2018). In situ examination of entangling and ingested debris and associated traumatic injuries is essential for revealing the pathologic impacts of fishing gear and debris on cetaceans. Changes can include laceration, amputation, and constriction-related injuries externally, and/or blockage, strangulation, ulceration, impaction, emaciation, and/or rupture internally. Evidence of chronic effects (e.g., emaciation) or prior trauma from entanglement and debris interaction, where material is no longer present, can also be obtained through careful clinical or postmortem examinations by scientists. Potential chemical exposure should also be evaluated, and it may or may not be accompanied by gross or histologic changes due to transfer of additives and sorbed priority pollutants from the plastic into the tissues (Fossi et al., 2016; Rochman et al., 2013). Based on the protocols developed at the IWC workshop, recommended procedures are given next for the analysis of marine litter impacts in stranded cetaceans.

Recommended Diagnostic Approach

Evaluating possible impacts due to entanglement and ingestion impacts should be done using a classical differential diagnostic approach, when possible, to enable the following:

1. Detection of trauma, chemical exposure, and other sequelae related to exposure;
2. Analysis of their roles in contributing to morbidity and mortality in the context of other potential causes, such as infectious or noninfectious disease, nutrition, and other possible etiologies.

In situations when a full differential diagnostic approach is not possible, efforts to document the presence of marine litter, both ingested and entangled, are still very important. Most studies focus on macroplastic, but efforts should also be taken to document microplastic ingestion. All necropsies of stranded cetaceans should include the following components, as appropriate:

1. Gross necropsy examination and report: This should include descriptions, sketches, images, measurements, collection, and preservation of entanglement/debris, and affected body part(s). The entire gastrointestinal tract should be opened and examined. Standard cetacean necropsy protocols should be followed (see Braco and Moore, 2013; McLellan et al., 2004; Pugliares et al., 2007). If the gastrointestinal tract is to be examined for microplastics, care must be taken to control for contamination (see Contamination Control section).
2. Item characterization: Material should be categorized (see Stranded Organisms section) as rope, net, floats, monofilament, braided line, hooks, packaging, cigarette butts, plastics, and other anthropogenic material. Size, shape (image analysis of digital photographs), mass, volume, and polymer type if plastic (e.g., Raman spectroscopy or Fourier transform infrared spectroscopy, FT-IR) should all be recorded, and all evidence should be identified as to source using established techniques (Browne et al., 2010) as practical and in collaboration with the relevant industries, to maximize the integration of data into these industries, such as plastics and fishing.

3. Confirmatory diagnostics: Further analyses as practical and indicated should be undertaken, such as histopathology, imaging, analytical chemistry, blood test, and organ function tests, to document presence of and the type of items ingested, as well as possible impacts to the animals. It would be useful to provide resources to develop techniques to identify particles of plastic in the tissues of animals. Criteria for the assignation of degree of confidence of findings (e.g., quality of data) of entanglement or ingestion contributing to or causing morbidity and mortality have been published and should be applied (Moore et al., 2013a). Chain of custody documentation should be maintained as required or possible.
4. Training designed for specific countries and regions and database maintenance would both enhance understanding of these problems.

Classification of Item Types

With regard to classification of the types of items ingested, categorization is essential for understanding sources, distribution, and impact on cetaceans. A two-part classification system is recommended to address this requirement. The first aspects should include characteristics adequate to understand the use, configuration, and other aspects of the ingested item while it is still in active use. Largely, these characteristics will map to the industrial function of the item: holding liquids, catching fish, or providing buoyancy. The second aspect of the classification system should focus on characteristics of the item after it has left human possession and that contribute to the harm the item might cause to cetaceans. For instance, this might include color (i.e., visibility), flexibility, sharp edges, size, strength, density, site in water, flexibility, shape/aspect ratio, and a host of other aspects that affect its ability to harm cetaceans.

Identification of Microplastics in the Gastrointestinal Tract

Gut contents (from mouth to anus) should be washed through a set of three nested sieves (118–250, 500, and 1000 μm), using prefiltered tap water. Intestines should be divided into equal separate pieces (20 pieces if it is possible) and washed individually, as well as the stomach chamber and esophagus. Material retained in the small mesh size sieve should be retained for microplastic analysis. For microplastic analysis, the retained samples should be dissolved using a premade solution of 10% KOH following Foekema et al. (2013), reviewed in Lusher et al. (2017), to remove the remaining organic matter. Three times the amount of solution (by mass) should be added to the suspension, and then left for approximately 3 weeks or until biologic material has dissolved. The remaining solution should then be filtered under vacuum. Particles retained on GF/C microfiber filter paper should then be identified under a microscope. Visual classification can be carried out using existing criteria (Lusher et al., 2014, 2015), and a subsample of plastics (10% of the overall total) should be retained for polymer classification using FT-IR spectroscopy, enabling the spectra to then be compared to a polymer library. The subsample retained should be chosen so it is representative of the overall particles analyzed, with proportions of colors and shapes that represent those in the total found. FT-IR determines the structure of molecules through the analysis of their absorption spectra (for more information, refer to Lusher et al., 2013). Polymers that match reference spectra with a Euclidian distance with a high level of certainty (>70%) can be accepted as microplastics, while those with a lower level of certainty (60%–70%) should be subjected to further visual examination of spectra characteristics before being accepted or rejected.

Contamination Control

To ensure minimal contamination, all work should be carried out under contamination-controlled conditions (following (Lusher et al., 2015)), with all personnel wearing clean cotton clothing and gloves, and minimal air movement within the laboratory. All gut sections should be covered prior to analysis and exposed to air for the minimum time possible. It is desirable to wash the gut externally using filter water before it is opened. This will reduce external contamination from the necropsy examination procedure. All manipulation instruments and equipment must be cleaned and checked under a microscope for contamination with airborne fibers before use, as previous studies have commented on the abundance of textile fibers found in biologic samples (e.g., Foekema et al., 2013). Blank samples should be processed in the same way as gut samples. To check for sufficient contamination control, recently published EU guidelines suggest that background levels of microplastics in control samples should be less than 10% of the overall microplastic average throughout all samples (MSFD Technical Subgroup on Marine Litter, 2013).

Free-Ranging Organisms

Plastic pollution and its effects in living organisms cannot be assessed by the direct quantification of plastics but can be carried out through monitoring of plastic additives and biologic responses to the exposure to plastic as a proxy of plastic ingestion. This approach will be extensively discussed later in this chapter.

IDENTIFICATION OF PLASTIC ADDITIVES IN TISSUE SAMPLES: PHTHALATES DETECTION

Ingestion of plastics and microplastics by marine organisms may have physical and toxicologic noxious effects, creating a serious threat to marine species.

Several authors (Baini et al., 2017; Fossi et al., 2016; Hardesty et al., 2015) report the need to develop methods to evaluate plastic exposure, through the detection of plastic tracers, in free-ranging marine wildlife, particularly related to plastic/microplastic ingestion. The presence of phthalates (bis-2-ethylhexyl phthalate [DEHP] and the mono-2-ethylhexyl phthalate [MEHP]) was explored for the first time in the Mediterranean fin whale (*Balaenoptera physalus*) by Fossi et al. (2012, 2014, 2016). DEHP and MEHP were analyzed following a method described by Takatori et al. (2004) with a few modifications and subsequently further modified by Baini et al. in 2017.

Sample Collection and Analysis

Samples (three subaliquots of 2 g each) of blubber and muscle (liver and kidney, if available from stranded organisms) obtained from stranded animals or remote dart sampling (skin biopsy) should be collected under the dorsal fin and put in aluminium foil (which has been previously cleaned with hexane) and maintained in a plastic-free container. Samples must be stored at least at −20°C (ideally at −80°C) until analysis. For analysis, the sample should be thawed, weighed (wet weight), and lyophilized. The percentage of water present in each sample must be calculated to determine the concentration of phthalates in terms of both wet weight and dry weight. The procedure for analysis of phthalate esters in organisms is similar to that for many organic contaminants and involves extraction, separation, purification, and instrumental detection as described in Baini et al. (2017). The method for the extraction of phthalates from solid samples has been kept as simple as possible, using mechanical shaking and ultrasonic extraction.

The solid–liquid extraction method coextracts lipids and organic compounds with the phthalates. In lipid-rich samples, such as blubber, it is therefore necessary to conduct a final clean-up of the extract. The clean-up methodology used is based upon the differential solubility of phthalate esters in sulphuric acid and the solvent used. The final extract must be concentrated under a gentle stream of nitrogen, for analysis by gas chromatography–mass spectrometry (GC–MS) with an internal standard. Blank analysis must be carried out for each set of samples to check for possible laboratory contamination.

Plastic Additives in Environmental and Cetacean Samples

Seven phthalate esters (PAEs) as plastic tracers were detected for the first time in superficial neustonic/planktonic samples and blubber collected from four cetacean species from the northwestern Mediterranean Sea (i.e., from within the Pelagos Sanctuary). Among all phthalates, the highest frequency of detection was obtained from DEHP (78.9%), followed in descending order by MBzP (57.9%), MEHP and BBzP (both 52.6%), DIOIP (47.4%), and MBP and DNHP (both 42.1%). All 19 superficial tows analyzed presented microplastic debris, confirming the widespread distribution of microplastics on the seawater surface in the Mediterranean Sea. The abundance of the smallest items (<0.5 mm) found in the samples was significantly correlated with MEHP, MBzP, and BBzP, while the abundance of 0.5–1 mm and 2.5–5 mm items were related to BBzP and MBP, respectively. Regarding the PAEs, detected in the neustonic/planktonic samples correlated with microplastics (MEHP, MBzP, BBzP, and MBP), appreciable levels of these compounds were found in all four cetacean species. The present results represent a step forward in the application of a new noninvasive analytical method to evaluate PAEs as tracers of plastic exposure/ingestion in marine wildlife.

The role of plastic and microplastics as vectors of environmental contaminants is still debated. Some studies demonstrate that fish and seabirds that have ingested plastics can accumulate significantly larger concentrations of halogenated organic compounds (HOC) (Bakir et al., 2016) and PBDEs, in particular BDE#s 183-209 (Rochman et al., 2014; Tanaka et al., 2015), and trace elements (Lavers et al., 2014) compared to the organisms that have low or any plastic items in their stomach or live in low-density plastic accumulation areas. Although, several modeling studies demonstrate that the capacity of plastic to sorb contaminants from the surrounding environment, the rate of adsorption depends on several factors (e.g., the type of polymer). Moreover, the HOCs bioaccumulated from natural prey typically overwhelm the flux from ingested microplastic, which implies that microplastic ingestion may not significantly increase the exposure to HOCs (Herzke et al., 2016; Koelmans et al., 2016). More research is needed to better understand the role of plastics as an additional source of contaminants to these organisms.

DEVELOPMENT OF BIOMARKERS TO EVALUATE THE EXPOSURE OF PLASTIC ADDITIVE IN FREE-RANGING CETACEANS

Currently, one of the main toxicologic issues in the management and conservation of the marine environment is the study of the persistent organic pollutants, including PCBs, and their impact on cetacean health and reproduction (Jepson et al., 2016), as well as the potential impact of compounds released from plastics, such as bisphenol A (BPA), industrial derivates, such as perfluoroalkyl substances (PFASs), and the halogenated flame retardant class polybrominated diphenyl ether (PBDEs). In this regard, the assessment of toxicologic risk in wildlife requires the development of sensitive biomarkers including those based on the use of in vitro systems. BPA is one of the most distributed compounds in the world, both in aquatic and terrestrial ecosystems, acting as agonist or antagonist for endocrine receptors (Rubin, 2011). The PFAS are used in the production of several repelling agents, fluoropolymers, pesticides, lubricants, paints, and firefighting foams due to their ability to repel both water and oils (Key et al., 1997; Prevedouros et al., 2006). Due to their thermic and chemical stability, they are persistent in the environment (Muir and de Wit, 2010) and can accumulate in some organisms, including marine mammals (Lam et al., 2016; Routti et al., 2016; Tartu et al., 2017).

Organotypic cultures from skin biopsies can represent a useful tool to develop gene expression biomarkers in cetaceans. Here, we present a case study based on the analysis of organotypic cultures of three odontocetes species (sperm whale, killer whale, and bottlenose dolphin) exposed to increasing concentrations of BPA and perfluorooctanoic acid (PFOA). The potential biomarker genes such as the peroxisome proliferator-activated receptors α and γ (*PPAR* α and γ) were analyzed. The PPARs belong to a superfamily of ligand-dependent nuclear receptor (*PPAR*α, β, and γ) that regulates physiologic processes of lipids homeostasis, inflammation, adipogenesis, and reproduction (Schupp and Lazar, 2010). *PPAR*α and *PPAR*γ seem to be modulated by the presence of BPA and PFOA, respectively (Buhrke et al., 2015; Mathieu-Denoncourt et al., 2015). The two genes of interest (*PPAR*α, *PPAR*γ) were sequenced in three odontocetes species. The mRNA levels were quantified in response to the two different treatments in the slice samples. Four genes (*PPAR*α, *PPAR*γ, *ER*α, and *E2F1*) are modulated by the treatments in all the three species. In particular, the results of this set of experiments revealed that the BPA treatments induce the expression of the genes *PPAR*α and *PPAR*γ, showing a dose-response trend (Fig. 6.2). Increasing

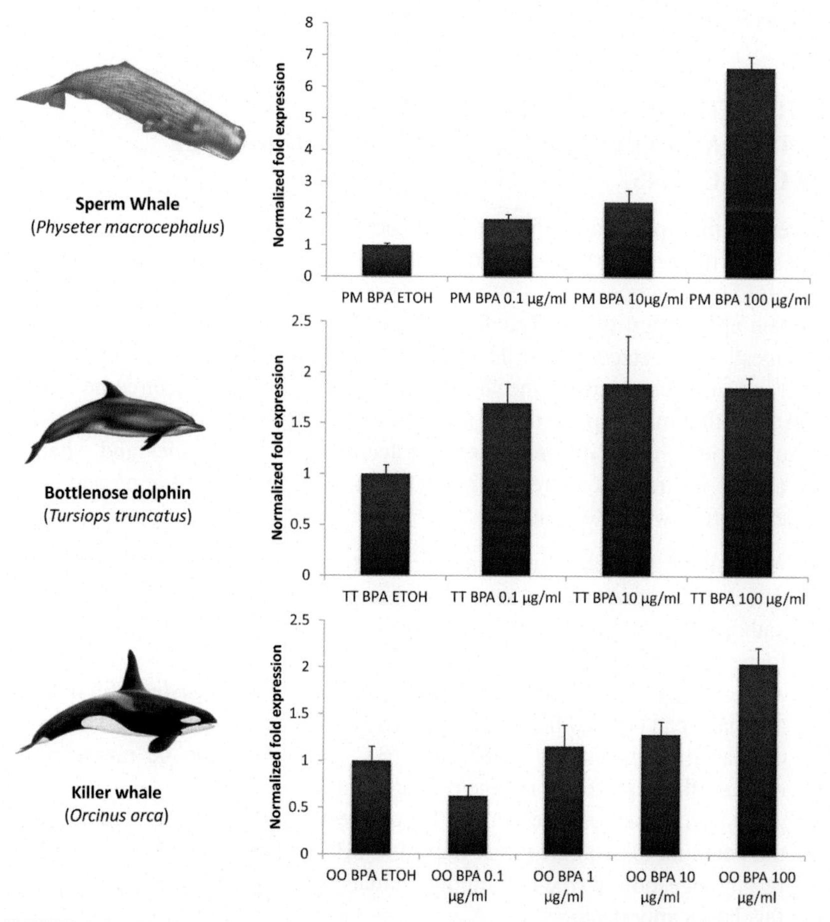

FIGURE 6.2 *PPAR*γ gene expression in sperm whale (*PM*), bottlenose dolphin (*TT*), and killer whale (*OO*) organotypic cell cultures treated with increasing concentration of bisphenol A (BPA) plus control (ethanol). mRNA levels were normalized to two reference genes, *GAPDH* and *YWHAZ*.

the BPA concentration increases the bottlenose dolphin, killer whale, and sperm whale mRNA levels, as well as for *E2F1* apart from the killer whale. Conversely, the PFOA exposure shows a downregulation of the *PPARα* and *PPARγ* both in sperm whale and killer whale slices, while *ERα* and *E2F1* are poorly induced by PFOA in both species.

With respect to the interspecific responses to the BPA treatment, the sperm whale exhibited the highest upregulation of the *PPARγ* gene compared to the other two species. On the other hand, the killer whale showed a dose-response of *ERα* gene expression after PFOA exposure, suggesting the role of the compound as an endocrine disruptor with estrogenic activity.

These data represent the first evidence of an effect of emerging contaminants on cetaceans, assessed using an in vitro experiment, and suggesting the potential use of these diagnostic markers as an early warning signal of exposure to plastic-released compounds and emerging contaminants in monitoring of marine wildlife.

A THREEFOLD MONITORING APPROACH TO DETECT MARINE LITTER INGESTION AND IMPACT IN CETACEANS

Given the multiple potential physical and ecotoxicologic effects of marine litter ingestion, the impact of litter on marine organisms should be assessed using a threefold approach.

The application of the threefold approach can elucidate not only the rate of ingestion in cetaceans, but also the multiple sublethal stresses that marine litter ingestion can cause in the short and long term. Each of the three investigation tools that make up the threefold approach can be applied independently or simultaneously using different methods according to the species and whether the animal is stranded or free ranging (Fig. 6.3). The threefold approach comprises the following elements:

- Analysis of gastrointestinal content: For stranded cetaceans, it is possible to detect the occurrence and rate of marine litter ingestion and any associated pathology through analysis of the gastrointestinal content, with a particular focus on plastics and microplastics.
- Analysis of the levels of plastic additives, as a proxy for ingestion: An indirect approach can be used for free-ranging as well as stranded animals. The levels of plastic additives and associated PBT compounds can be measured to evaluate the exposure to marine plastic pollution.
- Analysis of biomarker responses: Biomarker responses can be used to detect the potential toxicologic effect related to PBT and plastic additives related to plastic ingestion in free-ranging individuals or in stranded organisms up to a few hours after death.

i) Plastic detection

Analysis of the ingested marine litter/microplastics:

- Occurrence (%)
- Abundance (n°)
- Weight (g)
- Polymer analysis

ii) Plastic tracers detection

Analysis of plastic additives:

- Phthalates
- PBDEs
- Bisphenol A

- Analysis of PBT compounds:

- PCBs
- DDTs
- PAHs
- Mercury

iii) Biomarkers detection

Effects at molecular level:

- Measure of DNA damage
- Alterations of gene expression
- Alteration of proteins

- Effects at cellular level:

- Alteration of cell functions

- Effects at tissue level:

- Histological and histopathological alterations

FIGURE 6.3 The threefold monitoring approach to detect marine litter presence and impact in cetacean species (stranded and free-ranging organisms).

HOT SPOT AREAS OF MARINE LITTER FOR MARINE MAMMALS: THE CASE STUDY OF THE MEDITERRANEAN SEA

Monitoring of the impact of marine litter on cetacean species should include an understanding of likely exposure to marine litter in the area concerned. Here, we present the case study of the Mediterranean Sea.

The Mediterranean Sea is one of the areas most affected by marine plastic pollution in the world. Plastics and other polymer materials are the most common types of marine litter, representing some 80% of litter found. As larger pieces of plastic debris fragment into smaller pieces, the abundance of microplastics in marine habitats increases; 115,000–1,050,000 particles/km^2 are estimated to float in the Mediterranean Sea (Cózar et al., 2015). The marine litter problem in the Mediterranean is exacerbated by the basin's limited exchanges with other oceans, highly developed coastal tourism, densely populated coasts, busy offshore waters (with 30% of the world's maritime traffic), high temperatures accelerating litter degradation into secondary products that are difficult to collect or treat, and inputs of litter from very urbanized areas and large rivers.

Despite recent advances made within the framework of the Barcelona Convention Regional Plan for Marine Litter Management in the Mediterranean and the EU Marine Strategy Framework Directive (Descriptor 10), much work is still required to tackle marine litter in the Mediterranean and reduce the risks posed to Mediterranean marine wildlife. To-date the impacts of plastic pollution on marine biota, in particular on those species inhabiting Marine Protected Areas (MPAs) and Specially Protected Areas of Mediterranean Importance (SPAMI), are poorly understood.

Recent studies suggest that marine litter, including microplastics and plastic additives (phthalates), tends to accumulate in pelagic areas of the Mediterranean Sea, indicating a potential overlap between marine litter accumulation areas and the feeding grounds of some endangered species (Fossi et al., 2016). One of these areas is the Pelagos Sanctuary, one of the SPAMI (Notarbartolo-Di-Sciara et al., 2003) and located in the northwestern Mediterranean Sea. In this area, cetaceans coexist with high human pressure and are exposed to a considerable amount of plastic pollution, including microplastics (Collignon et al., 2012; Fossi et al., 2013, 2017).

Baleen whales are potentially exposed to microlitter ingestion as a result of their filter-feeding activity. The interactions between cetaceans and microplastics have been investigated recently in free-ranging fin whales (*Balaenoptera physalus*), comparing populations living in two semienclosed basins, the Mediterranean Sea and the Sea of Cortez (in the Gulf of California, Mexico) (Fossi et al., 2016). Fin whales, the only resident mysticete in the Mediterranean, aggregate during summer in the feeding grounds of the Pelagos Sanctuary and migrate to the southern Mediterranean Sea during winter. In the Sea of Cortez, fin whales are resident and genetically isolated from other Pacific populations. Fin whales forage on the dense aggregations of krill deep in the water column and near to the surface, engulfing an average of $71\,m^3$ of water per mouthful (Goldbogen et al., 2007). As a result, fin whales are exposed to a high potential risk of microplastic ingestion in their feeding grounds due to the ingestion of both contaminated prey and the direct ingestion of floating microplastics. This species can represent a critical indicator of the microplastic contamination in a whole basin (Fossi and Panti, 2017). In this case study, considerable abundance of microplastics and plastic additives was demonstrated in the superficial zooplankton samples from the Pelagos Sanctuary of the Mediterranean Sea (compared to those detected in Sea of Cortez). In addition, pelagic areas containing high densities of microplastics overlapped with whale feeding grounds, suggesting that whales are exposed to microplastics during foraging.

The abundance of microplastics in the Mediterranean environment, along with the high concentrations of PBT chemicals, plastic additives, and biomarker responses detected in the biopsies of Mediterranean whales (as compared to those in whales inhabiting the Sea of Cortez), suggest that exposure to microplastics through direct ingestion and consumption of contaminated prey poses a major threat to the health of fin whales in the Mediterranean Sea (Fig. 6.4).

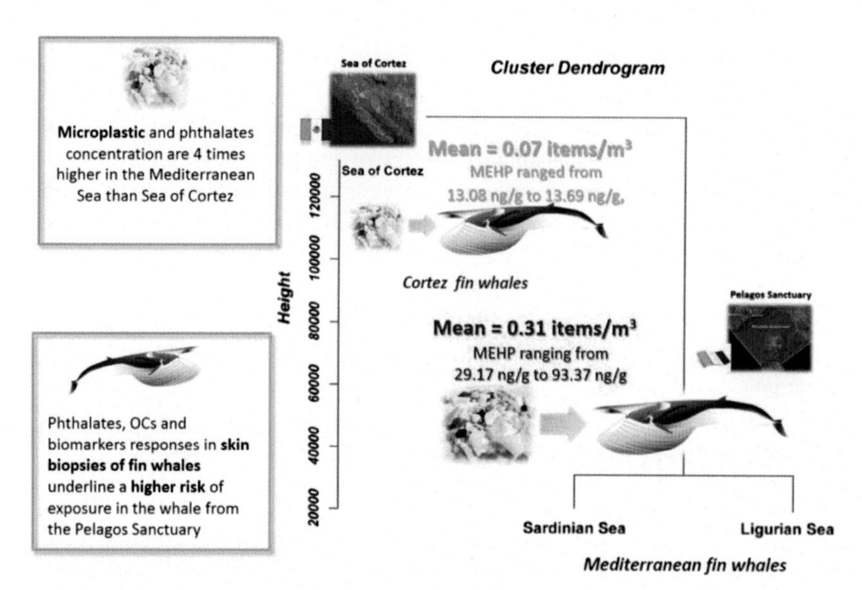

FIGURE 6.4 Comparison of microplastic abundance and fin whale ecotoxicologic status between Sea of Cortez (Mexico) and Mediterranean Sea by a cluster analysis.

The fin whale is a large-scale indicator of the impact of microplastics and related contaminants in pelagic environments, as well as a charismatic sentinel of the integrity of the marine food chain on the basin scale.

IDENTIFICATION OF THE SPECIES MOST THREATENED SPECIES BY MARINE LITTER

A key step forward in addressing this problem will be to identify the species most threatened by interaction with marine litter to implement effective mitigation measures. However, there is a need for much more data before it is possible to accurately assess the risk of morbidity and mortality from debris interactions and the conservation implications for different cetacean populations. Using the protocols described within this chapter, we propose that at all opportunities, necropsies include examination for ingestion or entanglement in marine litter and that stranding networks collate the following information to determine rates of interactions:

1. the total number of carcasses per species and region where a full necropsy was conducted (including examination of stomach contents);
2. the number of these in which ingestion of debris was detected;
3. the number of carcasses per species and region where a definitive cause of death was determined;
4. the number of mortalities attributed to debris;
5. the number with pathology attributed to debris.

Although in many cases it is difficult to irrefutably determine the cause of death or pathology, we propose that necropsy protocols following the guidelines set out herein and that published criteria for assigning a degree of confidence of findings should be applied (see Moore et al., 2013b). By reporting rates rather than just absolute numbers of interactions observed, it will be possible to circumvent the influence of geographic differences in shoreline coverage, carcass recovery rates, and necropsy effort and gain a more reliable assessment of the level of threat that marine litter poses to cetaceans. Such data should be made available and collated at the national level for analysis and reporting to appropriate fora, including under the MSFD, regional conservation agreements such as ASCOBANS and ACCOBAMS, and internationally to the IWC.

FUTURE DEVELOPMENT AND RECOMMENDATIONS

The impact of marine litter, in particular plastic pollution, on cetaceans is raising concern in the scientific community and public opinion (Germanov et al., 2018; Rochman et al., 2016). However, as underlined earlier in this chapter, the research and data available are still limited and sometimes lack consistency and comparability. Further nonlethal research targeted on understanding the actual impact on cetacean populations is therefore vital.

As a first priority, the harmonization/standardization of the protocols for the analysis of marine litter in stranded organisms is recommended and, in particular, regarding the analysis of microplastics and biologic and ecotoxicologic responses. This could be achieved by creating a strong community of marine litter experts who share knowledge, facilities, and samples.

One of the fundamental roles in this research topic is that played by stranding networks to collect and share samples for different marine litter analyses and to make such data available. Therefore, networks should be strongly encouraged to increase their study of the impacts of plastic pollution, with sufficient resources allocated to enable analysis. It is vital that such data be made available for analysis and collated at a national, regional, and international level through existing mechanisms, including regional and international agreements such as the IWC and the Convention on Migratory Species and its regional agreements ASCOBANS and ACCOBAMS.

Marine litter can affect different cetacean species in different ways, and different populations may be subject to differential pressures according to the occurrence and abundance of marine plastic pollution. The identification of accumulation areas and point sources of pollution could represent a key step in identifying the most threated species and populations worldwide.

With multiple stressors on cetacean populations, the additional risk posed by marine plastic pollution needs to be better understood. Given the existing evidence of severe welfare impacts on cetaceans, the potential conservation threat, as well as the known impacts on the wider marine environment and biodiversity, major reductions in consumption of plastics is one of the most urgent environmental issues, with a critical need for action by all stakeholders to eliminate marine plastic pollution.

ACKNOWLEDGMENTS

The authors would like to thank the Italian Ministry of Environment, Territory and Sea, which supported the fin whale Mediterranean project (prot n.39752/III-17). The authors want to thank all the researchers involved in the studies on cetaceans conducted in the Mediterranean Sea presented in the present paper.

REFERENCES

Adimey, N.M., Hudak, C.A., Powell, J.R., Bassos-Hull, K., Foley, A., Farmer, N.A., White, L., Minch, K., 2014. Fishery gear interactions from stranded bottlenose dolphins, Florida manatees and sea turtles in Florida, U.S.A. Marine Pollution Bulletin. 81, 103–115. https://doi.org/10.1016/j.marpolbul.2014.02.008.

Allen, R., Jarvis, D., Sayer, S., Mills, C., 2012. Entanglement of grey seals Halichoerus grypus at a haul out site in Cornwall, UK. Marine Pollution Bulletin. 64, 2815–2819. https://doi.org/10.1016/j.marpolbul.2012.09.005.

Arbelo, M., Espinosa de los Monteros, A., Herráez, P., Andrada, M., Sierra, E., Rodríguez, F., Jepson, P., Fernández, A., 2013. Pathology and causes of death of stranded cetaceans in the Canary Islands (1999–2005). Diseases of Aquatic Organism. 103, 87–99. https://doi.org/10.3354/dao02558.

Baini, M., Martellini, T., Cincinelli, A., Campani, T., Minutoli, R., Panti, C., Finoia, M.G., Fossi, M.C., 2017. First detection of seven phthalate esters (PAEs) as plastic tracers in superficial neustonic/planktonic samples and cetacean blubber. Analytical Methods. 9, 1512–1520. https://doi.org/10.1039/C6AY02674E.

Baird, R.W., Hooker, S.K., 2000. Ingestion of plastic and unusual prey by a juvenile harbour porpoise. Marine Pollution Bulletin. 40, 719–720. https://doi.org/10.1016/S0025-326X(00)00051-5.

Bakir, A., O'Connor, I.A., Rowland, S.J., Hendriks, A.J., Thompson, R.C., 2016. Relative importance of microplastics as a pathway for the transfer of hydrophobic organic chemicals to marine life. Environmental Pollution. 219, 56–65. https://doi.org/10.1016/j.envpol.2016.09.046.

Barros, N.B., Gasparini, J.L., Barbosa, L.A., Netto, R.F., Moreaes, C.S., 1997. Ingestao de plástico como provável causa mortis de una baleia pilto de peitorais curtas, *Globicephala macrorhynchus* Gray, 1846. In: No Litoral Do Estado De Espírito Santo, Anais Do 7o Congresso Nordestino de Ecologia, Ilhéus, Bahia, Brazil, p. 336.

Barros, N.B., Odell, D.K., Patton, G.W., 1990. Ingestion of plastic debris by stranded marine mammals from Florida. In: Shomura, R.S., Godfrey, M.L. (Eds.), Proceedings of the Second International Conference on Marine Debris NOAA Tech. Memo. NOAA-TM-NMFS-SWFSC-154.

Bassoi, M., 1997. Avaliação da dieta alimentar de toninha, *Pontoporia blainvillei* (Gervais and D' Orbigny, 1844), capturadas acidentalmente na pesca costeira de emalhe no sul do Rio Grande do Sul (Bachelor thesis). FURG, Rio Grande.

Bastida, R., Rivero, L., Rodríguez, D., 2000. Presencia inusual de elementos de origen antrópico en los contenidos estomacales de la franciscana (*Pontoporia blainvillei*). In: Technical Paper WP26 Presented to IV Workshop Para a Coordenação Da Pesquisa e Conservação Da Franciscana, *Pontoporia Blainvillei*, No AtlâNtico Sul Ocidental. 05–09 November, Porto Alegre.

Baulch, S., Perry, C., 2014. Evaluating the impacts of marine debris on cetaceans. Marine Pollution Bulletin. 80, 210–221. https://doi.org/10.1016/j.marpolbul.2013.12.050.

Baxter, A., May, 2009. Report on the Blue Whale Stranding Northwest Coast of the South Island, New Zealand. Department of Conservation, Nelson, New Zealand.

Berrow, S., Massett, N., O'Brien, J., 2007. Humpback whale *Megaptera novaeangliae* off Cos. Kerry and Galway. Irish Naturalists' Journal 28, 339–340.

Berrow, S., Ryan, C., O'Brien, J., 2010. Goose barnacle (*Conchoderma auritum* (L.)) attached to tooth of stranded Sowerby's beaked whale (Mesoplodon bidens Sowerby). Irish Naturalists' Journal 31, 136.

Besseling, E., Foekema, E.M., Van Franeker, J.A., Leopold, M.F., Kühn, S., Bravo Rebolledo, E.L., Heße, E., Mielke, L., IJzer, J., Kamminga, P., Koelmans, A.A., 2015. Microplastic in a macro filter feeder: humpback whale *Megaptera novaeangliae*. Marine Pollution Bulletin. 95, 248–252. https://doi.org/10.1016/j.marpolbul.2015.04.007.

Bogomolni, A., Pugliares, K., Sharp, S., Patchett, K., Harry, C., LaRocque, J., Touhey, K., Moore, M., 2010. Mortality trends of stranded marine mammals on Cape Cod and southeastern Massachusetts, USA, 2000 to 2006. Diseases of Aquatic Organism. 88, 143–155. https://doi.org/10.3354/dao02146.

Bossley, M., n.d. The Last Word: Tangled up in Blue. Nat. Aust. 28.

Braco, S., Moore, K.T., 2013. Protocol for examining marine mammals for signs of human interaction. NOAA Technical Document 1–93 NOAA-TM-NMFS-SWFSC-510.

Browne, M.A., Galloway, T.S., Thompson, R.C., 2010. Spatial patterns of plastic debris along estuarine shorelines. Environmental Science and Technology. 44, 3404–3409. https://doi.org/10.1021/es903784e.

Buhrke, T., Krüger, E., Pevny, S., Rößler, M., Bitter, K., Lampen, A., 2015. Perfluorooctanoic acid (PFOA) affects distinct molecular signalling pathways in human primary hepatocytes. Toxicology. 333, 53–62. https://doi.org/10.1016/j.tox.2015.04.004.

Caldwell, M.C., Caldwell, D.K., Sibenaler, J.B., 1965. Observation on captive and wild atlantic bottlenose dolphins (*Tursiops truncatus*) in the northeastern Gulf of Mexico. Los Angeles County Museum Contributions in Science 91, 1–10.

Carretta, J.V., Wilkin, S.M., Muto, M.M., Wilkinson, K., 2013. Sources of Human-Related Injury and Mortality for US Pacific West Coast Marine Mammal Stock Assessments, 2007–2011. US Dep. Commer. NOAA Tech. Memo. NMFS-SWFSC, p. 83. 514.

Cassoff, R., Moore, K., McLellan, W., Barco, S., Rotstein, D., Moore, M., 2011. Lethal entanglement in baleen whales. Diseases of Aquatic Organism. 96, 175–185. https://doi.org/10.3354/dao02385.

Cawthorn, M.W., 1985. Entanglement in and ingestion of plastic litter by marine mammals, sharks and turtles in New Zealand waters. In: Shomura, R.S., Yoshida, H.O. (Eds.), Proceedings of the Workshop on the Fate and Impact of Marine Debris. NMFS, Honolulu, Hawaii, pp. 336–343. U.S. Dep. Commer., NOAA Tech. Memo.

Ceccarelli, D.M., 2009. Impacts of plastic debris on Australian marine wildlife. In: C&R Consulting (Ed.), Report by C&R Consulting for the Department of the Environment, Water, Heritage and the Arts, p. 83.

Chatto, R., Warneke, R.M., 2000. Records of cetacean strandings in the northern territory of Australia. The Beagle Records of the Museum and Art Galleries of the Northern Territory 16, 163.

Citta, J.J., Burns, J.J., Quakenbush, L.T., Vanek, V., George, J.C., Small, R.J., Heide-Jørgensen, M.P., Brower, H., 2014. Potential for bowhead whale entanglement in cod and crab pot gear in the Bering Sea. Marine Mammal Science. 30, 445–459. https://doi.org/10.1111/mms.12047.

Coleman, F.C., Wehle, D.H.S., 1984. Plastic pollution: a worldwide oceanic problem. Parks 9, 9–12.

Collard, F., Gilbert, B., Eppe, G., Parmentier, E., Das, K., 2015. Detection of anthropogenic particles in fish stomachs: an isolation method adapted to identification by Raman spectroscopy. Archives of Environmental Contamination and Toxicology. 69, 331–339. https://doi.org/10.1007/s00244-015-0221-0.

Collignon, A., Hecq, J.-H., Glagani, F., Voisin, P., Collard, F., Goffart, A., 2012. Neustonic microplastic and zooplankton in the north western Mediterranean Sea. Marine Pollution Bulletin. 64, 861–864. https://doi.org/10.1016/j.marpolbul.2012.01.011.

Committee on Taxonomy, 2017. List of Marine Mammal Species and Subspecies. Society for Marine Mammalogy. [WWW Document]. Soc. Mar. Mammal www.marinemammalscience.org.

Cornish, V., Willing, J., Senner, S., 2011. Marine debris: more than a low-grade fever for marine mammals and sea turtles. In: Carswell, B., McElwee, K., Morison, S. (Eds.), Technical Proceedings of the Fifth International Marine Debris Conference. National Oceanic and Atmospheric Administration Technical Memorandum NOS-OR&R-38, Honolulu, Hawaii, pp. 604–608.

Cózar, A., Sanz-Martín, M., Martí, E., González-Gordillo, J.I., Ubeda, B., Gálvez, J.Á., Irigoien, X., Duarte, C.M., 2015. Plastic accumulation in the Mediterranean Sea. PLoS One. 10, e0121762. https://doi.org/10.1371/journal.pone.0121762.

Cuenca, G.C., Macusi, E.D., Abreo, N.A.S., Blatchley, D.D., 2016. First evidence of plastic ingestion by the Rare Deraniyagala's beaked whale (*Mesoplodon hotaula*). IAMURE International Journal of Ecology Conservation 19, 1.

Curran, E., Hernandez-Milian, G., Rogan, E., Whooley, P., 2014. Common Dolphins in the River Lee at Cork City, 33, p. 142.

De Pierrepont, J.F., Dubois, B., Desormonts, S., Santos, M.B., Robin, J.P., 2005. Stomach contents of English channel cetaceans stranded on the coast of Normandy. Journal of the Marine Biological Association of the United Kingdom. 85 (1539). https://doi.org/10.1017/S0025315405012762.

de Stephanis, R., Giménez, J., Carpinelli, E., Gutierrez-Exposito, C., Cañadas, A., 2013. As main meal for sperm whales: plastics debris. Marine Pollution Bulletin. 69, 206–214. https://doi.org/10.1016/j.marpolbul.2013.01.033.

Deaville, R., Jepson, P.D., Brownlow, A., Reid, R.J., Smith, B., Duffell, E.L., Sabin, R.C., Penrose, R., Perkins, M., 2010. Final Report for the Period 1st January 2005–31st December 2010. UK Strandings Investigation Programme, p. 98.

Degange, A.R., Newby, T.C., 1980. Mortality of seabirds and fish in a lost salmon driftnet. Marine Pollution Bulletin 11, 322–323.

Denuncio, P., Bastida, R., Dassis, M., Giardino, G., Gerpe, M., Rodríguez, D., 2011. Plastic ingestion in franciscana dolphins, *Pontoporia blainvillei* (Gervais and d'Orbigny, 1844), from Argentina. Marine Pollution Bulletin. 62, 1836–1841. https://doi.org/10.1016/j.marpolbul.2011.05.003.

Derraik, J.G.B., 2002. The pollution of the marine environment by plastic debris: a review. Marine Pollution Bulletin. 44, 842–852. https://doi.org/10.1016/S0025-326X(02)00220-5.

Desforges, J.-P.W., Galbraith, M., Ross, P.S., 2015. Ingestion of microplastics by zooplankton in the Northeast Pacific Ocean. Archives of Environmental Contamination and Toxicology. 69, 320–330. https://doi.org/10.1007/s00244-015-0172-5.

Di Beneditto, A.P.M., Awabdi, D.R., 2014. How marine debris ingestion differs among megafauna species in a tropical coastal area. Marine Pollution Bulletin. 88, 86–90. https://doi.org/10.1016/j.marpolbul.2014.09.020.

Di Beneditto, A.P.M., Ramos, R.M.A., 2014. Marine debris ingestion by coastal dolphins: what drives differences between sympatric species? Marine Pollution Bulletin. 83, 298–301. https://doi.org/10.1016/j.marpolbul.2014.03.057.

Eunomia Research, Consulting Ltd., 2016. Study to Support the Development of Measures to Combat a Range of Marine Litter Sources, Report for DG Environment of the European Commission.

Evans, K., Hindell, M., 2004. The diet of sperm whales (*Physeter macrocephalus*) in southern Australian waters. ICES Journal of Marine Science. 61, 1313–1329. https://doi.org/10.1016/j.icesjms.2004.07.026.

Fernández, R., Santos, M.B., Carrillo, M., Tejedor, M., Pierce, G.J., 2009. Stomach contents of cetaceans stranded in the Canary Islands 1996–2006. Journal of the Marine Biological Association of the United Kingdom. 89 (873). https://doi.org/10.1017/S0025315409000290.

Finley, K., 2001. Natural history and conservation of the Greenland whale or bowhead, in the Northwest Atlantic. Artic 54.

Foekema, E.M., De Gruijter, C., Mergia, M.T., van Franeker, J.A., Murk, A.J., Koelmans, A.A., 2013. Plastic in north sea fish. Environmental Science and Technology. 130711150255009 https://doi.org/10.1021/es400931b.

Fossi, M.C., Coppola, D., Baini, M., Giannetti, M., Guerranti, C., Marsili, L., Panti, C., de Sabata, E., Clò, S., 2014. Large filter feeding marine organisms as indicators of microplastic in the pelagic environment: the case studies of the Mediterranean basking shark (*Cetorhinus maximus*) and fin whale (*Balaenoptera physalus*). Marine Environmental Research. 100, 17–24. https://doi.org/10.1016/j.marenvres.2014.02.002.

Fossi, M.C., Marsili, L., Baini, M., Giannetti, M., Coppola, D., Guerranti, C., Caliani, I., Minutoli, R., Lauriano, G., Finoia, M.G., Rubegni, F., Panigada, S., Bérubé, M., Urbán Ramírez, J., Panti, C., 2016. Fin whales and microplastics: the Mediterranean Sea and the sea of Cortez scenarios. Environmental Pollution. 209, 68–78. https://doi.org/10.1016/j.envpol.2015.11.022.

Fossi, M.C., Panti, C., 2017. Sentinel Species of Marine Ecosystems. https://doi.org/10.1093/acrefore/9780199389414.013.110.

Fossi, M.C., Panti, C., Guerranti, C., Coppola, D., Giannetti, M., Marsili, L., Minutoli, R., 2012. Are baleen whales exposed to the threat of microplastics? A case study of the Mediterranean fin whale (*Balaenoptera physalus*). Marine Pollution Bulletin. 64, 2374–2379. https://doi.org/10.1016/j.marpolbul.2012.08.013.

Fossi, M.C., Panti, C., Marsili, L., Maltese, S., Spinsanti, G., Casini, S., Caliani, I., Gaspari, S., Muñoz-Arnanz, J., Jimenez, B., Finoia, M.G., 2013. The Pelagos Sanctuary for Mediterranean marine mammals: marine protected area (MPA) or marine polluted area? The case study of the striped dolphin (*Stenella coeruleoalba*). Marine Pollution Bulletin. 70, 64–72. https://doi.org/10.1016/j.marpolbul.2013.02.013.

Fossi, M.C., Romeo, T., Baini, M., Panti, C., Marsili, L., Campani, T., Canese, S., Galgani, F., Druon, J.-N., Airoldi, S., Taddei, S., Fattorini, M., Brandini, C., Lapucci, C., 2017. Plastic debris occurrence, convergence areas and fin whales feeding ground in the Mediterranean marine protected area Pelagos Sanctuary: a modeling approach. Frontiers in Marine Science. 4. https://doi.org/10.3389/fmars.2017.00167.

Frantzis, A., 2007. Fisheries interaction with cetacean species in Hellas. In: Papaconstantinou, C., Zenetos, A., Tserpes, G., Vassilopoulou, V. (Eds.), State of the Hellenic Fisheries. Hellenic Centre for Marine Research, Athens.

Frias, J.P.G.L., Otero, V., Sobral, P., 2014. Evidence of microplastics in samples of zooplankton from Portuguese coastal waters. Marine Environmental Research. 95, 89–95. https://doi.org/10.1016/j.marenvres.2014.01.001.

Galgani, F., Hanke, G., Maes, T., 2015. Global distribution, composition and abundance of marine litter. In: Bergmann, M., Gutow, L., Klages, M. (Eds.), Marine Anthropogenic Litter. Springer International Publishing, pp. 29–56.

Gassner, I., Rogan, E., Bruton, T., 2005. A live stranding of true's beaked whale *Mesoplodon mirus* true. Irish Naturalists' Journal 28, 170.

GEF Council Meeting, 2012. Global environmental facility, investing in our planet. In: GEF Council Meeting November 13-15 2012, Washington DC. GEF/STAP/C.43/Inf.04 October 15, 2012 Impact of Marine Debris on Biodiversity: Current Status and Potential Solutions.

Geise, L., Gomes, N., 1992. Occurencia De Plastico No Estomago De Um Golfinho, Sotalia guianensis (Cetacea, Delphinidae). In: Anais Da 3a Reuniao de Trabalho de Especialistas Em Mamiferos Aquaticos de America Do Sul, Montevideo, Uruguay, pp. 26–28.

Germanov, E.S., Marshall, A.D., Bejder, L., Fossi, M.C., Loneragan, N.R., 2018. Microplastics: no small problem for filter-feeding megafauna. Trends in Ecology and Evolution. https://doi.org/10.1016/j.tree.2018.01.005.

Gill, A., Fairbairns, B., Fairbairns, R., 2000. Some observations of minke whale (*Balaenoptera acutorostrata*) feeding behaviour and associations with seabirds in the coastal waters of the Isle of Mull, Scotland. European Research on Cetaceans 13, 61–64.

Goldbogen, J., Pyenson, N., Shadwick, R., 2007. Big gulps require high drag for fin whale lunge feeding. Marine Ecology Progress Series. 349, 289–301. https://doi.org/10.3354/meps07066.

Gomerčić., M.D., Galov, A., Gomerčić, T., Škrtić, D., Ćurković, S., Lucić, H., Vuković, S., Arbanasić, H., Gomerčić, H., 2009. Bottlenose dolphin (*Tursiops truncatus*) depredation resulting in larynx strangulation with gill-net parts. Marine Mammal Science. 25, 392–401. https://doi.org/10.1111/j.1748-7692.2008.00259.x.

Gorzelany, J.F., 1998. Unusual deaths of two free-ranging Atlantic bottlenose dolphins (*Tursiops truncatus*) related to ingestion of recreational fishing gear. Marine Mammal Science. 14, 614–617. https://doi.org/10.1111/j.1748-7692.1998.tb00748.x.

Haelters, J., Kerckhof, F., Doom, M., Evans, P.G.H., Van den Neucker, T., Jauniaux, T., 2017. First case of a narwhal (*Monodon monoceros*) in Belgium. In: Proceedings of the 31st Annual European Cetancean Society Conference, p. 173.

Hardesty, B.D., Holdsworth, D., Revill, A.T., Wilcox, C., 2015. A biochemical approach for identifying plastics exposure in live wildlife. Methods in Ecology and Evolution. 6, 92–98. https://doi.org/10.1111/2041-210X.12277.

Hare, M.P., Mead, J.G., 1987. Handbook for Determination of Adverse Human-Marine Mammal Interactions from Necropsis. U.S. Department of Commerce, Northwest and Alaska Fisheries Service, Seattle, p. 35.

Haynes, J.A., Limpus, C.J., 2000. Marine wildlife stranding and mortality database annual report, 2000: II cetacean and pinnipeds. In: Queensland Parks and Wildlife Service (Ed.), Research Coordination Unit Parks and Wildlife Strategy Division.

Henderson, J.R., 1988. Marine debris in Hawaii. In: Alverson, D.L., June, J.A. (Eds.), Proceedings of the North Pacific Rim Fishermen's Conference on Marine Debris. Natural Resources Consultants, Seattle.

Hernandez-Milian, G., 2014. Trophic Role of Small Cetaceans and Seals in Irish Waters (Ph.D. thesis). University College Cork, Ireland.

Herzke, D., Anker-Nilssen, T., Nøst, T.H., Götsch, A., Christensen-Dalsgaard, S., Langset, M., Fangel, K., Koelmans, A.A., 2016. Negligible impact of ingested microplastics on tissue concentrations of persistent organic pollutants in northern Fulmars off coastal Norway. Environmental Science and Technology. 50, 1924–1933. https://doi.org/10.1021/acs.est.5b04663.

Heyning, J.E., Lewis, T.D., 1990. Entanglements of baleen whales in fishing gear off southern California. In: 40th Report of the International Whaling Commission, Cambridge, pp. 427–431.

Humpback Whale Recovery Team, 1991. Final Recovery Plan for the Humpback Whale, *Megaptera novaeangliae*. National Marine Fisheries Service, Silverspring, M.D.

IWC, 2013. Report of the IWC Scientific Committee Workshop on Marine Debris.

Jackson, K., 2014. 2013 Alaska Region Marine Mammal Stranding Summary. National Marine Fisheries Service, Alaska Region, Protected Resources, Juneau, Alaska.

Jackson, K., 2013. 2012 Alaska Region Marine Mammal Stranding Summary. National Marine Fisheries Service, Alaska Region, Protected Resources, Juneau, Alaska.

Jackson, K., 2012. 2011 Alaska Region Marine Mammal Stranding Summary. National Marine Fisheries Service, Alaska Region, Protected Resources, Juneau, Alaska.

Jackson, K., 2011. 2010 Alaska Region Marine Mammal Stranding Summary. National Marine Fisheries Service, Alaska Region, Protected Resources, Juneau, Alaska.

Jackson, K., 2010. 2009 Alaska Region Marine Mammal Stranding Summary. National Marine Fisheries Service, Alaska Region, Protected Resources, Juneau, Alaska.

Jackson, K., Sternfeld, M., 2009. 2008 Alaska Region Marine Mammal Stranding Summary. National Marine Fisheries Service, Alaska Region, Protected Resources, Juneau, Alaska.

Jacobsen, J.K., Massey, L., Gulland, F., 2010. Fatal ingestion of floating net debris by two sperm whales (*Physeter macrocephalus*). Marine Pollution Bulletin. 60, 765–767. https://doi.org/10.1016/j.marpolbul.2010.03.008.

Jambeck, J.R., Geyer, R., Wilcox, C., Siegler, T.R., Perryman, M., Andrady, A., Narayan, R., Law, K.L., 2015. Plastic waste inputs from land into the ocean. Science. 347, 768–771. https://doi.org/10.1126/science.1260352.

Jepson, P.D., Deaville, R., Barber, J.L., Aguilar, À., Borrell, A., Murphy, S., Barry, J., Brownlow, A., Barnett, J., Berrow, S., Cunningham, A.A., Davison, N.J., ten Doeschate, M., Esteban, R., Ferreira, M., Foote, A.D., Genov, T., Giménez, J., Loveridge, J., Llavona, Á., Martin, V., Maxwell, D.L., Papachlimitzou, A., Penrose, R., Perkins, M.W., Smith, B., de Stephanis, R., Tregenza, N., Verborgh, P., Fernandez, A., Law, R.J., 2016. PCB pollution continues to impact populations of orcas and other dolphins in European waters. Scientific Reports. 6. https://doi.org/10.1038/srep18573.

Johnson, A., Salvador, G., Kenney, J., Robbins, J., Kraus, S., Landry, S., Clapham, P., 2005. Fishing gear involved in entanglements of right and humpback whales. Marine Mammal Science. 21, 635–645. https://doi.org/10.1111/j.1748-7692.2005.tb01256.x.

Jones, L.J., Ferrero, R.C., 1985. Observations of net debris and associated entanglements in the north pacific and Bering Sea, 1978–1984. In: Shomura, R.S., Yoshida, H.O. (Eds.), Proceedings of the Workshop on the Fate and Impact of Marine Debris. NMFS, Honolulu, Hawaii, pp. 183–196. U.S. Dep. Commer., NOAA Tech. Memo.

Kaladharan, P., Asokan, P.K., Mohammed Koya, K., Bhint, H.M., 2014. Plastic debris in the stomach of a Longman's beaked whale, *Indopacetus pacificus* (Longman, 1926) stranded off Sutrapada, Veraval, Saurashtra coast, India. Journal of the Marine Biological Association of India. 56, 92–94. https://doi.org/10.6024/jmbai.2014.56.2.01802-15.

Kasteleine, R.A., Lavaleye, M.S.S., 1992. Foreign bodies in the stomach of a female harbour porpoise (*Phococena phococena*) from the north sea. Aquatic Mammals 18, 40–46.

Katsanevakis, S., 2008. Marine debris, a growing problem: sources, distribution, composition and impact. In: Marine Pollution. New Research. Nova Science Publishers, pp. 53–100.

Key, B.D., Howell, R.D., Criddle, C.S., 1997. Fluorinated organics in the biosphere. Environmental Science and Technology. 31, 2445–2454. https://doi.org/10.1021/es961007c.

Kiessling, I., 2003. Finding Solutions: Derelict Fishing Gear and Other Marine Debris in Northern Australia. National Oceans Office, Darwin.

Knowlton, A., Hamilton, P., Marx, M., Pettis, H., Kraus, S., 2012. Monitoring North Atlantic right whale *Eubalaena glacialis* entanglement rates: a 30 yr retrospective. Marine Ecology Progress Series. 466, 293–302. https://doi.org/10.3354/meps09923.

Koelmans, A.A., Bakir, A., Burton, G.A., Janssen, C.R., 2016. Microplastic as a vector for chemicals in the aquatic environment: critical review and model-supported reinterpretation of empirical studies. Environmental Science and Technology. 50, 3315–3326. https://doi.org/10.1021/acs.est.5b06069.

Kraus, S.D., 1990. Rates and potential causes of mortality in north Atlantic right whales (*Eubalaena glacialis*). Marine Mammal Science 6, 278–291.

Kühn, S., Rebolledo, E.L.B., van Franeker, J.A., 2015. Deleterious effects of litter on marine life. In: Bergmann, M., Gutow, L., Klages, M. (Eds.), Marine Anthropogenic Litter. Springer International Publishing, pp. 75–116.

Laist, D.W., 1997. Impacts of marine debris: entanglement of marine life in marine debris including a comprehensive list of species with entanglement and ingestion records. In: Coe, J.M., Rogers, D.B. (Eds.), Marine Debris. Springer, New York, pp. 99–139.

Lam, J.C.W., Lyu, J., Kwok, K.Y., Lam, P.K.S., 2016. Perfluoroalkyl substances (PFASs) in marine mammals from the south China sea and their temporal changes 2002–2014: concern for alternatives of PFOS? Environmental Science and Technology. 50, 6728–6736. https://doi.org/10.1021/acs.est.5b06076.

Lavers, J.L., Bond, A.L., Hutton, I., 2014. Plastic ingestion by flesh-footed shearwaters (*Puffinus carneipes*): implications for fledgling body condition and the accumulation of plastic-derived chemicals. Environmental Pollution. 187, 124–129. https://doi.org/10.1016/j.envpol.2013.12.020.

Levy, A.M., Brenner, O., Scheinin, A., Morick, D., Ratner, E., Goffman, O., Kerem, D., 2009. Laryngeal snaring by ingested fishing net in a common bottlenose dolphin (*Tursiops truncatus*) off the Israeli shoreline. Journal of Wildlife Diseases. 45, 834–838. https://doi.org/10.7589/0090-3558-45.3.834.

Lowy, L., 1993. Foods and feeding ecology. In: Burns, J.J., Montague, J.J., Cowles, C.J. (Eds.), The Bowhead Whale. The Society for Marine Mammology, pp. 201–238.

Lusher, A.L., Burke, A., O'Connor, I., Officer, R., 2014. Microplastic pollution in the Northeast Atlantic ocean: validated and opportunistic sampling. Marine Pollution Bulletin. 88, 325–333. https://doi.org/10.1016/j.marpolbul.2014.08.023.

Lusher, A.L., Hernandez-Milian, G., Berrow, S., Rogan, E., O'Connor, I., 2018. Incidence of marine debris in cetaceans stranded and bycaught in Ireland: recent findings and a review of historical knowledge. Environmental Pollution. 232, 467–476. https://doi.org/10.1016/j.envpol.2017.09.070.

Lusher, A.L., Hernandez-Milian, G., O'Brien, J., Berrow, S., O'Connor, I., Officer, R., 2015. Microplastic and macroplastic ingestion by a deep diving, oceanic cetacean: the True's beaked whale *Mesoplodon mirus*. Environmental Pollution. 199, 185–191. https://doi.org/10.1016/j.envpol.2015.01.023.

Lusher, A.L., McHugh, M., Thompson, R.C., 2013. Occurrence of microplastics in the gastrointestinal tract of pelagic and demersal fish from the English channel. Marine Pollution Bulletin. 67, 94–99. https://doi.org/10.1016/j.marpolbul.2012.11.028.

Lusher, A.L., Welden, N.A., Sobral, P., Cole, M., 2017. Sampling, isolating and identifying microplastics ingested by fish and invertebrates. Analytical Methods. 9, 1346–1360. https://doi.org/10.1039/C6AY02415G.

MacLeod, C.D., 2007. Understanding the impacts of human activities on beaked whales: from individuals to species, and from local to global. In: Dolman, J.S., MacLeod, C.D., Evans, P.G.H. (Eds.), Proceedings of the ECS Workshop on Beaked Whale Research. ECS, San Sebastián, Spain.

Mann, J., Smolker, R.A., Smuts, B.B., 1995. Response to calf entanglement in free-ranging bottlenose dolphins. Marine Mammal Science 11, 100–106.

Marine Mammal Commission, 2006. Annual Report to Congress 2005. Bethesda, M.D.

Marine Mammal Commission, 1993. Annual Report to Congress 1992. Washington, D.C.

Martin, A.R., Clarke, M.R., 1986. The diet of sperm whales (*Physeter macrocephalus*) captured between Iceland and Greenland. Journal of the Marine Biological Association of the United Kingdom. 66 (779). https://doi.org/10.1017/S0025315400048426.

Mate, B.R., 1985. Incidents of marine mammal encounters with debris in active fishing gear. In: Shomura, R.S., Yoshida, H.O. (Eds.), Proceedings of the Workshop on the Fate and Impact of Marine Debris. NMFS, Honolulu, Hawaii, pp. 453–457. U.S. Dep. Commer., NOAA Tech. Memo.

Mathieu-Denoncourt, J., Wallace, S.J., de Solla, S.R., Langlois, V.S., 2015. Plasticizer endocrine disruption: highlighting developmental and reproductive effects in mammals and non-mammalian aquatic species. General and Comparative Endocrinology. 219, 74–88. https://doi.org/10.1016/j.ygcen.2014.11.003.

Mattila, D.K., Lyman, E., 2006. A note on the entanglement of large whales in marine debris. In: Unpublished Report to the Scientific Committee of the International Whaling Commission: SC/58/BC2.

Mauger, G., Kerleau, F., Robin, J.P., Dubois, B., De Pierrepont, J.F., De Meersman, P., Custers, I., 2002. Marine debris obstructing stomach of a young minke whale (*Balaenoptera acutorostrata*) stranded in Normany, France. In: Presented at the 8th International American Cetacean Society International Conference, Seattle W.A.

Mazzariol, S., Guardo, G.D., Petrella, A., Marsili, L., Fossi, C.M., Leonzio, C., Zizzo, N., Vizzini, S., Gaspari, S., Pavan, G., Podestà, M., Garibaldi, F., Ferrante, M., Copat, C., Traversa, D., Marcer, F., Airoldi, S., Frantzis, A., Quirós, Y.D.B., Cozzi, B., Fernández, A., 2011. Sometimes sperm whales (*Physeter macrocephalus*) cannot find their way back to the high seas: a multidisciplinary study on a mass stranding. PLoS One. 6, e19417. https://doi.org/10.1371/journal.pone.0019417.

McCauley, S., Bjomdal, K., 1999. Conservation implications of dietary dilution from debris ingestion: sublethal effects in post-hatchling loggerhead sea turtles. Conservation Biology 13, 925–929.

McLellan, W.A., Rommel, S.A., Moore, M.J., Pabst, D.A., 2004. Right whale necropsy protocol. In: Final Report to NOAA Fisheries for Contract # 40AANF112525 U.S. Department of Commerce, National Oceanic and Atmospheric Administration, Available from NOAA Fisheries Service, 1315 East West Highway, Silver Spring, MD 20910, USA 51 pp.

Moore, E., Lyday, S., Roletto, J., Litle, K., Parrish, J.K., Nevins, H., Harvey, J., Mortenson, J., Greig, D., Piazza, M., Hermance, A., Lee, D., Adams, D., Allen, S., Kell, S., 2009. Entanglements of marine mammals and seabirds in central California and the north-west coast of the United States 2001–2005. Marine Pollution Bulletin. 58, 1045–1051. https://doi.org/10.1016/j.marpolbul.2009.02.006.

Moore, M.J., Andrews, R., Austin, T., Bailey, J., Costidis, A., George, C., Jackson, K., Pitchford, T., Landry, S., Ligon, A., McLellan, W., Morin, D., Smith, J., Rotstein, D., Rowles, T., Slay, C., Walsh, M., 2013a. Rope trauma, sedation, disentanglement, and monitoring-tag associated lesions in a terminally entangled North Atlantic right whale (*Eubalaena glacialis*). Marine Mammal Science. 29, E98–E113. https://doi.org/10.1111/j.1748-7692.2012.00591.x.

Moore, M.J., van der Hoop, J., Barco, S., Costidis, A., Gulland, F., Jepson, P., Moore, K., Raverty, S., McLellan, W., 2013b. Criteria and case definitions for serious injury and death of pinnipeds and cetaceans caused by anthropogenic trauma. Diseases of Aquatic Organism. 103, 229–264. https://doi.org/10.3354/dao02566.

MSFD Technical Subgroup on Marine Litter, 2013. Guidance on Monitoring of Marine Litter in European Seas. Publications Office, Luxembourg.

Muir, D.C.G., de Wit, C.A., 2010. Trends of legacy and new persistent organic pollutants in the circumpolar arctic: overview, conclusions, and recommendations. The Science of the Total Environment. 408, 3044–3051. https://doi.org/10.1016/j.scitotenv.2009.11.032.

Murphy, F., Russell, M., Ewins, C., Quinn, B., 2017. The uptake of macroplastic & microplastic by demersal & pelagic fish in the Northeast Atlantic around Scotland. Marine Pollution Bulletin. 122, 353–359. https://doi.org/10.1016/j.marpolbul.2017.06.073.

Notarbartolo-Di-Sciara, G., Zanardelli, M., Jahoda, M., Panigada, S., Airoldi, S., 2003. The fin whale *Balaenoptera physalus* (L. 1758) in the Mediterranean Sea. Mammal Review. 33, 105–150. https://doi.org/10.1046/j.1365-2907.2003.00005.x.

Obbard, R.W., Sadri, S., Wong, Y.Q., Khitun, A.A., Baker, I., Thompson, R.C., 2014. Global warming releases microplastic legacy frozen in Arctic Sea ice. Earths Future. 2, 315–320. https://doi.org/10.1002/2014EF000240.

O'Brien, J., Berrow, S.D., 2006. Seaweed ingestion by a bottlenose dolphin. Irish Naturalists' Journal 28, 338–339.

Oliveira de Meirelles, A.C., Duarte do Rego Barros, H.M., 2007. Plastic debris ingested by a rough-toothed dolphin, *Stenobredanensis*, stranded alive in northeastern Brazil. Biotemas 20 (1), 127–131 Março 2007 20, 127–131.

Pellini, G., Gomiero, A., Fortibuoni, T., Ferrà, C., Grati, F., Tassetti, N., Polidori, P., Fabi, G., Scarcella, G., 2018. Characterization of microplastic litter in the gastrointestinal tract of Solea solea from the Adriatic Sea. Environmental Pollution. 234, 943–952. https://doi.org/10.1016/j.envpol.2017.12.038.

Philo, L.M., George, J.C., Albert, T.F., 1992. Rope entanglement of bowhead whales (*Balaena mysticetus*). Marine Mammal Science. 8, 306–311. https://doi.org/10.1111/j.1748-7692.1992.tb00414.x.

Pinedo, M.C., 1982. Análises Dos Conteúdos Estomacias De *Pontoporia blainvillei* (Gervais and D'orbigny, 1844) E *Tursiops gephyreus* (Lahille, 1908) (Cetacea, Platanistidae E Delphinidae). In: Na Zona Estuarial E Costeira Do Rio Grande, Rs, Brazil. Inuversidade Do Rio Grande, Rio Grande, Brazil, p. 95.

Poncelet, E., Van Canneyt, O., Boubert, J.-J., 2000. Considerable amount of plastic debris in the stomach of a Cuvier's beaked whale (*Ziphius cavirostris*) washed ashore on the French Atlantic coast. European Research on Cetaceans 14, 44–47.

Prevedouros, K., Cousins, I.T., Buck, R.C., Korzeniowski, S.H., 2006. Sources, fate and transport of perfluorocarboxylates. Environmental Science and Technology. 40, 32–44. https://doi.org/10.1021/es0512475.

Pugliares, K., Herzke, S., Bogomolni, A., Harry, C., Touhey, K., Moore, M., 2007. Marine mammal necropsy: an introductory guide for stranding responders and field biologists. In: Woods Hole Oceanographic Institution Technical Document 2007-06 117 pp.

Roberts, S.M., 2003. Examination of the stomach contents from a Mediterranean sperm whale found south of Crete, Greece. Journal of the Marine Biological Association of the United Kingdom. 83, 667–670. https://doi.org/10.1017/S0025315403007628h.

Rochman, C.M., Cook, A.-M., Koelmans, A.A., 2016. Plastic debris and policy: using current scientific understanding to invoke positive change: plastic debris and policy. Environmental Toxicology and Chemistry. 35, 1617–1626. https://doi.org/10.1002/etc.3408.

Rochman, C.M., Hoh, E., Kurobe, T., Teh, S.J., 2013. Ingested plastic transfers hazardous chemicals to fish and induces hepatic stress. Scientific Reports. 3 (3263). https://doi.org/10.1038/srep03263.

Rochman, C.M., Lewison, R.L., Eriksen, M., Allen, H., Cook, A.-M., Teh, S.J., 2014. Polybrominated diphenyl ethers (PBDEs) in fish tissue may be an indicator of plastic contamination in marine habitats. The Science of the Total Environment. 476–477, 622–633. https://doi.org/10.1016/j.scitotenv.2014.01.058.

Routti, H., Gabrielsen, G.W., Herzke, D., Kovacs, K.M., Lydersen, C., 2016. Spatial and temporal trends in perfluoroalkyl substances (PFASs) in ringed seals (*Pusa hispida*) from Svalbard. Environmental Pollution. 214, 230–238. https://doi.org/10.1016/j.envpol.2016.04.016.

Rubin, B.S., 2011. Bisphenol A: an endocrine disruptor with widespread exposure and multiple effects. The Journal of Steroid Biochemistry and Molecular Biology. 127, 27–34. https://doi.org/10.1016/j.jsbmb.2011.05.002.

Sadove, S.S., Morreale, S.J., 1990. Marine mammal and sea turtle encounters with marine debris in the New York Bight and the Northeast Atlantic. In: Shomura, R.S., Godfrey, M.L. (Eds.), Proceedings of the Second International Conference on Marine Debris. NMFS, Honolulu, Hawaii, pp. 562–570. U.S. Dep. Commer., NOAA Tech. Memo.

Santos, M.B., Martin, V., Arbelo, M., Fernández, A., Pierce, G.J., 2007. Insights into the diet of beaked whales from the atypical mass stranding in the Canary Islands in September 2002. Journal of the Marine Biological Association of the United Kingdom. 87 (243). https://doi.org/10.1017/S0025315407054380.

Schofield, T.D., Stamper, A., Whitaker, B., Geraci, J., 1994. A young whale, the victim of ocean-borne plastic. In: Paper Posters and Manuscripts from the Third International Conference on Marine Debris, p. 35.

Schupp, M., Lazar, M.A., 2010. Endogenous ligands for nuclear receptors: digging deeper. Journal of Biological Chemistry. 285, 40409–40415. https://doi.org/10.1074/jbc.R110.182451.

Schwartz, M.L., Hohn, A.A., Bernard, H.J., Chivers, S.J., Peltier, K.M., 1992. Stomach Contens of Beach-Cast Cetaceans Collected along the San Diego County Coast of California 1972–1991. U.S. Department of Commerce, Southwest Fisheries Science Center.

Scott, M.D., 2001. A note on the release and tracking of a rehabilitated pygmy sperm whale (*Kogia breviceps*). Journal of Cetacean Research and Management 3, 87–94.

Secchi, E.R., Zarzur, S., 1999. Plastic debris ingested by a Blainville's beaked whale, *Mesoplodon densirostris*, washed ashore in Brazil. Aquatic Mammals 25, 21–24.

Secretariat of the Convention on Biological Diversity, 2016. Marine debris: understanding, preventing and mitigating the significant adverse impacts on marine and coastal biodiversity, technical series. In: Secretariat of the Convention on Biological Diversity, Montreal.

Setälä, O., Fleming-Lehtinen, V., Lehtiniemi, M., 2014. Ingestion and transfer of microplastics in the planktonic food web. Environmental Pollution. 185, 77–83. https://doi.org/10.1016/j.envpol.2013.10.013.

Shoham-Frider, E., Amiel, S., Roditi-Elasar, M., Kress, N., 2002. Risso's dolphin (*Grampus griseus*) stranding on the coast of Israel (eastern Mediterranean). Autopsy results and trace metal concentrations. The Science of the Total Environment. 295, 157–166. https://doi.org/10.1016/S0048-9697(02)00089-X.

Simmonds, M.P., 2012. Cetaceans and marine debris: the great unknown, cetaceans and marine debris: the great unknown. Journal of Marine Biology. 2012, e684279. https://doi.org/10.1155/2012/684279.

Simmonds, M.P., 2011. Eating plastic: a preliminary evaluation of the impact on cetaceans of ingestion of plastic debris. In: Submission to the IWC Scientific Committee, pp. 1–14.

Slooten, E., Wang, J., Dungan, S., Forney, K., Hung, S., Jefferson, T., Riehl, K., Rojas-Bracho, L., Ross, P., Wee, A., Winkler, R., Yang, S., Chen, C., 2013. Impacts of fisheries on the critically endangered humpback dolphin *Sousa chinensis* population in the eastern Taiwan Strait. Endangered Species Research. 22, 99–114. https://doi.org/10.3354/esr00518.

Smiddy, P., Murphy, S., Ingram, S., 2002. Fin whale *Balaenoptera physalus* (L.). Irish Naturalists' Journal 27, 169.

Souza, S.P., Siciliano, S., Cuenca, S., Sanctis, B., 2005. A True's beaked whale (*Mesoplodon mirus*) on the coast of Brazil: adding a new beaked whale species to the Western Tropical Atlantic and South America. Latin American Journal of Aquatic Mammals. 4. https://doi.org/10.5597/lajam00077.

Spence, L., 1994. The death of a whale: using a stranding incident for public education. In: Clary, J.C. (Ed.), Poster Abstracts and Manuscripts from the Third International Conference on Marine Debris, May 8–13, 1994, 4, Miami, Florida, p. 108 U.S. Dep. Commer., NOAA Tech. Memo. NMFSAFSC-51.

Stamper, M.A., Whitaker, B.R., Schofield, T.D., 2006. Case study: morbidity in a pygmy sperm whale *Kogia breviceps* due to ocean-bourne plastic. Marine Mammal Science. 22, 719–722. https://doi.org/10.1111/j.1748-7692.2006.00062.x.

Sternfeld, M., 2006. Draft 2005 Alaska Region Marine Mammal Stranding Summary. National Marine Fisheries Service, Alaska Region, Protected Resources, Juneau, Alaska.

Sternfeld, M., 2005. Draft 2004 Marine Mammal Strandings as Reported to National Marine Fisheries Service, Alaska Region. National Marine Fisheries Service, Alaska Region Protected Resources, Juneau, Alaska.

Sternfeld, M., 2004. Marine Mammal Strandings as Reported to National Marine Fisheries Service, Alaska Region: 2003. National Marine Fisheries Service, Alaska Region Protected Resources, Juneau, Alaska.

Takatori, S., Kitagawa, Y., Kitagawa, M., Nakazawa, H., Hori, S., 2004. Determination of di(2-ethylhexyl)phthalate and mono(2-ethylhexyl)phthalate in human serum using liquid chromatography-tandem mass spectrometry. Journal of Chromatography B. 804, 397–401. https://doi.org/10.1016/j.jchromb.2004.01.056.

Tanaka, K., Takada, H., Yamashita, R., Mizukawa, K., Fukuwaka, M., Watanuki, Y., 2015. Facilitated leaching of additive-derived PBDEs from plastic by seabirds' stomach oil and accumulation in tissues. Environmental Science and Technology. 49, 11799–11807. https://doi.org/10.1021/acs.est.5b01376.

Tarpley, R.J., 1990. Plastic ingestion in a pygmy sperm whale, *Kogia breviceps*. In: Shomura, R.S., Godfrey, M.L. (Eds.), Proceedings of the Second International Conference on Marine Debris. NMFS, Honolulu, Hawaii. U.S. Dep. Commer., NOAA Tech. Memo NNFS, NOM-TN-NHFS-SUFSC-154.

Tarpley, R.J., Marwitz, S., 1993. Plastic debris ingestion by cetaceans along the texas coast: two case reports. Aquatic Mammals 19, 93–98.

Tartu, S., Bourgeon, S., Aars, J., Andersen, M., Lone, K., Jenssen, B.M., Polder, A., Thiemann, G.W., Torget, V., Welker, J.M., Routti, H., 2017. Diet and metabolic state are the main factors determining concentrations of perfluoroalkyl substances in female polar bears from Svalbard. Environmental Pollution. 229, 146–158. https://doi.org/10.1016/j.envpol.2017.04.100.

Tonay, A.M., Dede, A., Öztürk, A.A., Öztürk, B., 2007. Stomach content of harbour porpoises (*Phocoena phocoena*) from the Turkish Western Black sea in spring and early summer. In: Rapport Commission Internationale Pour l'Exploration Scientifique de La Mer Méditerranée 38, p. 616.

Unger, B., Herr, H., Benke, H., Böhmert, M., Burkhardt-Holm, P., Dähne, M., Hillmann, M., Wolff-Schmidt, K., Wohlsein, P., Siebert, U., 2017. Marine debris in harbour porpoises and seals from German waters. Marine Environmental Research. 130, 77–84. https://doi.org/10.1016/j.marenvres.2017.07.009.

Unger, B., Rebolledo, E.L.B., Deaville, R., Gröne, A., IJsseldijk, L.L., Leopold, M.F., Siebert, U., Spitz, J., Wohlsein, P., Herr, H., 2016. Large amounts of marine debris found in sperm whales stranded along the North Sea coast in early 2016. Marine Pollution Bulletin. 112, 134–141. https://doi.org/10.1016/j.marpolbul.2016.08.027.

van der Hoop, J., Moore, M., Fahlman, A., Bocconcelli, A., George, C., Jackson, K., Miller, C., Morin, D., Pitchford, T., Rowles, T., Smith, J., Zoodsma, B., 2014. Behavioral impacts of disentanglement of a right whale under sedation and the energetic cost of entanglement. Marine Mammal Science. 30, 282–307. https://doi.org/10.1111/mms.12042.

van Franeker, J.A., Bravo Rebolledo, E.L., Hesse, E., IJsseldijk, L.L., Kühn, S., Leopold, M., Mielke, L., 2018. Plastic ingestion by harbour porpoises *Phocoena phocoena* in The Netherlands: establishing a standardised method. Ambio. https://doi.org/10.1007/s13280-017-1002-y.

Vitale, D., Varneau, N., Tison, Y., 1992. Stomach obstruction in a sperm whale beached on the Lavezzi islands: macropollution in the Mediterranean. Journal de Recherche Océanographique. Paris 16, 100–102.

Walker, W.A., Coe, J.M., 1990. Survey of marine debris ingestion by *Odontocete cetaceans*. In: Shomura, R.S., Godfrey, H.L. (Eds.), Proceedings of the Second International Conference on Marine Debris, 2–7 April 1989. Honolulu, Hawaii U.S. Dep. Comer., NOAA Tecli. Memo. NNFS. NOM-TH-NHFS-SWFSC-154. 1990.

Walker, W.A., Hanson, M.B., 1999. Biological observations on Stejneger's beaked whale, *Mesoplodon stejnegeri*, from strandings on Adak Island, Alaska. Marine Mammal Science. 15, 1314–1329. https://doi.org/10.1111/j.1748-7692.1999.tb00893.x.

Waring, G., Pace, R., Quintal, J., Fairfield, C., Maze-Foley, K., 2004. US Atlantic and Gulf of Mexico marine mammal stock assessments 2003. In: NOAA Technical Memorandum NMFS-NE, pp. 182–287.

Woodall, L.C., Sanchez-Vidal, A., Canals, M., Paterson, G.L.J., Coppock, R., Sleight, V., Calafat, A., Rogers, A.D., Narayanaswamy, B.E., Thompson, R.C., 2014. The deep sea is a major sink for microplastic debris. Royal Society Open Science. 1 (140317). https://doi.org/10.1098/rsos.140317.

World Economic Forum, Ellen MacArthur Foundation, McKinsey & Company, 2016. The New Plastics Economy – Rethinking the Future of Plastics.

Yamada, T.K., Yatabe, A., Pitman, R.L., Brownell Jr., R.L., 2012a. Review of current knowledge on *Indopacetus pacificus* including identification of knowledge gaps and suggestions for future research. In: 64th Meeting of the International Whaling Commission. Panama City, Panama. SC/64/SM26.

Yamada, T.K., Yatabe, A., Pitman, R.L., Brownell Jr., R.L., 2012b. Review of current knowledge on Hubbs' beaked whale, *Mesoplodon carlhubbsi*, from the Seas around Japan and data from the North America. In: 64th Meeting of the International Whaling Commission. Panama City, Panama. SC/64/SM27.

Chapter 7

Persistent Organic Pollutants in Cetaceans Living in a Hotspot Area: The Mediterranean Sea

Letizia Marsili[1], Begoña Jiménez[2], Asunción Borrell[3]
[1]University of Siena, Siena, Italy; [2]Institute of Organic Chemistry (IQOG-CSIC), Madrid, Spain; [3]University of Barcelona, Barcelona, Spain

INTRODUCTION

The Mediterranean Sea is the largest semienclosed European sea, with a single major opening, represented by the Strait of Gibraltar. The Mediterranean Sea is situated between Europe to the north, Africa to the south, and Asia to the east. The Siculo-Tunisian separates two distinct zones, i.e., the western and the eastern basins, and acts as a geographic and hydrologic frontier. The eastern basin is connected to the Red Sea and the Indian Ocean through the artificial Suez Canal in the south, and to the Black Sea through the Sea of Marmara and the Dardanelles Strait in the north (Fig. 7.1A).

It is estimated that water takes over one century to be completely renewed through the Strait of Gibraltar. As evaporation exceeds precipitation and river runoff, the water of the Mediterranean Sea is saltier than the water in other European seas. In fact, the salinity in the basin is 2%–3% higher than the salinity of open ocean waters. This sea is one of the most oligotrophic marine systems in the world, i.e., it is rich in oxygen and poor in nutrients. Oligotrophy increases from west to east. However, some areas, such as the Corso-Ligurian Basin and the Gulf of Lion, are characterized by high levels of primary productivity related to the upwelling of nutrients.

The Mediterranean coasts support a high density of inhabitants; more than 507 million people live in the region, and it is expected that the population of the coastal states of the Mediterranean will rise to 700 million by the end of the 21st century (Blue Plan, 2008). Furthermore, this basin is economically important: fishing and oil and gas extraction are among the main activities, but it is also used for commercial shipping, recreational boating, and tourism (UNEP, 1996; Shaltout and Omstedt, 2014). Approximately 30% of all international

Marine Mammal Ecotoxicology. https://doi.org/10.1016/B978-0-12-812144-3.00007-3

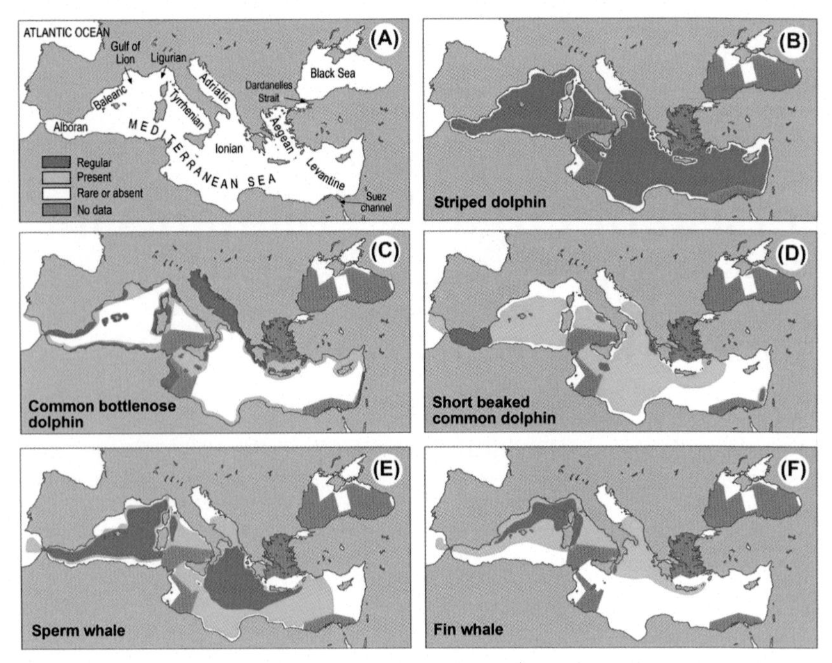

FIGURE 7.1 Range of distribution of the five most common cetacean species (B) striped dolphin; (C) common bottlenose dolphin; (D) short beaked common dolphin; (E) sperm whale; (F) fin whale) in the Mediterranean Sea (A). *(Figure adapted from Cagnolaro, L., Cozzi, B., Notarbartolo di Sciara, G., Podestà, M., 2015. In: Calderini (Ed.), Fauna d'Italia Vol. XLIX - Mammalia IV – Cetacea, pp. 3–375.)*

maritime traffic passes through its waters, meaning that over 200,000 vessels cross the Mediterranean Sea annually (Vaes and Druon, 2013). These characteristics, combined with an increase in the coastal industrial economy, have a great impact on its waters, and sewage and industrial and incidental discharges are an important source of persistent organic pollutants (POPs).

The Mediterranean Sea is a marine biodiversity hot spot. Coll et al. (2010) listed approximately 17,000 marine species; additionally, undescribed species will be added in the future. Furthermore, the invasion of alien species, most of them entering through the Suez Channel, is a crucial factor that will continue to change the biodiversity of the Mediterranean. In these waters, the other principal threats for marine life are overexploitation, habitat loss, and degradation, which are mostly due to human activities, such as fisheries, ship traffic, water pollution, and coastal anthropization.

A high percentage of the Mediterranean marine species are endemic (Bianchi and Morri, 2000). There are several unique and endangered habitats, including the endemic *Posidonia oceanica* and coralliferous assemblages (Green and Short, 2003; Ballesteros, 2006), as well as a set of emblematic species of conservation concern, such as sea turtles, several cetaceans, and the critically endangered Mediterranean monk seal (*Monachus monachus*).

Regarding the cetaceans, a dozen species habitually inhabit the basin, but only eight are consistently found in the basin when we exclude the Black Sea. The most common species are the striped dolphin (*Stenella coeruleoalba*), common bottlenose dolphin (*Tursiops truncatus*), short-beaked common dolphin (*Delphinus delphis*), sperm whale (*Physeter macrocephalus*), and fin whale (*Balaenoptera physalus*). Thus, these five species were used as case studies in this chapter. Their distributional ranges are shown in Fig. 7.1B–F. Less abundant species include the Cuvier's beaked whale (*Ziphius cavirostris*), long-finned pilot whale (*Globicephala melas*), and Risso's dolphin (*Grampus griseus*). The killer whale (*Orcinus orca*) and rough-toothed dolphin (*Steno bredanensis*) are considered local species only in small areas of the Mediterranean Sea or in some areas linked to the Mediterranean (Cagnolaro et al., 2015). All of these species have been included in the IUCN Red List of Threatened Species (2017). Chemical (and noise) pollution, incidental mortality caused by fishing gear, overfishing, and habitat degradation are the main threats affecting cetacean populations in the Mediterranean Sea.

Among the chemical pollutants, POPs are the main group of concern due to their highly toxic properties, persistence, bioaccumulation, and long-range transport potential (Jones and de Voogt, 1999). Currently, a total of 23 chemicals or groups of halogenated compounds, comprising pesticides, industrial chemicals, and unintentional products are included on the list of POPs regulated under the Stockholm Convention (UNEP, 2009).

Given that POPs are bioaccumulative and magnify through food chains, mobile—though resident—top predators, such as dolphins, have been proposed as potential indicator species (Aguilar et al., 2002). Moreover, they feed at high trophic levels, have a small body size—and thus, a high metabolic rate—and their body contains a large amount of fat that is capable of retaining lipophilic compounds. These three biologic traits synergistically place cetaceans in an ecologic situation where they are more likely to be adversely affected by POPs (Aguilar and Borrell, 1994a). Furthermore, their lower capacity for degrading these chemicals (Tanabe et al., 1988; Watanabe et al., 1989) means these compounds accumulate more readily than in other mammals or birds with comparable biologic traits; as a result, the toxic effects of the compounds are exacerbated (Aguilar and Borrell, 2005).

In this chapter, the threats, time trends, and geographic variations of POPs are reviewed specifically for cetaceans in the Mediterranean Sea. Most of the available literature addresses legacy organochlorine contaminants (OCs), including polychlorinated biphenyls (PCBs), dichlorodiphenyltrichloroethanes (DDTs), hexachlorocyclohexanes (HCHs), and hexachlorobenzene (HCB), most of which were banned in the late 1970s in many countries. The more recent polybrominated diphenyl ethers (PBDEs) were highly used in a wide variety of industrial applications until they started being regulated under the

Stockholm Convention (UNEP, 2009). To date, the group of compounds imposing more emerging concern are the per- and polyfluoroalkylated substances (PFASs). Some of these compounds (i.e., PFOS and its salts) are already listed under Annex B of restricted substances under the UNEP Stockholm Convention on POPs, and others are candidate substances (i.e., PFOA, its salts, and related compounds). The main families of chlorinated, brominated, and fluorinated POPs are reviewed for each cetacean species within the Mediterranean basin.

CASE STUDIES

Striped Dolphin

Currently, the striped dolphin is the most abundant cetacean, and it is present throughout the Mediterranean Sea, both in the eastern and western basins (Fig. 7.1B). However, it is not found at uniform densities. The striped dolphin typically shows a preference for highly productive, open waters beyond the continental shelf (Aguilar, 2000). The Mediterranean striped dolphin is opportunistic, feeding on a wide variety of pelagic and bathypelagic oceanic organisms that form large and dense aggregates in the water column (Aguilar and Gaspari, 2012).

Due to its broad distribution (Fig. 7.1B) and large population size, the striped dolphin has traditionally been the most studied odontocete in terms of its role as a pollutant bioindicator in the Mediterranean. Thus, a series of studies, proportionally distributed along time, have emerged about POP contamination in Mediterranean striped dolphins, but all these studies focused on the western basin and specifically reported levels and effects in the French, Spanish, and Italian subpopulations. To our knowledge, such studies have not been conducted in the eastern basin. In these studies, blubber has mainly been selected as the most representative tissue for monitoring organochlorine levels, though other tissues have also been analyzed. This fact facilitates comparisons between surveys and areas of available data in the literature. However, available data are spread over a wide region and have had distinct sampling procedures for more than 35 years. Moreover, analytical techniques have changed considerably during this period and have differed in accuracy between laboratories; additionally, the results are reported in different units, i.e., lipid weight (lw), wet weight (ww), or dry weight (dw). This wide geographic range and variation in analytical techniques adds uncertainty to understanding the variation in pollutant loads between localities and years.

Organochlorine Compounds

Temporal and Spatial Trends

During the 1990s, information regarding POPs in striped dolphins was mainly focused on DDT and PCB concentrations in different regions. The first study that reported levels of organochlorines in the Mediterranean striped dolphin was conducted by Alzieu and Duguy (1979), who analyzed PCBs and DDTs in 80 stranded odontocetes on the Atlantic and Mediterranean French coasts. They found extremely high levels of both PCBs (172–260 mg/kg dw) and dichloro-diphenyldichloroethylene (DDE) (347–455 mg/kg dw) in the Mediterranean striped dolphins. The concentrations in blubber were between 5 and 10 times higher in individuals from the Mediterranean than in those from the Atlantic. Since then, and as a result of several subsequent studies, it has been known that the striped dolphin population in the Mediterranean waters has been exposed to high levels of OC pollutants, and consequently, the OC levels in their tissues are extremely high.

After this study, the concentrations of both DDTs and PCBs in stranded dolphins along the French coast decreased consistently; the mean concentrations in blubber in 2002–03 ($n=3$) were as follows: PCBs = 70 and DDTs = 24 mg/kg lw (Wafo et al., 2005), and those in 2007–09 ($n=37$) were as follows: PCBs = 57 and DDTs = 16 mg/kg lw (Wafo et al., 2012a). Moreover, in a further study, other groups of pesticides were detected (in order of abundance): dieldrin, endrin, heptachlor-epoxide, endosulfan, lindane, aldrin, and heptachlor. Compared to DDT, these pesticides showed very low concentration: all were below 0.3 mg/kg lw (Wafo et al., 2012b).

Since 1987, 8 years after the first French publication, striped dolphins from the Mediterranean waters off central and northwestern Spain began to be controlled by means of free-ranging dolphin blubber biopsies collected by butterfly valve darts (Aguilar and Nadal, 1984; Aguilar and Borrell, 2005). Between 1987 and 1989, PCB levels in blubber ($n=87$) were typically between 100 and 500 (median = 282), and DDTs were between 50 and 250 (median = 120) mg/kg lw. After 1991, concentrations of both groups of compounds slowly decreased due to the ban on the use of DDT and PCB compounds that was implemented around the end of the 1970s and early 1980s. However, the decline in PCBs (i.e., from 342 mg/kg lw in 1987 to 76 mg/kg lw in 2002) was steeper than that of DDTs (i.e., from 198 mg/kg lw in 1987 to 55 mg/kg lw in 2002) (Aguilar and Borrell, 2005). PCB and DDT levels seemed to stabilize from 2003 to 2009, though concentrations still consistently exceeded all mammalian toxicity thresholds (Castrillon et al., 2010; Jepson et al., 2016). In fact, the concentrations during the 2000s were similar to those found in France (Wafo et al., 2005, 2012a,b).

Dolphins from the Alboran Sea showed PCB concentrations that were half of the concentrations in dolphins from the northwestern Mediterranean; however, both groups had similar DDT concentrations. This indicates that countries surrounding the Alboran Sea may be more agricultural than industrialized, and the data agree with those of Dachs et al. (1996), who found that concentrations and fluxes of these compounds in sinking particles collected in the Alboran Sea were characteristic of a rather pristine environment due to the relative remoteness of this region; specifically, the area was far from pollution sources, such as river discharges, and was rather dissimilar from the northwestern Mediterranean. Despite this, a declining trend was also observed during the period between

1992 and 1994 ($n=29$) and 2007 ($n=15$). Total PCBs decreased by approximately 50%, from 68 to 35 mg/kg lw, and DDT decreased by approximately 67%, from 79 to 25 mg/kg lw (Aguilar and Borrell, 2005; Fossi et al., 2013).

Meanwhile, the Italian free-ranging subpopulation was also being monitored through biopsy sampling (Marsili and Focardi, 1996; Marsili, 2000; Fossi et al., 2004, 2013). From 1991 to 1993, 89 blubber biopsies were collected from the Ligurian, Tyrrhenian, and Ionian seas, and they had a mean PCB concentration of 60 mg/kg dw (range = 15.5–86) and a mean DDT concentration of 40 mg/kg dw (range: 15.6–63.5) (Marsili and Focardi, 1996). Moreover, the striped dolphins biopsied in 1993 in the Ligurian Sea showed higher concentrations (i.e., $n=24$, PCBs = 50, and DDTs = 30 mg/kg dw) than those from the Tyrrhenian (i.e., $n=22$; PCBs = 40, and DDTs = 20 mg/kg dw) and Ionian seas (i.e., $n=6$, PCBs = 20, and DDTs = 20 mg/kg dw); this result was related to the greater industrialized area and higher abundance of intensive farming and horticulture found in Liguria (Marsili and Focardi, 1996; Marsili, 2000).

Marsili et al. (2011) reported a slight decrease in DDT and PCB levels in both stranded and free-ranging dolphins from 1996 to 2003, followed by an increase since 2005. Accordingly, available data about biopsies sampled during this period show a decrease in the different Italian seas. Thus, during the period between 1993 ($n=22$) and 2002 ($n=10$), PCBs in Tyrrhenian dolphins decreased approximately 65%, from 40 to 14 mg/kg dw, and DDTs decreased approximately 50%, from 20 to 10 mg/kg dw (Marsili, 2000; Fossi et al., 2004). However, concentrations in dolphins from the Ligurian Sea seemed to have decreased less; during the period between 1993 ($n=24$) and 2007 ($n=20$), both PCBs and DDTs decreased by approximately 30%, from 70 to 50 mg/kg lw and from 44 to 30 mg/kg lw, respectively (Marsili, 2000; Fossi et al., 2013). Additionally, the concentrations in dolphins from the Ionian Sea seemed to have remained stable during the period between 1993 ($n=6$) and 2007 ($n=12$); PCBs ranged from 27 to 30 mg/kg lw, and DDTs ranged from 27 to 23 mg/kg lw (Marsili, 2000; Fossi et al., 2013). Storelli et al. (2012) reported a mean PCB level of 55 mg/kg lw in 17 stranded striped dolphins collected in different coastal areas of the Adriatic Sea during the period between 1999 and 2004.

In summary, data from striped dolphins showed that during the 1990s, concentrations of DDTs and PCBs in the Balearic Sea and the Gulf of Lion were even higher than those in the Italian and Alboran Seas. Within the Italian waters, Liguria was the most polluted region. However, for these compounds, a decreasing contamination trend was found from the northwestern basin to the eastern and southern areas. These spatial data, as a whole, provide evidence that the Ebro and Rhone rivers that flow into the northwestern basin might have been the primary vector for contamination by POPs. This is consistent with findings in other organisms, which showed that POP levels in the western Mediterranean were exceptionally high, particularly in the northwestern basin, as a consequence of its proximity to the highly populated and industrialized European coasts (Fowler, 1987). However, in later years—probably due the regulation of DDTs and PCBs for extended periods of time (>30–40 years) in many countries—the concentrations appeared to have homogenized throughout the entire area, to the point that a definitive spatial trend is no longer apparent.

Epizootics

Starting in the summer of 1990, an epizootic that was identified as an infection produced by a morbillivirus (Domingo et al., 1990) caused a massive striped dolphin die-off in the Mediterranean Sea (Aguilar and Raga, 1993). In 1991 and 1992, secondary outbursts affected the coasts of southern Italy, Sicily (Bortolotto et al., 1992), the Greek Islands (Cebrian, 1995), and northern Turkey. Because this species inhabits offshore waters, only a small proportion of the affected dolphins were stranded, but it is speculated that the total population declined to just one-third of its initial level (Aguilar and Raga, 1993).

The levels of OC pollutants found in the dolphins examined during the epizootic along the Spanish coasts were found to be between two- and three-fold higher than the levels commonly found in the healthy population ($n=72$, median in blubber; PCBs = 778 and DDTs = 339 mg/kg lw) (Kannan et al., 1993; Aguilar and Borrell, 1994b). It is necessary to emphasize the role of non-ortho substituted congeners in exerting toxic impact because the observed levels were among the highest values reported in the literature (Kannan et al., 1993; Borrell et al., 1996). Moreover, the high concentrations of DDTs that were also present in these animals might have contributed to additive toxic effects. Based on these high levels and on the knowledge that organochlorine compounds depress the immune system of mammals (Loose et al., 1977; De Swart et al., 1996; Busbee et al., 1999), it was suggested that PCBs caused higher susceptibility to the viral disease and played an important role in the development of the epizootic (Kannan et al., 1993; Aguilar and Borrell, 1994b).

Similarly, stranded dolphins along the Italian coasts (Tyrrhenian, Adriatic and Ionian) between 1988 and 1994 were analyzed for HCB, DDTs and PCBs (Marsili et al., 1997). The PCB and DDT levels in dolphins stranded in 1990 and 1991 ($n=12$ median in blubber; PCBs = 165; DDTs = 92 mg/kg lw) were several times higher than the levels in dolphins stranded in previous ($n=4$, median in blubber PCBs = 25; DDTs = 9 mg/kg lw) and posterior years ($n=8$, median in blubber PCBs = 88; DDTs = 36 mg/kg lw). Moreover, in 1990, stranded dolphins in Liguria showed higher levels ($n=16$, median PCBs = 241; DDTs = 162 mg/kg lw) than those in the other three Italian seas (Marsili and Focardi, 1997), confirming that the Ligurian Sea was the most polluted sea in Italy.

These high OC concentrations found in diseased dolphins confirmed the results obtained in Spain from dolphins that died during the epizootic (Marsili et al., 1997). However, the contribution to toxic equivalents (TEQs) of non-ortho substituted congeners only accounted for <1% of the total PCB TEQs in deceased dolphins ($n=6$) (Reich et al., 1999); in contrast, in Spain, after calculations were corrected using the toxic equivalent factors proposed by Ahlborg et al. (1994), non-ortho congeners contributed between 14% and 32%

(Kannan et al., 1993; Borrell et al., 1996). On the other hand, the livers of striped dolphins, among other cetaceans, stranded in 1990 along the coast of the Tyrrhenian ($n=3$) and Ligurian ($n=2$) seas were analyzed for polychlorinated dibenzo-p-dioxins (PCDDs) and polychlorinated dibenzofurans (PCDFs) (Jiménez et al., 2000), to ascertain if PCDDs and PCDFs influenced the 1990–92 mortality event. Additionally, 2,3,7,8-substituted PCDDs and PCDFs were found in all analyzed samples (mean concentration PCDD/Fs=79 ng/kg ww). However, the contribution of PCDDs and PCDFs to overall TEQs was always lower than the contributions of mono- and non-ortho-PCBs for such dolphins (Reich et al., 1999; Jiménez et al., 2000), discarding these compounds as factors contributing to the mortality event (Jiménez et al., 2000). Years later, the blubber of biopsied dolphins that was collected in 2002 in the south Tyrrhenian Sea showed total PCDD/F levels of 73.91 ng/kg ww (Fossi et al., 2004). However, making comparisons between the data sets is impracticable, as the results were reported as wet weight values for different tissues, and these compounds are highly lipophilic. In these samples, coplanar PCBs were found at higher levels, with a mean value of 372 ng/kg ww, than PCDD/Fs.

Overall, the levels of both PCBs and DDTs detected in the stranded dolphins in the Italian seas in the 1990s were approximately 3–4 times lower than the values detected in dolphins in the Catalonian-Balearic basin and the French area; however, they were more similar to the values detected in the Alboran Sea.

In 2007 and 2008, 17 years after the 1990 epizootic, dozens of dead striped dolphins suddenly washed up on Mediterranean beaches, displaying the same symptoms observed during the epizootic in 1990; researchers believed this indicated a new bud of the virus (Raga, 2008). However, the levels of PCBs and DDTs in the blubber of dolphins found stranded in the Balearic Sea ($n=18$, PCBs=57 mg/kg lw; DDTs=63 mg/kg lw) were similar to the samples collected from dolphins stranded previously and afterward (Castrillon et al., 2010), indicating that this new epizootic event was unlikely to be enhanced by PCB concentrations, which, despite being high, did not seem to have affected the immune systems of the dolphins. In fact, this second outbreak was much lighter than the outbreak of 1990, as fewer dolphins died, at least in Catalonian waters (Castrillon et al., 2010).

Brominated Compounds

PBDEs and methoxylated PBDEs (MeO-PBDEs) were analyzed in the livers of five striped dolphins, which among other odontocete species, were stranded in 1990 on the Tyrrhenian ($n=3$) and Ligurian ($n=2$) coasts. The striped dolphins showed the highest concentrations (means: total PBDEs=3625 μg/kg lw; MeO-PBDEs=104 μg/kg lw) compared with the other odontocetes (range total PBDEs=85–2917 μg/kg lw) (Pettersson et al., 2004). The Mediterranean dolphins, despite showing some slightly higher concentrations, displayed the same PBDE congener pattern displayed by cetaceans from the Atlantic Ocean and the North Sea. After this first study, at least two more modern studies on PBDEs and MeO-PBDEs appeared, reporting levels in the 2000s.

The PBDE levels detected in blubber biopsies collected in 2007 from free-ranging striped dolphins were compared among three locations: the Ligurian Sea ($n=20$; PBDEs = 150 μg/kg lw), the western Ionian Sea ($n=12$; PBDEs = 165 μg/kg lw), and the Strait of Gibraltar ($n=15$; PBDEs = 75 μg/kg lw). The PBDE concentrations were higher in the two Italian subpopulations, whereas the Gibraltar samples had lower average concentrations (Fossi et al., 2013). These dolphins had concentrations that were 20 times lower than the concentrations that were detected in dolphins that died during the epizootic events (Pettersson et al., 2004).

The last study that reported PBDEs and MeO-PBDEs in Mediterranean striped dolphins is very recent (Barón et al., 2015a). Blubber was analyzed from 11 stranded dolphins, which were sampled on the Alboran Sea coastline between 2004 and 2011. The following concentrations were detected: PBDEs = 940 μg/kg lw and MeO-PBDE = 870 μg/kg lw. The PBDEs were one order of magnitude higher than the PBDE concentrations in the biopsied dolphins from the Strait of Gibraltar. In this study, other halogenated flame retardants were analyzed, such as hexabromobenzene (HBB), dechlorane 602 (Dec 602), dechlorane 603 (Dec 603), and dechlorane plus (DP), but the detected concentrations were all very low. Additionally, the brain was analyzed, and both halogenated natural products (HNPs) and halogenated flame retardants (HFRs) were detected in brain samples, implying that these compounds are able to surpass the blood-brain barrier and reach the brain. Moreover, some compounds (i.e., 2,5,7-tribromo-4a-bromomethyl-1,1-dimethyl-2,3,4,4a,9,9a-hexahydro-1H-xanthene (TetraBHD), BDE-153, and HBB) presented higher levels in the brain than in the blubber.

Fluorinated Compounds

PFOS was measured in the livers of four striped dolphins (stranded in 1991 along the Tyrrhenian and south Adriatic coasts) by Kannan et al. (2002), who found low levels (26 μg/kg ww).

Common Bottlenose Dolphin

The common bottlenose dolphin is the most common cetacean on the continental shelf of the Mediterranean Sea (Fig. 7.1C) (Notarbartolo di Sciara, 2016), and it is often aggregated in small nuclei and resident in only relatively small areas (Gonzalvo et al., 2016; Bearzi et al., 2012). Each of these nuclei should be considered particularly fragile due to their isolation. In this sense, it is feared that genetic deterioration could have strong negative consequences for the maintenance of the species, as the population is suspected to have declined at least 30% over the last 60 years (Bearzi et al., 2012). As a result, in the Mediterranean Sea, the common bottlenose dolphin is classified as vulnerable on the IUCN Red List of Threatened Species criteria A2cde. The opportunistic common bottlenose dolphin feeds on most available preys, and their diet is mainly composed of demersal fish and cephalopods (Bearzi et al., 2008; Blanco et al., 2001).

As this dolphin is at the top of the trophic chain and lives very close to the coast, it is particularly vulnerable to anthropic contamination. Thus, it is an excellent bioindicator of environmental quality, especially in coastal areas closely linked with land-based activities. Several studies have been carried out on the presence and accumulation of various legacy and emerging POPs in common bottlenose dolphins around the world (e.g., Table 7.1). Regarding the Mediterranean Sea, most of the studies that exist on the contamination of POPs in this species address chlorinated POPs of both agricultural and industrial origin.

Organochlorine Compounds

Many studies concerning organochlorine levels in different tissues of stranded specimens or in subcutaneous blubber obtained using biopsy techniques on free-ranging specimens have been published. The first study that reported levels of PCBs in the Mediterranean common bottlenose dolphin was conducted by Alzieu and Duguy (1979), who analyzed these xenobiotics in the blubber of a stranded specimen along the French coast. The PCBs were present at extremely high levels, i.e., 321 mg/kg dw. Similar PCB levels were found by Corsolini et al. (1995a) in the blubber of one stranded specimen on Spain's Mediterranean coast (PCBs = 234 mg/kg ww and DDTs = 24 mg/kg ww). A review by Jepson et al. (2016) on PCBs in blubber of bottlenose dolphins from the western Mediterranean Sea (i.e., Spain's coastal areas) reported PCB concentrations from stranded/biopsied individuals sampled in 1990–2012; males ($n = 8$) showed mean levels of 182.70 mg/kg lw, and females ($n = 11$) showed mean levels of 192.23 mg/kg lw.

In the Italian seas (Tyrrhenian, Adriatic, and Ionian seas), Marsili and Focardi (1997) and Marsili (2000) reported mean levels of HCB, DDTs, and PCBs found in different tissues (melon [$n = 2$]; blubber [$n = 8$]; liver [$n = 13$]; muscle [$n = 12$]; brain [$n = 2$]; kidney [$n = 3$]; heart [$n = 4$]; and milk [$n = 1$]) of 14 common bottlenose dolphins, which were stranded between 1987 and 1992. HCB showed the lowest concentrations (range: 0.108 mg/kg lw in heart to 1.433 mg/kg lw in liver). Differences in xenobiotic levels were not found between the different seas or in the three tissues most frequently analyzed (i.e., liver, muscle, and blubber). In the blubber, the levels of HCB, DDTs, and PCBs were 0.473 mg/kg lw, 11.59 mg/kg lw, and 41.65 mg/kg lw, respectively.

Borrell and Aguilar (2007) reported the trend of DDT and PCB levels in the blubber of 36 specimens stranded in the northwestern Mediterranean Sea (Catalonia, Valencia, and Balearic Islands) between 1978 and 2002. The mean concentration of OCs (in blubber) measured between 1978 and 2002 showed a declining trend for both DDTs and PCBs. In fact, the DDT levels were 303 in 1978, 194 in 1987, and 13 mg/kg lw in 2002; PCB levels were 708 in 1978, 486 in 1987, and 116 mg/kg lw in 2002. In one specimen collected in 2001 on the French coast, only the liver, kidney, and muscle were analyzed (Wafo et al., 2005). The muscle had the highest level of PCBs (1.38 mg/kg dw), while the kidney showed the highest level of DDTs (0.34 mg/kg dw). In nine

TABLE 7.1 Comparison of Blubber PCB Levels in Common Bottlenose Dolphin in the Mediterranean and in Other Areas of the World

Ocean/ Sea	Area	Year of Collection	*n*	Sex	PCBs (mg/kg lw)			References
					Mean	Max.	Min.	
Northern Hemisphere								
Mediter-ranean	Ligurian sea	2007–09	9	9U	367.90			Lauriano et al. (2014)
	North Adriatic Sea	2011	6	5M 1F	110.46	293.23	28.64	Jepson et al. (2016)
	North Adriatic Sea	2000–05	13	5F 6M 2U	97.00	494.00	2.00	Romanić et al. (2014)
	Balearic Sea	2000–02	9	2M 2F 5U	149.84	271.60	63.10	Borrell and Aguilar (2007)
	Gulf of Ambracia	2013	14	9M 5F	26.77	108.70	3.00	Gonzalvo et al. (2016)
NE Atlantic	W England and Wales	2004–07	10	6M 4F	93.86	446.55	4.11	Jepson et al. (2016)
	Scotland	2004–12	8	5M 3F	71.32	125.08	8.53	Jepson et al. (2016)
	W Ireland	2000	8	6M 2F	23.90	59.60	8.14	Berrow et al. (2002)
	Portugal/Galicia	2004–11	16	8M 8F	85.67	382.20	5.08	Jepson et al. (2016)
	Gulf of Cadiz		23	13M 10F	203.67	446.22	3.69	Jepson et al. (2016)
	Canary Islands	2001–05	6	4M 2F	36.54	136.68	1.07	Carballo et al. (2008)

Continued

TABLE 7.1 Comparison of Blubber PCB Levels in Common Bottlenose Dolphin in the Mediterranean and in Other Areas of the World—cont'd

Ocean/ Sea	Area	Year of Collection	*n*	Sex	PCBs (mg/kg lw)			References
					Mean	Max.	Min.	
NW Atlantic	Florida	2000–07	77	77M	159.72	658.00	16.00	Kucklick et al. (2011)
	Gulf of Mexico	2000–07	151	151M	68.94	109.00	24.00	Kucklick et al. (2011)
	Bermudas	2000–07	3	3M	38.80	86.10	17.00	Kucklick et al. (2011)
	Gulf of Mexico, Texas	2010–12	108	108U	47.70			Balmer et al. (2015)
Mid Pacific	Hawaii	2009–11	3	1M 1F 1U	11.80	20.30	7.49	Bachman et al. (2014)
W Pacific	Taiwan	2000–01	6	4F 2M	5.36	10.41	2.15	Chou et al. (2004)
	China	2005	1	1U	4.90			Huang et al. (2007)
Southern Hemisphere								
SW Atlantic	Rio de Janeiro	2000–05	2	2M	11.80	10.10	13.50	Lailson-Brito et al. (2012)
SW Indian	Tanzania	2000–2002	18	10M 8F	0.00	0.00	0.00	Mwevura et al. (2010)
	Reunion Island	2010–11	32	32U	5.20	67.50	0.10	Dirtu et al. (2016)
SE Indian	Swan River, Australia	2009	5	3F 2M	22.34	53.94	10.26	Holyoake et al. (2010)
	Bunbury Australia	2008–09	4	2M 2F	8.46	25.33	2.37	Holyoake et al. (2011)
SW Pacific	Victoria, Australia	2004–08	10	7M 3F	5.46	0.37	11.51	Monk et al. (2014)

U, unknown.

specimens stranded along the northwestern Italian coastline between 2007 and 2009, Lauriano et al. (2014) reported mean blubber PCB concentrations of 367.9 mg/kg lw and mean DDT concentrations of 143.7 mg/kg lw

Moving to the eastern part of the basin, Corsolini et al. (1995b) reported PCB and DDT levels for seven specimens (five males and two females) stranded along the coast of the Adriatic Sea in 1992. The mean PCB level was 1000 mg/kg lw (males = 1024; females = 590), and the mean DDT level was 330 mg/kg lw (males = 399; females = 140).

In 13 common bottlenose dolphins stranded between 2000 and 2005 in the northern part of the Adriatic Sea (Croatian coast), the DDT and PCB levels were analyzed in several tissues (i.e., blubber, kidney, lung, muscle, liver, and heart). The PCBs were found at higher concentrations than DDTs in all of the analyzed tissues. In the blubber, the mean PCB concentration was 22.05, and the mean DDT concentration was 11.31 mg/kg ww (Romanić et al., 2014). On the Apulian coasts (South Adriatic Sea), Storelli and Marcotrigiano (2000) analyzed the brain, heart, lung, liver, intestine, and kidney, but not the blubber, of three bottlenose dolphins stranded in 1996–97. Therefore, these data could not be used for a bibliographic comparison. The liver showed the highest levels, with a range between 5.8 and 130.3 mg/kg lw. The same authors published two studies on the accumulation of PCBs in bottlenose dolphins stranded in the same area during the 1999–2000 ($n=9$) and 2001–02 ($n=12$) periods (Storelli and Marcotrigiano, 2003; Storelli et al., 2007). The results regarding concentrations in the blubber were only reported in the first paper, which included data on nine specimens and reported a mean PCB level of 32.7 mg/kg lw (Storelli and Marcotrigiano, 2003). In the 12 specimens stranded in 2001–02, the PCB concentrations in the liver (44.73 mg/kg lw) and in the kidney (39.38 mg/kg lw) were comparable, but they were significantly higher than those observed in the muscle and lung, which presented similar levels to one another (19.86 and 16.10 mg/kg lw, respectively). DDTs and PCBs were measured in tissues of six common bottlenose dolphins collected along Israel's Mediterranean coast during 2004–06. The DDT and PCB concentrations were highest in the blubber, with a wide concentration range of 0.92–142 and 4.7–7.9 mg/kg ww, respectively (Shoham-Frider et al., 2009). These unique representatives of the eastern coast of the Mediterranean show that the eastern basin is much less contaminated by PCBs than the western basin, probably due to lower inputs of industrial waste.

Few studies have been conducted in the Mediterranean Sea that uniquely focused on free-ranging bottlenose dolphins. In the summers between 1994 and 1998, subcutaneous blubber samples were obtained from seven bottlenose dolphins in the Ionian Sea (Fossi et al., 2003, 2007). Concentrations of HCB, DDTs, and PCBs in these dolphins were 0.1, 17, and 19 mg/kg ww, respectively. Fossi et al. (2000) reported the DDT and PCB levels for one specimen of common bottlenose dolphin, which was sampled in the Ionian Sea. The DDT concentration was approximately 8 mg/kg dw, while the PCB concentration was approximately 18 mg/kg dw.

Jepson et al. (2016) reported mean PCB levels in the subcutaneous blubber of six common bottlenose dolphins sampled using biopsy techniques in Slovenia; the PCB value was 126.83 mg/kg lw in males and 28.64 mg/kg lw in females. More recently, Gonzalvo et al. (2016) reported OC levels from common bottlenose dolphins present in the semiclosed waters of the Gulf of Ambracia in western Greece. In 2013, blubber samples were collected from 14 samples using a biopsy technique. The mean blubber concentrations of HCB, DDTs, and PCBs (in mg/kg lw with the SD in brackets) were 0.041 (0.020), 62.85 (99.92), and 26.77 (28.27), respectively. The DDT values were the highest for all the analyzed animals, with the exception of two dolphins.

In Table 7.1, a comparison of PCB levels in the Mediterranean and in other areas of this widely distributed species is shown. Only specimens collected and analyzed after the year 2000 are displayed; this was done to avoid differences in concentrations caused by the collection year. The levels shown are based on blubber lipid. In this table, we can appreciate the high levels found in the Mediterranean specimens compared to the levels found in specimens in other parts of the world. The Northern Hemisphere is much more contaminated than the Southern Hemisphere, and the Atlantic Ocean is the most contaminated ocean, including the Mediterranean Sea.

Brominated Compounds

There are only three studies about brominated POPs in resident common bottlenose dolphins in the Mediterranean Sea (Pettersson et al., 2004; Barón et al., 2015a,b). Pettersson et al. (2004) collected liver samples from four specimens stranded along the Tyrrhenian ($n=1$) and Adriatic ($n=3$) coasts between 1990 and 1992 and found concentrations of PBDEs from 66 to 518 μg/kg lw and concentrations of MeO-PBDEs from <15 to 97 μg/kg lw; in fact, these were the lowest levels found among the cetacean species analyzed in the study (including Risso's dolphin, fin whale, pilot whale, and striped dolphin).

Brominated compound levels in blubber biopsies collected from free-ranging common bottlenose dolphins in 2012 were compared between two locations, the Strait of Gibraltar ($n=20$; PBDEs=1184; MeO-PBDEs=813; HBCD=300 μg/kg lw) and the Gulf of Cadiz ($n=20$; PBDEs=830; MeO-PBDEs=475; HBCD=51 μg/kg lw) (Barón et al., 2015a). Moreover, the blubber and brain were analyzed from one male stranded between 2004 and 2011 on the Alboran Sea coastline (Barón et al., 2015b). The PBDE level was 850, and the MeO-PBDE level was 940 μg/kg lw in the blubber; in the brain, the PBDE level was 330, and the MeO-PBDE level was 93.0 μg/kg lw. Additionally, high levels of HNPs and HFRs (2290 μg/kg lw and 880 μg/kg lw, respectively) were found in the blubber.

Fluorinated Compounds

PFOS, FOSA, PFHxS, and PFOA were measured in liver samples from stranded specimens collected along the Tyrrhenian ($n=4$) and Adriatic ($n=2$) seas in 1991

(Kannan et al., 2002). The mean liver concentration of PFOS was 84.5 μg/kg ww, which was higher than in the striped dolphin but lower than in the common dolphin and cetaceans from coastal waters of Florida (Kannan et al., 2001). The levels of FOSA were similar to those of PFOS, averaging 78.5 μg/kg ww

Short-Beaked Common Dolphin

The short-beaked common dolphin seems to have started a recession in the 1960s. Initially, it was believed that this species occupied the entire western Mediterranean, but by 1975, it had disappeared from the northern one-third of the basin; today, its distribution is mostly restricted to the waters of Andalusia, North Africa, and Sicily (Bearzi et al., 2003, Fig. 7.1D). Since this decline was one of the most notorious declines observed in cetaceans in this region, the species has been included on the list of strictly protected species of the Berne Convention and classified as endangered on the IUCN Red List (Bearzi, 2012). Sparse information exists about the common dolphin's diet. It appears to consume a wide range of prey, including predominantly epipelagic and mesopelagic fish species as well as crustaceans and cephalopods (Bearzi et al., 2003).

Organochlorine Compounds

Published data on POPs in common dolphins of the Mediterranean are limited. Few surveys have been conducted, and sample sizes are usually too small to enable the proper assessment of trends and geographic differences among subpopulations. Information on organochlorine levels, mainly on PCBs and DDTs, suggest levels are in the mid to low end of the range of concentrations detected in other *Delphinidae* species from the same region. This variability may be the result of a subtle, but meaningful, difference in diet composition, a dissimilar ability of the different species to handle pollutants, or a combination of both factors.

During the swordfish fishery seasons of 1992, 1993, and 1994, 27 dolphins in the Alboran Sea (next to the Strait of Gibraltar) became entangled in driftnets; samples from these dolphins were analyzed for PCBs (33 mg/kg lw) and DDTs (33 mg/kg lw). The concentrations of PCBs were similar, and the concentrations of DDTs were twice as high as those of common dolphins from the Atlantic Ocean (northwestern Spain) (Borrell et al., 2001); additionally, when compared to striped dolphins from the same area and years, common dolphins showed concentrations of both groups of compounds that were half of those found for striped dolphins (Borrell and Aguilar, 2005). Although slightly less polluted (i.e., DDTs = 15; PCBs = 19 mg/kg dw), similar differences between the two species were found by Fossi et al. (2000), who analyzed three biopsies from common dolphins and 18 biopsies from striped dolphins collected in 1993 in the Ligurian Sea. The fact that the striped dolphin is situated somewhat higher in the trophic web could explain the differences in concentrations between the species (Borrell and Aguilar, 2005).

Thirteen common dolphins biopsied in the Ionian Sea between 1994 and 1998 showed even lower concentrations (mean: HCB ~ 70 ng/kg ww; DDTs ~ 12 mg/kg ww; PCBs ~ 8 mg/kg ww) (Fossi et al., 2003), and again, the concentrations were half of those found in the striped dolphins ($n = 8$, mean: HCB ~ 175 ng/kg ww; DDTs ~ 19 mg/kg ww; PCBs ~ 18 mg/kg ww). These authors emphasized that the correlations between OCs with endocrine disruption capacities (i.e., ppDDE, opDDT and PCB153) and BPMO activity found in these dolphins were signals of a potential toxicological stress in this species, even though the total levels of these OCs were lower than in other odontocetes in the area.

The highest levels reported in Mediterranean common dolphins were found in a large male stranded on the North Adriatic coast in 2004 (PCBs = 138; HCB = 0.4; tHCH = 0.9; DDTs = 105.9 mg/kg lw; Lazar et al., 2012); however, the fact that it was only one individual that was a very old male (length = 2.2 m) might misrepresent the average concentrations found in the population.

Brominated Compounds

Classical (PBDEs) and alternative (HNs, HBB, PBEB, DBDPE, and HBCD) flame retardants, together with naturally produced MeO-PBDEs, were studied in short-beaked common dolphins, among other species (i.e., bottlenose dolphin and pilot whale), which were biopsied in 2012 in the Alboran Sea (Barón et al., 2015b). Several PBDEs were detected in the species, while the emerging brominated flame retardants analyzed (i.e., HBB, PBEB, and DBDPE) were not detected in any sample. In terms of contaminants analyzed, common dolphins had the lowest levels of most anthropogenic and naturally produced contaminants. The PBDEs ($n = 15$; 203 μg/kg lw) and MeO-PBDEs ($n = 15$; 225 μg/kg lw) in common dolphins were lower than those in bottlenose dolphins and pilot whales; moreover, their levels were lower than the reported mean level of PBDEs in common dolphins from the NE Atlantic (422–758 μg/kg lw; Alonso et al., 2014). Similarly, HBCD levels (38 μg/kg lw) in common dolphins were lower than those found in any other population in other water masses (Law et al., 2014). In fact, concentrations were below the upper threshold limit level (i.e., 1500 μg/kg lw) associated with endocrine disruption in grey seals (Hall et al., 2003).

Additionally, HNPs (MHC-1, TriBHD, TetraBHD, MeO-PBDEs, Q1, and related PMBPs) and HFRs (PBDEs, HBB, Dec 602, Dec 603, and DP) were determined in blubber and brain samples collected from 10 stranded common dolphins on the Alboran Sea coastline between 2004 and 2011 (Barón et al., 2015a). As has been showed in striped dolphins, contaminant levels in stranded animals do not necessarily reflect the contaminant burdens of the entire population; rather, they may be biased toward higher pollutant loads because heavily contaminated animals may be more likely to suffer morbidity or mortality. Thus, the levels in stranded specimens (PBDEs = 1000 μg/kg lw and MeO-PBDE = 690 μg/kg lw) were more than 3 times higher than those from biopsied dolphins in the same area. Similar to reports on the striped dolphin, some compounds, such as HBB or TetraBHD, presented higher concentrations in the brain than in the blubber.

Fluorinated Compounds

PFOS, FOSA, PFHxS, and PFOA were measured in the liver of one common dolphin (stranded in 1998 in the Tyrrhenian Sea), among other stranded cetaceans along the Italian coast (Kannan et al., 2002). The highest concentration of PFOS was found in the common dolphin (940 μg/kg ww), while the mean concentration of PFOS in the livers of other species (i.e., bottlenose dolphin, striped dolphin, and long-finned pilot whale) were one order of magnitude lower. FOSA, PFHxS, and PFOA were under detection levels, with the exception of FOSA in the common dolphin (878 μg/kg ww). The result of that study suggested the widespread occurrence of PFOS in marine mammals in the Mediterranean (Kannan et al., 2002).

Sperm Whale

Found throughout the Mediterranean basin, the sperm whale is more abundant in the western and central basins and is uncommon in the eastern Mediterranean (Fig. 7.1E). They prefer deep continental waters close to the continental slope, where mesopelagic cephalopods, the species' preferred prey, are most abundant (Praca and Gannier, 2008), although they can also be found in deeper offshore waters (Notarbartolo di Sciara and Birkun, 2010). The most serious threats to sperm whales in the Mediterranean are bycatch in high-seas swordfish and tuna driftnets, disturbance from intense marine traffic and collisions with large vessels, and underwater noise from mineral prospecting (seismic air guns), military operations, and illegal dynamite fishing gear (Notarbartolo di Sciara and Birkun, 2010). Based on these threats and their consequences, in the Mediterranean, the sperm whale is classified as endangered on the IUCN Red List (Notarbartolo di Sciara et al., 2012).

Organochlorine Compounds

Few papers report the POP toxicologic status of Mediterranean sperm whales, and these studies primarily assessed DDTs and PCBs.

In December 2009, a pod of seven sperm whale males was stranded on the Adriatic coast of southern Italy (Mazzariol et al., 2011). According to genetic and photoidentification studies, these specimens belonged to the Mediterranean Sea population. Complete necropsies were performed only on the three animals of better conservation status; however, biologic material (i.e., blubber, muscle, and liver) was collected from all seven specimens for contaminant analysis (i.e., HCB, DDTs, PCBs) (Marsili et al., 2014). Among the different OCs, HCB was the compound with the lowest level, with similar values among individuals and tissues. In the blubber, the levels of HCB, DDTs, and PCBs (mean ± SD) were 0.522 ± 1.04, 157.2 ± 268.4, and 148.7 ± 270.4 mg/kg lw, respectively.

Praca et al. (2011) reported the average levels of DDTs and PCBs in 14 specimens (12 biopsies and 2 stranded) sampled in the northwestern Mediterranean between 2006 and 2009. The PCB concentrations were 107.81 ± 108.72 mg/kg lw, and the DDT concentrations were 115.98 ± 112.35 mg/kg lw. Moreover, skin and blubber biopsies were sampled from 61 sperm whales (14 females and 47 males) in the northwestern Mediterranean Sea during eight WWF campaigns between 2006 and 2013 (Pinzone et al., 2015). The DDT and PCB levels (mean ± SD), split by sex, were as follows. In males, DDTs = 44.17 ± 41.91 and PCBs = 24.24 ± 17.42 mg/kg lw; in females, DDTs = 17.37 ± 11.12 and PCBs = 16.88 ± 7.24 mg/kg lw. These results show a clear difference between sexes. In 15 of these biopsies, PCDDs (0.36 ± 0.31 μg/kg lw) and PCDFs (0.16 ± 0.17 μg/kg lw) were also analyzed. Table 7.2 shows a comparison of PCB levels in the Mediterranean and in other areas where this widely distributed species is found. Only specimens collected and analyzed after 1998 are displayed; this was done to avoid differences in concentrations caused by the collection year. The levels shown are based on blubber lipid basis. In this table, we can appreciate the high levels found in Mediterranean specimens compared to the levels found in specimens in other parts of the world.

Brominated Compounds

In the specimens studied by Pinzone et al. (2015), PBDE levels were 0.382 ± 0.176 mg/kg lw in males and 0.248 ± 0.106 mg/kg lw in females.

Fin Whale

Fin whales in the Mediterranean show a steady population, with likely more than 5000 individuals throughout the Mediterranean (Fig. 7.1F) (Notarbartolo-di-Sciara et al., 2003). Fin whales concentrate in the Tyrrhenian-Ligurian-Provençal basin in late spring and summer (Forcada et al., 1996), where they feed on their main prey, the euphausiids *Meganyctiphanes norvegica* (Aguilar, 2009), which is abundant due to cold water outcrops in summer. This mysticete is thought to be mainly threatened because of collisions with ships (Panigada et al., 2006) and microplastics pollution (Fossi et al., 2016). In the Mediterranean, the fin whale is classified as vulnerable on the IUCN Red List of Threatened Species (Panigada and Notarbartolo di Sciara, 2012). Though POPs have been extensively studied in other cetacean species in the Mediterranean basin, studies on fin whales and POPs are scarce.

Organochlorine Compounds

The first data on OC levels in fin whale blubber from the Mediterranean were reported by Alzieu and Duguy (1979), who reported 14.8 and 4.96 mg/kg ww

TABLE 7.2 Comparison of Blubber PCB Levels in Sperm Whale in the Mediterranean and in Other Areas of the World

Ocean/Sea	Area	Year of Collection	*n*	Sex	PCBs (mg/kg lw)			References
					Mean	Max.	Min.	
Northern Hemisphere								
Mediterranean	South Adriatic Sea	2009	7	7M	148.70	950.85	19.67	Mazzariol et al. (2011) and Marsili et al. (2014)
	Ligurian Sea/Gulf of Lions	2006–09	14	1M 13U	107.81			Praca et al. (2011)
	Corso Ligurian Basin	2006–13	61	47M 14F	24.24 16.88	68.48 27.24	1.58 8.73	Pinzone et al. (2015)
E Pacific	Gulf of California	1999	10	2M 8F	1.42			Godard-Codding et al. (2011)
Equator								
E Pacific	Galapagos	2000	10	10M	1.32			Godard-Codding et al. (2011)
Pacific	West Galapagos	2000	10	3M 7F	0.75			
W Pacific	Kiribati	2000	10	10F	7.56			
W Pacific	Papua New Guinea	2001	10	2M 8F	1.14			
Southern Hemisphere								
SW Pacific	Tasmania Australia	1998	37	5M 32F	0.90	3.30	0.30	Evans et al. (2004)

U, unknown.

of total DDTs and PCBs, respectively, in immature specimens. Furthermore, Viale (1981) reported 0.80 and 0.30 mg/kg ww of total DDTs and PCBs, respectively, in the blubber of one specimen of fin whale from the Mediterranean Sea. The markedly lower concentrations in fin whales than in the odontocetes were expected because of the fact that they feed mostly on little contaminated planktonic crustaceans and are thus situated low in the food web, contrarily to the majority of odontocetes.

During the summers between 1990 and 1993, Marsili et al. (1996) sampled subcutaneous blubber from 68 specimens of fin whales. In the Ligurian Sea, PCBs and DDTs ranged from 5.5 to 7.1 mg/kg ww and 4.2 to 9.5 mg/kg ww, respectively. The measured organochlorine concentrations in this study indicated that DDTs decreased significantly between 1990 and 1991, while PCBs reached a maximum in 1991; additionally, the PCB congeners 153, 138, 187, 180, and 170 showed the highest percentages. Later, Marsili et al. (1998) concluded that the relationship between OCs and mixed function oxidase activity in skin biopsy samples of male Mediterranean fin whales emphasized the validity of their nondestructive approach in the study of marine mammals. The study by Fossi et al. (2003) showed high concentrations of DDT metabolites and PCB congeners in this mysticete. Moreover, a statistically significant correlation was found between BPMO activity and OC levels in skin biopsies of male fin whales.

The most recent study on fin whales (Pinzone et al., 2015) reported data on 70 individuals (35 males and 35 females), which were collected in the northwestern Mediterranean Sea during eight WWF campaigns between 2006 and 2009. Among all the whale species studied by Pinzone et al. (2015), fin whales presented the lowest values, in accordance with its trophic position, with PCB concentrations of (mean $\pm$ SD) 5.72 $\pm$ 5.18 mg/kg lw, and DDT concentrations of 6.64 $\pm$ 5.55 mg/kg lw. The study from Pinzone et al. (2015) demonstrated an important exposure to pollutants in Mediterranean cetaceans, which often surpassed the estimated toxicity threshold value, i.e., 17 mg/kg lw, for blubber in marine mammals; this threshold was set by Jepson et al. (2005) and Kannan et al. (2000), and at levels greater than this threshold, the specimen may experience deleterious health effects. However, the study by Pinzone et al. (2015) regarding fin whales did not appear to show PCB concentrations higher than the threshold of 17 mg/kg lw. It is remarkable that Pinzone et al. (2015), in the main conclusions from their study, state that the concentrations found in their whales were much higher than not only those in the Southern Hemisphere but also countries of the North Atlantic Ocean.

Brominated Compounds

PBDEs and MeO-PBDEs were analyzed in the liver of one fin whale female specimen that was stranded in 1990 on the Tyrrhenian coast; among other odontocete species, this female showed the highest levels of total PBDEs

(3625 μg/kg lw) and MeO-PBDEs (104 μg/kg lw) compared to the other odontocetes (total PBDEs = 886 μg/kg lw) (Pettersson et al., 2004).

Among all the species studied by Pinzone et al. (2015), fin whales presented the lowest PBDE concentrations (PBDEs: 177 ± 208 μg/kg lw), in accordance with its trophic position.

Fluorinated Compounds

Concentrations of PFOS, FOSA, PFHxS, and PFOA were measured in tissues collected from stranded fin whales from the Italian coasts of the Mediterranean Sea (i.e., the Tyrrhenian Sea). PFOS, FOSA, PFOA, and PFHxS, measured in the muscle of one specimen, showed levels of <19, <19, <38, and <19 μg/kg ww, respectively (Kannan et al., 2002).

CONCLUSIONS AND RESEARCH NEEDS

The present review conducted on cetaceans from the Mediterranean Sea reveals an important exposure to POPs in Mediterranean cetaceans, which often surpasses the estimated toxicity threshold value. However, for some legacy POPs, levels seem to decrease due to their ban and regulation, and the main remaining problem exists in terms of emerging and/or not yet regulated compounds, for which levels are high and are showing increasing tendencies.

Given the particular susceptibility of Mediterranean cetaceans to anthropogenic pressures, specifically to contamination by legacy and emerging POPs, further research is needed in marine mammal ecotoxicology that focuses on multiple-stress exposure. Ecotoxicologic studies on marine mammals are of special importance not only to conserve local populations but also to predict the human exposure to these hazardous substances in the marine environment. It is remarkable that recently published literature raises attention on the possibility of human health problems caused by the common consumption of fish in the Mediterranean basin. In fact, the levels of dioxin-like compounds found by Pinzone et al. (2015) were much higher than the 2 pg WHO-TEQ/g lw limit for human lifetime intake, which is the limit that can be ingested without appreciable health risk (Charnley and Doull, 2005).

REFERENCES

Aguilar, A., 2000. Population biology, conservation threats and status of Mediterranean striped dolphins (*Stenella coeruleoalba*). Journal of Cetacean Research and Management 2 (1), 17–26.

Aguilar, A., 2009. Fin whale *Balaenoptera physalus*. In: Perrin, W.F., Würsig, B., Thewissen, J.G.M. (Eds.), Encyclopedia of Marine Mammals, second ed. Academic Press, Amsterdam, pp. 433–437.

Aguilar, A., Borrell, A., 1994a. Assessment of organochlorine pollutants in cetaceans by means of skin and hypodermic biopsies. In: Fossi, M.C., Leonzio, C. (Eds.), Nondestructive Biomarkers in Vertebrates. Lewis Publisher, Florida, pp. 245–267.

Aguilar, A., Borrell, A., 1994b. Abnormally high polychlorinated biphenyl levels in striped dolphins (*Stenella coeruleoalba*) affected by the 1990–1992 Mediterranean epizootic. The Science of the Total Environment 154 (2–3), 237–247.

Aguilar, A., Borrell, A., 2005. DDT and PCB reduction in the western Mediterranean from 1987 to 2002, as shown by levels in striped dolphins (*Stenella coeruleoalba*). Marine Environmental Research 59 (4), 391–404.

Aguilar, A., Borrell, A., Reijnders, P.J.H., 2002. Geographical a temporal variation in levels of organochlorine contaminants in marine mammals. Marine Environmental Research 53, 425–452.

Aguilar, A., Gaspari, S., 2012. *Stenella coeruleoalba* (Mediterranean subpopulation). The IUCN Red List of Threatened Species 2012: e.T16674437A16674052.

Aguilar, A., Nadal, J., 1984. Obtencion de biopsies hipodérmicas de cetáceos en libertad. Investigación Pesquera 48, 23–29.

Aguilar, A., Raga, J.A., 1993. The striped dolphin epizootic in the Mediterranean Sea. Ambio 524–528.

Ahlborg, U.G., Becking, G.C., Birnbaum, L.S., Brouwer, A.A., Derks, H.J.G.M., Feeley, M., Golor, G., Hanberg, A., Larsen, J.C., Liem, A.K.D., Safe, S.H., Schlatter, C., Waern, F., Younes, M., Yrjänheikki, E., 1994. Toxic equivalency factors for dioxin-like PCBs: report on WHO-ECEH and IPCS consultation, December 1993. Chemosphere 28 (6), 1049–1067.

Alonso, M.B., Azevedo, A., Torres, J.P.M., Dorneles, P.R., Eljarrat, E., Barceló, D., Lailson-Brito Jr., J., Malm, O., 2014. Anthropogenic (PBDE) and naturally-produced (MeO-PBDE) brominated compounds in cetaceans—a review. The Science of the Total Environment 481, 619–634.

Alzieu, C., Duguy, R., 1979. Teneurs en composés organochlorés chez les cetacés et pinnipèdes fréquentant les côtes françaises. Oceanologica Acta 2 (1), 107–120.

Bachman, M.J., Keller, J.M., West, K.L., Jensen, B.A., 2014. Persistent organic pollutant concentrations in blubber of 16 species of cetaceans stranded in the Pacific Islands from 1997 through 2011. Science of the Total Environment 488, 115–123.

Ballesteros, E., 2006. Mediterranean coralligenous assemblages: a synthesis of present knowledge. Oceanography and Marine Biology - an Annual Review 44, 123–195.

Balmer, B.C., Ylitalo, G.M., McGeorge, L.E., Baugh, K.A., Boyd, D., Mullin, K.D., Rosel, P.E., Sinclair, C., Wells, R.S., Zolman, E.S., Schwacke, L.H., 2015. Persistent organic pollutants (POPs) in blubber of common bottlenose dolphins (*Tursiops truncatus*) along the northern Gulf of Mexico coast, USA. Science of the Total Environment 527, 306–312.

Barón, E., Hauler, C., Gallistl, C., Giménez, J., Gauffier, P., Castillo, J.J., Fernández-Maldonado, C., de Stephanis, R., Vetter, W., Eljarrat, E., Barceló, D., 2015a. Halogenated natural products in dolphins: brain–blubber distribution and comparison with halogenated flame retardants. Environmental Science and Technology 49 (15), 9073–9083.

Barón, E., Giménez, J., Verborgh, P., Gauffier, P., De Stephanis, R., Eljarrat, E., Barceló, D., 2015b. Bioaccumulation and biomagnification of classical flame retardants, related halogenated natural compounds and alternative flame retardants in three delphinids from Southern European waters. Environmental Pollution 203, 107–115.

Bearzi, G., 2012. *Delphinus delphis*. The IUCN Red List of Threatened Species 2012: e.T6336A16236707.

Bearzi, G., Reeves, R.R., Notarbartolo-Di-Sciara, G., Politi, E., Canadas, A.N.A., Frantzis, A., Mussi, B., 2003. Ecology, status and conservation of short-beaked common dolphins *Delphinus delphis* in the Mediterranean Sea. Mammal Review 33 (3–4), 224–252.

Bearzi, G., Fortuna, C., Reeves, R.R., 2008. Ecology and conservation of common bottlenose dolphins *Tursiops truncatus* in the Mediterranean Sea. Mammal Review 39 (2), 92–123.

Bearzi, G., Fortuna, C., Reeves, R., 2012. *Tursiops truncatus*. The IUCN Red List of Threatened Species 2012- Vulnerable.

Berrow, S.D., McHugh, B., Glynn, D., McGovern, E., Parsons, K.M., Baird, R.W., Hooker, S.K., 2002. Organochlorine concentrations in resident bottlenose dolphins (*Tursiops truncatus*) in the Shannon estuary. Ireland. Marine Pollution Bulletin 44 (11), 1296–1303.

Bianchi, C.N., Morri, C., 2000. Marine biodiversity of the Mediterranean Sea: situation, problems and prospects for future research. Marine Pollution Bulletin 40, 367–376.

Blanco, C., Salomón, O., Raga, J.A., 2001. Diet of the bottlenose dolphin (*Tursiops truncatus*) in the western Mediterranean Sea. Journal of the Marine Biological Association of the United Kingdom 81 (6), 1053–1058.

Blue Plan, 2008. The Blue Plan's Sustainable Development Outlook for the Mediterranean. UNEP Blue Plan Activity Centre, Sophia Antipolis, France.

Borrell, A., Aguilar, A., 2005. Differences in DDT and PCB residues between common and striped dolphins from the Southwestern Mediterranean. Archives of Environmental Contamination and Toxicology 48 (2), 501–508.

Borrell, A., Aguilar, A., 2007. Organochlorine concentrations declined during 1987-2002 in western Mediterranean bottlenose dolphins, a coastal top predator. Chemosphere 66 (2), 347–352.

Borrell, A., Aguilar, A., Corsolini, S., Focardi, S., 1996. Evaluation of toxicity and sex-related variation of PCB levels in Mediterranean striped dolphins affected by an epizootic. Chemosphere 32 (12), 2359–2369.

Borrell, A., Cantos, G., Pastor, T., Aguilar, A., 2001. Organochlorine compounds in common dolphins (*Delphinus delphis*) from the Atlantic and Mediterranean waters of Spain. Environmental Pollution 114 (2), 265–274.

Bortolotto, A., Casini, L., Stanzani, L.A., 1992. Dolphin mortality along the Southern Italian Coast (June-September 1991). Aquatic Mammals 18, 56–60.

Busbee, D., Tizard, I., Scott, J., Ferrick, D., Ott-Reeves, E., 1999. Environmental pollutants and marine mammal health: the potential impact of hydrocarbons and halogenated hydrocarbons on immune system dysfunction. Journal of Cetacean Research and Management (Special Issue 1), 223–248.

Cagnolaro, L., Cozzi, B., Notarbartolo di Sciara, G., Podestà, M., 2015. In: Calderini (Ed.), Fauna d'Italia Vol. XLIX - Mammalia IV – Cetacea, pp. 3–375.

Carballo, M., Arbelo, M., Esperon, F., Mendez, M., de la Torre, A., Munoz, M.J., 2008. Organochlorine residues in the blubber and liver of bottlenose dolphins (*Tursiops truncatus*) stranded in the Canary Islands, North Atlantic Ocean. Environmental Toxicology 23 (2), 200–210.

Castrillon, J., Gomez-Campos, E., Aguilar, A., Berdié, L., Borrell, A., 2010. PCB and DDT levels do not appear to have enhanced the mortality of striped dolphins (*Stenella coeruleoalba*) in the 2007 Mediterranean epizootic. Chemosphere 81 (4), 459–463.

Cebrian, D., 1995. The striped dolphin *Stenella coeruleoalba* epizootic in Greece, 1991–1992. Biological Conservation 74 (2), 143–145.

Charnley, G., Doull, J., 2005. Human exposure to dioxins from food, 1992-2002. Food and Chemical Toxicology 43, 671–679.

Chou, C.C., Chen, Y.N., Li, C.S., 2004. Congener-specific polychlorinated biphenyls in cetaceans from Taiwan waters. Archives of Environmental Contamination and Toxicology 47 (4), 551–560.

Coll, M., Piroddi, C., Steenbeek, J., Kaschner, K., Lasram, F.B.R., Aguzzi, J., Ballesteros, E., Bianchi, C.N., Corbera, J., Dailianis, T., Danovaro, R., Estrada, M., Froglia, C., Galil, B.S., Gasol, J.M., Gertwagen, R., Gil, J., Guilhaumon, F., Kesner-Reyes, K., Kitsos, M.S., Koukouras, A., Lampadariou, N., Laxamana, E., López-Fé de la Cuadra, C.M., Lotze, H.K., Martin, D., Mouillot, D., Oro, D., Raicevich, S., Rius-Barile, J., Saiz-Salinas, J.I., San Vicente, C., Somot, S., Templado, J., Turon, X., Vafidis, D., Villanueva, R., Voultsiadou, E., 2010. The biodiversity of the Mediterranean Sea: estimates, patterns, and threats. PLoS One 5 (8), e11842.

Kannan, K., Koistinen, J., Beckmen, K., Evans, T., Garzelany, J., Hansen, K.J., Jones, P.D., Giesy, J.P., 2001. Environmental Science and Technology 35, 1593–1598.

Kannan, K., Corsolini, S., Falandysz, J., Oehme, G., Focardi, S., Giesy, J.P., 2002. Perfluorooctanesulfonate and related fluorinated hydrocarbons in marine mammals, fishes, and birds from coasts of the Baltic and the Mediterranean Seas. Environmental Science and Technology 36 (15), 3210–3216.

Kucklick, J., Schwacke, L., Wells, R., Hohn, A., Guichard, A., Yordy, J., Hansen, L., Zolman, E., Wilson, R., Litz, J., Nowacek, D., Rowles, T., Pugh, R., Balmer, B., Sinclair, C., Rosel, P., 2011. Bottlenose dolphins as indicators of persistent organic pollutants in the western North Atlantic Ocean and northern Gulf of Mexico. Environmental Science and Technology 45 (10), 4270–4277.

Lailson-Brito, J., Dorneles, P.R., Azevedo-Silva, C.E., Bisi, T.L., Vidal, L.G., Legat, L.N., Azevedo, A.F., Torres, J.P.M., Malm, O., 2012. Organochlorine compound accumulation in delphinids from Rio de Janeiro State, southeastern Brazilian coast. Science of the Total Environment 433, 123–131.

Lauriano, G., Di Guardo, G., Marsili, L., Maltese, S., Fossi, M.C., 2014. Biological threats and environmental pollutants, a lethal mixture for mediterranean cetaceans? Journal of the Marine Biological Association of the United Kingdom 1–5 First view article.

Law, R.J., Covaci, A., Harrad, S., Herzke, D., Abdallah, M.A.E., Fernie, K., Toms, L.M., Takigami, H., 2014. Levels and trends of PBDEs and HBCDs in the global environment: status at the end of 2012. Environment International 65, 147–158.

Lazar, B., Holcer, D., Mackelworth, P., Klincic, D., Romanic, S.H., 2012. Organochlorine contaminant levels in tissues of a short-beaked common dolphin, *Delphinus delphis*, from the northern Adriatic Sea. Natura Croatica: Periodicum Musei Historiae Naturalis Croatici 21, 391–401.

Loose, L.D., Pitman, K.A., Bentitz, K.F., Silkworth, J.B., 1977. Polychlorinated biphenyl and hexachlorobenzene induced humoral immunosuppression. Journal of the Reticuloendothelial Society 22, 253–271.

Marsili, L., 2000. Lipophilic contaminants in marine mammals: review of the results of ten years work at the Department of Environmental Biology, Siena University (Italy). The Control of Marine Pollution: Current Status and Future Trends, Special Issue International Journal of Environment and Pollution 13 (1–6), 416–452.

Marsili, L., Focardi, S., 1996. Organochlorine levels in subcutaneous blubber biopsies of fin whales (*Balaenoptera physalus*) and striped dolphins (*Stenella coeruleoalba*) from the Mediterranean Sea. Environmental Pollution 91 (1), 1–9.

Marsili, L., Focardi, S., 1997. Chlorinated hydrocarbon (HCB, DDTs and PCBs levels in cetaceans stranded along the Italian coasts: an overview. Environmental Monitoring and Assessment 45 (2), 129–180.

Marsili, L., Fossi, M.C., Notarbartolo Di Sciara, G., Zanardelli, M., Focardi, S., 1996. Organochlorine levels and mixed function oxidase activity in skin biopsy specimens from Mediterranean cetaceans. Fresenius Environmental Bulletin 5 (9/10), 723–728.

Marsili, L., Casini, C., Marini, L., Regoli, A., Focardi, S., 1997. Age, growth and organochlorines (HCB, DDTs and PCBs) in Mediterranean striped dolphins *Stenella coeruleoalba* stranded in 1988-1994 on the coasts of Italy. Marine Ecology Progress Series 151, 273–282.

Marsili, L., Fossi, M.C., Notarbartolo Di Sciara, G., Zanardelli, M., Nani, B., Panigada, S., Focardi, S., 1998. Relationship between organochlorine contaminants and mixed function oxidase activity in skin biopsy specimens of Mediterranean fin whales (*Balaenoptera physalus*). Chemosphere 37 (8), 1501–1510.

Marsili, L., Maltese, S., D Coppola, L.C., Fossi, M.C., 2011. Trend temporale della contaminazione da organoclorurati in esemplari di stenella striata (*Stenella coeruleoalba*) spiaggiati e free-ranging del Mar Mediterraneo. Biologia Marina Mediterranea 18 (1), 188–189.

Marsili, L., Maltese, S., Coppola, D., Carletti, L., Mazzariol, S., Fossi, M.C., 2014. Ecotoxicological status of seven sperm whales (*Physter macrocephalus*) stranded along the Adriatic coast of Southern Italy. Aquatic Conservation: Marine and Freshwater Ecosystems 24 (1), 103–118.

Mazzariol, S., Di Guardo, G., Petrella, A., Marsili, L., Fossi, M.C., Leonzio, C., Zizzo, N., Vizzini, S., Gaspari, S., Pavan, G., Podestà, M., Garibaldi, F., Ferrante, M., Copat, C., Traversa, D., Marcer, F., Airoldi, S., Frantzis, A., De Bernaldo Quiròs, Y., Cozzi, B., Fernàndez, A., 2011. Sometimes sperm whales (*Physeter macrocephalus*) cannot find their way back to the high seas: a multidisciplinary study on a mass stranding. PLoS One 6 (5), e19417.

Monk, A., Charlton-Robb, K., Buddhadasa, S., Thompson, R.M., 2014. Comparison of mercury contamination in live and dead dolphins from a newly described species, *Tursiops australis*. PloS One 9 (8), e104887.

Mwevura, H., Amir, O.A., Kishimba, M., Berggren, P., Kylin, H., 2010. Organohalogen compounds in blubber of Indo-Pacific bottlenose dolphin (*Tursiops aduncus*) and spinner dolphin (Stenella longirostris) from Zanzibar, Tanzania. Environmental Pollution 158 (6), 2200–2207.

Notarbartolo di Sciara, G., 2016. Marine mammals in the Mediterranean Sea: an overview. Advances in Marine Biology 75, 1–36.

Notarbartolo-di-Sciara, G., Zanardelli, M., Jahoda, M., Panigada, S., Airoldi, S., 2003. The fin whale *Balaenoptera physalus* (L. 1758) in the Mediterranean Sea. Mammal Review 33 (2), 105–150.

Notarbartolo di Sciara, G., Birkun, A.J.R., 2010. Conserving Whales and Dolphins in the Mediterranean and Black Seas. An ACCOBAMS Status Report. ACCOBAMS, Monaco, p. 212.

Notarbartolo di Sciara, G., Frantzis, A., Bearzi, G., Reeves, R., 2012. *Physeter macrocephalus*. The IUCN Red List of Threatened Species 2012: e.T41755A2955634.

Panigada, S., Pesante, G., Zanardelli, M., Capoulade, F., Gannier, A., Weinrich, M.T., 2006. Mediterranean fin whales at risk from fatal ship strikes. Marine Pollution Bulletin 52, 1287–1298.

Panigada, S., Notarbartolo di Sciara, G., 2012. *Balaenoptera physalus*. The IUCN Red List of Threatened Species 2012: e.T2478A2787161.

Pettersson, A., van Bavel, B., Engwall, M., Jimenez, B., 2004. Polybrominated diphenylethers and methoxylated tetrabromodiphenylethers in cetaceans from the Mediterranean Sea. Archives of Environmental Contamination and Toxicology 47 (4), 542–550.

Pinzone, M., Budzinski, H., Tasciotti, A., Ody, D., Lepoint, G., Schnitzler, J., Scholl, G., Thomé, J.P., Tapie, N., Eppe, G., Das, K., 2015. POPs in free-ranging pilot whales, sperm whales and fin whales from the Mediterranean Sea: influence of biological and ecological factors. Environmental Research 142, 185–196.

Praca, E., Gannier, A., 2008. Ecological niches of three teuthophageous odontocetes in the north-western Mediterranean Sea. Ocean Science 4 (1), 49–59.

Praca, E., Laran, S., Lepoint, G., Thomé, J.P., Quetglas, A., Belcari, P., Sartor, P., Dhermain, F., Ody, D., Tapie, N., Budzinski, H., Das, K., 2011. Toothed whales in the northwestern Mediterranean: insight into their feeding ecology using chemical tracers. Marine Pollution Bulletin 62 (5), 1058–1065.

Raga, J., 2008. Dolphin morbillivirus epizootic resurgence, Mediterranean Sea. Emerging Infectious Diseases 14 (3), 471–473.

Reich, S., Jimenez, B., Marsili, L., Hernández, L.M., Schurig, V., González, M.J., 1999. Congener specific determination and enantiomeric ratios of chiral polychlorinated biphenyls in striped dolphins (*Stenella coeruleoalba*) from the Mediterranean Sea. Environmental Science and Technology 33 (11), 1787–1793.

Romanić, S.H., Holcer, D., Lazar, B., Klinčić, D., Mackelworth, P., Fortuna, C.M., 2014. Organochlorine contaminants in tissues of common bottlenose dolphins *Tursiops truncatus* from the northeastern part of the Adriatic Sea. Environmental Toxicology and Pharmacology 38 (2), 469–479.

Shaltout, M., Omstedt, A., 2014. Recent sea surface temperature trends and future scenarios for the Mediterranean Sea. Oceanologia 56, 411–443.

Shoham-Frider, E., Kress, N., Wynne, D., Scheinin, A., Roditi-Elsar, M., Kerem, D., 2009. Persistent organochlorine pollutants and heavy metals in tissues of common bottlenose dolphin (*Tursiops truncatus*) from the Levantine Basin of the Eastern Mediterranean. Chemosphere 77, 621–627.

Storelli, M.M., Marcotrigiano, G.O., 2000. Environmental contamination in bottlenose dolphin (*Tursiops truncatus*): relationship between levels of metals, methylmercury, and organochlorine compounds in an adult female, her neonate, and a calf. Bulletin of Environmental Contamination and Toxicology 64 (3), 333–340.

Storelli, M.M., Marcotrigiano, G.O., 2003. Levels and congener pattern of polychlorinated biphenyls in the blubber of the Mediterranean bottlenose dolphins *Tursiops truncatus*. Environment International 28, 559–565.

Storelli, M.M., Barone, G., Piscitelli, G., Storelli, A., Marcotrigiano, G.O., 2007. Tissue-related polychlorinated biphenyls accumulation in Mediterranean cetaceans: assessment of toxicological status. Bulletin of Environmental Contamination and Toxicology 78, 206–210.

Storelli, M.M., Barone, G., Giacominelli-Stuffler, R., Marcotrigiano, G.O., 2012. Contamination by polychlorinated biphenyls (PCBs) in striped dolphins (*Stenella coeruleoalba*) from the Southeastern Mediterranean Sea. Environmental Monitoring and Assessment 184 (9), 5797–5805.

Tanabe, S., Watanabe, S., Kan, H., Tatsukawa, R., 1988. Capacity and mode of PCB metabolism in small cetaceans. Marine Mammal Science 4, 103–124.

UNEP, 1996. State of the Marine and Coastal Environment in the Mediterranean Region. MAP Technical Report Series No. 100. UNEP, Athens.

UNEP, 2009. In: U. N. E. P. Report of the Conference of the Parties of the Stockholm Convention on Persistent Organic Pollutants on the Work of its Fourth Meeting; United Nations Environment Programme: Stockholm Convention on Persistent Organic Pollutants: Geneva, p. 112.

Vaes, T.J., Druon, N., 2013. Mapping of Potential Risk of Ship Strike with Fin Whales in the Western Mediterranean Sea. Report of the Joint Research Centre of the European Commission.

Viale, D., 1981. Lung pathology in stranded cetaceans on the Mediterranean coasts. Aquatic Mammals 8, 96–100.

Wafo, E., Sarrazin, L., Diana, C., Dhermain, F., Schembri, T., Lagadec, V., Rebouillon, P., 2005. Accumulation and distribution of organochlorines (PCBs and DDTs) in various organs of *Stenella coeruleoalba* and a *Tursiops truncatus* from Mediterranean littoral environment (France). The Science of the Total Environment 348 (1–3), 115–127.

Wafo, E., Mama, C., Risoul, V., Schembri, T., Dhermain, F., Portugal, H., 2012a. Chlorinated pesticides in the bodies of dolphins of the French Mediterranean coastal environment. Advances in Environmental Sciences - International Journal of the Bioflux Society 4 (1), 29–35.

Wafo, E., Risoul, V., Schembri, T., Lagadec, V., Dhermain, F., Mama, C., Portugal, H., 2012b. PCBs and DDTs in *Stenella coeruleoalba* dolphins from the French Mediterranean coastal environment (2007–2009): current state of contamination. Marine Pollution Bulletin 64 (11), 2535–2541.

Watanabe, S., Shimada, T., Nakamura, S., Nishiyama, N., Yamashita, N., Tanabe, S., Tatsukawa, R., 1989. Specific profile of liver microsomal cytochrome P-450 in dolphin and whales. Marine Environmental Research 27, 51–65.

FURTHER READING

NOAA - National Oceanic and Atmospheric Administration, 2009. Large Marine Ecosystems. The Mediterranean LME. Web page: http://www.lme.noaa.gov/.

Chapter 8

Pollutants in Tropical Marine Mammals of the Galápagos Islands, Ecuador: an Ecotoxicological Quest to the Last Eden

Juan José Alava[1,2,3], **Peter S. Ross**[3]
[1]*Institute for the Oceans and Fisheries, University of British Columbia, Vancouver, BC, Canada;*
[2]*Fundación Ecuatoriana para el Estudio de Mamíferos Marinos (FEMM), Guayaquil, Ecuador;*
[3]*Ocean Pollution Research Program, Coastal and Ocean Research Institute, Ocean Wise Conservation Association, Vancouver, BC, Canada*

INTRODUCTION

The Galápagos Islands are the last tropical Eden, where the genesis of evolution, shaped in their endemic species, inspired Darwin theory's "*On the Origin of Species*" in 1859. Because of its unique biodiversity and endemism that provides strong evidence of evolutionary theory such as natural selection, adaptation, speciation, and radiation processes, the Galápagos Islands, harboring both the Galápagos National Park (GNP) and the Galápagos Marine Reserve (GMR), were designated a UNESCO-World Natural Heritage Site and Biosphere of the Earth (Alava et al., 2014). However, the Galápagos Islands were recently enlisted as a heritage in risk in 2007 due to the rising number of invasive species, emergent human population growth, and increasing tourism, despite still conserving 95% of its biodiversity (Watkins and Cruz, 2007).

The human and ecologic footprint on the Galápagos Islands is unraveled as the geographic opening of the islands in terms of "continentalization," defined as an anthropogenic process reducing the degree of isolation of this fragile ecoregion due to the ongoing reliance on and massive influx of energy, fuel, and materials transported from continental Ecuador, jeopardizing the long-term preservation of the islands (Charles Darwin Foundation, 2010; Grenier, 2010). Despite the oceanic remoteness of the islands and being a renowned

Marine Mammal Ecotoxicology. https://doi.org/10.1016/B978-0-12-812144-3.00008-5

World Heritage Site, the Galápagos are not immune to the global and regional pollution by ubiquitous contaminants in marine biota (Alava et al., 2014). This has been demonstrated by the exposure to persistent organic pollutants (POPs) in Galápagos sea lion (*Zalophus wollebaeki*) (Alava et al., 2009, 2011a, 2014; Alava, 2011a; Alava and Gobas, 2012), marine fish (Alava and Gobas, 2012; Alava et al., 2014), and sperm whales (*Physeter macrocephalus*) (Godard-Codding et al., 2011), as well as metals (mercury) in yellowfin tuna (*Thunnus albacares*) sampled around Galápagos waters (Muñoz-Abril, 2016).

Due to the remoteness and isolation of the Galápagos Islands relative to other geographic areas, the Galápagos Island species offer a special opportunity to undertake research related to contaminants in tropical environments. The moderate population levels and enforcement of environmental control and management practices on the islands ensures that local pollutant sources are in most cases low compared to global sources (Alava et al., 2009, 2011a,b; Alava and Gobas, 2012; Alava et al., 2014). These conditions provide a unique mesocosm to study the behavior of global pollutants in marine mammalian species and their food chains. Studies of the bioaccumulation, biomagnification, and food web transport of POPs in tropical systems such as remote islands around the equatorial Pacific Ocean are limited (Alava, 2011b; Alava and Gobas, 2012).

For the purpose of this chapter, two sentinel species of endemic marine mammals, including the Galápagos sea lion (*Z. wollebaeki*) and the Galápagos fur seal (*Arctocephalus galapagoensis*), are used as study cases to provide insights of the state of local food webs (Alava and Salazar, 2006; Alava et al., 2009, 2011a,b, 2014, 2017a; Alava, 2011a,b; Alava and Gobas, 2012). In this context, these endemic pinnipeds represent "*the canaries in the coal mine*" at sea to investigate ecotoxicological impacts by pollutants in the Galápagos. These species are also affected by other factors, both natural and human-induced environmental stressors in the GMR. Chemical assaults and biologic contamination have emerged in recent times from anthropogenic sources such as urbanized areas (i.e., sewage from homes and run off from impervious surfaces), agriculture (i.e., nonpoint sources and run off from pesticides application), and maritime traffic (i.e., oil spills and chronic hydrocarbon emissions and ballast water from fishing and tourism boats) (see Alava et al., 2014).

CHEMICAL POLLUTION IN THE GALÁPAGOS MARINE RESERVE: AN INCONVENIENT TRUTH

Historically, the Galápagos Islands represented a strategic oceanic island location occupied by the US military forces from 1941 to 1946 during World War II (Woram, 2005) with a military base on Baltra Island (near the semiurbanized Santa Cruz Island) in 1943 (González et al., 2008). This military presence exacted a toll on the terrestrial and coastal-marine environments in the Galápagos, where the organochlorine pesticide DDT was used to eliminate introduced rats (e.g., black rats, *Rattus rattus*) in the islands (Alava et al., 2011a,b, 2014).

Ironically, the effort to eradicate rats was unsuccessful, as the invasive rodents were not eliminated in the long-term, but the legacy of the past use of DDT persisted in the marine environment of the islands, as demonstrated in the Galápagos sea lion (Alava et al., 2011a; Alava and Gobas, 2012). However, global environmental transport of POPs from continental sources cannot be ruled out as a pathway for contaminants in this remote region (Alava et al., 2009, 2011a; Alava and Gobas, 2012).

Recently, sperm whales (*P. macrocephalus*), sampled in Galápagos waters, revealed the highest expression levels for cytochrome P450 1A1 (CYP1A1), an enzyme used as a biomarker to assess exposure to organic pollutants such as polycyclic aromatic hydrocarbons and polychlorinated biphenyls (PCBs), in comparison to other studied regions of the Pacific (Godard-Codding et al., 2011).

Coastal development, burgeoning tourism, escalating human population growth, fisheries overexploitation, and chemical and biologic pollution are identified as the major threats and impacts to the GMR (Merlen, 1995; MacFarland and Cifuentes, 1996; Bensted-Smith et al., 2002; Carr et al., 2013; Alava et al., 2014; Schiller et al., 2015).

Moreover, operational and accidental releases of hydrocarbons (e.g., oil, diesel, gas) occur regularly around the islands from ships, with the former occurring in the long term, causing chronic degradation in the marine environment (Kingston et al., 2003; Lessmann, 2004), and the latter (i.e., oil spills) resulting in acute impacts to the marine endemic fauna (Wikelski et al., 2001, 2002; Romero and Wikelski, 2002). While oil spills imprint the most visible example of chemical assaults on sea life, less evident and more critical global toxicants of concern involve POPs, which have recently been assessed in a few organisms in the Galápagos (Alava and Gobas, 2012; Alava et al., 2014).

GALÁPAGOS PINNIPEDS SAMPLING AND STUDY AREA

Galápagos sea lions and fur seals are endemic pinnipeds to the Galápagos Islands, where these two species play a crucial role as key predators and sentinel species for the functioning, services, and health of the marine ecosystem (Alava and Salazar, 2006; Alava et al., 2014, 2017a). Both species are endangered and are the only tropical pinnipeds biogeographically found at and near the equator (latitude 0°) (Alava, 2017). This was the central rationale to select these species as indicators of contamination in the Galápagos.

Fieldwork was conducted in 2005 and 2008 to collect biopsy samples from live captured, free-ranging Galápagos pinnipeds from several rookeries around islands from the GMR, as illustrated in Fig. 8.1. A total of 41 Galápagos sea lion pups were sampled (i.e., $n=21$ in 2005; $n=20$ in 2008) following the field sampling and methods described elsewhere (Alava et al., 2009, 2011a). Furthermore, contaminant data for three Galápagos fur seals sampled in Fernandina Island (i.e., Cabo Hammond) and Isabela Island (i.e., Cabo Marshall) in 2005 (Alava et al., 2017a) are reported in this chapter. Similarly, concentration data

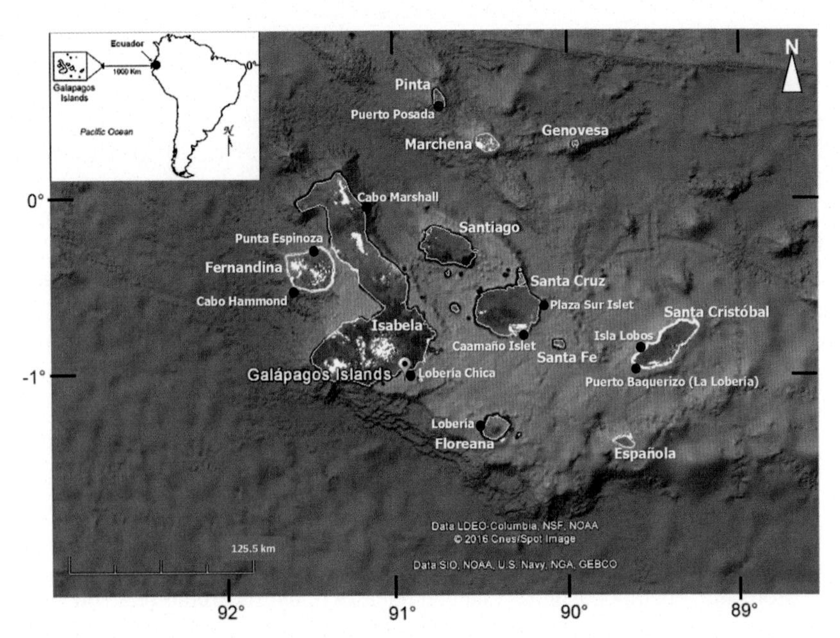

FIGURE 8.1 The Galápagos Islands comprise an archipelago with 13 major volcanic islands, situated at 1000 km (01°40′N–01°25′S and 89°150′W–92°00′W) from Ecuador's mainland coast, in the Southeastern Pacific ocean. *The red circles* depict the sampling sites (pinniped rookeries) in the Galápagos Marine Reserve in 2005 and 2008.

for POPs detected in marine fish species part of the Galápagos sea lions' diet prey (i.e., mullets, *Mugil curema*; Galápagos thread herrings, *Ophistonema berlangai*) are also provided here.

POP ANALYSIS

Analytical methods for organic contaminant analytes, quality assurance/quality control procedures, and data analysis are described elsewhere (Alava et al., 2009, 2011a). The POPs analyzed included PCBs, polybrominated diphenyl ethers (PBDEs), dioxins (i.e., polychlorinated dibenzo-*p*-dioxins, PCDDs), furans (i.e., polychlorinated furans, PCDFs), dichlorodiphenyltrichloroethanes (DDTs), and several organochlorine pesticides (i.e., mirex, dieldrin, beta-hexachlorocyclohexane (β-HCH), chlordanes).

LEGACY PCBs VERSUS EMERGING PBDEs

Galápagos sea lions were found to have low concentrations of PCBs and PBDEs (Alava et al., 2009; Alava and Gobas, 2012). PCB concentrations (mean ± standard error of the mean [SEM]) in male and female pups were 122 ± 84 μg/kg lipid weight (lw) and 93 ± 59 μg/kg lw in 2005 and 91 ± 30 μg/kg lw and 136 ± 32 μg/kg lw

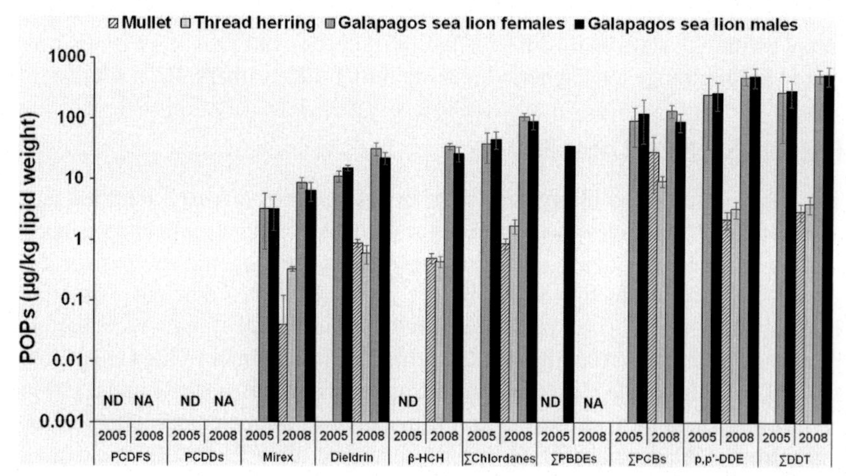

FIGURE 8.2 Mean concentrations of POPs (µg/kg lipid weight) measured in Galápagos sea lions pups (*Z. wollebaeki*), thread herring (*O. berlangai*), and mullets (*M. curema*) highlight that both prey and predator (sea lions) are exposed to several POPs in the remote Galápagos Islands. PCDFs and PCDDs were nondetected in any Galápagos sea lion sample collected in 2005, as quantified concentrations were below the method of detection limit (MDL) (Alava et al., 2009). The highest MDL was 146 pg/g ww, for octachlorodibenzo-p-dioxin, whereas the lowest MDL was 51.4 pg/g ww, for 1,2,3,4,7,8 hexachlorodibenzofuran (Alava et al., 2009). The pesticide *β*-HCH was also below the MDL in sea lion samples collected in 2005 (Alava et al., 2014). Total concentrations of chlordanes (∑Chlordanes) were calculated as the sum of *trans*-chlordane, *cis*-chlordane, *trans*-nonachlor, and *cis*-nonachlor. The total concentration of PBDEs (∑PBDE) was calculated as the sum of the concentrations of four congeners (i.e., BDEs 47, 49, 66, and 183) above the MDL and detected in a single sea lion (male) sampled in 2005 (Alava et al., 2009). Total PCB concentrations (∑PCBs) were calculated as the sum of 72 congeners consistently detected at concentrations above the MDL in 2005 (Alava et al., 2009); in 2008, ∑PCBs were calculated as the sum of 20 PCB congeners consistently detected in samples and above the MDL (Alava and Gobas, 2012). Total DDTs (∑DDT) were calculated as the sum of *o,p′*-DDE, *p,p′*-DDE, *o,p′*-DDD, *p,p′*-DDD, *o,p′*-DDT, and *p,p′*-DDT for Galápagos sea lions (Alava et al., 2011a). Samples for mullets and thread herring were not collected in 2005. *ND* indicates that a given POP was nondetected, as the contaminant quantified was below detections limits; *NA* means that a particular contaminant was not analyzed in a given year. *Error bars* are standard error of the mean (SEM).

in 2008, respectively. While PCB concentrations were among the lowest levels detected among pinniped species, these concentrations were still higher than those detected in southern elephant seal pups (*Mirounga leonina*) from Antarctica (Miranda-Filho et al., 2007). Based on our biomagnification assessment (Alava and Gobas, 2012), PCB concentrations in 2008 were significantly higher in pups relative to prey for adult individuals (i.e., mullets, *Mugil curema*; Galápagos thread herrings, *Ophistonema berlangai*), as shown in Fig. 8.2. Conversely, PBDE concentrations were negligible, with a single individual exhibiting detectable concentrations (i.e., 35 µg/kg lw; Fig. 8.2) above the detection limit (Alava et al., 2009).

No differences were found in PCB levels detected in pups sampled from different colonies (Alava et al., 2009). Thus, the detection of PCBs and PBDEs was linked to long-range atmospheric transport rather than primary, local sources.

DIOXINS AND FURANS

The incineration of solid organic waste in open dumps or rustic landfills and application of pesticides in the agricultural zones of human-inhabited islands in the Galápagos may have indirectly served as anthropogenic sources for the generation and introduction of dioxins (i.e., polychlorinated dibenzo-*p*-dioxins, PCDDs) and furans (i.e., polychlorinated furans, PCDFs) into the coastal-marine environment (Alava et al., 2014). In the human center of the Galápagos, most of the solid waste is organic matter, ranging from 60% to 70% in the 1990s and from 40% to 86% in 2008, and it is disposed of in open areas assigned for this purpose (Alava et al., 2014).

While quantified concentrations of PCDDs and PCDFs were below the detection limits in pup samples collected in 2005 (Fig. 8.2; Alava et al., 2009), questions remain about uncharacterized risks to the terrestrial and aquatic biota in the human centers of the islands, as current practices do not prevent the unintentional by-production of PCDDs and PCDFs (i.e., burning of both organic matter in open dumps and pesticide-contaminated agricultural soils and vegetation).

THE RAISING OF DDTs: EXPOSURE AND BIOMAGNIFICATION

Contrasting the low concentrations of PCBs and PBDEs, and lack of detection of dioxins in sea lions, DDTs dominated the set of POPs found in Galápagos sea lion pups (Figs. 8.2 and 8.3). The mean concentration (±SEM) of $\sum$DDTs in male and female pups was 293 ± 135 μg/kg lw and 274 ± 233 μg/kg lw in 2005 and 533 ± 183 μg/kg lw and 516 ± 125 μg/kg lw in 2008, respectively, as shown in Fig. 8.2 and Table 8.1 (Alava et al., 2011a, 2014). Concentrations of *p,p′*-DDE (i.e., the major metabolite of DDT and a potent antiandrogenic chemical) in male and female pups were 266 ± 130 μg/kg lw and 252 ± 221 μg/kg lw in 2005 and 505 ± 180 μg/kg lw and 408 ± 120 μg/kg lw in 2008. These concentrations were comparable to higher to concentrations found in other marine or aquatic mammal species from the Southern Hemisphere, although lower relative to other species in the northern hemisphere (Alava et al., 2011a). Similar to PCBs, DDT biomagnification was confirmed in the Galápagos sea lion food chain, wherein concentrations were significantly greater than those observed in low trophic level prey, i.e., *M. curema* and *O. berlangai* (Alava and Gobas, 2012).

Similar to Galápagos sea lions, contaminant data for POPs in three fur seals (two males and one female) sampled in 2005 revealed that this species was exposed to DDTs, as illustrated in Fig. 8.3 and Table 8.1 (Alava et al., 2017a). Total concentrations of DDT ($\sum$DDT) in fur seals ranged from 277 to 1890 μg/kg lw;

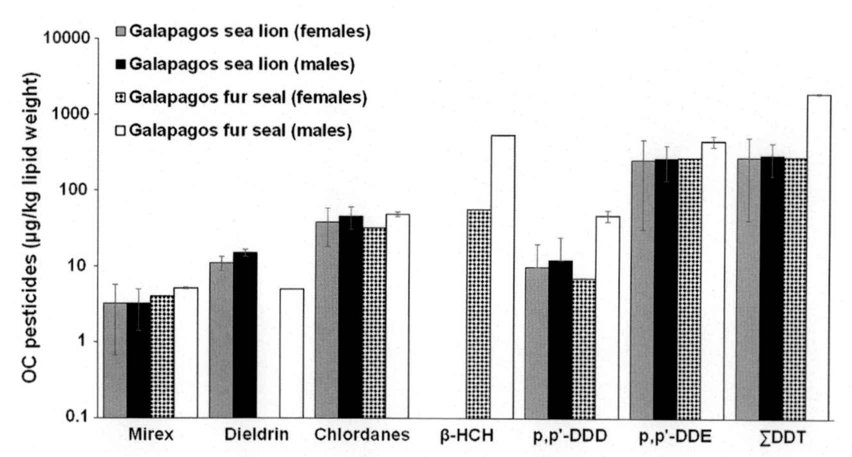

FIGURE 8.3 Mean concentrations of OC pesticides (μg/kg lipid weight) measured in pups of Galápagos fur seal (*A. galapagoensis*) and Galápagos sea lion (*Z. wollebaeki*) sampled in 2005. The concentration patterns observed in these two endemic marine mammal species revealed the ubiquitous exposure to these group of POPs in the Galápagos Islands. In fur seals, chlordane concentrations are represented by the detected concentrations of *trans*-nonachlor (Alava et al., 2017a), while for sea lions, ∑Chlordanes is the sum of *trans*-chlordane, *cis*-chlordane, *trans*-nonachlor, and *cis*-nonachlor. Total DDTs (∑DDT) were calculated as the sum of *o,p′*-DDE, *p,p′*-DDE, *o,p′*-DDD, *p,p′*-DDD, *o,p′*-DDT, and *p,p′*-DDT for Galápagos sea lions (Alava et al., 2011a), while for fur seals, ∑DDT concentrations are mainly the sum of *p,p′*-DDE, *p,p′*-DDD, and *p,p′*-DDT, as the rest of analytes were below the MDL (Alava et al., 2017a). The pesticide *β*-HCH was below the MDL in Galápagos sea lion samples collected in 2005 (Alava et al., 2014), but it was detected in two fur seals (Alava et al., 2017a). Dieldrin was detected in all sea lions (Alava et al., 2014) and in one fur seal male (Alava et al., 2017a). Only three fur seals (two males and a female) were sampled in 2005. *Error bars* are standard error of the mean (SEM).

and, the metabolites *p,p′*-DDE and *p,p′*-DDD in fur seal males ($n=2$) exhibited a mean ± SEM concentration of 455 ± 75 μg/kg lw and 47 ± 7.8 μg/kg lw, respectively. As for the single fur seal female sampled, concentrations of ∑DDT, *p,p′*-DDE, and *p,p′*-DDD were 277 μg/kg lw, 270 μg/kg lw, and 7 μg/kg lw, respectively (Fig. 8.3; Table 8.1). While the small sample size in fur seals precluded a robust statistical comparison with sea lions, ∑DDT concentrations in fur seal males appear to be higher when compared to sea lions (Fig. 8.3; Table 8.1). Differences in dietary preferences, trophic level, and foraging strategies between sea lions and fur seals may explain this distinction in ∑DDT concentrations.

Despite the limited empirical contaminant data for other diet items and most marine fauna of the Galápagos Islands, Galápagos sea lions pups exhibited significantly higher ∑DDT concentrations relative to mullets and thread herrings (Fig. 8.4) (Alava and Gobas, 2012). Moreover, recent food web-bioaccumulation modelling work in progress also demonstrated that DDT biomagnifies in the marine food web of Galápagos sea lions (Alava, 2011b; J.J. Alava, unpublished data).

TABLE 8.1 Concentrations of OC Pesticides (μg/kg Lipid Weight) in Galápagos Sea Lion and Fur Seal Pups Sampled in the Galápagos Marine Reserve

	Galápagos Sea Lion Pups			
	Year	**Males**	**Females**	
	2005	**(*n* = 8)**	**(*n* = 13)**	
OC Pesticides	**2008**	**(*n* = 10)**	**(*n* = 10)**	**References**
∑DDT	2005	293 (51.0–1200)	274 (16.0–3070)	Alava et al. (2011a)
	2008	533 (16.3–1666)	516 (71.2–1230)	Alava et al. (2011a)
				Alava and Gobas (2012)
p,p′-DDE	2005	266 (18.1–1140)	252 (0.150–2900)	Alava et al. (2011a)
	2008	505 (13.6–1650)	480 (65.4–1183)	Alava and Gobas (2012)
Mirex	2005	3.20 (0.55–7.70)	3.22 (0.11–13.0)	Alava et al. (2014)
	2008	6.40 (0.85–24.0)	8.60 (2.50–21.0)	Alava and Gobas (2012)
Dieldrin	2005	15.0 (1.30–60.0)	11.3 (1.15–103)	Alava et al. (2014)
	2008	22.0 (9.00–63.0)	31.0 (9.00–83.0)	Alava and Gobas (2012)
β-HCH	2005	ND	ND	Alava et al. (2014)
	2008	26.0 (7.75–78.0)	34.2 (18.3–52.0)	Alava and Gobas (2012)
Chlordanes	2005	45.5 (16–123)	38 (2.35–382)	Alava et al. (2014)
	2008	90.5(18.8–255)	107(48.0–180)	Alava and Gobas (2012)

	Year	Galápagos Fur Seal Pups		
OC Pesticides	**2005**	**Males ($n=2$)**	**Female ($n=1$)**	**References**
∑DDT	2005	1890–1970.5	277	Alava et al. (2017a)
p,p′-DDE	2005	380–530	270	Alava et al. (2017a)
Mirex	2005	5.0–5.30	4.0	Alava et al. (2017a)
Dieldrin	2005	<MDL[a]–5.0	<MDL	Alava et al. (2017a)
β-HCH	2005	<MDL–540	56	Alava et al. (2017a)
trans-Nonachlor	2005	45–52	32	Alava et al. (2017a)

For Galápagos sea lions, the mean and ranges are reported, while for fur seals, the concentration range for two males and data for one female are shown. *NC* indicates that samples for this species were not collected; *ND* means that the chemical compound/analyte was not detected in the organism; *NA* indicates that the chemical compound was not analyzed.

[a]MDL, *method of detection limit, which is calculated as the mean of procedural blanks run during the sample analysis plus three times the standard deviations of the procedural blanks (Alava et al., 2011a). The MDL is then subtracted from the quantified contaminant data, and only concentrations above the MDL were reported (Alava et al., 2017a).*

Adapted from data reported by Alava, J.J., Palomera, C., Bendell, L., Ross, P.S., 2014. Pollution as a threat for the conservation of the Galapagos Marine Reserve: environmental impacts and management perspectives. In: Vinueza, L., Denkinger, J. (Eds.), The Galapagos Marine Reserve: A Dynamic Socio- Ecological System. ©Springer Science and Business Media, New York, pp. 247–283; Alava, J.J., Denkinger, J., Jimenez, P.J., Carvajal, R., Salazar, S., 2017a. Population Status, anthropogenic stressors and conservation of the Galapagos fur seal (*Arctocephalus galapagoensis*): an overview. In: Alava, J.J. (Ed.), Tropical Pinnipeds: Bio-Ecology, Threats and Conservation. CRC Press and Taylor & Francis Group, Boca Raton, FL, USA, pp. 120–131.

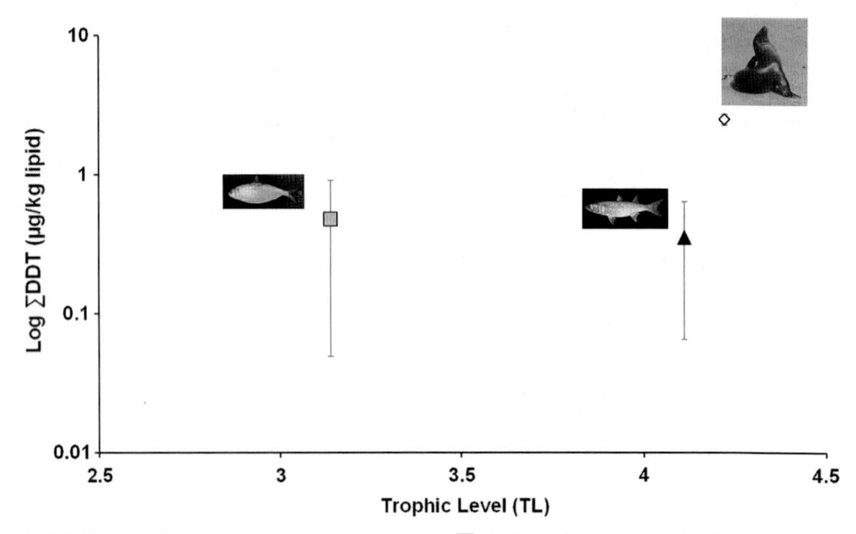

FIGURE 8.4 The relationship between mean Log ∑DDT concentrations (μg/kg lipid weight) in thread herring (TL = 3.14; *gray square*), mullet (TL = 4.11; *black triangle*), and Galápagos sea lion (TL = 4.22; *white jack*) versus trophic level shows that Galápagos sea lions, the top predator in the food chain, contained significant higher concentrations of DDTs compared to those concentrations detected in two of its prey items. *Error bars* are 95% CI; *TL*, trophic level. *(Data for ∑DDT and TL were based on Alava, J.J., Gobas, F.A.P.C., 2012. Assessing biomagnification and trophic transport of persistent organic pollutants in the food chain of the Galápagos sea lion (*Zalophus wollebaeki*): conservation and management implications. In: Romero, A., Keith, E.O. (Eds.), New Approaches to the Study of Marine Mammals. InTech, Croatia. ISBN: 979-953-307-948-5, pp. 77–108.)*

OTHER ORGANOCHLORINE PESTICIDES

In addition to DDTs, several other organochlorine (OC) pesticides were detected in both Galápagos sea lions and fur seals. Galápagos sea lions sampled in 2005 and 2008 were exposed to an important number of OC pesticides, including mirex, dieldrin, beta-hexachlorocyclohexane (β-HCH), and chlordanes (i.e., *trans*-chlordane, *cis*-chlordane, *trans*-nonachlor, and *cis*-nonachlor) (Alava and Gobas, 2012; Alava et al., 2014). The mean concentration and ranges of OC pesticides in Galápagos sea lion pups are reported in Table 8.1.

Galápagos fur seals (two males and one female) sampled in 2005 revealed that this species was also exposed to organochlorine pesticides (OC pesticides), including mirex, dieldrin, beta-hexachlorocyclohexane (β-HCH), and *trans*-nonachlor, as illustrated in Fig. 8.3 and Table 8.1 (Alava et al., 2017a; J.J. Alava, unpublished data). The pesticides mirex and *trans*-nonachlor were detected in all samples, while β-HCH was detected in two (i.e., in one male and a female) of the three fur seals (range: 56–540 μg/kg lw; Table 8.1) and dieldrin in only one male (Fig. 8.3). The OC pesticide concentrations detected in Galápagos fur seals are similar or within the range of those measured in Galápagos sea lions (Fig. 8.3).

ROLE OF OCEANIC AND ATMOSPHERIC TRANSPORT OF POPs

Although local sources may contribute to POP contamination in the Galápagos, the long-range atmospheric transport of POPs to this remote archipelago likely plays an important role in delivering pollutants to biota in this region (Alava et al., 2009; Alava and Gobas, 2012). For instance, trans-Pacific flow of contaminants from tropical Asia to the eastern Pacific is a well-established process (Iwata et al., 1993; Wilkening et al., 2000). However, the local use of DDT in the past in the Galápagos (i.e., DDT use to kill introduced rats) and adjacent areas in Latin America likely explains part of our observations (Alava et al., 2011a,b, 2014). The major use of DDT on the latitudinal band between 6°N and 6°S, engulfing part of the tropics and the equator (i.e., latitude 0°), occurred from 1945 to 1965 (Schenker et al., 2008).

The regional atmospheric-oceanic system in and around the Galápagos Islands, including the confluence of the NE and SE trade winds (i.e., the Inter-Tropical Converge Zone-ITCZ), winds from the west and the Panama (El Niño) and Humboldt oceanographic currents from the east as well as the Equatorial undercurrent or Cromwell current coming from the west, may well contribute to the distribution of these contaminants in this particular region of the Southeastern Pacific Ocean (Alava et al., 2011a). Both long-range environmental transport and/or regional fractionation or distillation (i.e., The Grasshopper Effect) of DDT emissions resulting from both past and recent use to control malaria along midlatitudes of the Pacific coast of South America and subsequently redistributed and transferred to the Galápagos were suggested (Alava et al., 2011a; Alava and Gobas, 2012) as pathways of DDT contamination in the Galápagos marine environment.

POP HEALTH RISKS

In general, POP concentrations were detected at low levels in both pinniped species, implying moderate health risks (Alava et al., 2009, 2011a, 2017a; Alava and Gobas, 2012). Despite PCB concentrations in Galápagos sea lion pups being lower than the PCB-health effect risk threshold of 1300 μg/kg lw (Mos et al., 2010), these contaminants may cause low-level disruption of endocrine and immune systems (Ross et al., 1995; Mos et al., 2006; Tabuchi et al., 2006; Hall and Thomas, 2007), affecting the ability of marine mammals to combat disease (Ross, 2002; Alava et al., 2014). The potential for interplay between contaminant-associated immunotoxicity and the emergence of infectious diseases is troubling. For example, canine distemper virus and *Leptospira* have already been detected in Galápagos sea lion pups from the urbanized San Cristóbal Island (Denkinger et al., 2017), where interactions with domestic animals (e.g., dogs) and introduced species (e.g., rats), and contact with humans were reported (Denkinger et al., 2015; Alava et al., 2014).

Similar to PCBs, DDT concentrations in Galápagos sea lion pups were close to levels expected to be associated with impacts on the endocrine systems in males, as it has been previously reported that 1% of the male pups exceeded the *p,p′*-DDE toxic effect concentration (i.e., 64 μg/kg wet weight [ww]; Kelce et al., 1995) associated with potent antiandrogenic effects (Alava et al., 2011a). Adult male Galápagos sea lions can be expected to exhibit DDT concentrations that are higher than those in pups as DDTs accumulate throughout the animal's life because they are unable to offload contaminants during reproduction (Alava et al., 2011a; Alava and Gobas, 2012).

The interplay of POPs and emerging infectious diseases can be especially critical during periods of nutritional stress due to lack of prey supply during the strongest El Niño episodes or El Niño Southern Oscillation events, when mass mortality occurs and populations often approach the critical tipping point of extinction (Alava et al., 2011a, 2014, 2017a,b).

CONTAMINANTS OF EMERGING CONCERN IN THE GALÁPAGOS ISLANDS

Contaminants or chemicals of emerging concern (CECs) in the Galápagos include current-use pesticides (CUPs), pharmaceuticals and personal care products (PPCPs), antifouling biocides/paints (organotins, OT), and microplastics (MPs), as well as mercury, as previously reported in anthropogenic threat assessments conducted for the Galápagos (Alava and Salazar, 2006; Alava, 2011a; Alava et al., 2014). The selection of these chemicals of emerging concern was based on the context of the existing literature, weight of evidence, and best judgment experts for these contaminants classes (for a review, see Alava and Salazar, 2006; Alava et al., 2014). An updated rapid assessment and overview about the potential risks and implications of these CECs in Galápagos marine mammals is discussed next.

CUPs

In the Galápagos, there is evidence that CUPs were used (Alava et al., 2014). While CUPs include organophosphate and carbamate classes, a wide variety of other groups of pesticides seems to have been applied in the Galápagos, including insecticides such as neonicotinoid (Acetamiprid) and synthetic pyrethroids (PYR) (i.e., Deltamethrin to control the biovector of dengue, the mosquito *Aedes aegypti*) as well as herbicides such as glyphosate (Rodeo or Roundup) and paraquat (Gramoxone) to eliminate weeds and invasive vegetation (Alava et al., 2014). Though CUPs are generally less persistent and bioaccumulative than legacy POPs, some can be acutely toxic pesticides to animal vertebrates, including fish (see Johannessen and Ross, 2002; Tierney et al., 2008).

In the Galápagos, agriculture has occurred or occurs on all four human-inhabited islands (Santa Cruz, Santa Cristóbal, Floreana, and Isabela), mainly in the highlands where CUPs may have been used (Alava et al., 2014). While the impact of CUPs is basically unknown in Galápagos marine mammals and their prey (e.g., marine fish), it is known that pyrethroids (e.g., permethrin and cypermethrin) can bioaccumulate and be maternally transferred in marine mammals, as recently found in Franciscana dolphins (*Pontoporia blainvillei*) from Brazil, despite the notion that these insecticides are converted to nontoxic metabolites by hydrolysis in mammals (Alonso et al., 2012, 2015).

PPCPs

Because of the marked presence of a human population inhabiting the Galápagos (i.e., >30,000 individuals by 2017; see Alava et al., 2014) and about 200,000 tourists visiting the islands annually (Epler, 2007; GNP, 2010), mainly in the islands harboring different degrees of urbanization and rural areas (i.e., Santa Cruz, San Cristóbal, Isabela, Floreana) with lack of wastewater treatment plants (Alava et al., 2014), the use and release of PPCPs such as antibiotics, drugs, disinfectants, and fragrances, as well as the use of UV filters (sunscreens) in public beaches into the coastal-marine environment cannot be ruled out. Overflow from rudimentary septic tanks (i.e., latrines or cesspools) and runoff from human sewage discharges from these islands threaten the water quality near urbanized centers and increase the risk of fecal contamination in coastal waters (Moir and Armijos, 2007; Stumpf et al., 2013; Alava et al., 2014).

For instance, a recent study found that sewage-impacted sites at San Cristóbal Island exhibited higher levels of antibiotic resistant *E. coli*, suggesting that human activities are increasing the levels of resistance due to the use of antibiotics by the local populations as well as tourists carrying and using antibiotics (Overbey et al., 2015).

Triclosan represents a risk for marine mammals, as this substance was detected in plasma (9.0 ng/g ww) of a captive killer whale (*Orcinus orca*) fed with a diet of Pacific herring (*Clupea pallasii*) harvested from the coast of British Columbia (Bennett et al., 2009). Likewise, a previous study also confirmed for first time the bioaccumulation of triclosan, with plasma concentration ranging 0.025–0.27 ng/g ww, in bottlenose dolphins (*Tursiops truncatus*) from the southeast coast (South Carolina and Florida) of the United States (Fair et al., 2009). Whether this can be found in Galápagos species is unclear, highlighting the need for future research.

Meanwhile, tourists using sunscreen around the Galápagos may contribute to the release of these substances in the marine environment, and questions linger on whether Galápagos sea lions and fur seals are exposed to these substances. For example, a major ingredient of sunscreens, octocrylene (OCT), was found for first time in Franciscana dolphins (*P. blainvillei*) in Brazil (Gago-Ferrero et al., 2013); subsequently, several UVFs (i.e., 4-methylbenzylidene camphor (4MBC), 2-ethyl-hexyl-4-trimethoxycinnamate (EHMC), 2-ethyl-hexyl-4-dimethyl-aminobenzoate and OCT) were found to be maternally transferred through placenta in both Franciscana and Guiana dolphins (*Sotalia guianensis*) from Brazil (Alonso et al., 2015), possessing a new looming threat for tropical marine mammals.

Antifouling Paints

The antifouling biocide of concern in the Galápagos is tributyltin (TBT), which is toxic and readily bioaccumulates in marine mammals and humans around the world (Tanabe, 1999; Linley-Adams, 1999; Antizar-Ladislao, 2008). Other antifouling paints can include phenyltin (PTs) compounds, including triphenyltin (TPT), as well as organotin (OT)-free antifouling agents such as Irgarol and Diuron.

An assessment of TBT with associated health effects (i.e., endocrine disruption of gonad development in marine invertebrates, defined as imposex) on muricide gastropod species (i.e., *Thais* spp.; *Plicopurpura* spp.) collected from Santa Cruz and San Cristóbal islands in the Galápagos revealed the occurrence of this biocide and imposex incidence in the Galápagos (Rodríguez-Grimón et al., 2016). Incidence of imposex levels in muricide gastropods and TBT concentrations in sediments from the Gulf of Guayaquil (Ecuador) have also been reported (Castro et al., 2012). In the 1970s, most vessels or ships in the world carried OT-based antifouling compounds (i.e., TBT) painted on their hulls to reduce or prevent the immersed solid surfaces from biofouling by colonizing marine organisms (IMO, 2002); thus it cannot be ruled out that tourism cruise ships and cargo vessels carrying TBT-based antifouling paints on their hulls and arriving and anchoring in the Galápagos have impacted local marine fauna, including invertebrates and marine mammals. However, it is unknown if endemic pinnipeds of the Galápagos are exposed to TBT, as assessments to investigate this pollutant have not yet been conducted in marine mammals (Alava et al., 2014).

Microplastics

In the Galápagos, the impact of marine debris, especially plastic materials, particularly causes concern because no appropriate solid waste management programs exists on board vessels (i.e., fishing boats, merchant-transportation ships, and recreational-tourism cruise ships), although the level of municipal waste collection is high and fairly organized in the islands (Alava et al., 2014). Of particular concern are microplastics (MPs), which are defined by size as particles <5 mm (i.e., <5000 μm), resulting from two basic categories: primary MPs, which are deliberately manufactured (e.g., microbeads in cosmetics, industrial cleaners, or virgin resin pellets for manufacturing and nurdles), and secondary MPs, which are break-down products of larger plastic (i.e., polymers > 5 mm) such as clothing, ropes, bags, and bottles (Moore, 2008; GESAMP, 2010; Browne et al., 2007; Andrady, 2011; Duis and Coors, 2016).

These micropollutants are considered global pollutants and part of a much bigger ocean pollution problem because of their insidious and ubiquitous nature, becoming more evident in recent times (Browne et al., 2007; Andrady, 2011; Ross and Morales-Caselles, 2015). Most MPs likely originate from land-based sources (e.g., household and industrial waste and wastewater), although these pollutants can also be released from sea-based sources (e.g., fishing, shipping).

Although research to assess the extent of the impact of plastics in marine mammals of Galápagos Islands has yet to be conducted, it is known that 26 species of cetaceans can ingest plastic bags, fishing line, and other plastic (Moore, 2008). In fact, while ingestion of MPs has been found in the stomach content of a deep diving, oceanic cetacean species (True's beaked whale, *Mesoplodon mirus*) stranded on the coast of Ireland (Lusher et al., 2015), recent research shows that fin whales (*Balaenoptera physalus*) from the Mediterranean Sea are exposed to MPs due to direct ingestion and consumption of contaminated prey (Fossi et al., 2016).

Mercury

Empirical evidence of the legacy of anthropogenic mercury (i.e., amalgamation and cinnabar mining) during Inca, colonial, and preindustrial times (i.e., increased preindustrial mercury pollution beginning between 1400 and 1600 AD) in the South American Andes was found in sediment cores of El Junco Lake on San Cristóbal Island (Galápagos), indicating and corroborating long-range atmospheric transport of preindustrial mercury emissions from South America (Cooke et al., 2013).

Ecotoxicological assessments of mercury and its methylated form, methylmercury (MeHg), in marine organisms are very limited around the Galápagos Islands, with only one study on mercury in muscles of yellowfin tuna (*T. albacares*) collected in Galápagos waters (Muñoz-Abril, 2016). The mean concentration of mercury in this species was 0.5 mg/kg ww, ranging from 0.09 to 6.8 mg/kg ww, with the latter one of the highest concentrations reached in the Southeastern Pacific (Muñoz-Abril, 2016), surpassing well above international consumption safety levels or guidelines for mercury (e.g., 0.5 mg/kg, 1.0 mg/kg ww) (FAO/WHO, 2016). Mercury have also been reported in yellowfin tuna (*T. albacares*) and common dolphinfish or "dorado" (*Coryphaena hippurus*) landed at Ecuador's mainland coast, where these pelagic fish species exhibited mean concentrations of 1.4 (range, <0.005–6.0 mg/kg ww) and 1.6 (range, 0.04–5.3 mg/kg ww), respectively (Araújo and Cedeño-Macías, 2016). No mercury studies in pinnipeds or other marine mammals have been conducted thus far in the Galápagos, and questions remain on whether marine apex predators are contaminated with MeHg.

POLLUTANT MANAGEMENT AND CONSERVATION IMPLICATIONS IN THE LAST EDEN

Ecuador is a signatory country of the Stockholm Convention on POPs, which was ratified on June 7, 2004. The National Plan for the Implementation of the POP Management in Ecuador has since been enacted. While regulations have been implemented through the Ecuadorian Guidelines for the Control and Management of Environmental Pollution as well as commitments to pursuing the mandate of the Stockholm Convention on POPs in Ecuador, the control and management of environmental pollution in the Galápagos Islands warrants additional efforts.

The World Health Organization (WHO) reactivated the use of the pesticide DDT to control and eliminate the malaria mosquito in tropical countries because of growing cases of malaria (WHO, 2006). While the concentrations of DDT and associated health risks in wildlife are generally believed to be declining, this may no longer be the case in tropical countries from the Southern Hemisphere, where DDT is increasingly used and can biomagnify in food chains (Alava and Gobas, 2012). A renewed use of DDT to combat malaria is likely to increase DDT concentrations in the Southern Hemisphere and in particular put marine mammal populations at greater risk because of the biomagnification of these substances in their food webs (Alava and Gobas, 2012). DDT is included on Schedule 2 of the Stockholm Convention because of its harmful health effects in human and wildlife populations. Continuation of this initiative will help to control DDT contamination in the Galápagos.

Of additional attention are the OT-based antifouling compounds (e.g., TBTs), which were recognized by the International Maritime Organization (IMO, 2017a) in the late 1980s (IMO, 2002). In 1990, the International Maritime Organization's (IMO) Marine Environment Protection Committee (MEPC) adopted a resolution recommending that governments should adopt measures to eliminate the use of antifouling paint containing TBT on nonaluminum hulled vessels (<25 m in length) and eliminate the use of antifouling paints with a leaching rate of more than 4 μg of TBT per day. On January 2003, the resolution called for a global prohibition on the application of organotin compounds by January 1, 2003, and a complete prohibition by January 1, 2008. Thus, the International Convention on the Control of Harmful Anti-fouling Systems (AFS) on ships was adopted on October 5, 2001 and entered in force on September 17, 2008. The AFS Convention prohibits the use of harmful organotins in antifouling paints used on ships and establishes a mechanism to prevent the potential future use of other harmful substances in antifouling systems (IMO, 2017a). While the Galápagos Islands were designated as a particularly Sensitive Sea Area by the IMO in 2005 under Resolution MEPC-135(53) to prevent marine pollution by spills and hazardous contamination coming from ships (Alava et al., 2014), there are no restrictions on antifouling application on ships (including TBT) since Ecuador is not a signatory of the AFS Convention (Rodríguez-Grimón et al., 2016). Enforcement of this convention and implementation of regulations to foster pollution control and management of antifouling paints are urgently required in the Galápagos.

Likewise, Ecuador signed (2013) and ratified (2016) the Minamata Convention on Mercury, which recently entered in force in August 16, 2017, to control, reduce, and eliminate mercury emissions, and enact regulations of gold mines (UNEP, 2017). However, artisanal and small-scale gold mining is still a severe source of mercury pollution in Ecuador's mainland coast, mainly in the Gulf of Guayaquil.

Finally, the London Convention/Protocol on the Prevention of Marine Pollution by Dumping of Wastes and Other Matter (The London Protocol) is an additional global instrument to promote policies for the effective control of all sources of marine pollution to protect the marine environment from human activities (IMO, 2017b). This protocol should be adopted by Ecuador's government (Ecuador is not yet a party state) to implement management actions to address and mitigate waste releases and plastic debris in the Galápagos.

ACKNOWLEDGMENTS

Special thanks to the Charles Darwin Foundation for the Galápagos Islands, the Galápagos National Park, and the Santa Barbara Marine Mammal Center for their support and logistic field assistance. We thank Dr. S. Villegas-Amtmann and Dr. D. P. Costa for collecting and shipping the biopsy samples for contaminant analysis in 2005. We are indebted to Dr. Frank Gobas, S. Salazar, G. Jimenez-Uzcátegui, M. Cruz, P. Martinez, G. Merlen, J. Geraci, P. Howorth, and the volunteers from the Marine Mammal Center in Santa Barbara (E. Stetson, C. Powell, D. Noble, N. Stebor, D. Storz, and S. Crane) for their assistance in the field work and live capture of pups during the 2008 expedition as part of the Project Health Status, Genetic and Rescue Techniques of Galápagos Pinnipeds of the Charles Darwin Foundation, and the Galápagos National Park Service (Servicio Parque Nacional Galápagos). We also thank Dr. Michael Ikonomou and Cory Dubetz from the Institute of Ocean Sciences, Department of Fisheries and Ocean Canada (DFO) for assisting and supporting the chemical analysis of contaminants. Funding to develop this chapter was provided by Mitacs-SSHRC Joint Initiative and *OceanCanada* Partnership at University of British Columbia (Institute for the Oceans and Fisheries) and Ocean Pollution Research Program at the Ocean Wise Conservation Association. Official permits for carrying out this research and exporting of samples in our previous research were given by the Galápagos National Park.

REFERENCES

Alava, J.J., Salazar, S., 2006. Status and conservation of Otariids in Ecuador and the Galapagos Islands. In: Trites, A.W., Atkinson, S.K., DeMaster, D.P., Fritz, L.W., Gelatt, T.S., Rea, L.D., Wynne, K.M. (Eds.), Sea Lions of the World. Alaska Sea Grant College Program, University of Alaska, Fairbanks, AK.

Alava, J.J., Ikonomou, M.G., Ross, P.S., Costa, D.P., Salazar, S., Aurioles-Gamboa, D., Gobas, F.A.P.C., 2009. Polychlorinated biphenyls and polybrominated diphenyl ethers in Galapagos sea lions (*Zalophus wollebaeki*). Environmental Toxicology and Chemistry 28 (11), 2271–2282.

Alava, J.J., 2011a. Bioaccumulation of Pollutants in Galapagos Sea Lions and Marine Mammals from British Columbia, Canada (Ph.D. thesis). School of Resource and Environmental Management, Faculty of Environment, Simon Fraser University, BC, Canada.

Alava, J.J., 2011b. Modeling the bioaccumulation of DDT in the Galapagos sea lion food web: conservation implications. In: Abstracts, 19th Biennial Conference on the Biology of Marine Mammals. Cumulative Effects of Threats to Marine Mammals: Challenges to Animals, Scientists, and Managers. 27 November–2 December, 2011. Tampa, Florida, USA, p. 9.

Alava, J.J., Ross, P.S., Ikonomou, M.G., Cruz, M., Jimenez-Uzcategui, G., Salazar, S., Costa, D.P., Villegas-Amtmann, S., Howorth, P., Gobas, F.A.P.C., 2011a. DDT in endangered Galapagos sea lions (*Zalophus wollebaeki*). Marine Pollution Bulletin 62, 660–671.

Alava, J.J., Salazar, S., Cruz, M., Jimenez-Uzcategui, G., Villegas-Amtmann, S., Paez-Rosas, D., Costa, D.P., Ross, P.S., Ikonomou, M.G., Gobas, F.A.P.C., 2011b. DDT strikes back: Galapagos sea lions face increasing health risks. Ambio 40, 425–430.

Alava, J.J., Gobas, F.A.P.C., 2012. Assessing biomagnification and trophic transport of persistent organic pollutants in the food chain of the Galapagos sea lion (*Zalophus wollebaeki*): conservation and management implications. In: Romero, A., Keith, E.O. (Eds.), New Approaches to the Study of Marine Mammals. InTech, Croatia. ISBN: 979-953-307-948-5, pp. 77–108.

Alava, J.J., Palomera, C., Bendell, L., Ross, P.S., 2014. Pollution as a threat for the conservation of the Galapagos Marine Reserve: environmental impacts and management perspectives. In: Vinueza, L., Denkinger, J. (Eds.), The Galapagos Marine Reserve: A Dynamic Socio-Ecological System. ©Springer Science and Business Media, New York, pp. 247–283.

Alava, J.J., Denkinger, J., Jimenez, P.J., Carvajal, R., Salazar, S., 2017a. Population Status, anthropogenic stressors and conservation of the Galapagos fur seal (*Arctocephalus galapagoensis*): an overview. In: Alava, J.J. (Ed.), Tropical Pinnipeds: Bio-Ecology, Threats and Conservation. CRC Press and Taylor & Francis Group, Boca Raton, FL, USA, pp. 120–131.

Alava, J.J. (Ed.), 2017. Tropical Pinnipeds: Bio-Ecology, Threats and Conservation. CRC Press and Taylor & Francis Group, Boca Raton, FL, USA, p. 325.

Alava, J.J., Cheung, W.W.L., Ross, P.S., Sumaila, R.U., 2017b. Climate change-contaminant interactions in marine food webs: towards a conceptual framework. Global Change Biology 23, 3984–4001. https://doi.org/10.1111/gcb.13667.

Alonso, M.B., Feo, M.L., Corcellas, C., Vidal, L.G., Bertozzi, C.P., Marigo, J., Secchi, E.R., Bassoi, M., Azevedo, A.F., Dorneles, P.R., Torres, J.P.M., 2012. Pyrethroids: a new threat to marine mammals? Environment International 47, 99–106.

Alonso, M.B., Feo, M.L., Corcellas, C., Gago-Ferrero, P., Bertozzi, C.P., Marigo, J., Flach, L., Meirelles, A.C.O., Carvalho, V.L., Azevedo, A.F., Torres, J.P.M., 2015. Toxic heritage: maternal transfer of pyrethroid insecticides and sunscreen agents in dolphins from Brazil. Environmental Pollution 207, 391–402.

Andrady, A.L., 2011. Microplastics in the marine environment. Marine Pollution Bulletin 62, 1596–1605. https://doi.org/10.1016/j.marpolbul.2011.05.030.

Antizar-Ladislao, B., 2008. Environmental levels, toxicity and human exposure to tributyltin (TBT)-contaminated marine environment. A review. Environment International 34, 292–308.

Araújo, C.V.M., Cedeño-Macías, L.A., 2016. Heavy metals in yellowfin tuna (*Thunnus albacares*) and common dolphinfish (*Coryphaena hippurus*) landed on the Ecuadorian coast. The Science of the Total Environment 541, 149–154.

Bennett, E.R., Ross, P.S., Huff, D., Alaee, M., Letcher, R.J., 2009. Chlorinated and brominated organic contaminants and metabolites in the plasma and diet of a captive killer whale (*Orcinus orca*). Marine Pollution Bulletin 58 (7), 1078–1083.

Bensted-Smith, R., Powel, G., Dinerstein, E., 2002. Planning for the ecoregion. In: Bensted-Smith, R. (Ed.), A Biodiversity Vision for the Galápagos Islands. Charles Darwin Foundation and World Wildlife Fund, Puerto Ayora, Santa Cruz, Galápagos.

Browne, M.A., Galloway, T., Thompson, R., 2007. Microplastic—an emerging contaminant of potential concern? Integrated Environmental Assessment and Management 3, 559–566. https://doi.org/10.1002/ieam.5630030215.

Carr, L.A., Stier, A.C., Fietz, K., Montero, I., Gallagher, A.J., Bruno, J.F., 2013. Illegal shark fishing in the Galápagos Marine Reserve. Marine Policy 39, 317–321.

Charles Darwin Foundation, 2010. The human footprint on Galápagos. In: Charles Darwin Foundation Annual Report 2010. Puerto Ayora, Santa Cruz Island, Galapagos Islands, Ecuador.

Castro, Í.B., Arroyo, M.F., Costa, P.G., Fillmann, G., 2012. Butyltin compounds and imposex levels in Ecuador. Archives of Environmental Contamination and Toxicology 62 (1), 68–77.

Cooke, C.A., Hintelmann, H., Ague, J.J., Burger, R., Biester, H., Sachs, J.P., Engstrom, D.R., 2013. Use and legacy of mercury in the Andes. Environmental Science and Technology 47 (9), 4181–4188.

Denkinger, J., Gordillo, L., Montero-Serra, I., Murillo, J.C., Guevara, N., Hirschfeld, M., Fietz, K., Rubianes, F., Dan, M., 2015. Urban life of Galápagos sea lions (*Zalophus wollebaeki*) on San Cristobal Island, Ecuador: colony trends and threats. Journal of Sea Research 105, 10–14.

Denkinger, J., Guevara, N., Ayala, S., Murillo, J.C., Hirschfeld, M., Montero-Serra, I., Fietz, K., Goldstein, T., Ackermann, M., Barragán, V., Cabrera, F., 2017. Pup mortality and evidence for pathogen exposure in Galapagos sea lions (*Zalophus wollebaeki*) on San Cristobal Island, Galapagos, Ecuador. Journal of Wildlife Diseases 53 (3). https://doi.org/10.7589/2016-05-092.

Duis, K., Coors, A., 2016. Microplastics in the aquatic and terrestrial environment: sources (with a specific focus on personal care products), fate and effects. Environmental Sciences Europe 28, 1–25. https://doi.org/10.1186/s12302-015-0069-y.

Epler, B., 2007. Tourism, the Economy and Population Growth and Conservation in Galápagos. Fundación Charles Darwin, Puerto Ayora, Santa Cruz Island, Galápagos Islands, Ecuador.

Fair, P.A., Lee, H.B., Adams, J., Darling, C., Pacepavicius, G., Alaee, M., Bossart, G.D., Henry, N., Muir, D., 2009. Occurrence of triclosan in plasma of wild Atlantic bottlenose dolphins (*Tursiops truncatus*) and in their environment. Environmental Pollution 157 (8–9), 2248–2254.

FAO/WHO, April, 2016. Discussion paper on maximum levels for methylmercury in fish. Joint FAO/WHO Food Standards Programme, Codex Committee on Contaminants in Foods. Codex Alimentarius Commission. Rotterdam, The Netherlands, 4-8. CX/CF 16/10/15. 19pp.

Fossi, M.C., Marsili, L., Baini, M., Giannetti, M., Coppola, D., Guerranti, C., Caliani, I., Minutoli, R., Lauriano, G., Finoia, M.G., Rubegni, F., 2016. Fin whales and microplastics: the Mediterranean Sea and the Sea of Cortez scenarios. Environmental Pollution 209, 68–78.

Gago-Ferrero, P., Alonso, M.B., Bertozzi, C.P., Marigo, J., Barbosa, L., Cremer, M., Secchi, E.R., Azevedo, A., Lailson-Brito Jr., J., Torres, J.P., Malm, O., 2013. First determination of UV filters in marine mammals. Octocrylene levels in Franciscana dolphins. Environmental Science and Technology 47 (11), 5619–5625.

Galapagos National Park, 2010. Informe de Ingreso de Turistas. Galapagos National Park, Puerto Ayora, Santa Cruz, CA.

GESAMP, 2010. In: Bowmer, T., Kershaw, P.J. (Eds.), Proceedings of the GESAMP International Workshop on Plastic Particles as a Vector in Transporting Persistent, Bio-Accumulating and Toxic Substances in the Oceans. IMO/FAO/UNESCO-IOC/UNIDO/WMO/IAEA/UN/UNEP Joint Group of Experts on the Aspects of Marine Environmental Protection. GESAMP Rep. Stud. No. 82, 68 pp.

Godard-Codding, C.A.J., Clark, R., Fossi, M.C., Marsili, L., Maltese, S., West, A.G., Valenzuela, L., Rowntree, V., Polyak, I., Cannon, J.C., Pinkerton, K., Rubio-Cisneros, R., Mesnick, S.L., Cox, S.B., Kerr, I., Payne, R., Stegeman, J.J., 2011. Pacific Ocean–wide profile of CYP1A1 expression, stable carbon and nitrogen isotope ratios, and organic contaminant burden in sperm whale skin biopsies. Environmental Health Perspectives 119, 337–343.

González, J.A., Montes, C., Rodríguez, J., Tapia, W., 2008. Rethinking the Galapagos Islands as a complex social-ecological system: implications for conservation and management. Ecology and Society. 13 (2), 13. http://www.ecologyandsociety.org/vol13/iss2/art13/.

Grenier, C., 2010. The Geographic opening of Galápagos. In: Galápagos Report 2009–2010. CDF, GNP, and Governing Council of Galápagos, Puerto Ayora, Galápagos, Ecuador.

Hall, A.J., Thomas, G.O., 2007. Polychlorinated biphenyls, DDT, polybrominated diphenyl ethers and organic pesticides in United Kingdom harbor seals (*Phoca vitulina*)—mixed exposures and thyroid homeostasis. Environmental Toxicology and Chemistry 26, 851–861.

IMO, 2002. Antifouling Systems. Focus on IMO. International Maritime Organization, London, United Kingdom, p. 31.

IMO, 2017a. International Convention on the Control of Harmful Anti-Fouling Systems on Ships. International Maritime Organization (IMO). http://www.imo.org/en/About/Conventions/ListOfConventions/Pages/International-Convention-on-the-Control-of-Harmful-Anti-fouling-Systems-on-Ships-(AFS).aspx.

IMO, 2017b. Convention on the Prevention of Marine Pollution by Dumping of Wastes and Other Matter. International Maritime Organization (IMO). http://www.imo.org/en/OurWork/Environment/LCLP/Pages/default.aspx.

Iwata, H., Tanabe, S., Sakai, N., Tatsukawa, R., 1993. Distribution of persistent organochlorines in the oceanic air and surface seawater and the role of ocean on their global transport and fate. Environmental Science and Technology 27 (6), 1080–1098.

Johannessen, D.I., Ross, P.S., 2002. Late-Run Sockeye at Risk: An Overview of Environmental Contaminants in Fraser River Salmon Habitat Can. Tech. Rep. Fish. Aquat. Sci. 2429.

Kelce, W.R., Stone, C.R., Laws, S.C., Gray, L.E., Kemppainen, J.A., Wilson, E.M., 1995. Persistent DDT metabolite p,p-DDE is a potent androgen receptor antagonist. Nature 375, 581–585.

Kingston, P.F., Runciman, D., McDougall, J., 2003. Oil contamination of sedimentary shores of the Galápagos Islands following the wreck of the Jessica. Marine Pollution Bulletin 47, 303–312.

Lessmann, R.P., 2004. Current protections on the Galapagos Islands are inadequate: the International Maritime Organization should declare the islands a particularly sensitive sea area. Colorado Journal of International Environmental Law and Policy 15, 117–151.

Linley-Adams, G., 1999. The Accumulation and Impact of Organotins on Marine Mammals, Seabirds and Fish for Human Consumption Report for World Wildlife Fund -UK, WWF-UK Project No 98054. 26 p.

Lusher, A.L., Hernandez-Milian, G., O'Brien, J., Berrow, S., O'Connor, I., Officer, R., 2015. Microplastic and macroplastic ingestion by a deep diving, oceanic cetacean: the True's beaked whale, *Mesoplodon mirus*. Environmental Pollution 199, 185–191. https://doi.org/10.1016/j.envpol.2015.01.023.

MacFarland, C., Cifuentes, M., 1996. Case study: Galapagos, Ecuador. In: Dompka, V. (Ed.), Human Population, Biodiversity and Protected Areas: Science and Policy Issues Report of a Workshop, American Association for the Advancement of Science (AAAS), Washington, DC, 20–21 April 1995.

Merlen, G., 1995. Use and misuse of the seas around the Galapagos Archipelago. Oryx 29, 99–106.

Miranda-Filho, K.C., Metcalfe, T.L., Metcalfe, C.D., Metcalfe, R.B., Robaldo, R.B., Muelbert, M.M.C., Colares, E.P., Martinez, P.E., Bianchini, A., 2007. Residues of persistent organochlorine contaminants in southern elephant seals (*Mirounga leonina*) from Elephant Island, Antarctica. Environmental Science and Technology 41, 3829–3835.

Moir, F.C., Armijos, E., 2007. Integrated water supply and wastewater solutions for the town of Puerto Ayora on the island of Santa Cruz–the Galapagos Islands–Ecuador. Proceedings of the Water Environment Federation 70, 4816–4842.

Moore, C.J., 2008. Synthetic polymers in the marine environment: a rapidly increasing, long-term threat. Environmental Research 108, 131–139.

Mos, L., Morsey, B., Jeffries, S.J., Yunker, M.B., Raverty, S., De Guise, S., Ross, P.S., 2006. Chemical and biological pollution contribute to the immunological profiles of free-ranging harbor seals. Environmental Toxicology and Chemistry 25, 3110–3117.

Mos, L., Cameron, M., Jeffries, S.J., Koop, B., Ross, P.S., 2010. Risk-based analysis of PCB toxicity in harbour seals. Integrated Environmental Assessment and Management 6 (4), 631–640.

Muñoz-Abril, L.J., 2016. Ecología trófica, diversidad genética y contaminación por mercurio del atún aleta amarilla (*Thunnus albacares*) en la Reserva Marina de Galápagos y el continente ecuatoriano. Trabajo de Titulación en Ecología. Universidad San Francisco de Quito, Quito, Ecuador. 96 p. http://repositorio.usfq.edu.ec/bitstream/23000/5742/1/126989.pdf.

Overbey, K.N., Hatcher, S.M., Stewart, J.R., 2015. Water quality and antibiotic resistance at beaches of the Galápagos Islands. Frontiers in Environmental Science 3, 64. https://doi.org/10.3389/fenvs.2015.00064.

Rodríguez-Grimón, R.O., Osorio, M.F.A., de Freitas, D.M., Castro, Í.B., 2016. Tributyltin impacts in Galapagos Islands and Ecuadorian shore: marine protected areas under threat. Marine Policy 69, 24–31.

Romero, L.M., Wikelski, M., 2002. Severe effects of low-level oil contamination on wildlife predicted by the corticosterone-stress response: preliminary data and a research agenda. Spill Science and Technology Bulletin 7, 309–313.

Ross, P.S., De Swart, R.L., Reijnders, P.J.H., Van Loveren, H., Vos, J.G., Osterhaus, A.D.M.E., 1995. Contaminant-related suppression of delayed type hypersensitivity and antibody responses in harbor seals fed herring from the Baltic Sea. Environmental Health Perspectives 103, 162–1677.

Ross, P.S., 2002. The role of immunotoxic environmental contaminants in facilitating the emergence of infectious diseases in marine mammals. Human and Ecological Risk Assessment 8, 277–292.

Ross, P.S., Morales-Caselles, C., 2015. Out of sight, but no longer out of mind: microplastics as a global pollutant. Integrated Environmental Assessment and Management 11 (4), 721–722.

Schiller, L., Alava, J.J., Grove, J., Reck, G., Pauly, D., 2015. The demise of Darwin's fishes: evidence of fishing down and illegal shark finning in the Galápagos Islands. Aquatic Conservation: Marine and Freshwater Ecosystems 25 (3), 431–446.

Stumpf, C.H., Gonzalez, R.A., Noble, R.T., 2013. Investigating the coastal water quality of the Galápagos Islands, Ecuador. In: Walsh, S.J., Mena, C.F. (Eds.), Science and Conservation in the Pollution as an Emerging Threat for the Conservation of the Galápagos Marine. Galapagos Islands: Frameworks and Perspectives, Social and Ecological Interactions in the Galápagos Islands. Springer, New York.

Schenker, U., Scheringer, M., Hungerbühler, K., 2008. Investigating the global fate of DDT: model evaluation and estimation of future trends. Environmental Science and Technology 42, 1178–1184.

Tabuchi, M., Veldhoen, N., Dangerfield, N., Jeffries, S., Helbing, C., Ross, P., 2006. PCB related alteration of thyroid hormones and thyroid hormone receptor gene expression in free-ranging harbor seals (*Phoca vitulina*). Environmental Health Perspectives 114, 1024–1031.

Tanabe, S., 1999. Butyltin contamination in marine mammals-a review. Marine Pollution Bulletin 39 (1–12), 62–72.

Tierney, K.B., Sampson, J.L., Ross, P.S., Sekela, M.A., Kennedy, C.J., 2008. Salmon olfaction is impaired by an environmentally realistic pesticide mixture. Environmental Science and Technology 42 (13), 4996–5001. https://doi.org/10.1021/es800240u.

UNEP, 2017. Minamata Convention on Mercury. Retrieved from: http://www.mercuryconvention.org/Convention/tabid/3426/language/en-US/Default.aspx.

Watkins, G., Cruz, F., 2007. Galápagos at Risk: A Socioeconomic Analysis of the Situation in the Archipelago. Charles Darwin Foundation, Puerto Ayora, Province of Galápagos, Ecuador.

Wikelski, M., Romero, L.M., Snell, H.L., 2001. Marine iguanas oiled in the Galapagos. Science 292, 437–438.

Wikelski, M., Wong, V., Chevalier, B., Rattenborg, N., Snell, H.L., 2002. Marine iguanas die from trace oil pollution. Nature 417, 607–608.

Wilkening, K.E., Barrie, L.A., Engle, M., 2000. Trans-pacific air pollution. Science 290 (5489), 65–67.

Woram, J., 2005. Charles Darwin Slept Here. Rockville Press, NY.

World Health Organization, 2006. WHO: WHO Gives Indoor Use of DDT, a Clean Bill of Health for Controlling Malaria: WHO Promotes Indoor Residual Spraying with Insecticides as One of Three Main Interventions to Fight Malaria. World Health Organization, Washington, DC. http://www.who.int/mediacentre/news/releases/2006/pr50/en/index.html.

Section II

Effects of Toxicological and Cumulative Stress on Marine Mammal Health

Chapter 9

Field Sampling Techniques and Ecotoxicologic Biomarkers in Cetaceans

Céline A.J. Godard-Codding[1], Maria Cristina Fossi[2]
[1]The Institute of Environmental and Human Health Texas Tech University, Lubbock, TX, United States; [2]University of Siena, Siena, Italy

INTRODUCTION

Pressures on dolphins and whales in ocean and sea waters worldwide are higher than they have ever been in the last decades and have different origins. In the last 30 years, there has been growing concern about hazards to cetaceans occasioned by multiple stress factors, due to bioaccumulation and effects of anthropogenic contaminants combined with the impact of infectious diseases, marine litter including microplastics, climate change, food depletion (overfishing), trapping, noise, ship collisions, whale watching activities, and genetic erosion (Notarbartolo di Sciara et al., 2003; Bossart, 2011; Fossi et al., 2013, 2018). The simultaneous combination of some or all of these pressures for different cetacean species may produce dramatic effects on population stability. Exposure to contaminants combined with other human impacts can affect survival, recruitment, reproductive success, mutation rates, and migration, and it may play a significant role in the partitioning of genetic variation among populations exposed to different levels of stress (Whitehead et al., 2003).

While the levels of some legacy contaminants may be declining in the marine environment, those of others such as polybrominated diphenyl ethers (PBDEs) and emerging contaminants including plastic additives seem to be increasing (Petterson et al., 2004; Baini et al., 2017). The need for sensitive markers of both legacy and emerging contaminants in cetaceans and for alternative ex vivo models to explore susceptibilities to these compounds led the scientific community to develop several biomarker techniques applicable to skin biopsies or tissues collected from stranded animals as a new nonlethal investigative tool (Godard et al., 2004; Fossi et al., 2006; Panti et al., 2011; Godard-Codding et al., 2011). This approach was born in the early 1990s (Fossi et al., 1992), using specifically nonlethal methods, and it resulted in the first molecular

Marine Mammal Ecotoxicology. https://doi.org/10.1016/B978-0-12-812144-3.00009-7

biomarkers determination (induction of CYP1A1) in skin biopsies of free-ranging cetaceans (fin whale *Balaenoptera physalus* and striped dolphin *Stenella coeruleoalba*) sampled in the Mediterranean Sea. The first published record of cetacean cell culture was probably by Kniazeff and Groyon in 1966, but the potential of such an in vitro model for research on the impact of pollutants and related biomarker studies was initiated in the mid-1990s and started gaining popularity in the 2000s (Carvan et al., 1994, 1995; Gauthier et al., 1998, 1999; Fossi et al., 2006; Godard et al., 2006). In 2004, Godard et al., initiated the first organotypic model in cetaceans using skin biopsy slices from sperm whales (*Physeter macrocephalus*). This model allowed for the validation of CYP1A1 expression as a biomarker of exposure in cetaceans by demonstrating a causal relationship between chemical exposure and CYP1A1 induction. In the last 30 years, the literature published on this topic has increased exponentially, both applying this methodology in different species distributed worldwide and using increasingly sophisticated techniques. More recently, contaminants of emerging concern (CEC), e.g., pharmaceuticals, personal care products, nanomaterials, and plastics, are increasingly being detected in surface water. Several are classified as endocrine disruptor chemicals (EDCs), for they can alter the normal hormone functions affecting reproduction, development, and metabolism (Casals-Casas and Desvergne, 2011). EDCs place the entire marine ecosystem at risk, at every trophic food level, with the top predators, such as cetaceans, being the most vulnerable. A recent study identified gene markers specific for different CEC exposure in the skin of cetaceans (Mancia et al., 2018).

In this context the main aim of this chapter is to do the following: (1) describe the most suitable sampling techniques applicable to nonlethal ecotoxicologic studies in cetaceans; (2) identify biomarkers that can help determine which pollutants are of particular concern in cetacean species; (3) identify, among the various categories of selected biomarker techniques, the series of markers that best allows us to diagnose the presence of a specific class of pollutants, or a mixture of them; (4) describe case studies with a special focus on interspecific variability of biomarker responses, the application of the biomarkers approach in cetaceans inhabiting hot spot areas, and in worldwide monitoring of sentinel species. Among the body of literature on ecotoxicologic investigations conducted on cetaceans through the use of biomarker techniques, this chapter will focus essentially on the skin matrix and on two study areas (as main topics of investigations for the authors of this chapter): the Mediterranean Sea and the Pacific Ocean.

FIELD SAMPLING TECHNIQUES IN CETACEANS

The development of a series of nondestructive techniques to evaluate residue levels and biomarker responses is recommended, in place of the lethal approach, for hazard assessment and conservation of endangered species of aquatic mammals. In the last decades, much interest has been shown internationally toward skin biopsies as a sensitive nonlethal technique for hazard assessment of free-ranging cetaceans exposed to POPs and emerging contaminants (Godard et al., 2004; Fossi et al., 2006; Panti et al., 2011; Godard-Codding et al., 2011). The biomarker and residue analysis approaches can be used in marine mammals with skin samples collected from either freshly stranded or free-ranging animals. While other sample matrices valuable for ecotoxicologic research can be collected noninvasively or minimally invasively from wild cetaceans, this chapter will focus primarily on the skin matrix.

Stranded Animals for Ecotoxicologic Investigation

In theory, all stranded cetaceans in a good state of conservation can be used to some extent in ecotoxicologic investigation. Skin as well as blood and all internal organs and tissues can be suitable materials for a wide range of analyses. Different sample matrices may be more suited to particular biomarker or residue analyses than others, depending on the contaminant or class of chemicals of interest, the time between death and sampling, and the sample storage condition. Geraci and Lounsbury (1993) and Mazzariol et al. (2011) thoroughly covered specimen and data collection in the case of marine mammal strandings. Whenever possible and before postmortem examinations, the stranded cetaceans need to be measured for total length and weighed. When weight measurement is not possible, the total length may be used to estimate weight according to formulas specific to the species investigated. Ideally, sample collection for all organs and tissues needs to be performed by necropsy-authorized personnel for later microscopic examinations (i.e., histopathology, immunohistochemistry, and ultrastructural investigations), as well as virologic and microbiologic investigations. Teeth and baleens can also be sampled for age determination (two teeth for each animal in odontocetes), while gastric contents can be collected during opening of the stomach complexes for the examination of organic, inorganic (including marine litter), and parasitic loads. Skin samples can be collected for genetic, stable isotope, fatty acid, hormone, biomarker, and chemical analyses, as well as cell and organotypic cultures. Most analyses are suitable with frozen storage conditions, but some require storage at various temperature or in various media, buffers, or solutions.

Ecotoxicologic Investigation in Free-Ranging Cetaceans: The Skin Biopsy Approach

Several international institutions, such as the International Whaling Commission (IWC) have encouraged research on panels of sensitive nonlethal biomarkers, combined with analyses of persistent, bioaccumulative, and toxic (PBT) residues in skin biopsies of free-ranging animals. This approach can help define the health status of cetacean species with respect to multiple threats and supports IWC projects such as Pollution 2000+ and Pollution 2020. From an ecotoxicologic perspective, it is preferable to obtain samples from live free-ranging animals with collection methodologies that pose no or minimum disturbance. The most useful samples for nondestructive studies in cetaceans are skin biopsy specimens, obtained remotely by dart. The biopsy dart method has been used successfully on a range of cetacean species worldwide and is considered relatively benign (Noren and Mocklin, 2012). The response of cetaceans to skin biopsy collection is considered low in odontocetes and low to moderate in mysticetes, while wound healing appears to be rapid, with no apparent adverse health effects (Noren and Mocklin, 2012).

Cetacean skin biopsies are suitable for hazard assessment of free-ranging cetaceans (Fossi et al., 1992, 2013, 2014, 2016, 2018; Godard et al., 2004; Godard-Codding et al., 2011). Various dart methodologies have been used successfully, as reviewed in Noren and Mocklin (2012). Skin biopsies (epidermis and dermis/blubber) from free-ranging dolphins (such as *Tursiops truncatus, Stenella coeruleoalba*) can be obtained using an aluminum pole armed with biopsy tips (e.g., 0.7 cm ø, 3.0 cm length) or with a crossbow and darts. Skin biopsies from large odontocete (*Physeter macrocephalus*) or mysticete (such as *Balaenoptera physalus* or other baleen whale) (Fig. 9.1) species can be obtained with a crossbow or air gun and darts armed with tips (e.g., 0.9 cm ø, 4.0 cm length). Several models of crossbows (such as a Barnett Wildcat II crossbow with a 150-pound test bow), air guns, and darts (preferably untethered and with or without prongs of different angles in the tip) are available. To avoid the possibility of infection, the bolt tip needs to be sterilized before deployment. Biopsy samples can be taken between the dorsal fin and the upper part of the caudal peduncle upon approaching the animal at a suitable distance and speed as specifically permitted for the species and research project. The skin biopsy needs to be stored immediately in the proper conditions required for intended analyses. Common storage conditions include frozen, as is, in liquid

FIGURE 9.1 Skin biopsy sampling in fin whale (*Balaenoptera physalus*) in the Mediterranean Sea. Remote sampling of integument biopsies from mysticete can be obtained using a crossbow and darts armed with tips (0.9 cm ø, 4.0 cm length).

nitrogen, dry ice, or −80 and −20C freezers or stored either cold or at room temperature in cell medium, buffer, or specific reagents.

Skin biopsy is a powerful tool for ecotoxicologic studies for the following reasons: (1) it allows collection of a large number of samples across a wide geographic range; (2) it allows collection of sequential samples from the same animal if identified by photo identification or genetics; (3) it is suitable for residue analysis of many contaminants including dioxin-group chemicals (suitable for calculation of TEQs), other halogenated aromatic hydrocarbons, polycyclic aromatic hydrocarbons (PAHs), plastic additives (phthalates), and heavy metals; (4) it is suitable for several biomarker analyses (see the next section) and cell and organ culture. A number of successful studies show that cetacean skin biopsies are a powerful nonlethal tool for assessing ecotoxicologic risk in marine mammals and aspects of feeding ecology and food preferences (Fig. 9.1).

SKIN BIOMARKERS IN THE CETACEAN'S WORLD

The United Nations Intergovernmental Oceanographic Commission (IOC) (including the Global Ocean Observing System), the IWC, and the marine mammal scientific community have encouraged research on panels of sensitive biomarkers combined with contaminant residue analyses when assessing the health status of cetacean species as sentinels of ocean integrity (Pollution2000+ (IWC), IOC 2001, Knap et al., 2002, Godard-Codding et al., 2011, Parsons et al., 2015). Cetacean skin biopsies are suitable for hazard assessment of free-ranging cetaceans (Fossi et al., 1992, 2013, 2014, 2016, 2018; Godard et al., 2004; Godard-Codding et al., 2011) obtainable with minimum disturbance for animals. The biomarkers highlighted in this chapter are measured in skin biopsies as either general or specific diagnostic signals and at different hierarchic levels (see Fig. 9.2 and Table 9.1).

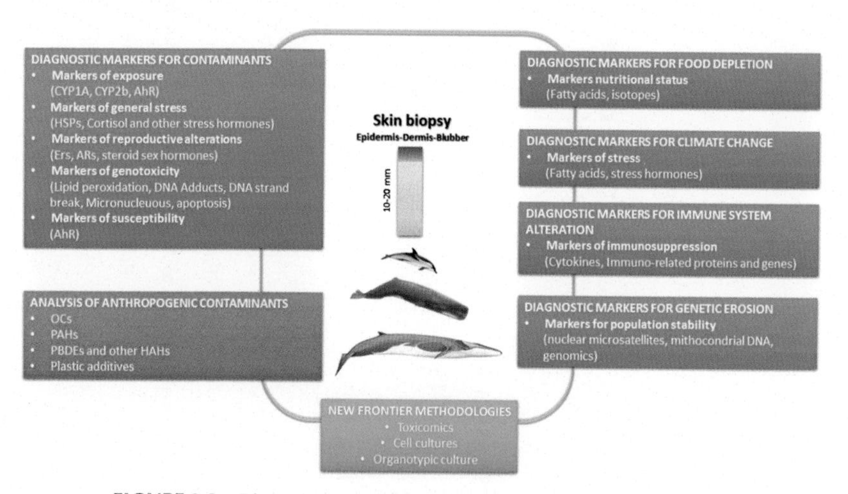

FIGURE 9.2 Biomarker and residue analyses relevant to the skin matrix.

TABLE 9.1 Skin Biopsy for Diagnosis of Anthropogenic Threats in Cetaceans (Anthropogenic Contaminants, Markers, Biologic Responses, and Analytical Methodology)

Marker	Response	Methodology	Reference
Detection of Anthropogenic Contaminants			
HCB, DDTs, PCBs	Organochlorine contaminants (OCs)	GC/MS: concentration in blubber	Godard-Codding et al., 2011 Fossi et al. (1992, 2003, 2010, 2013, 2014, 2016)
HCB, DDTs, PCBs	Organochlorine contaminants (OCs)	HRCGC: concentration in blubber	Marsili and Focardi (1997)
PBDEs	Flame retardants: polybrominated diphenyl ethers	GC/MS: concentration in blubber	Petterson et al. (2004)
PAEs	Phthalate esters	GC/MS: concentration in blubber	Baini et al. (2017)
Markers of Exposure to Anthropogenic Contaminants			
CYP1A1	Protein induction: enzyme induction	Enzyme induction in skin biopsy	Fossi et al. (1992, 2003)
CYP1A1, CYP2B	Protein induction, substrate-inducible, and substrate-specific (OCs, PAHs, PBDEs, PCDDs)	WB: protein concentration in skin biopsy	Fossi et al. (2008, 2010, 2013, 2014, 2016)

CYP1A1	Upregulation (DDTs, HCHs)	qRT-PCR: gene expression in skin cultures	Jia et al. (2015)
	Upregulation (PCBs)	qRT-PCR: gene expression in skin samples	Noel et al. (2014)
	Protein expression	Immunohistochemistry in skin biopsy	Godard-Codding et al., 2011
	Dose-response expression by AhR agonist	Immunohistochemistry in skin organotypic culture	Godard et al. (2004)
	Upregulation of mRNA CYP1A1 (PAHs, PCBs, PCDDs)	qRT-PCR: gene expression in skin biopsy	Panti et al. (2011)
	Upregulation of mRNA CYP1A1 (PAHs, PCBs, PCDDs)	qRT-PCR: gene expression in skin biopsy	Fossi et al. (2010, 2013, 2014, 2016)
CYP1B1	Gene expression	Gene sequencing	Godard et al. (2000)
AhR	Upregulation (HCHs, HCB)	qRT-PCR: gene expression in skin cultures	Jia et al. (2015)
	Upregulation (PCBs)	qRT-PCR: gene expression in skin samples	Noel et al. (2014)
	Upregulation mRNA of nuclear transcription factor (PAHs, PCBs, PCDDs)	qRT-PCR: gene expression in skin biopsy	Panti et al. (2011)
	Upregulation mRNA of nuclear transcription factor (PAHs, PCBs, PCDDs)	qRT-PCR: gene expression in skin biopsy	Fossi et al. (2010, 2013, 2014)
Steroid hormone receptors (ERs)	Dysregulated in the presence of endocrine disrupting chemicals (EDCs)	qRT-PCR: gene expression in skin biopsy	Panti et al. (2011)
	Dysregulated in the presence of endocrine disrupting chemicals (EDCs)	qRT-PCR: gene expression in skin biopsy	Fossi et al. (2010, 2013, 2014, 2016)

Continued

TABLE 9.1 Skin Biopsy for Diagnosis of Anthropogenic Threats in Cetaceans (Anthropogenic Contaminants, Markers, Biologic Responses, and Analytical Methodology)—cont'd

Marker	Response	Methodology	Reference
Marker of General Stress			
Cell death	Cytotoxicity upon PAH exposure	Cell culture in skin and other organs	Godard et al. (2006)
HSP70	Upregulation (DDTs)	qRT-PCR: gene expression in skin cultures	Jia et al. (2015)
	Stress-induced transcription factor (chemicals, general stress)	qRT-PCR: gene expression in skin biopsy	Panti et al. (2011)
	Stress-induced transcription factor (chemicals, general stress)	qRT-PCR: gene expression in skin biopsy	Fossi et al. (2010, 2013, 2014, 2016)
E2F-1	Stress-induced apoptosis signal (chemicals, general stress)	qRT-PCR: gene expression in skin biopsy	Panti et al. (2011)
	Stress-induced apoptosis signal (chemicals, general stress)	qRT-PCR: gene expression in skin biopsy	Fossi et al. (2010, 2013, 2014, 2016)
Markers of Genetic Erosion (Individual Heterozygosity)			
Population genetic variability	Within-individual genetic variation related to health/resilience. Calculation of heterozygosity by assigning a score to each locus, weighted by the average heterozygosity at that locus size/stability	Standardized heterozygosity observed (st.het.Obs)	Coltman and Slate (2003)
Population genetic variability	Within-individual genetic variation related to health/resilience	Standardized heterozygosity observed (st het.Obs)	Fossi et al. (2013)

Protein Expression in Cetaceans' Skin Biopsies

Each biomarker described in this chapter represents a different diagnostic signal (Fossi et al., 2014). The cytochrome P450 family is the most important metabolic/detoxifying enzyme system in mammals. CYP1A1 induction is widely used as a biomarker of exposure to aryl hydrocarbon (AhR) agonists in animal species (Stegeman et al., 1992; Fossi et al., 2003, 2013, 2014, 2016). Among the few studies of cytochrome P450s in cetaceans, several have examined the metabolism of foreign chemicals in hepatic microsomes or cell culture (Goksøyr et al., 1986; Murk et al., 1994; White et al., 1994; Boon et al., 1998; White et al., 2000), and a CYP1A1 gene has been identified in several species (Teramitsu et al., 2000). Correlations between nonortho and mono-ortho PCB burdens in blubber and hepatic CYP1A1 content and activity have been observed in beluga whales (White et al., 1994). The use of skin biopsies for measuring CYP1A1 activity in marine mammals has been advocated as a valid nondestructive method since the early 1990s (Fossi et al., 1992, 2003). CYP1A1 was confirmed as a biomarker of exposure in cetaceans in 2004, when a causal relationship between chemical exposure and CYP1A1 induction was established ex vivo using sperm whale skin biopsy slices exposed to various concentrations of an AhR agonist (Godard et al., 2004). Overall, CYP1A and CYP2B have been detected in cetacean skin, and induction of these isoforms was found after exposure to various AhR agonists, including lipophilic contaminants such as OCs, PAHs, and brominated flame retardant both ex vivo and in field studies (Godard et al., 2004; Fossi et al., 2006, 2008; Hooker et al., 2008; Montie et al., 2008; Fossi et al., 2010, 2013, 2014, 2016; Godard-Codding et al., 2011). The induction of CYP isoforms can thus be considered a powerful biomarker of exposure. Furthermore, basal expression and upregulation of extrahepatic CYP enzymes can significantly affect local disposition of xenobiotics or endogenous compounds in peripheral tissues (Shimada et al., 1996; Wilson et al., 2007) and thus modify their pharmacologic and toxicologic effects or affect absorption of xenobiotics into the systemic circulation.

Gene Expression in Cetaceans' Skin Biopsies

Skin samples collected from live animals or immediately after death are amenable to gene expression studies. RNA starts degrading very rapidly (within minutes), and the length of time between sample collection and proper storage conditions is an important factor to take into account when considering gene expression studies. Storage in liquid nitrogen is ideal but may be challenging in field conditions. Storage in freezers is adequate (at −80 C preferred over −20 C), and some solutions are now commercially available that can be used to ensure RNA stability at room temperature or on ice for a limited time. A limited number of ecotoxicologic studies have been conducted in cetaceans so far. Panti et al. (2011) were the first to evaluate the mRNA levels of CYP1A1, E2F, heat shock protein HSP70, and estrogen receptor ER alpha in cetaceans. They measured the gene expression of these five putative biomarkers by quantitative real-time PCR in skin biopsies of Mediterranean striped dolphins and observed correlations between contamination levels and some of the genes examined. A similar study was conducted by Fossi et al. (2014) in striped dolphin skin biopsy slices exposed to various pollutants, including OCs, PBDEs, and PAHs, where the sensitivity and selectivity of the resulting biomarker responses were assessed. The same gene expression biomarkers were evaluated in fin whale in the Mediterranean Sea (*Balaenoptera physalus*) and in several cetacean species in the Sea of Cortez (Fossi et al., 2010, 2014, 2016). Noel et al. (2014) measured the levels of 13 health-related gene transcripts and their possible association with POP burdens in artic belugas (*Delphinapterus leucas*). They observed a positive correlation between PCBs and both CYP1A1 and AhR gene expression. Van Dolah et al. (2015) investigated the skin transcriptome of common bottlenose dolphins (*Tursiops truncatus*) from the Northern Gulf of Mexico before and after the Deepwater Horizon oil spill of 2010. The skin transcriptomes were characterized using a dolphin microarray. Interestingly, global gene expression did not correlate strongly with persistent organic pollutant burdens in blubber but was influenced by seasonal effects. Jia et al. (2015) examined the expression of CYP1A1, AhR, and HSP70 in Indo-Pacific humpback dolphin (*Sousa chinensis*) skin fibroblast cultures exposed to various POPs and other AhR agonists. Positive correlations between contaminant exposure and biomarker expression were observed. Together, these studies illustrate the power of gene expression analyses for ecotoxicologic investigation in cetaceans.

The Relevance of In Vitro (Cell Cultures) and Ex Vivo (Organotypic Cultures) Models in Ecotoxicology

Both cell and organotypic cultures can be established from skin samples obtained by biopsy or from stranded animals shortly after death. The recent development of these cultures for ecotoxicology research in cetaceans opens critical avenues of scientific inquiries in species for which in vivo experimentation is prohibited. The first published account of skin cetacean cell culture appears to be a 1981 publication by Jarell and Árnason focused on chromosome staining in belugas (*Delphinapterus leucas*). Since then, over 20 studies have been published on cetacean skin cultures (Gauthier et al., 1998, 1999; Marsili et al., 2000; Fossi et al., 2006; Godard et al., 2006). The in vitro skin model has allowed for the investigation of CYP1A1 and CYP2B biomarker expression (Fossi et al., 2006, 2008, Jia et al., 2015), the cytotoxicity and genotoxicity of a variety of contaminants (Gauthier et al., 1998, 1999; Fossi et al., 2006; Godard et al., 2006; Chen et al., 2009, 2012; Frenzilli et al., 2014; Jia et al., 2015), and transcriptomics of cetacean skin cells exposed to hexavalent chromium (Pabuwal et al., 2013) and methylmercury (Ellis et al., 2010). The fields of medicine, pharmacology, and terrestrial mammalian toxicology rely heavily on in vitro methodology, and its possible applications in cetacean toxicology have only begun to be explored.

Godard et al. (2006) compared the cytotoxicity of benzo[a]pyrene in skin, testis, and lung cell lines in the right whale (*Eubalaena glacialis*). This study showed greater toxicity in testis cells than in skin or lung cells after 48- and 72-h exposure to benzo[a]pyrene. It was the first study comparing the in vitro effects of a toxicant in multiple organs of the same cetacean, providing information on the relative sensitivity of somatic versus reproductive organs. Frenzilli et al., in 2014, compared the genotoxic potential of nanosized TiO_2 anatase and microsized rutile in bottlenose dolphin (*Tursiops truncatus*), mouse, and human skin cell cultures and observed some species differences. This illustrates the potential for in vitro studies to provide a path to conducting meaningful toxicity comparisons across taxa that are relevant to cetacean ecotoxicology.

In 2004, Godard et al., initiated the first organotypic model in cetaceans using skin biopsy slices from sperm whales (*Physeter macrocephalus*). This ex vivo model allowed for the validation of CYP1A as biomarker of exposure to AhR agonist in cetaceans. Fossi et al. (2014) used a similar model for exposing skin samples collected from a female striped dolphin within 4h after death. The skin slices were exposed to organochlorine compounds, polybrominated diphenyl ethers, and PAHs, and a series of biomarkers were investigated including CYP1A1, CYP2B, heat shock protein 70, estrogen receptor alpha, and E2F transcription factor expression. These studies illustrate the considerable potential of ex vivo methodology for cetacean ecotoxicology, and we recommend its use. The organotypic model has been shown to more closely simulate in vivo conditions than cell culture in toxicology assessments and is widely used in pharmacology (Parrish et al., 1995; Boess et al., 2003; Vickers and Fisher, 2004; Pampaloni et al., 2007). The adoption of precision-cut slices and the validation of slice viability in cetacean ex vivo studies will further increase their relevance and scope of applications.

The Relevance of Hormone Analyses in a Multidisciplinary Biomarker Approach to Ecotoxicologic Investigation

Hormone analyses can provide information on stress and reproductive fitness that is critical for the conservation and management of cetaceans (Parsons et al., 2015; De Mello and De Oliveira, 2016). For example, in various species, progesterone and glucocorticosteroid analyses have successfully been used to detect pregnancy and stress status, respectively, while testosterone analyses may be useful as markers of male reproductive status (Mansour et al., 2002; Kellar et al., 2006, 2009, 2013, 2015; Pérez et al., 2011; Trego et al., 2013; Fair et al., 2014; Thompson et al., 2014; Bechshoft et al., 2015; Trana et al., 2015, 2016; Clark et al., 2016; Champagne et al., 2017). When paired with residue analyses, especially of EDCs, or with biomarkers of exposure or effects to these chemicals, hormone analyses have the potential to greatly increase our understanding of the impact of toxicologic threats to this taxon. Hormone analyses have been conducted in a variety of cetacean species and sample matrices, including blubber, serum, urine, feces, saliva, muscle, earwax, baleen, and blow (Yoshioka et al., 1994; Mansour et al., 2002; Rolland et al., 2005; Hunt et al., 2006; Kellar et al., 2006; Biancani et al., 2009; Hogg et al., 2009; Kellar et al., 2009; Pérez et al., 2011; Kellar et al., 2013; Trego et al., 2013; Trumble et al., 2013; Hunt et al., 2014a,b; Thompson et al., 2014; Bechshoft et al., 2015; Kellar et al., 2015; Trana et al., 2015; Clark et al., 2016; Trana et al., 2016; Champagne et al., 2017; Hayden et al., 2017). To date, most published studies on the topic have been conducted on blubber samples collected primarily from freshly dead animals with only a limited but growing number of studies reporting on analyses of skin biopsies. Historically, hormone analyses performed in blubber samples have relied on immunoassays, but Hayden et al. (2017) and Boggs et al. (2017) developed a liquid chromatography-tandem mass spectrometry (LC-MS/MS) methodology that allows for the simultaneous detection of multiple hormones including progesterone, 17-hydroxyprogesterone, testosterone, androstenedione, cortisol, 11-deoxycortisol, cortisone, and 11-deoxycorticosterone in a single sample. The sample mass requirement for hormone analyses by LC-MS/MS is small (50 mg minimum), which allows for the concurrent analyses of residues or biomarkers in the same biopsy. Hence, hormone analyses are a highly relevant fit to a multidisciplinary biomarker approach to ecotoxicologic investigation.

APPLICATION IN CASE STUDIES

Interspecific Variability

Fossi et al., in 2014 applied for the first time a suite of diagnostic biomarkers to seven cetacean species (*B. physalus, B. edeni, B. musculus, T. truncatus, D. capensis P. microcephalus,* and *O. orca*) to evaluate the role of the feeding habits and migratory behavior in the toxicologic status of these species from the Gulf of California, Mexico. The authors investigated the interspecific differences in cytochrome P450 1A1 and 2B (CYP1A1 and CYP2B, respectively), aryl hydrocarbon receptor, and E2F transcription factor 1 and the contaminants levels [organochlorine compounds, polybrominated diphenyl ethers (PBDEs) and PAHs] in four odontocete species (common bottlenose dolphin, long-beaked common dolphin, sperm whale, and killer whale) and three mysticete species (blue whale, fin whale, and Bryde's whale) using skin biopsy. Differences in contaminant levels and molecular biomarker responses between odontocete and mysticete species were observed. The canonical discriminant analysis on principal component analysis factors, performed to reveal clustering variables, showed that odontocetes are characterized by the highest levels of lipophilic contaminants compared to mysticetes, with the highest levels of polychlorinated biphenyls, dichlorodiphenyltrichloroethanes and PBDEs detected in killer whale and the lowest levels in Bryde's whale.

The biomarker data showed interspecific differences among the seven species, revealing highest CYP1A and CYP2B protein levels in the mysticete

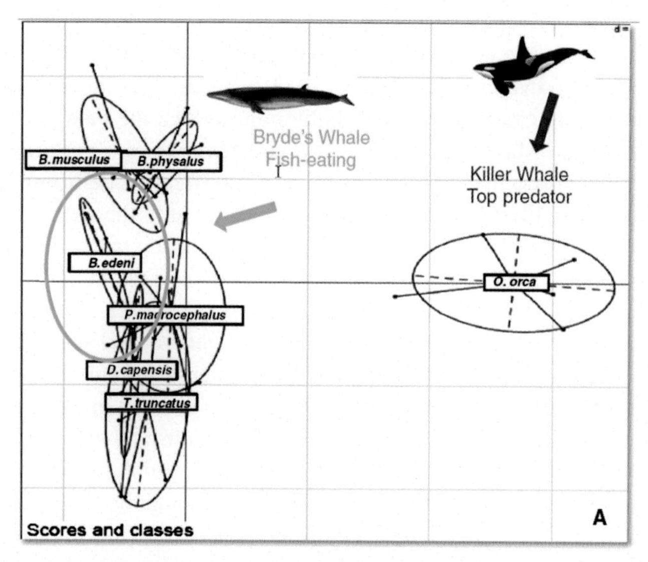

FIGURE 9.3 Correlation between contaminant (total OCs, PAHs, and PBDEs) and biomarkers (CYP1A1, CYP2B, AhR, E2F1) variables and discriminant functions in all the seven species studied. *(Reprinted with permission from Fossi, M.C., Casini, S., Maltese, S., Panti, C., Spinsanti, G., Marsili, L., 2014. An "ex vivo" model to evaluate toxicologic responses to mixtures of contaminants in cetaceans: integumentum biopsy slices. Environmental Toxicology 29 (10), 1107–1121.)*

fish-eating species (Bryde's whale). These data suggested that several factors can influence the different toxicologic hazard observed among the cetacean species inhabiting the Gulf of California: (1) the position in the food chain (odontocete show the highest POP concentrations and biomarker responses as a signal of toxicologic exposure); (2) the inductive ability linked to the presence of planar (CYP1A1 induction) and globular (CYP2B induction) halogenated compounds in the blubber of blue whale; (3) the migratory/resident behavior of the species (the two baleen whales, blue whale, and fin whale, although being at the same levels of the food web, show different levels of contaminants and biomarker responses due to the different feeding grounds); and (4) the role of evolutionary pressure related to the different dietary habits of the species on CYP enzymatic systems (zooplankton-eating species fin whale and blue whale vs. the fish-eating species Bryde's whale). In evaluating the toxicologic status, the authors' final aim was to identify among the seven species analyzed those that are most threatened by the toxicologic hazard (killer whale), while analyzing the concurring factors influencing the level of hazard (Fig. 9.3).

Biomarker Approach in Hot Spot Areas

In one of the most biodiverse area of the world, the Mediterranean Sea, cetaceans coexist with high human pressure and are subject to a considerable amount of anthropogenic pollutants. The concurrence of man-made pressures on cetaceans in the Mediterranean Sea is potentially affecting population stability and marine biodiversity. Fossi et al. (2013) reported this finding for the only pelagic marine protected area in the Mediterranean Sea: the Pelagos Sanctuary for Mediterranean Marine Mammals. The authors applied a multidisciplinary tool, using diagnostic markers elaborated in a statistical model to rank toxicologic stress in Mediterranean cetaceans. As a case study the authors analyzed PBT chemicals combined with a wide range of diagnostic markers of exposure to anthropogenic contaminants and genetic variation as a marker of genetic erosion in striped dolphin (*Stenella coeruleoalba*) skin biopsies. In that paper the markers measured in skin biopsies served as both general and specific diagnostic signals at different hierarchic levels. The biomarkers were subdivided into (1) markers of exposure to anthropogenic contaminants, (2) markers of general stress, and (3) markers of genetic erosion. Finally, a statistical model was applied to obtain a complete toxicologic profile of the striped dolphin in the Pelagos Sanctuary and other Mediterranean areas (Ionian Sea and Strait of Gibraltar).

The set of diagnostic tools in skin biopsies, proposed in the 2013 Fossi et al., paper (Fig. 9.4), provided evidence of toxicologic stress in striped dolphin living in the Pelagos Sanctuary, and it underlined differences in PBT chemicals and molecular biomarker responses in the three striped dolphin populations investigated. The highest toxicologic stress in the Pelagos population was highlighted by high PBT chemical levels, combined with correlated biomarker responses. The results supported an association between genetic diversity and toxicologic stress, confirming that genetic variability is linked to resilience. Individuals with lower st.het_Obs reported significantly higher contaminant loads (50% of the dolphins of the Pelagos Sanctuary).

The final application of the classification model (STatistical Risk Elaborating System in Stenella, ST.R.E.S.S) provided an outline of the toxicologic status of striped dolphin populations and represents a potential tool for the monitoring and conservation of cetacean biodiversity. The complete data set confirmed that striped dolphins in the Pelagos Sanctuary are subject to greater toxicologic stress than other Mediterranean populations. Particular concern arises from the evidence that 50% of the striped dolphins from the Pelagos Sanctuary were classified in the high toxicologic hazard group. In conclusion, by applying a set of diagnostic biomarkers in skin biopsies of striped dolphin elaborated in the classification model, the authors provided the first evidence of toxicologic stress in cetaceans living in the only pelagic MPA of the Mediterranean Sea, the SPAMI (Specially Protected Areas and Biological Diversity in the Mediterranean, Pelagos Sanctuary).

Biomarker Approach in Sperm Whale Worldwide

Godard-Codding et al., in 2011, conducted a large-scale monitoring study incorporating biomarker analyses and contaminant residues in skin biopsies of the threatened sperm whale (*Physeter macrocephalus*) (Fig. 9.5). They explored the

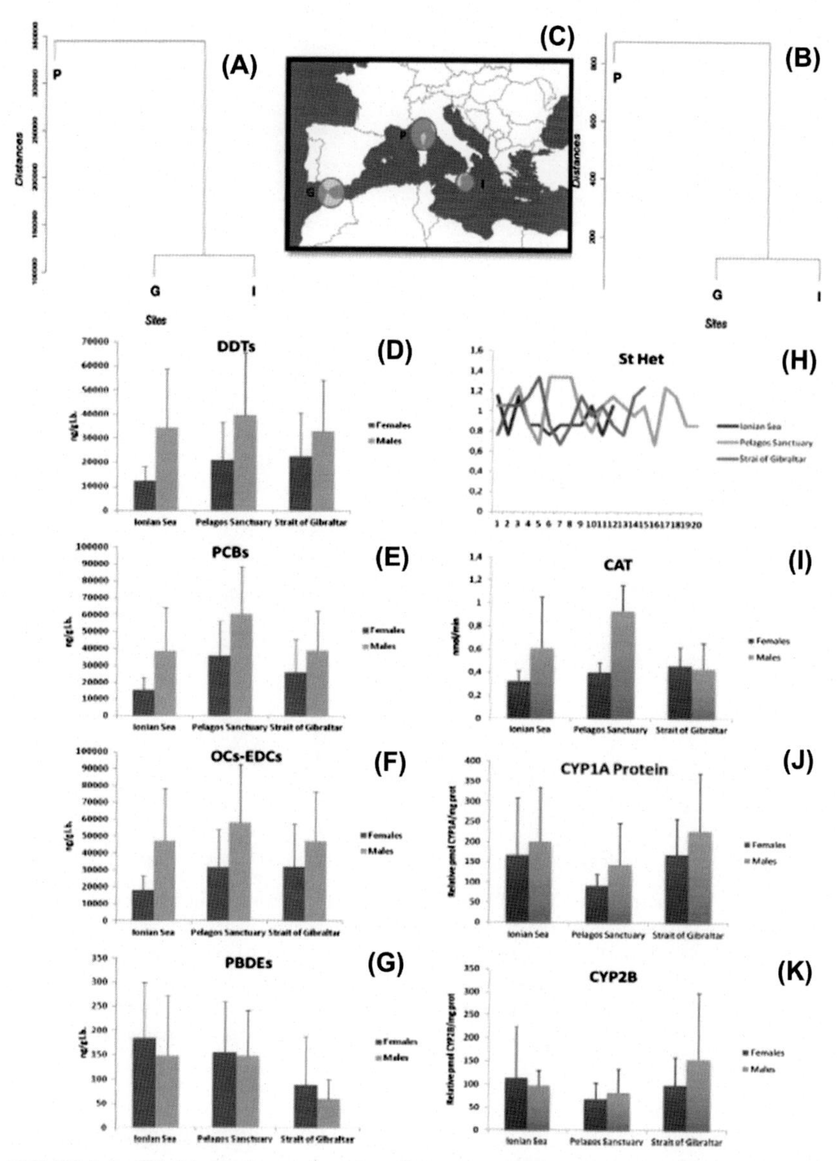

FIGURE 9.4 Skin biopsy diagnostic tool applied to three Mediterranean striped dolphin populations. Dendrograms of classification of the three populations studied: (A) sites grouped by contaminants (OCs and PBDEs). (B) Sites grouped by biomarkers. (C) Study areas: Pelagos Sanctuary (P), in Ionian Sea (I), and the Strait of Gibraltar (G). PBT detection in subcutaneous blubber: (D) DDT levels, (E) PCB levels, (F) OC–EDC levels, and (G) PBDE levels. Markers of genetic erosion: (H) st.het_Obs. Markers of general stress: (I) catalase activity (CAT). Markers of contaminants exposure: protein levels of (J) CYP1A1, (K) CYP2B. For all set of data, mean value ± SD are reported. *(Reprinted with permission from Fossi, M.C., Panti, C., Marsili, L., Maltese, S., Spinsanti, G., Casini, S., Caliani, I., Gaspari, S., Munoz-Arnanz, J., Jimenez, B., Finoia, M.G., 2013. The Pelagos Sanctuary for Mediterranean marine mammals: Marine Protected Area (MPA) or marine polluted area? The case study of the striped dolphin (*Stenella coeruleoalba*). Marine Pollution Bulletin 70 (1–2), 64–72.)*

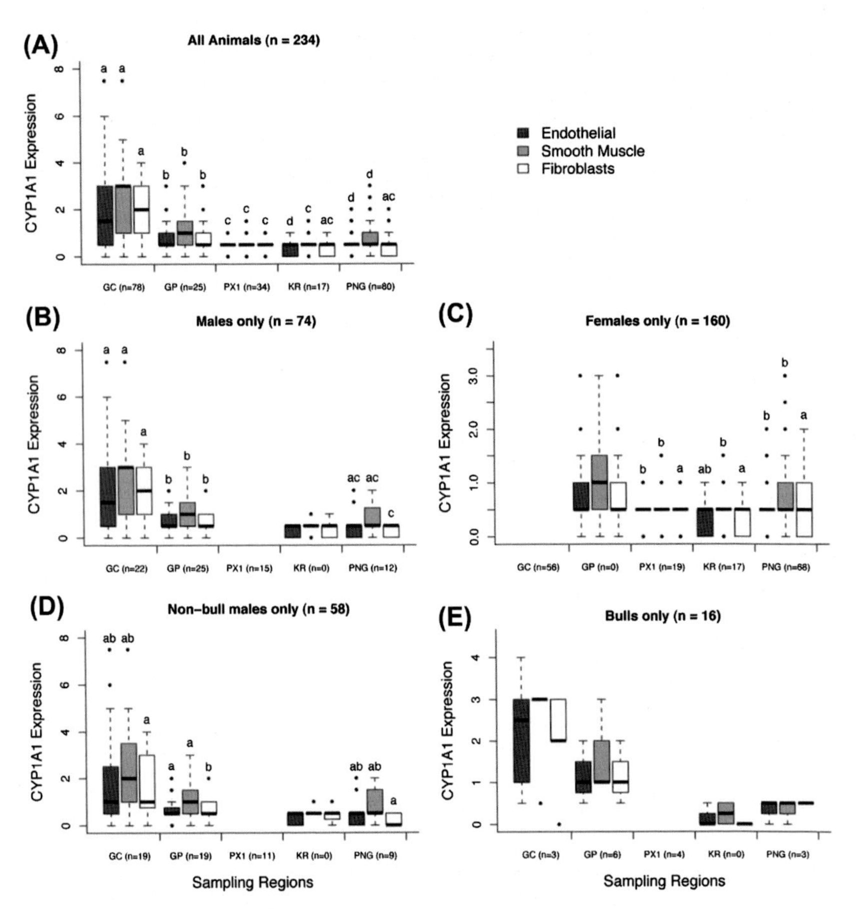

FIGURE 9.5 Mean cell-specific CYP1A1 expression in sperm whale skin biopsies among sampling locations: (A) all animals, (B) males, (C) females, (D) nonbull males only, and (E) bulls only. Note: all Galapagos animals were males, and all Kiribati animals were females. Error bars represent the standard error of the mean. Different letters indicate statistically different means within a specific cell type ($P<.05$). Regional differences were detected in bulls ($P<.05$), but small sample size precluded statistical analysis and thus ranking of means. *(Reprinted with permission from Godard-Codding et al., (2011).)*

hypothesis that this multidisciplinary approach could reveal geographic trends in exposure on an ocean-wide scale. The analyses included cytochrome P450 1A1 expression, stable nitrogen and carbon isotope ratios, and contaminant burdens for 234 biopsies collected in five regions of the Pacific Ocean (waters of the Gulf of California, Galapagos, Pacific Crossing, Kiribati, and Papua New Guinea). Sex of all animals sampled was determined by genetic analyses. CYP1A1 provided information on exposure to AhR agonists, isotope ratios served as general indicators of trophic position and latitude, and residue analyses quantified blubber burdens for PAH, PCBs, and PHAHs. Significant regional differences were revealed by the biomarker analyses, where whales from the sampling sites farthest away from continents exhibited the lowest CYP1A1 expression, while those from the Galapagos, a United Nations Educational, Scientific, and Cultural Organization World Heritage marine reserve, exhibited the highest. Sex, diet, or regional variation did not appear to have influenced CYP1A1 expression. No significantly correlation was observed with contaminant burdens in blubber, but the power to detect significant associations was limited by small sample sizes. This large-scale monitoring study provided an important baseline for this widely used biomarker of exposure to aryl hydrocarbon receptor agonists and was successful at identifying regional differences on an ocean scale. The authors concluded that future biomarker studies conducted in cetacean skin biopsy at an ocean scale are warranted, as they may help identify whether globally distributed chemicals occur at biochemically relevant concentrations in sensitive ecosystems and provide a measure of ocean integrity.

CONCLUSION

A new era of discoveries in cetacean ecotoxicology began in 1992, with the first detection of the CYP 450 in the skin biopsy of Mediterranean cetaceans. In 2018, more than 130 papers (Pab Med data, June 2018) have been published on the application of biomarkers in cetaceans, and many of these papers are based on the use of innovative molecular biology methodologies in skin biopsies. The Mediterranean and Pacific case studies presented here emphasize the need to consider multiple sources of stressors when a multidisciplinary biomarker approach is applied in ecotoxicologic investigation. Cetaceans worldwide are exposed to a variety of adversities that potentially decrease their survival or reproductive success. These include weather, food shortages, competitors, parasites, disease, and human-induced effects (Fossi et al., 2018). Each factor affects marine mammals in a different way, but more importantly, factors can also interact and create impacts far greater than any one factor alone. The case studies reported in this chapter underline that the same anthropogenic pressures are potentially affecting the population stability of cetaceans and other large marine organisms in "hot spot areas." Analyzing persistent bioaccumulative and toxic chemicals, diagnostic markers of exposure to anthropogenic contaminants, and the genetic variation using microsatellite markers, a statistical model revealed that, among subpopulations of the same cetacean species, an association between genetic diversity and toxicologic stress exists, confirming genetic variability is linked to resilience (Panti et al., 2011; Fossi et al., 2013).

In conclusion, these findings, obtained by the application of the multidisciplinary biomarker approach in skin biopsy, underline that in areas where several anthropogenic activities place pressure on populations, top predators are exposed to multiple stressors including plastic pollution (see Chapter 6), and these species may function as useful sentinels of the consequences for the food chain and human health.

ACKNOWLEDGMENTS

The authors want to thank all the authors involved in the studies on cetaceans conducted in the Mediterranean Sea and Pacific Ocean presented in the present paper. The authors should like to give special thanks to Dr. Cristina Panti, for her great contribution to the development of gene expression biomarkers.

REFERENCES

Baini, M., Martellini, T., Cincinelli, A., Campani, T., Minutoli, R., Panti, C., Finoia, M.G., Fossi, M.C., 2017. First detection of seven phthalate esters (PAEs) as plastic tracers in superficial neustonic/planktonic samples and cetacean blubber. Analytical Methods 9, 1512–1520.

Bechshoft, T., Wright, A.J., Weisser, J.J., Teilmann, J., Dietz, R., Hansen, M., Björklund, E., Styrishave, B., 2015. Developing a new research tool for use in free-ranging cetaceans: recovering cortisol from harbour porpoise skin. Conservation Physiology 3 (1), cov016.

Biancani, B., Da Dalt, L., Lacave, G., Romagnoli, S., Gabai, G., 2009. Measuring fecal progestogens as a tool to monitor reproductive activity in captive female bottlenose dolphins (*Tursiops truncatus*). Theriogenology 72, 1282–1292.

Boess, F., Kamber, M., Romer, S., Gasser, R., Muller, D., Albertini, S., Suter, L., 2003. Gene expression in two hepatic cell lines, cultured primary hepatocytes, and liver slices compared to the in vivo liver gene expression in rats: possible implications for toxicogenomics use of in vitro systems. Toxicological Sciences 73 (2), 386–402.

Boggs, A.S., Schock, T.B., Schwacke, L.H., Galligan, T.M., Morey, J.S., McFee, W.E., Kucklick, J.R., 2017. Rapid and reliable steroid hormone profiling in *Tursiops truncatus* blubber using liquid chromatography tandem mass spectrometry (LC-MS/MS). Analytical and Bioanalytical Chemistry 409 (21), 5019–5029.

Boon, J., et al., 1998. The use of microsomal in vitro assay to study phase I biotransformation of chlorobornanes (Toxaphene) in marine mammals and birds. Possible consequences of biotransformation for bioaccumulation and genotoxicity. Comparative Biochemistry and Physiology C 121 (1-3), 385–403.

Bossart, G.D., 2011. Marine mammals as sentinel species for oceans and human health. Veterinary Pathology Online 48, 676–690.

Carvan III, M.J., Santostefano, M., Safe, S., Busbee, D., 1994. Characterization of a bottlenose dolphin (*Tursiops truncatus*) kidney epithelial cell line. Marine Mammal Science 10 (1), 52–69.

Carvan III, M.J., Flood, L.P., Campbell, B.D., Busbee, D.L., 1995. Effects of benzo(a)pyrene and tetrachlorodibenzo(p)dioxin on fetal dolphin kidney cells: inhibition of proliferation and initiation of DNA damage. Chemosphere 30 (1), 187–198.

Casals-Casas, C., Desvergne, B., 2011. Endocrine Disruptors: From Endocrine to Metabolic Disruption. Annual Review of Physiology 73 (73), 135–162.

Champagne, C.D., Kellar, N.M., Crocker, D.E., Wasser, S.K., Booth, R.K., Trego, M.L., Houser, D.S., 2017. Blubber cortisol qualitatively reflects circulating cortisol concentrations in bottlenose dolphins. Marine Mammal Science 33 (1), 134–153.

Chen, T.L., Wise, S.S., Holmes, A., Shaffiey, F., Wise Jr., J.P., Thompson, W.D., Kraus, S., Wise Sr, J.P., 2009. Cytotoxicity and genotoxicity of hexavalent chromium in human and north Atlantic right whale (*Eubalaena glacialis*) lung cells. Comparative Biochemistry and Physiology Part C 150 (4), 487–494.

Chen, T.L., Lacerte, C., Wise, S.S., Holmes, A., Martino, J., Wise Jr., J.P., Thompson, D.W., Wise Sr, J.P., 2012. Comparative cytotoxicity and genotoxicity of particulate and soluble hexavalent chromium in human and sperm whale (*Physeter microcephalus*) skin cells. Comparative Biochemistry and Physiology Part C 155, 143–150.

Clark, C.T., Fleming, A.H., Calambokidis, J., Kellar, N.M., Allen, C.D., Catelani, K.N., Robbins, M., Beaulieu, N.E., Steel, D., Harvey, J.T., 2016. Heavy with child? Pregnancy status and stable isotope ratios as determined from biopsies of humpback whales. Conservation Physiology 4 (1), 1–3.

Coltman, D.W., Slate, J., 2003. Microsatellite measures of inbreeding: A meta-analysis. Evolution 57 (5), 971–983.

De Mello, D.M.D., De Oliveira, C.A., 2016. Biological matrices for sampling free-ranging cetaceans and the implications of their use for reproductive endocrine monitoring. Mammal Review 46 (2), 77–91.

Ellis, B.C., Gattoni-Celli, S., Kindy, S., 2010. The impact of methylmercury on 1,25-dihydroxyvitamin D3-induced transcriptomic responses in dolphin skin cells. Journal of Biological Chemistry 391, 245–258.

Fair, P.A., Schaefer, A.M., Romano, T.A., Bossart, G.D., Lamb, S.V., Reif, J.S., 2014. Stress response of wild bottlenose dolphins (*Tursiops truncatus*) during capture–release health assessment studies. General and Comparative Endocrinology 206, 203–212.

Fossi, M.C., Marsili, L., Leonzio, C., Notarbartolo Di Sciara, G., Zanardelli, M., Focardi, S., 1992. The use of nondestructive biomarker in Mediterranean Cetaceans: preliminary data on MFO activity in skin biopsy. Marine Pollution Bulletin 24, 459–461.

Fossi, M.C., Marsili, L., Neri, G., Natoli, A., Politi, E., Panigada, S., 2003. The use of a non-lethal tool for evaluating toxicological hazard of organochlorine contaminants in Mediterranean cetaceans: new data 10 years after the first paper published in MPB. Marine Pollution Bulletin 46, 972–982.

Fossi, M.C., et al., 2006. Development of new-tools to investigate toxicological hazard due to endocrine disruptor organochlorines and emerging contaminants in Mediterranean cetaceans. Marine Environmental Research 62, S200–S204.

Fossi, M.C., et al., 2008. First detection of CYP1A1 and CYP2B induction in Mediterranean cetacean skin biopsies and cultured fibroblasts by Western blot analysis. Marine Environmental Research 66 (1), 3–6.

Fossi, M.C., et al., 2010. A multi-trial diagnostic tool in fin whale (Balaenoptera physalus) skin biopsies of the Pelagos Sanctuary (Mediterranean Sea) and the Gulf of California (Mexico). Marine Environmental Research 69, S17–S20.

Fossi, M.C., Panti, C., Marsili, L., Maltese, S., Spinsanti, G., Casini, S., Caliani, I., Gaspari, S., Munoz-Arnanz, J., Jimenez, B., Finoia, M.G., 2013. The Pelagos Sanctuary for Mediterranean marine mammals: Marine Protected Area (MPA) or marine polluted area? The case study of the striped dolphin (*Stenella coeruleoalba*). Marine Pollution Bulletin 70 (1–2), 64–72.

Fossi, M.C., Casini, S., Maltese, S., Panti, C., Spinsanti, G., Marsili, L., 2014. An "ex vivo" model to evaluate toxicological responses to mixtures of contaminants in cetaceans: integumentum biopsy slices. Environmental Toxicology 29 (10), 1107–1121.

Fossi, M.C., Marsili, L., Baini, M., Giannetti, M., Coppola, D., Guerranti, C., Caliani, I., Minutoli, R., Lauriano, G., Finoia, M.G., Rubegni, F., Panigada, S., Bérubé, M., Urbán, R.J., Panti, C., 2016. Fin whales and microplastics: the Mediterranean Sea and the Sea of Cortez scenarios. Environmental Pollution 209, 68–78.

Fossi, M.C., Panti, C., Baini, M., Lavers, J.L., 2018. A review of plastic-associated pressures: cetaceans of the Mediterranean sea and Eastern Australian Shearwaters as case studies. Frontiers in Marine Science 5, 173.

Frenzilli, G., et al., 2014. Effects of in vitro exposure to titanium dioxide on DNA integrity of bottlenose dolphin (*Tursiops truncatus*) fibroblasts and leukocytes. Marine Environmental Research 100, 68–73.

Gauthier, J.M., Dubeau, H., Rassart, É., 1998. Mercury-Induced micronuclei in skin fibroblasts of beluga whales. Environmental Toxicology and Chemistry 17 (12), 2487–2493.

Gauthier, J.M., Dubeau, H., Rassart, E., 1999. Induction of micronuclei in vitro by organochlorine compounds in beluga whale skin fibroblasts. Mutation Research/Genetic Toxicology and Environmental Mutagenesis 439 (1), 87–95.

Geraci, J., Lounsbury, V., 1993. Marine Mammals Ashore: A Field Guide for Strandings. Texas A&M Sea Grant Publication.

Godard, C.A., 2000. Novel cytochrome P450 1 genes in marine vertebrates: characterization and expression of multiple CYP1B genes in the fish *Stenotomus chrysops* and of CYP1A1 and CYP1B1 in the cetacean *Stenella coeruleoalba*. Environmental Toxicology. Lubbock, Texas Tech University, p. 151.

Godard, C.A., Smolowitz, R.M., Wilson, J.Y., Payne, R.S., Stegeman, J.J., 2004. Induction of cetacean cytochrome P4501A1 by β-naphthoflavone exposure of skin biopsy slices. Journal of Toxicological Sciences 80 (2), 268–275.

Godard, C.A.J., Wise, S.S., Kelly, R.S., Goodale, B., Kraus, S., Romano, T., O'Hara, T., Wise, J.P., 2006. Benzo [a] pyrene cytotoxicity in right whale (*Eubalaena glacialis*) skin, testis and lung cell lines. Marine Environmental Research 62, S20–S24.

Godard-Codding, C.A., Clark, R., Fossi, M.C., Marsili, L., Maltese, S., West, A.G., Valenzuela, L., Rowntree, V., Polyak, I., Cannon, J.C., Pinkerton, K., 2011. Pacific Ocean–wide profile of CYP1A1 expression, stable carbon and nitrogen isotope ratios, and organic contaminant burden in sperm whale skin biopsies. Environmental Health Perspectives 119 (3), 337–343.

Goksøyr, A., et al., 1986. Initial characterization of the hepatic microsomal cytochrome P-450 system of the piked whale (minke) *Balaenoptera acutorostrata*. Marine Environmental Research 19, 185–203.

Hayden, M., Bhawal, R., Escobedo, J., Harmon, C., O'Hara, T., Klein, D., San-Francisco, S., Zabet-Moghaddam, M., Godard-Codding, C.A., 2017. Nano-LC-MS/MS analysis of steroids from gray whale blubber. Rapid Communications in Mass Spectrometry 31 (13), 1088–1094.

Hogg, C.J., Rogers, T.L., Shorter, A., Barton, K., Miller, P.J., Nowacek, D., 2009. Determination of steroid hormones in whale blow: it is possible. Marine Mammal Science 25 (3), 605–618.

Hooker, S.K., et al., 2008. Changes in persistent contaminant concentration and CYP1A1 protein expression in biopsy samples from northern bottlenose whales, Hyperoodon ampullatus, following the onset of nearby oil and gas development. Environmental Pollution 152 (1), 205–216.

Hunt, K.E., Rolland, R.M., Kraus, S.D., Wasser, S.K., 2006. Analysis of fecal glucocorticoids in the North Atlantic right whale (*Eubalaena glacialis*). General and Comparative Endocrinology 148 (2), 260–272.

Hunt, K.E., Stimmelmayr, R., George, C., Hanns, C., Suydam, R., Brower, H., Rolland, R.M., 2014a. Baleen hormones: a novel tool for retrospective assessment of stress and reproduction in bowhead whales (Balaena mysticetus). Conservation Physiology 2 (1).

Hunt, K.E., Rolland, R.M., Kraus, S.D., 2014b. Detection of steroid and thyroid hormones via immunoassay of North Atlantic right whale (*Eubalaena glacialis*) respiratory vapor. Marine Mammal Science 30 (2), 796–809.

IOC, 2001. The Strategic Plan for the Health of the Ocean Panel for GOOS. Intergovernmental oceanographic commission, UNESCO, Paris.

Jarrell, G.H., Árnason, Ú., 1981. Banded karyotypes of a Belukha whale, *Delphinapterus leucas*. Hereditas 93, 37–41.

Jia, K., Ding, L., Zhang, L., Zhang, M., Yi, M., Wu, Y., 2015. In vitro assessment of environmental stress of persistent organic pollutants on the indo-pacific humpback dolphin. Toxicology in Vitro 30 (1), 529–535.

Kellar, N.M., Trego, M.L., Marks, C.I., Dizon, A.E., 2006. Determining pregnancy from blubber in three species of delphinids. Marine Mammal Science 22 (1), 1–6.

Kellar, N.M., Trego, M.L., Marks, C.I., Chivers, S.J., Danil, K., Archer, F.I., 2009. Blubber testosterone: a potential marker of male reproductive status in short-beaked common dolphins. Marine Mammal Science 25 (3), 507–522.

Kellar, N.M., Keliher, J., Trego, M.L., Catelani, K.N., Hanns, C., George, J.C., Rosa, C., 2013. Variation of bowhead whale progesterone concentrations across demographic groups and sample matrices. Endangered Species Research 22 (1), 61–72.

Kellar, N.M., Catelani, K.N., Robbins, M.N., Trego, M.L., Allen, C.D., Danil, K., Chivers, S.J., 2015. Blubber cortisol: a potential tool for assessing stress response in free-ranging dolphins without effects due to sampling. PLoS One 10 (2), e0115257.

Knap, A., Dewailly, E., Furgal, C., Galvin, J., Baden, D., Bowen, R.E., et al., 2002. Indicators of ocean health and human health: developing a research and monitoring framework. Environmental Health Perspectives 110 (9), 839–845.

Kniazeff, A.J., Groyon, R.M., 1966. Virus susceptibility of cell cultures derived from several species of aquatic and terrestrial mammals. In Vitro 2, 132.

Mancia, A., et al., 2018. Impact of marine contaminants of emerging concern on the cetacean transcriptome. Isj-Invertebrate Survival Journal 15, 129–129.

Mansour, A.A., Mkay, D.W., Lien, J., Orr, J.C., Banoub, J.H., ØIen, N., Stenson, G., 2002. Determination of pregnancy status from blubber samples in minke whales (*Balaenoptera acutorostrata*). Marine Mammal Science 18 (1), 112–120.

Marsili, L., Focardi, S., 1997. Chlorinated hydrocarbon (HCB, DDTs and PCBs) levels in cetaceans stranded along the Italian coasts: an overview. Environmental Monitoring and Assessment 45, 129–180.

Marsili, L., et al., 2000. Skin biopsies for cell cultures from Mediterranean free-ranging cetaceans. Marine Environmental Research 50, 523–526.

Mazzariol, S., Guardo, G.D., Petrella, A., Marsili, L., Fossi, C.M., Leonzio, C., et al., 2011. Sometimes sperm whales (*Physeter macrocephalus*) cannot find their way back to the high seas: a multidisciplinary study on a mass stranding. PLoS One 6, e19417.

Montie, E.W., et al., 2008. Cytochrome P4501A1 expression, polychlorinated biphenyls and hydroxylated metabolites, and adipocyte size of bottlenose dolphins from the Southeast United States. Aquat Toxicol 86 (3), 397–412.

Murk, A., et al., 1994. In vitro metabolism of 3,3',4,4'-tetrachlorobiphenyl in relation to ethoxyresorufin-O-deethylase activity in liver microsomes of some wildlife species and rat. European Journal of Pharmacology 270, 253–261.

Noel, M., et al., 2014. PCBs are associated with altered gene transcript profiles in arctic Beluga Whales (*Delphinapterus leucas*). Environmental Science and Technology 48 (5), 2942–2951.

Noren, D.P., Mocklin, J.A., 2012. Review of cetacean biopsy techniques: factors contributing to successful sample collection and physiological and behavioral impacts. Marine Mammal Science 28 (1), 154–199.

Notarbartolo di Sciara, G., Zanardelli, M., Jahoda, M., Panigada, S., Airoldi, S., 2003. The fin whale *Balaenoptera physalus* (L. 1758) in the Mediterranean Sea. Mammal Review 33 (2), 105–150.

Pabuwal, V., et al., 2013. Transcriptomic analysis of cultured whale skin cells exposed to hexavalent chromium [Cr(VI)]. Aquatic Toxicology 134, 74–81.

Pampaloni, F., Reynaud, E.G., Stelzer, E.H., 2007. The third dimension bridges the gap between cell culture and live tissue. Nature Reviews Molecular Cell Biology 8 (10), 839.

Panti, C., et al., 2011. Ecotoxicological diagnosis of striped dolphin (*Stenella coeruleoalba*) from the Mediterranean basin by skin biopsy and gene expression approach. Ecotoxicology 20 (8), 1791–1800.

Parrish, A.R., Gandolfi, A.J., Brendel, K., 1995. Precision-cut tissue slices: applications in pharmacology and toxicology. Life Sciences 57 (21), 1887–1901.

Parsons, E.C., Baulch, S., Bechshoft, T., Bellazzi, G., Bouchet, P., Cosentino, A.M., Godard-Codding, C.A., Gulland, F., Hoffmann-Kuhnt, M., Hoyt, E., Livermore, S., 2015. Key research questions of global importance for cetacean conservation. Endangered Species Research 27 (2), 113–118.

Pérez, S., García-López, Á., De Stephanis, R., Giménez, J., García-Tiscar, S., Verborgh, P., Mancera, J.M., Martínez-Rodriguez, G., 2011. Use of blubber levels of progesterone to determine pregnancy in free-ranging live cetaceans. Marine Biology 158 (7), 1677–1680.

Petterson, A., van Bavel, B., Engwall, M., Jimenez, B., 2004. Polybrominated diphenylethers and metoxylated tetrabromodiphenylethers on cetaceans from the Mediterranean Sea. Archives of Environmental Contamination and Toxicology 47, 542–550.

Rolland, R.M., Hunt, K.E., Kraus, S.D., Wasser, S.K., 2005. Assessing reproductive status of right whales (*Eubalaena glacialis*) using fecal hormone metabolites. General and Comparative Endocrinology 142 (3), 308–317.

Shimada, T., et al., 1996. Activation of chemically diverse procarcinogens by human cytochrome P-450 1B1. Cancer Research 56, 2979–2984.

Stegeman, J., et al., 1992. Molecular responses to environmental contamination: enzyme and protein systems as indicators of chemical exposure and effects. In: Huggett, R., Kimerle, R., Mehrle, P., Bergman, H. (Eds.), Biomarkers. Biochemical, physiological, and histological markers of anthropogenic stress. Lewis Publishers. SETAC Special Publications Series, Boca Raton, pp. 235–335.

Teramitsu, I., et al., 2000. Identification of novel cytochrome P450 1A genes from five marine mammal species. Aquatic Toxicology 51, 145–153.

Thompson, L.A., Spoon, T.R., Goertz, C.E., Hobbs, R.C., Romano, T.A., 2014. Blow collection as a non-invasive method for measuring cortisol in the beluga (*Delphinapterus leucas*). PLoS One 9 (12), e114062.

Trana, M.R., Roth, J.D., Tomy, G.T., Anderson, W.G., Ferguson, S.H., 2015. Influence of sample degradation and tissue depth on blubber cortisol in beluga whales. Journal of Experimental Marine Biology and Ecology 462, 8–13.

Trana, M.R., Roth, J.D., Tomy, G.T., Anderson, W.G., Ferguson, S.H., 2016. Increased blubber cortisol in ice-entrapped beluga whales (*Delphinapterus leucas*). Polar Biology 39 (9), 1563–1569.

Trego, M.L., Kellar, N.M., Danil, K., 2013. Validation of blubber progesterone concentrations for pregnancy determination in three dolphin species and a porpoise. PLoS One 8 (7), e69709.

Trumble, S.J., Robinson, E.M., Berman-Kowalewski, M., Potter, C.W., Usenko, S., 2013. Blue whale earplug reveals lifetime contaminant exposure and hormone profiles. Proceedings of the National Academy of Sciences of the United States of America 110 (42), 16922–16926.

Van Dolah, F.M., et al., 2015. Seasonal variation in the skin transcriptome of common bottlenose dolphins (*Tursiops truncatus*) from the northern Gulf of Mexico. PLoS One 10 (6), e0130934.

Vickers, A.E.M., Fisher, R.L., 2004. Organ slices for the evaluation of human drug toxicity. Chemico-Biological Interactions 150 (1), 87–96.

White, R., et al., 1994. Catalytic and immunochemical characterization of hepatic microsomal cytochromes P450 in beluga whale (*Delphinapterus leucas*). Toxicology and Applied Pharmacology 126, 45–57.

White, R., et al., 2000. In vitro metabolism of polychlorinated biphenyl congeners by beluga whale (*Delphinapterus leucas*) and pilot whale (*Globicephala melas*) and relationship to cytochrome P450 expression. Comparative Biochemistry and Physiology C 126 (3), 267–284.

Whitehead, A., Anderson, S.L., Kuivila, K.M., Roach, J.L., May, B., 2003. Genetic variation among interconnected populations of *Catostomus occidentalis*: implications for distinguishing impacts of contaminants from biogeographical structuring. Molecular Ecology 12, 2817–2833.

Wilson, J.Y., et al., 2007. Correlates of cytochrome P450 1A1 expression in bottlenose dolphin (*Tursiops truncatus*) integument biopsies. Toxicological Sciences 97 (1), 111–119.

Yoshioka, M., Okumura, T., Aida, K., Fujise, Y., 1994. A proposed technique for quantifying muscle progesterone content in minke whales (*Balaenoptera acutorostrata*). Canadian Journal of Zoology 72 (2), 368–370.

FURTHER READING

Fossi, M.C., Panti, C., Guerranti, C., Coppola, D., Giannetti, M., Marsili, L., Minutoli, R., 2012. Are baleen whales exposed to the threat of microplastics? A case study of the Mediterranean fin whale (*Balaenoptera physalus*). Marine Pollution Bulletin 64, 2374–2379.

Fossi, M.C., Baini, M., Panti, C., Galli, M., Jiménez, B., Muñoz-Arnanz, J., Marsili, L., Finoia, M.G., Ramírez-Macías, D., 2017. Are whale sharks exposed to persistent organic pollutants and plastic pollution in the Gulf of California (Mexico)? First ecotoxicological investigation using skin biopsies. Compartive Biochemistry and Physiology Part C 4 (167), 1–16.

White, N.D., Godard-Codding, C.A.J., Webb, S.J., Bossart, G.D., Fair, P.A., 2017. Immunotoxic effects of in vitro exposure of dolphin lymphocytes to Louisiana sweet crude oil and Corexit. Journal of Applied Toxicology 37 (6), 676–682.

Chapter 10

Ecotoxicological Biomarkers and Accumulation of Contaminants in Pinnipeds

Kristina Lehnert[1], Jean-Pierre Desforges[2], Krishna Das[3], Ursula Siebert[1]
[1]University of Veterinary Medicine Hannover, Büsum, Germany; [2]Aarhus University, Roskilde, Denmark; [3]University of Liege, Liege, Belgium

INTRODUCTION

Pinnipeds are long-lived predators with amphibious lifestyles, often residing in remote environments. Due to their position at the top of the food chain, they can accumulate high levels of environmental contaminants and pollutants (Das et al., 2003; Weijs et al., 2009). Some species are vulnerable, and their ecosystems are increasingly threatened by anthropogenic activities (Siebert et al., 2012a). Pollutant exposure and other stressors are known to affect marine mammal health status and increase their susceptibility to infectious disease (De Swart et al., 1996; Ross et al., 1996a; Beineke et al., 2005). Environmental contaminants as well as chronic and acute stress due to anthropogenic impacts into their habitat are known to affect the immune and endocrine system in seals (Das et al., 2008; Dupont et al., 2016).

Different approaches have been used to assess the effects of contaminants on the health of marine mammals and better understand the impact on their physiology (see Desforges et al., 2016 for a review). To evaluate detrimental changes potentially caused by xenobiotics on wildlife health, parameters of biologic processes with prognostic or diagnostic explanatory power need to be found (Fossi and Marsili, 1997; Barron et al., 2003). These so-called biomarkers are characteristic biologic traits that can be measured and objectively evaluated to indicate normal or pathogenic processes within the organism associated with any stress including environmental pollution or diseases (Zelikoff, 1998; Nyman et al., 2003). Biomarkers can relate to cells, genes, gene products, or hormones and metabolites (Hall et al., 1998; Simms and Ross, 2000) or behavioral changes (Ungherese and Ugolini, 2009; Capowiez et al., 2010). Additionally, complex organ functions (reproductive success, thyroid cell functionality) or pathologic changes in biologic tissues (reproductive organs, endocrine system) may be

Marine Mammal Ecotoxicology. https://doi.org/10.1016/B978-0-12-812144-3.00010-3

used as medical biomarkers (Brouwer et al., 1989; Helle et al., 1976; Beineke et al., 2005). Biomarkers have been used in many scientific fields, and they can be specific or integral, be markers of exposure and of effect. Increasingly, they have become instruments of ecotoxicology research in marine mammals (Fossi and Marsili, 1997; Kim and Hahn, 2002; Lehnert et al., 2017).

Advancements and adaptations of analytical techniques have facilitated the correlation of exposure to environmental metals and persistent organic pollutants (POPs) with alterations of cellular and humoral immunity in marine mammals (Desforges et al., 2016). However, cause and effect relationships are difficult to detect in free-ranging, protected, and long-lived mammals. Study design in live pinnipeds is restricted by legal as well as ethical constraints, and since new chemical compounds enter the global market constantly and are detected and analyzed in the environment permanently, researchers are facing many challenges in finding approaches to investigate effects of single compounds or environmentally relevant concentrations of mixtures in exposure studies (Weijs and Zaccaroni, 2016; Levin et al., 2016; Desforges et al., 2017) and to pinpoint causal relationships in the wild. The use of single cells in culture in experimental exposure studies is also limited in comparison to whole organisms or organ culture, where architecture of the tissue is maintained and pollutant kinetics as well as physiologic realism is improved. Ultimately, this research can inform managers, and the information derived may guide species, habitat, and population assessments as well as conservation plans.

In pinnipeds, pollutant exposure was correlated with tissue lesions and damage in functionality of endocrine organs, cell populations, and reduced reproductive success (Roos et al., 2012). Feeding studies in the field (De Swart et al., 1994, 1996; Reijnders 1986; Brouwer et al., 1989) as well as in vitro exposure experiments with immune cells (Kakuschke et al., 2008; Das et al., 2008; Dupont et al., 2016; Desforges et al., 2017), transcriptomic assays, and gene expression studies have been developed (Fonfara et al., 2008; Mancia et al., 2014; Brown et al., 2017) for marine mammals. While recent papers reviewed toxicologic research and immunotoxicity in marine mammals in general (Desforges et al., 2016; Weijs and Zaccaroni, 2016), this chapter summarizes the mechanisms of bioaccumulation in pinnipeds and collates the biomarkers currently used to assess effects on their organism in ecotoxicological research. A case study from the North Sea is given as an example of the impact pollution has on marine mammals in a heavily pressured habitat, followed by a presentation of pathways of contaminant accumulation in seals and an overview about ecotoxicological biomarkers applied in pinniped research.

MARINE MAMMALS IN THE NORTH SEA: CASE STUDY OF A POLLUTED AREA

Marine mammals from the North Sea are under continuous pressure of anthropogenic activities such as fisheries, ship traffic, oil exploration, and chemical and noise pollution (Richardson et al., 1995; Siebert et al., 1999, 2012a; Jepson et al., 2005, 2016; Lucke et al., 2009). This has prompted many studies over the past decades to evaluate the effects of cumulative stress on the health of these marine mammals.

Reproductive failure had been described in harbor seals from the Dutch Wadden Sea nearly three decades ago and was related to feeding on fish from that polluted area (Reijnders, 1986). This was the first demonstration of a causal relationship between naturally occurring levels of pollutants and a physiologic response in marine mammals. It was suggested that increased PCB levels impaired the reproduction of harbor seals (Reijnders 1980, 1986). PCBs, DDE, and methyl sulfone (MSF) metabolites have high affinity for binding a receptor protein in the uterus (uteroglobin, UG) and the lung. Hence, selective bioaccumulation of MSFs occurs in the uterus, which can lead to implantation failure or abortion (Troisi et al., 2001). Moreover, uteri changes such as stenosis or occlusion in phocids (ringed seals, grey seal, and harbor seals) were associated with higher levels of DDT and PCB (Helle et al., 1976; Olsson et al., 1994). Sixty-four percent of female grey seals had uterine leiomyomas (or fibrosis), mainly in the uterine corpus, in the study period 1975–97 (Bäcklin et al., 2003). Additionally, 65% of the animals with leiomyomas contained no functional corpora in the ovaries. Leiomyomas seem to be associated directly or indirectly with organochlorines (Bäcklin et al., 2003) and during recent time also with acanthocephalans infections in the cecum and colon and colonic ulcers and tunica muscularis hypertrophy (Bergman, 2007; Bäcklin et al., 2013).

Systematic investigations on diseases and causes of death of marine mammals within the North Sea showed that animals suffered frequently from severe parasitic and bacterial infections, particularly those of the respiratory tract (Clausen and Andersen, 1988; Baker and Martin, 1992; Jepson et al., 2000; Siebert et al., 2001, 2007; Jauniaux et al., 2002; Lehnert et al., 2005). Harbor porpoises from the North and Baltic Seas showed a significantly higher incidence of severe bacterial infections than those from less polluted waters around Greenland, Iceland, and Norway (Wünschmann et al., 2001; Siebert et al., 2006, 2009). Further investigations on the immune and endocrine systems of harbor porpoises indicated that animals from the North Sea showed lymphoid depletion in thymus and spleen, as well as a replacement of thyroid follicles by connective tissue indicative of reduced overall health status (Beineke et al., 2005; Das et al., 2006).

In 1988/89 and 2002 the harbor seal population in the North Sea was hit by a large-scale mortality event linked to Phocine Distemper Virus, which caused the death of tens of thousands individuals (Härkönen et al., 2006). It was discussed if higher chemical pollutant burden in seal populations may have resulted in greater susceptibility to viral diseases via impaired immune status (De Swart et al., 1996; Ross et al., 1996a).

The concentration of non-essential metals, such as mercury (Hg), lead (Pb), and cadmium (Cd) has been measured in marine mammals from German waters in different studies (Siebert et al., 1999, 2012a; Das et al., 2008, 2004; Kakuschke et al., 2006; Griesel et al., 2008). Concentration of non-essential metals depends upon the contamination of the environment but also on many other factors such as the species, the diet, the position in the food web, and the body size (Das et al. 2003, 2008). Harbor seals from the southern North Sea displayed higher Hg concentration in blood compared (Das et al., 2008; Dupont et al., 2013) to grey seals sampled in Scotland (Habran et al., 2013) (Fig. 10.1), underlying the highest exposure of harbor seals.

Organic substances like PCBs, PBDEs, PFOS, and DDT have also been reported in seals and harbor porpoises from the German North and Baltic Sea (e.g., Siebert et al., 2002; Thron et al., 2004; Das et al., 2006; Weijs et al., 2009; Ahrens et al., 2009; Huber et al., 2012).

Harbor porpoises from the North Sea showed extremely high PCB values (up to 199 μg/g lw) for the sum of seven ICES markers, which might be due to local pollution or discharges (Covaci et al., 2002). Resuspension and run-off from old deposits of PCBs and DDTs are possibly higher in locations closer to industrialized regions (Covaci et al., 2002). The high concentrations can also be related to the bad health condition of the animals (Jepson et al., 1999).

Σ_{21}PCB concentrations in blubber tissue ranged between 2.2–172 μg/g lw and 1.3–126 μg/g lw for harbor seals and harbor porpoises, respectively (Weijs et al., 2009). The large range of values displayed by both species underlies the numerous biotic factors involved in PCB lipid accumulation (e.g., age, gender, and body condition).

Concentrations of CB 153 were higher in harbor seals and harbor porpoises from the North Sea (Weijs et al., 2009) than in similar species from other seas and oceans (Table 10.1), indicating that the Southern part of the North Sea is still highly contaminated with PCBs, in agreement with previous published studies (Covaci et al., 2002; Weijs et al., 2009). Hazardous substances found in grey and harbor seals are numerous (Tables 10.2 and 10.3).

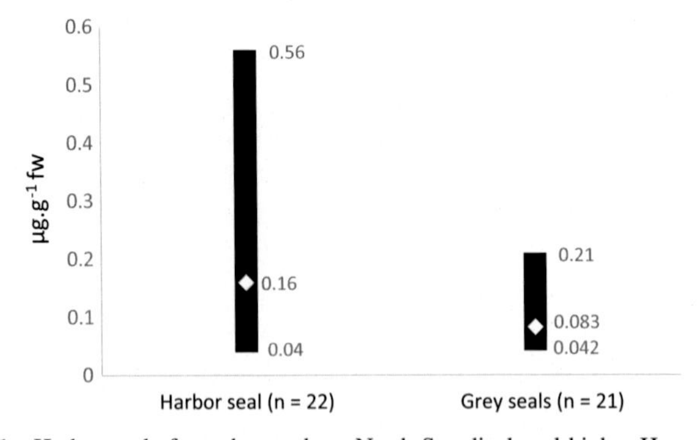

FIGURE 10.1 Harbor seals from the southern North Sea displayed higher Hg concentration in blood compared (Das et al., 2008; Dupont et al., 2013) to grey seals sampled in Scotland (Habran et al., 2013). Mean: 0.16 $\mu g.g^{-1}fw$ = 172 μg $Hg.L^{-1}$- 1 μM.

TABLE 10.1 Mean Concentrations (Standard Deviation) (μg/g Lipid Weight) of CB 153 in Blubber Tissue of Harbor Seals (*Phoca vitulina*) and Harbor Porpoises (*Phocoena phocoena*)

Species	Location	Year	*n*	CB 153 (μg/g Lipid Weight)				References
				AM	JM	AF	JF	
Harbor seal	Canada	1996–2000	8	10.6 (5.1)				Hobbs et al. (2002)
	Norway		6-4	0.61		0.12		Wolkers et al. (2004)
	Southern North Sea	1999–2004	8-8-2-8	28.9 (23.3)	7.2 (2.4)	4.3 (4.3)	10.3 (10.8)	Weijs et al. (2009)
Harbor porpoise	Baltic Sea	1985–1993	4-13	20 (13)	6.6 (3.6)			Berggren et al. (1999)
	Kattegat-Skagerrak	1988–1990	7-10	5.7 (2.3)	4.8 (2.5)			Berggren et al. (1999)
		1978–1981	5	19 (12)				Berggren et al. (1999)
	Norway	1988–1990	8	5.6 (4.6)				Berggren et al. (1999)
	United Kingdom	1999–2004	16-18-8-15	3.7 (3.2)	4.4 (7.4)	2.2 (1.5)	2.9 (1.6)	Law et al. (2006)
	Southern North Sea	1999–2004	8-11-4-10	28.7 (12.0)	3.9 (3.0)	1.7 (0.6)	3.7 (4.1)	Weijs et al. (2009)

Modified after Weijs, L., Dirtu, C.A., Das, K., Gheorghe, A., Reijnders, J.H.P., Neels, H., Blust, R., Covaci, A., 2009. Inter-species differences for polychlorinated biphenyls and polybrominated diphenyl ethers in marine top predators from the Southern North Sea: part 1. Accumulation patterns in harbour seals and harbour porpoises. Environmental Pollution, Elsevier, 437–444.

TABLE 10.2 Literature About Hazardous Substances in Harbor Seals

Hazardous Substances	Samples	Region of Sampling	Year	References
PCB, DDT	B	DE (N)	1974–1976	Drescher et al. (1977)
PFOS	Li, Ki	BE, FR; NL	1995–2000	Van de Viejver et al. (2003)
PCB, DDT	B, Li, K, Br	NL (N)	n.r.	Duinker et al. (1979)
PCB, DDT	B	NO	1988	Skaare et al. (1990)
PCB, DDT, Dioxin	B	DE (N)	1988–1989	Beck et al. (1990)
PCB, DDT	Bl	DE, DK (N)	2006–2008	Weijs et al. (2009)
PCB	B	USA	n.r.	Shaw et al. (2005)
PFOS	Li	DE	2007	Ahrens et al. (2009)
PFOS	Li	USA	2000–2007	Shaw et al. (2009)
PFOS	Li, K, Mu, S	NL	2002	Van de Viejver et al. (2005)
PBDEs, PCB	B	BE, NL, DE (N)	1999–2004	Weijs et al. (2009)
PCB, DDT	B, Li	DK, DE (N)	2001–2002	Siebert et al. (2012b)
Hg	Mu, Li	DE (N)	n.r.	Harms et al. (1978)
Hg, Cd, Pb	Li	GB	1988–1989	Law et al. (1991)
Hg	Bl	DE (N)	1997–2004	Das et al. (2008)
Hg, Cd, Pb	Ha, Sk	DE (N)	1988	Wenzel et al. (1993)

TABLE 10.2 Literature About Hazardous Substances in Harbor Seals—cont'd

Hazardous Substances	Samples	Region of Sampling	Year	References
Hg, Cd, Pb	Li, Br, K	DE (N)	1974–1976	Drescher et al. (1977)
Cd, Pb, As	Bl	DE (N), DK	2003–2004	Griesel et al. (2008) and Kakuschke et al. (2005)
Cd, Pb	Mu, Li, K	DE (N)	n.r.	Harms et al. (1978)
Pb, Cd	B, Li, K, Br, S, H, P	NL	n.r.	Duinker et al. (1979)
Hg	Li, K, Br	DE, DK, NL	1975–1976	Reijnders (1980)
As	U, G, Bl	DE	n.r.	Kuenstl et al. (2009)
Hg, Pb, As	Li	PO	2002	Ciesielski et al. (2006)

Hazardous substances: *As*, arsenic; *Cd*, cadmium; *Hg*, mercury; *Pb*, lead. Samples: *B*, blubber; *Bl*, blood; *Br*, brain; *G*, gastric juice; *H*, heart; *Ha*, hair; *K*, kidney; *Li*, liver; *Mi*, milk; *Mu*, muscle; *P*, placenta; *S*, spleen; *Sk*, skin; *U*, urine. Regions: *B*, Baltic Sea; *BE*, Belgium; *DE*, Germany; *DK*, Denmark; *GB*, Great Britain; *N*, North Sea; *n.r.*, not reported; *NL*, Netherlands; *NO*, Norway; *PO*, Poland; *USA*, United States of America; *Year*, year of sampling.

ENVIRONMENTAL TOXICOLOGY: WHERE ECOLOGY MEETS TOXICOLOGY

Apart from the environment and proximity to sources, biologic and ecologic factors play an important role in the pollutant burden of an individual. These include biologic factors such as age, body condition, metabolic rate, and diet (De Swart et al., 1996; Das et al., 2003; Thron et al., 2004). Sex of the animal is a major factor for different contaminants and their levels, as reported for Cd in harbor seal fur (Wenzel et al., 1993) or PCBs, DDT, and CHLs in blubber (Shaw et al., 2005). In these cases, females had lower burdens (and concentrations) than males due to contaminant transfer/offloading to their offspring during gestation and lactation (Shaw et al., 2005). Maternal transfer of contaminants has been shown in several pinniped species (Debier et al., 2003; Habran et al., 2013). Indeed, transfer of lipophilic contaminants in high-fat milk is a major exposure route for nursing seal pups, resulting in exposures to similar concentrations as adult animals.

TABLE 10.3 Literature About Hazardous Substances in Grey Seals

Hazardous Substances	Samples	Region of Sampling	Year	References
PCB	Bl, B, Mi	GB	1998–2000	Debier et al. (2003)
DDT	B, K	GB	1988	Law et al. (1989)
PFOS	Li, K	BE, FR, NL	1995–2000	Van de Viejver et al. (2003)
PFOS	Li	FI	n.r.	Kannan et al. (2002)
TBT	Li	PO	1996–2003	Ciesielski et al. (2004)
Hg, Cd, Pb	Li	GB	1988–1989	Law et al. (1991)
Hg	Ha, Bl, Mi	GB	2008	Habran et al. (2013)
Hg, Cd, Pb	Li	DE (N)	n.r.	Harms et al. (1978)
Pb, As, Cd	Bl	DE (N), GB	2003	Kakuschke et al. (2006)
Pb, Cd	Bl, Mi, B, Ha	GB	2008	Habran et al. (2013)
Hg, Cd, Pb	B, Mu, Li, K	GB	1988	Morris et al. (1989)
Hg, Cd, Pb, As	Li	PO	1996–2003	Ciesielski et al. (2004)

Hazardous substances: *As*, arsenic; *Cd*, cadmium; *Hg*, mercury; *Pb*, lead. Samples: *B*, blubber; *Bl*, blood; *Br*, brain; *H*, heart; *Ha*, hair; *K*, kidney; *Li*, liver; *Mi*, milk; *Mu*, muscle; *S*, spleen. Regions: *B*, Baltic Sea; *BE*, Belgium; *DE*, Germany; *Fi*, Finland; *FR*, France; *GB*, Great Britain; *N*, North Sea; *n.r.*, not reported; *NL*, Netherland; *PO*, Poland; *Year*, year of sampling.

Several POPs increase with age in males. A positive correlation between DDT and PCB concentrations and the age of male harbor seals was found, which could not be seen in females (Helle et al., 1976). Concentrations of non-ortho PCBs in immature porpoises were similar or higher than in mature ones, reflecting maternal transfer of selected congeners (Berggren et al., 1999). In younger harbor seals (<7 months) C6–C8 PFSAs, perfluorododecanoic acid (PFDoDA), and FOSA were significantly higher than in older seals (Ahrens et al., 2009), while it was contrary in PFDA (Ahrens et al., 2009). The median

PCB concentrations in the blubber of harbor porpoises from the Danish North Sea was 11 mg/kg ww (female) and 8.8 mg/kg ww (male) in young animals (age < 1 year) and 8.1 mg/kg (female) and 23 mg/kg ww (male) in older (>3 years) animals (Granby and Kinze, 1991). Moreover, different elements can influence each other; hepatic Hg is positively correlated with selenium (Se) and, additionally, with age (Das et al., 2004). Thron et al. (2004) found a positive correlation in harbor porpoises between $\Sigma_6$toxaphene and age. A significant correlation can also be found between mercury/methylmercury and age in harbor porpoises (Siebert et al., 1999; Ciesielski et al., 2004). In summary, age dependency varies with contaminant type and congeners (Thron et al., 2004).

Grey seal pups get an intake of different trace elements (Ca, Fe, Zn, Cu, Ni, Pb, Hg, Cr, V) through gestation and milk (Habran et al., 2013). During the lactation period the contaminant concentration in the blood of mature females varied significantly: while some elements decreased, others increased (Habran et al., 2013). In pups, most elements increase during lactation. Furthermore, the placental transfer of elements should be taken into consideration (Habran et al., 2013). All these examples show that a comparison of contaminant levels in marine mammals has to take into account the animals age and reproductive state.

Lipophilic contaminants will accumulate preferentially in blubber of harbor porpoises and harbor seals from the North Sea (Weijs et al., 2009). However, the liver of harbor porpoises contains also high concentrations of organohalogenated contaminants (such as PCB and DDT) (Covaci et al., 2002), of which p,p′-DDE is the main contributor to the total DDT (Berggren et al., 1999; Covaci et al., 2002). The inconsistent types of sampling (e.g., region, year, species, tissue, age) and the varying focus on certain contaminants make a comparison of the data difficult. Furthermore, in seals the timing of the molt is an important factor for the distribution of metals in the hair (Wenzel et al., 1993) and needs to be taken into account when using fur for analyses.

Because of different factors that may influence the concentration of hazardous substances, Thron et al. (2004) recommended to focus on subadult animals so that the contaminant intake through nursing has ceased and is replaced by a normal prey intake. Furthermore, the sex should not play a role in subadult animals, since no pregnancies have altered contaminant levels. Body condition is another important factor to consider when comparing contaminants; poor condition during starvation can lead to mobilization of stored contaminants in fat depots (Thron et al., 2004). For further studies, we recommend taking these factors into consideration, so a comparison between different regions as well as temporal trends can be created. It is important to know the actual pollutant burden in these animals to draw conclusions about the environmental status and ecosystem health.

BIOMARKERS

Biomarkers measure health or immune endpoints in the target animal and can depict physiologic processes within the organism, e.g., the detoxification of contaminants by quantifying receptors on the cellular level. Biomarkers for ecotoxicology must be evaluated in terms of their ability to reflect exposure and/or effects. Their specificity in terms of the toxic compound investigated as well as a strong correlation with a health effect are the properties most sought when trying to establish a biomarker in marine mammal research. Additionally, reproducibility, time trends, and cost effective study design are desirable. Traditionally, liver tissue or hepatic cells are used to assess exposure or effects of xenobiotics, because these organs are where compounds are metabolized and detoxified. In vulnerable wildlife, access to liver tissue is only feasible from dead stranded or diseased moribund animals (Behr et al., 2008; Hellwig, 2011). Current in vivo research focuses on minimally invasive sampling, including blood or blubber biopsy samples, which are best suited for free-ranging live animals and for which captive animals can be trained (Fossi et al., 1997). Urine, saliva, feces, and fur are also used as minimally invasive sampling matrices (Constable et al., 2006). The advantage of blood, blubber, and skin samples is that they yield cells to be used in ex vivo/in vitro approaches (Marsili et al., 2000; Weijs and Zaccaroni, 2016). In general, pinnipeds are more accessible for sampling than cetaceans due to their amphibious life style.

Endocrine Dysfunction

As previously mentioned, Bergman and Olsson (1986) revealed the presence of a disease complex in grey and ringed seals from the Baltic Sea in the 1960s and 1970s that was caused by organochlorine interference with the endocrine system. Health investigations revealed several alterations in different organ systems, including uterus leiomyoma, stenosis, occlusion, loss of bone structure, colon ulcera, claw lesions, and intestinal ulcers (Bergman, 2007; Bergman and Olsson, 1986).

In a following study (1977–96), the number of claw lesions decreased over the study period. The adrenocortical hyperplasia and gynecological health possibly also showed this decreasing trend. Contrary, colonic ulcers increased (Bergman, 1999). In 1960–69 and 1971–85, grey seal skulls were collected and compared to seals found before 1950. An increase of skull bone lesions compared to those from before 1950 could be observed (Bergman et al., 1992). In light of this, bone mineral density of males was studied in three different groups (very low organochlorine, high organochlorine, and decreasing organochlorine in the environment). The lowest trabecular bone mineral density was found in the years with high organochlorine burdens. The cortical bone mineral density increased over the years, which may be related to decreasing contaminant levels in the Baltic (Lind et al., 2003).

Brouwer et al. (1989) showed a decrease of total and free thyroxin (TT4 and FT4) and triiodothyronine (TT3) as a sign of change in thyroid tissue in common seals, fed with PCB-contaminated fish from the Wadden Sea. In addition to organochlorines, metallic trace elements have an effect on the endocrine system; together with PCBs, MeHg, As, Cd, and Se were shown to alter the in vitro biosynthesis of steroid hormones in grey seals (Freeman and Sangalan, 1977). Vitamin A (retinol) is a dietary hormone required for immune function, reproduction, growth, and development. Circulating vitamin A was correlated with PCB exposure in free-ranging harbor seals and indicated a contaminant-related disruption of hormone levels in plasma and blubber (Mos et al., 2006). A study in live captured harbor seal pups found circulatory retinol levels positively correlated with contaminant levels in the blubber of nonnursing animals, indicating a mobilization of liver vitamin A stores into circulation following exposure to milk-derived contaminants (Simms and Ross, 2000). Thyroid hormone serum levels as biomarkers were also associated with PCB exposure in grey seals (Hall et al., 1998) and analyzed together with vitamin A and hepatic POPs and their circulating hydroxyl metabolites in ringed and harbor seals (Routti et al., 2010a). The studies showed that circulating vitamin A and thyroid hormones are influenced by molting, lactating, and fasting in seals, and these metabolic processes have to be taken into account when interpreting results (Routti et al., 2010b).

Liver Function

A study using primary hepatocytes from harbor seals investigated the impact of PCBs on cell viability (Korff et al., 2009). In a two-step biopsy perfusion method (Reese and Byard, 1981; Clement et al., 2001), hepatocytes were isolated from fresh liver tissue of harbor seals from the North Sea. During cultivation, the hepatocytes were exposed to environmentally relevant contaminants with known (hepato-) toxic qualities (PFOS and an Aroclor mixture) in concentrations corresponding to those found in wild harbor seal tissue. Cell viability and maintenance was measured by evaluating the activity of mitochondrial dehydrogenases (XTT assay), the membrane integrity (LDH release), and urea synthesis (Korff et al., 2010). Although the main aspects of cell viability and the specific metabolism of primary seal hepatocytes were not reduced by the investigated pollutants, urea synthesis decreased slightly during cultivation (Korff et al., 2010). This finding shows that primary hepatocytes may be a cell culture model in which effects of pollutant exposure can be monitored closely under controlled experimental conditions. In subsequent proteome analyses, ten proteins from approximately 160/gel modified their protein expression levels (Behr et al., 2008). This showed that protein expression patterns enable discriminating between pollutant-incubated cells and negative control cells. Some of the upregulated proteins were shown to belong to the cytochrome P450 enzyme group (Behr et al., 2008).

Disease Surveillance

Pollutants are likely to promote disease and mortality by compromising the immune system (Wenzel et al., 1993). Mercury, cadmium, and lead, in particular, affect immune functions due to their high toxicity and accumulation characteristics (Wenzel et al., 1993). This was demonstrated in harbor porpoises, showing that those animals that died due to infections had higher concentrations of mercury, selenium, zinc, and an increased Hg:Se molar ratio than animals that died as a result of physical trauma (Bennett et al., 2001). While **pathology and histopathology** were mentioned previously, immune histopathology so far has been investigated only in harbor porpoises and polar bears (Beineke et al., 2005; Kirkegaard et al., 2005). Recently, also, **parasitology** data, e.g., acanthocephalan infections and intestinal lesions, were used as a biomarker for contaminant exposure and effects (Bergman, 2007). Parasitic infections, their prevalence, and intensity, as well as emerging species have been taken into account as bioindicators not only for the health or ecology of marine animals but also as markers for exposure and detrimental health effects of xenobiotics (Sures, 2004; Pascual and Abollo, 2005; Marcogliese and Pietrock, 2011; Siebert et al., 2006; Lehnert et al., 2014b). In the future, emerging infectious diseases and microparasites like viruses may be used as markers for effects of xenobiotics on the health of pinnipeds, as their role also as indicators for global change has become evident (Van Bressem et al., 2009).

It has been shown that environmental factors seem to play a role in the emergence and pathogenicity of morbillivirus epidemics, lobomycosis/LLD, toxoplasmosis, poxvirus-associated tattoo skin disease, and infectious diseases of multifactorial etiology in cetaceans worldwide (Van Bressem et al., 2009). While in the 80s two PDV epidemics were associated with contaminant loads and subsequent susceptibility to infectious disease in harbor seals (De Swart et al., 1996; Ross et al., 1996a), a recent study used in vitro techniques to show that Aroclor mixtures may increase PDV virus replication in harbor seal lymphocytes (Bogomolni et al., 2016). Moreover, different feeding experiments with fish from diverse contaminated areas revealed an immunotoxic risk and a higher susceptibility for virus infections in seals (Ross et al., 1996a; De Swart et al., 1996).

Hematology and Blood Chemistry

Hematology and blood clinical parameters have been used to assess health status and immune parameters in several studies investigating pinnipeds (Bossart, 2011; Hasselmeier et al., 2008) and correlating with tissue contaminant levels. Parameters like circulating leukocytes (by assessing total blood cell numbers or percent) or acute phase proteins and hormone levels can reflect infectious disease, immune system development, and stress (Beineke et al., 2010; De Swart et al., 1996; Hall et al., 1997; Mos et al., 2006). The number of granulocytes, neutrophils, and basophils were positively correlated with PCBs in harbor and northern fur seals (*Callorhinus ursinus*) (Beckmen et al., 2003; De Swart et al., 1994). However, mixed results were found for circulating blood cells in harbor seals from the North Sea linked to PCBs or metals (De Swart et al., 1994;

Reijnders, 1988; Weirup et al., 2013; Kakuschke et al., 2011) and northern fur seals (Beckmen et al., 2003) (see Desforges et al., 2016; Table 1). The effect of pollutants on immune cells and its extent remain uncertain and hamper the interpretation of hematology results as clear indicators of toxic effects in pinnipeds (Desforges et al., 2016). Elevated serum haptoglobin has been reported in PCB- and PFC-exposed harbor seals and northern fur seals (Beckmen et al., 2003; Zenteno-Savin et al., 1997), but due to confounding variables, those studies were unable to conclusively report that contaminants were the primary cause of the observed changes.

Immune System

Lymphocyte monitoring is the most often used in vitro biomarker technique in pinniped studies assessing effects of pollutants (Neale et al., 2002; Sørmo et al., 2009; Dupont et al., 2013). Lymphocytes are active immune cells of adaptive immunity, and their ability to proliferate shows their ability to mount an immune response. Usually, lymphocytes are isolated from blood samples of live or freshly dead animals, cultivated, and exposed to mitogens/contaminants to induce proliferation. Proliferation as well as protein expression of lymphocytes have been used as two biomarkers for immunotoxicity (Kakuschke et al., 2009a; Beineke et al. unpublished data). Immuno stimulation by environmental exposure in free-ranging marine mammals has been found in two studies of harbor seals (PCBs: Levin et al., 2005; metals: Kakuschke et al., 2005). Mostly proliferation in association with contaminant exposure was found to be suppressed in seals (suppl. Table 1, from Desforges et al., 2016; Fig. 10.2). In harbor seal lymphocytes from juveniles a higher susceptibility to the toxic effect of different Hg compounds compared to lymphocytes from adults was found. The proliferation seemed more inhibited from organic than from the inorganic compound (Kakuschke et al., 2009a). Even at low concentrations, methylmercury (0.2 and 1 μM) was shown to have immunosuppressive effects in in vitro tests with harbor seal cells (Das et al., 2008). In newborn harbor seals, elevated metallic trace elements concentrations induced an inhibition of lymphocyte proliferation and otherwise a hypersensitivity (Kakuschke et al., 2008) based on a lymphocyte proliferation test (LTT).

Lymphocyte ultrastructure was shown to be affected by MeHgCl exposure in an in vitro approach using morphologic characteristics with electron microscopy (Dupont et al., 2016). On the other hand, no detrimental effect of PCB in vitro exposure or in relation with PCB blubber burden was observed in ringed seal lymphocytes from Greenland (Levin et al., 2016).

Desforges et al. (2016) reviewed 50 studies that associated immunity with pollutants, the majority of which ($n = 29$) targeted pinnipeds and mostly linked organic pollutants, especially PCBs. A **meta-analysis** of published data showed a clear dose–response relationship for the effects of PCBs and heavy metals on lymphocyte proliferation, with higher pollutant concentrations having stronger

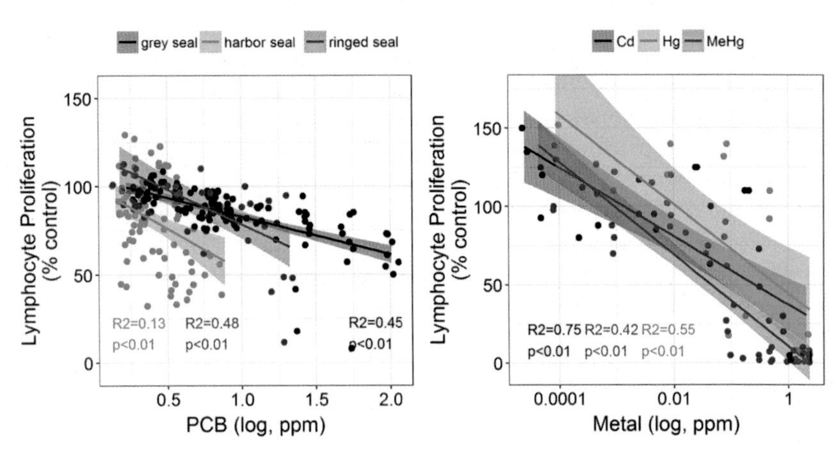

FIGURE 10.2 PCB and metal exposure reduces the lymphoproliferative response in marine mammals. Dose–response relationships are shown for different seal species (PCBs) or for all species grouped together but for different metals. Details on the data and processing are available in Desforges et al., (2016). *(Modified from Desforges, J.P.W., Sonne, C., Levin, M., Siebert, U., De Guise, S., Dietz, R., 2016. Immunotoxic effects of environmental pollutants in marine mammals. Environment International 86, 126–139.)*

effects: this was most obvious for PCBs, and impacts were species-specific (Desforges et al., 2016). **Modeling/in silico studies** on toxic effects in pinnipeds are pending and have been performed only in bioaccumulation studies (Weijs et al., 2014 only for POPs).

Neutrophils are phagocytic cells that engulf pathogens and can be used as markers for bacterial infections. **Phagocytosis** can be measured by flow cytometry and fluorescence. In vitro studies have been performed with samples of free-ranging harbor (Hammond et al., 2005; Pillet et al., 2000; Frouin et al., 2010; Mos et al., 2006) and grey seals (Hammond et al., 2005; Lalancette et al., 2003; Pillet et al., 2000) showing suppressive effects in association with metals (Pillet et al., 2000; Lalancette et al., 2003) and PCBs (Hammond et al., 2005; Frouin et al., 2010). Mos et al. (2006) found suppressive effects of PCBs in vivo, and Frouin et al. (2010) observed suppressed phagocytosis in captive harbor seal samples exposed to metals.

Oxidative burst is the rapid release of reactive oxygen species and an important reaction that occurs in phagocytes to degrade internalized particles and bacteria. **Leukocyte respiratory burst** modulation after stimulation with PCBs was found in harbor and grey seals with mixed effects. Stimulative effects in vivo (Mos et al., 2006) and in vitro (Frouin et al., 2010) were found in free-ranging harbor seals. Suppressive effects (Hammond et al., 2005) were observed in vitro in harbor seals from the wild. In grey seal samples, in vitro exposure to PCBs yielded no respiratory burst modulation (Hammond et al., 2005).

Two studies measured **natural killer (NK) cell activity** when exposed to PCBs in samples of harbor and grey seals. NK cells are granular lymphocytes that deploy their cytotoxic activity to destroy pathogens, thus acting as non-specific defenders in the early phase of viral infections (Ross et al., 1996b). One in vivo study observed suppressive effects (Ross et al., 1996a) in harbor seals; the other in vitro study in harbor and grey seals (Hammond et al., 2005) reported no effects of PCBs on NK cell activity. Desforges et al. (2017) exposed harp and hooded seal NK cells to a mixture of contaminants derived from blubber and observed NK stimulation and no effect, respectively, for each species.

Immunoglobulins (Igs) are antibodies with multiple isotypes differing in structure and function (De Guise, 2004; Ross and De Guise, 2007) that can antagonize bacterial infections and induce phagocytosis and NK cell activity. Two pinniped studies of immunoglobulins have looked at IgG levels, the most abundant isotype, and specific antibodies after immunization with different viruses or antigens. Ross et al. (1995) looked at the effects of PCBs on immunoglobulins after immunization with ovalbumin, tetanus toxoid, polio-, and rabies-virus in harbor seals and found suppressive effects in association with ovalbumin only. In Northern fur seals, immunization with tetanus toxoid had suppressive effects on circulating Igs in association with PCBs (Beckmen et al., 2003). Serum levels of IgG or total Ig are not pathogen-specific and their suppression by pollutants may indicate reduced antibody production; however, contrary effects in cetaceans have been observed (Fair et al., 2013), so any effects have to be interpreted with regard to natural factors such as age, sex, or species, as well as naturally occurring toxic compounds. Because antibody production is a process involving many compartments of the immune system (Beckmen et al., 2003), multiple parameters may influence the modulation by pollutants (Desforges et al., 2016). Mixed responses between studies and species limit the value of antibody modulation as a biomarker in marine mammals.

Gene Expression

To gain insight into immunological health-related effects and toxicity on the physiology of marine mammals, new tools to assess the early cellular biologic impacts were recently developed (Mancia et al., 2014; Weirup et al., 2013). In a molecular approach, **gene transcription** induced by xenobiotics as well as by inflammatory disease and stress has been used widely in wildlife. Mostly cytokine gene transcription has been measured using quantitative PCR (Fonfara et al., 2008; Neale et al., 2005; Kakuschke et al., 2009a) to establish mRNA levels or total copy numbers of selected markers in blood of pinnipeds. RNA is typically isolated from full blood samples or specific cell populations, transcribed into DNA, and transcripts of target DNA of selected proteins are quantified. But also hepatic tissue has been used to assess effects of xenobiotics; fresh tissue of preferably live animals is needed, and to not depict postmortem artifacts in gene expression (Harrison et al., 1995; Tomita et al., 2004), samples are conserved in liquid nitrogen or RNAlater. Hepatic cells of freshly dead seals were isolated and analyzed (Kim et al., 2005; Ishibashi et al., 2008) or harvested for in vitro exposure studies in cell culture (Behr et al., 2008) in Baikal and harbor seals. Cytokines like interferons (IFNs), interleukins (ILs), and various growth and stimulating factors act as messengers between the many different compartments of the immune system within the organism. Cytokines IL-2 and IL-10 are mediators of the immune system and have been widely used in marine mammals (Beineke et al., 2010; Fonfara et al., 2008) to detect susceptibility to infectious disease and immune suppression (Das et al., 2008; Müller et al., 2013; Weirup et al., 2013). IL-2 is known as a T cell growth factor but also is involved in the induction and termination of inflammatory immune responses (Malek, 2003). IL-10 has its function in preventing inflammatory and autoimmune pathologies, by suppressing macrophage and T cell functions (Sabat et al., 2010). Cell receptors and molecules involved in the detoxification of contaminants are also used to assess effects of contaminants in pinnipeds. The aryl hydrocarbon receptor (AHR) and the aryl hydrocarbon receptor nuclear translocator (ARNT) are known to modulate immunotoxic pollutants (Stevens et al., 2009) and are invoked by environmental contaminants. AHR and ARNT transcription were shown to be associated with pollutant (PHAHs) exposure in Baikal seal (*Pusa sibirica*) liver tissue (Kim et al., 2005). The transcription factor peroxisome proliferator-activated receptor (PPARα) mediates detoxifying enzymes (Shizu et al., 2013) and was found to be induced by poly- and perfluoroalkyl substances (PFCs) in Baikal seal hepatic tissue samples (Ishibashi et al., 2008). PPARα has been also used as a marker for fatty acid catabolism and effects of dietary lipids (De Vogel-van den Bosch et al., 2008), as well as lipid metabolism and homeostasis (Kersten et al., 2000) in model organisms and may be a future marker for metabolic processes influenced by toxins in marine mammals (Lehnert et al., 2017). Several receptors like estrogen receptor alpha (Esr1) and thyroid hormone receptor alpha (Thra), among others, have been established to depict changes of the endocrine and reproductive system on the molecular level (Tabuchi et al., 2006). The acute phase protein, heat-shock-protein HSP70, is important in assessing stress and immune reactions (De Maio, 2011), as it is involved in the protein folding and under stress responsible for prevention of the aggregation of unfolding proteins (Mayer and Bukau, 2005). HSP70 transcription has been used to evaluate habituation to handling and rehabilitation in harbor seal pups (Weirup et al., 2013) and grey seals (*Halichoerus grypus*) in human care (Lehnert et al., 2014a).

Blood samples of phocid seals and harbor porpoises (*Phocoena phocoena*) from the North and Baltic Seas (Müller et al., 2013; Weirup et al., 2013; Lehnert et al., 2014a) were used to establish minimally invasive health markers in vulnerable wildlife looking at targeted gene transcription. The selected markers were chosen to both characterize immunological status and stress levels. mRNA transcription of six immuno-relevant biomarkers was analyzed in 13 abandoned harbor seal pups from the North Sea, fostered at the Seal Centre Friedrichskoog, Germany. RNAlater blood samples were taken at admission, day 22, and before release, and analyzed using RT-qPCR. Significant differences in HSP70, cytokine IL-2, and xenobiotic biomarkers AHR, ARNT, and PPARα transcription were found between admission, during rehabilitation, and before release. Highest levels at admission may result from dehydration, handling, transport, and contaminant exposure via lactation. The significant decrease is linked to health improvement, feeding on a supplementary milk product, and adaptation to human care. The increase before release is suspected to be due to mounting infection pressure when seal pups were transferred to a joint and larger pool before release and contaminant exposure from feeding on fish (Weirup et al., 2013).

In blood of grey seal pups and adults in rehabilitation and permanent care, mRNA transcription of AHR, ARNT, PPARα, IL-2, and HSP70 as well as hematology values were measured and compared to rehabilitating harbor seal pups. In pups the highest levels at admission in xenobiotic biomarker, HSP70, and cytokine transcription may show contaminant exposure via lactation, stress during abandonment, and dehydration. The significant decrease may be linked to diet, health improvement, and adaptation. Adults showed higher levels and more variation in biomarker transcription, and clear species-specific differences between harbor and grey seal pups were found (Lehnert et al., 2014a).

In free-ranging harbor seals from North Sea health assessments, normalized transcript copy numbers were correlated to hematology and POP concentration in blood and trace metals in blood and fur. Significant interrelationships between markers and POP compounds, as well as with season, weight, and hematology values, indicate that biomarkers reflect pollutant exposure and effects. A significant relationship between cortisol levels and heat-shock protein expression was observed, indicating stress experienced during restraint of the seals. Interleukin-10 transcription showed significant correlations with trace elements in fur, pointing toward immune regulatory effects of metal exposure. The connection between IL-2, xenobiotic markers, and pollutants may indicate immune suppression in animals exposed to contaminants with subsequent susceptibility to inflammatory disease (Lehnert et al., 2016). IL-2 was highly correlated with AHR, which may be due to immune-modulatory effects. Increased AHR transcription was probably caused by exposure to environmental pollutants. This is in accordance with preliminary results with primary harbor seal hepatocytes, which were incubated with PCB mixtures (Behr et al., 2008; Hellwig, 2011; Korff et al., 2010). HSP70 and cortisol showed a positive correlation in adult harbor seals, probably reflecting stress experienced by the seals during capture (Lehnert et al., 2016). In previous studies on harbor and grey seal pups, HSP70 transcription decreased after admission and throughout rehabilitation, which probably reflects habituation and underlines the use of this acute phase protein as a stress marker in pinnipeds (Fonfara et al., 2008; Weirup et al., 2013; Lehnert et al., 2014a).

In free-ranging harbor seal pups, biopsy samples were used to analyze PCB contaminant burden and gene expression of several target genes associated with health or pollutant exposure (Noël et al., 2017). From eight selected markers chosen for their involvement in reproduction, growth, development, stress response, and metabolism, estrogen receptor alpha (Esr1), thyroid hormone receptor alpha (Thra), and glucocorticoid receptor (Nr3c1) showed significant correlations with PCB loads in blubber, indicating that PCBs are affecting the physiology of individual harbor seals and could impair their ability to respond to other stressors such as environmental changes. The results also showed that skin and inner and outer blubber yielded different gene transcription results, and therefore the response to PCBs was tissue-specific (Noël et al., 2017).

In another study using blood and biopsy samples in harbor seal pups, gene transcription and hematology were combined. Gene expression of thyroid hormone receptor *TR*-α in skin/blubber biopsy samples increased with contaminant load, and a concomitant decrease in circulating total thyroxine concentrations was found, indicating adverse effects on growth and development, as well as on energy metabolism in seals (Tabuchi et al., 2006).

In crabeater (*Lobodon carcinophaga*), Weddell (*Leptonychotes weddellii*), and Ross seals (*Ommatophoca rossii*), minimally invasive blood and fur samples were used to assess immune-relevant and xenobiotic effects on the health status of these vulnerable marine mammal species. Crabeater seals displayed the highest gene transcription in xenobiotic markers, where POP congeners in Antarctic krill or naturally occurring fluorine and cadmium may be responsible. A relationship between Hg concentration and HSP70 transcription indicated that Hg as a toxic metal probably caused modulation of HSP70 expression as a stress response in all three seal species (Lehnert et al., 2017). Differences in the feeding activities between the species and fasting and molting in crabeater seals (Desvergne et al., 2006) may have influenced the high PPARα transcription and may indicate PPARα as marker for diet in crabeater seals.

In another study targeting liver tissue, the release of approximately 260 kg of PCBs by a military radar facility over a 30-year period (1970–2000) contaminated some local marine biota, including the ringed seal (*Pusa hispida*). The abundance profiles of eight health-related gene transcripts were evaluated in liver samples collected from 43 ringed seals in the affected area. In ringed seals the mRNA transcript levels of five gene targets, including AHR, interleukin-1 β (Il1b), estrogen receptor α (Esr1), insulin-like growth factor receptor 1 (Igf1), and glucocorticoid receptor α (Nr3c1), correlated with increasing levels of blubber PCBs (Brown et al., 2014).

Transcriptomics were used in ringed seal blubber sample to identify molecular health indicators associated with PCB exposure. Five target genes were found associated with energy metabolism imbalance in local ringed seals. Those proteins and receptor transcripts correlated with higher blubber PCB burden and may represent indicators of PCB exposure and effects in pinnipeds (Brown et al., 2017).

OUTLOOK

Biomarker research in ecotoxicology studies on pinnipeds is a diverse and interdisciplinary field, in which veterinary medicine, toxicology, and wildlife ecology converge to develop and validate new health and immune endpoints and techniques to help understand the pressures of contaminants on marine mammals. The parameters used vary from the big picture (organ system, reproductive success) to the small puzzle piece (cell culture, transcriptomics) and on combinations of both (lymphocyte proliferation plus gene expression). Single compounds as well as mixtures are investigated, and matrices vary from single cells to tissues, organs, or organisms. Many species with species-specific traits or special habitats are targeted. Many markers show traits that can limit the interpretation of results to one pinniped species and may not allow extrapolation to other species. Without these ideal characteristics, which are difficult to meet, the use of biomarkers as a predictor of detrimental effects of toxins has certain limitations. But well-established techniques like lymphocyte proliferation have since delivered solid data for many species and enable researchers now to perform meta-analyses, while new markers are developed, established, and tested for their viability constantly. New techniques like NGS and transcriptomics help identify genes of interest and can guide scientists to find the markers best suited for their questions. However, many markers used so far have clearly shown that xenobiotics have detrimental effects on the health status and immune function in pinnipeds. The new biomarkers also increasingly allow more detailed insight into specific pathway mechanisms and characteristic effects of single pollutants. It seems that biomarkers for toxicology effects research in vulnerable marine wildlife is a very topical and still young field that will continue to thrive. Molecular biomarkers seem to be a sensitive, noninvasive tool to evaluate health and pollutant exposure and useful to serve as early warning indicators, monitoring, and a case-by-case tool for seal species in human care and the wild. All collected data will constitute an important reference point for future studies, and the selected markers seem useful to indicate biologic effects of stress on the cellular level. Ecotoxicological biomarkers may contribute to the optimization of an effect-oriented monitoring strategy concerning the influence of pollutants on marine mammals.

REFERENCES

Ahrens, L., Siebert, U., Ebinghaus, R., 2009. Temporal trends of polyfluoroalkyl compounds in harbour seals *(Phoca vitulina)* from the German Bight, 1999-2008. Chemosphere 76 (2), 151–158.

Bäcklin, B.M., Bredhult, C., Olovsson, M., 2003. Proliferative effects of estradiol, progesterone, and two CB congeners and their metabolites on gray seal (*Halichoerus grypus*) uterine myocytes in vitro. Toxicological Sciences 75 (1), 154–160.

Bäcklin, B.-M., Moraeus, C., Kauhala, K., Isomursu, M., HELCOM SEAL Expert Group, 2013. Pregnancy Rates of the Marine Mammals — Particular Emphasis on Baltic Grey and Ringed Seals HELCOM Core Indicator Report Available at: http://www.helcom.fi/Core%20Indicators/HELCOM-CoreIndicator-Preg-nancy_rates_of_marine_mammals.pdf.

Baker, J.R., Martin, A.R., 1992. Causes of mortality and parasites and incidental lesions in harbour porpoises (*Phocoena phocoena*) from British waters. The Veterinary Record 130, 554–558.

Barron, M.G., Heintz, R., Krahn, M.M., 2003. Contaminant exposure and effects in pinnipeds: implications for Steller sea lion declines in Alaska. The Science of the Total Environment 311 (1), 111–133.

Beck, H., Breuer, E.M., Droß, A., Mathar, W., 1990. Residues of PCDDs, PCDFs, PCBs and other organochlorine compounds in harbour seals and harbour porpoise. Chemosphere 20 (7–9), 1027–1034.

Beckmen, K.B., Blake, J.E., Ylitalo, G.M., Stott, J.L., O'Hara, T.M., 2003. Organochlorine contaminant exposure and associations with hematological and humoral immune functional assays with dam age as a factor in free-ranging northern fur seal pups (*Callorhinus ursinus*). Marine Pollution Bulletin 46, 594–606.

Behr, A., Wargel, A., Siebert, U., Hellwig, V., 2008. Analysis of polychlorinated biphenyl (PCB)-induced modification of protein expression in primary hepatocytes from harbour seals. Organohalogen Compounds 70, 1079–1082.

Beineke, A., Siebert, U., Mclachlan, M., Bruhn, R., Thron, K., Failing, K., Muller, G., Baumgartner, W., 2005. Investigations of the potential influence of environmental contaminants on the thymus and spleen of harbor porpoises (*Phocoena phocoena*). Environmental Science and Technology 39, 3933–3938.

Beineke, A., Siebert, U., Wohlsein, P., Baumgärtner, W., 2010. Immunology of whales and dolphins. Veterinary Immunology and Immunopathology 133 (2), 81–94.

Bennett, P.M., Jespon, P.D., Law, R.J., Jones, B.R., Kuiken, T., Baker, J.R., Rogan, E., Kirkwood, J.K., 2001. Exposure to heavy metals and infectious disease mortality in harbour porpoises from England and Wales. Environmental Pollution 112, 33–40.

Bergman, A., Olsson, M., 1986. Pathology of Baltic grey seal and ringed seal females with special reference to adrenocortical hyperplasia: Is environmental pollution the cause of a widely distributed disease syndrome? Finnish Game Research 44, 47–62.

Bergman, A., Olsson, M., Reiland, S., 1992. Skull-bone lesions in the baltic grey seal (*Halichoerus grypus*). Ambio 21, 517–519.

Bergman, A., 1999. Health condition of the Baltic grey seal (*Halichoerus grypus*) during two decades: gynaeco- logical health improvement but increased prevalence of colonic ulcers. Acta Pathologica Microbiologica et Immunologica Scandinavica 107, 270–282.

Bergman, A., 2007. Pathological Changes in Seals in Swedish Waters: The Relation to Environmental Pollution 2007 (131). (Doctoral thesis). Swedish University of Agricultural Sciences, Uppsala.

Berggren, P., Ishaq, R., Zebqhr, Y., Näf, C., Bandh, C., Broman, D., 1999. Patterns and levels of organochlorines (DDTs, PCBs, non-ortho PCBs and PCDD/Fs) in male harbour porpoises (*Phocoena phocoena*) from the Baltic Sea, the Kattegat-Skagerrak Seas and the west coast of Norway. Marine Pollution Bulletin 38 (12), 1070–1084.

Bogomolni, A., Frasca, S., Levin, M., Matassa, K., Nielsen, O., Waring, G., De Guise, S., 2016. In vitro exposure of harbor seal immune cells to Aroclor 1260 alters phocine distemper virus replication. Archives of Environmental Contamination and Toxicology 70 (1), 121–132.

Bossart, G.D., 2011. Marine mammals as sentinel species for oceans and human health. Veterinary Pathology 48, 676–690 Online.

Brouwer, A., Reijnders, P.J.H., Koeman, J.H., 1989. Polychlorinated biphenyl (PCB)-contaminated fish induces vitamin A and thyroid hormone deficiency in the common seal (*Phoca vitulina*). Aquatic Toxicology 15, 99–106.

Brown, T.M., Ross, P.S., Reimer, K.J., Veldhoen, N., Danger, N.J., Fisk, A.T., Helbing, C.C., 2014. PCB related effects thresholds as derived through gene transcript profiles in locally contaminated ringed seals (*Pusa hispida*). Environmental Science and Technology 48, 12952–12961.

Brown, T.M., Hammond, S.A., Behsaz, B., Veldhoen, N., Birol, I., Helbing, C.C., 2017. De novo assembly of the ringed seal (*Pusa hispida*) blubber transcriptome: a tool that enables identification of molecular health indicators associated with PCB exposure. Aquatic Toxicology 185, 48–57.

Capowiez, Y., Dittbrenner, N., Rault, M., Triebskorn, R., Hedde, M., Mazzia, C., 2010. Earthworm cast production as a new behavioural biomarker for toxicity testing. Environmental Pollution 158 (2), 388–393.

Ciesielski, T., Wasik, A., Kuklik, I., Skóra, K., Namiesnik, J., Szefer, P., 2004. Organtin compounds in the liver tissue of marine mammals from the Polish coast of the Baltic Sea. Environmental Science and Technology 2004 (38), 1415–1420.

Ciesielski, T., Bertenyi, Szefer Zs, Kuklik, I., Skóra, K., Namiesnik, J., Fodor, P., 2006. Interspecific distribution and co-associations of chemical elements in the liver tissue of marine mammals from the Polish Economical Exclusive Zone, Baltic Sea. Environment International 32, 524–532.

Clausen, B., Andersen, S., 1988. Evaluation of by-catch and health status of the harbour porpoise (*Phocoena phocoena*) in Danish waters. Danish Review of Game Biology 13, 1–20.

Clement, B., Christiansen, K., Girreser, U., 2001. Phase 2 metabolites of *N*-hydroxylated amidines (amidoximes): synthesis, *in vitro* formation by pig hepatocytes, and mutagenicity testing. Chemical Research in Toxicology 14, 319–326.

Constable, S., Parslow, A., Dutton, G., Rogers, T., Hogg, C., 2006. Urinary cortisol sampling: a non-invasive technique for examining cortisol concentrations in the Weddell seal, *Leptonychotes weddellii*. Zoo Biology 25 (2), 137–144.

Covaci, A., Jorens, P., Jacquemyn, Y., Schepens, R., 2002. Distribution of PCBs and organochlorine pesticides in umbilical cord and maternal serum. The Science of the Total Environment 298 (1–3), 45–53.

Das, K., Debacker, V., Pillet, S., Bouquegneau, J.-M., 2003. Heavy metals in marine mammals. In: Vos, J.V., Bossart, G.D., Fournier, M., O'Shea, T. (Eds.), Toxicology of Marine Mammals. Taylor and Francis Publishers, Washington, DC, pp. 135–167.

Das, K., Siebert, U., Fontaine, M., Jauniaux, T., Holsbeek, L., Bouquegneau, J.M., 2004. Ecological and pathological factors related to trace metal concentrations in harbour porpoises *Phocoena phocoena* from the North Sea and adjacent areas. Marine Ecology Progress Series 281, 283–295.

Das, K., Vossen, A., Tolley, K., Víkingsson, G., Thron, K., Müller, G., et al., 2006. Interfollicular fibrosis in the thyroid of the harbour porpoise: an endocrine disruption? Archives of Environmental Contamination and Toxicology 51 (4), 720–729.

Das, K., Siebert, U., Gillet, A., Dupont, A., Di-Poi, C., Fonfara, S., Mazzucchelli, G., De Pauw, E., De Pauw-Gillet, M.C., 2008. Mercury immune toxicity in harbour seals: links to in vitro toxicity. Environmental Health 7 (52).

Debier, C., Pomeroy, P.P., Dupont, C., Joiris, C., Comblin, V., Le Boulenge, E., Larondelle, Y., Thome, J.P., 2003. Dynamics of PCB transfer from mother to pup during lactation in UK grey seals *Halichoerus grypus*: differences in PCB profile between compartments of transfer and changes during the lactation period. Marine Ecology Progress Series 247, 249–256.

De Guise, S., 2004. Immunotoxicology in marine mammals. In: Tryphonas, H., Fournier, M., Blakley, B.R., Smits, J.E.G., Brousseau, P. (Eds.), Investigative Immunotoxicology. Taylor & Francis, New York, USA.

De Maio, A., 2011. Extracellular heat shock proteins, cellular export vesicles, and the Stress Observation System: a form of communication during injury, infection, and cell damage. Cell Stress and Chaperones 16 (3), 235–249.

Desforges, J.P.W., Sonne, C., Levin, M., Siebert, U., De Guise, S., Dietz, R., 2016. Immunotoxic effects of environmental pollutants in marine mammals. Environment International 86, 126–139.

Desforges, J.P., Levin, M., Jasperse, L., De Guise, S., Eulaers, I., Letcher, R.J., Acquarone, M., Nordøy, E., Folkow, L.P., Hammer Jensen, T., Grøndahl, C., Bertelsen, M.F., Leger, J.St., Almunia, J., Sonne, C., Dietz, R., 2017. Environmental Science and Technology 51 (19), 11431–11439. https://doi.org/10.1021/acs.est.7b03532.

Desvergne, B., Michalik, L., Wahli, W., 2006. Transcriptional regulation of metabolism. Physiological Reviews 86, 465–514.

De Swart, R.L., Ross, P.S., Vedder, L.J., Timmerman, H.H., Heisterkamp, S., Loveren, H.K., Vos, J.G., Reijnders, P.J.H., Osterhaus, A.D.M.E., 1994. Impairment of immune function in harbour seals (*Phoca vitulina*) feeding on fish from polluted waters. Ambio 23, 155–159.

De Swart, R.L., Ross, P.S., Vos, J.G., Osterhaus, A.D.M.E., 1996. Impaired immunity in harbour seals (*Phoca vitulina*) exposed to bioaccumulated environmental contaminants: review of a long-term feeding study. Environmental Health Perspectives 104, 823–828.

De Vogel-van den Bosch, H.M., Bünger, M., de Groot, P.J., Bosch-Vermeulen, H., Hooiveld, G.J.E.J., Müller, M., 2008. PPARalpha-mediated effects of dietary lipids on intestinal barrier gene expression. BMC Genomics 9, 231.

Drescher, H.E., Harms, U., Huschenbeth, E., 1977. Organochlorines and heavy metals in the harbour seal *Phoca vitulina* from the German North Sea Coast. Marine Biology 41 (1), 99–106.

Duinker, J.C., Hillebrand, M.T.J., Nolting, R.F., 1979. Organochlorines and metals in harbour seals (Dutch Wadden Sea). Marine Pollution Bulletin 10, 360–364.

Dupont, A., Siebert, U., Covaci, A., Weijs, L., Eppe, G., Debier, C., De Pauw-Gillet, M.-C., Das, K., 2013. Relationships between in vitro lymphoproliferative responses and levels of contaminants in blood of free-ranging adult harbour seals (*Phoca vitulina*) from the North Sea. Aquatic Toxicology 142–143, 210–220.

Dupont, A., De Pauw-Gillet, M.C., Schnitzler, J., Siebert, U., Das, K., 2016. Effects of methylmercury on harbour seal peripheral blood leucocytes in vitro studied by electron microscopy. Archives of Environmental Contamination and Toxicology 70 (1), 133–142.

Fair, P.A., Romano, T., Schaefer, A.M., Reif, J.S., Bossart, G.D., Houde, M., Muir, D., Adams, J., Rice, C., Hulsey, T.C., Peden-Adams, M., 2013. Associations between perfluoroalkyl compounds and immune and clinical chemistry parameters in highly exposed bottlenose dolphins (*Tursiops truncatus*). Environmental Toxicology and Chemistry 32, 736–746.

Fonfara, S., Kakuschke, A., Rosenberger, T., Siebert, U., Prange, A., 2008. Cytokine and acute phase protein expression in blood samples of harbour seal pups. Marine Biology 155 (3), 337–345.

Fossi, C., Marsili, L., 1997. The use of non-destructive biomarkers in the study of marine mammals. Biomarkers 2 (4), 205–216.

Fossi, M.C., Marsili, L., Junin, M., Castello, H., Lorenzani, J.A., Casini, S., et al., 1997. Use of non-destructive biomarkers and residue analysis to assess the health status of endangered species of pinnipeds in the south-west Atlantic. Marine Pollution Bulletin 34 (3), 157–162.

Freeman, H.C., Sangalan, G.B., 1977. A study on the effects of methylmercury, cadmium, arsenic, selenium, and a PCB (Arochlor 1254) on adrenal and testicular steroidogeneses *in vitro*, by the gray seal *Halichoerus grypus*. Archives of Environmental Contamination and Toxicology 5, 369–383.

Frouin, H., Lebeuf, M., Hammill, M., Masson, S., Fournier, M., 2010. Effects of individual polybrominated diphenyl ether (PBDE) congeners on harbour seal immune cells in vitro. Marine Pollution Bulletin 60, 291–298.

Granby, K., Kinze, K., 1991. Organochlorines in Danish and West Greenland harbour porpoises. Marine Pollution Bulletin 22, 458–462.

Griesel, S., Kakuschke, A., Siebert, U., Prange, A., 2008. Trace element concentrations in blood of harbour seals (*Phoca vitulina*) from the Wadden Sea. The Science of the Total Environment 392 (2–3), 313–323.

Habran, S., Pomeroy, P., Debier, C., Das, K., 2013. Changes in trace elements during lactation in a marine top predator, the grey seal. Aquatic Toxicology 126, 455–466.

Hall, A., Pomeroy, P., Green, N., Jones, K., Harwood, J., 1997. Infection, haematology and biochemistry in grey seal pups exposed to chlorinated biphenyls. Marine Environmental Research 43, 81–98.

Hall, A.J., Green, N.J., Jones, K.C., Pomeroy, P.P., Harwood, J., 1998. Thyroid hormones as biomarkers in grey seals. Marine Pollution Bulletin 36 (6), 424–428.

Hammond, J.A., Hall, A.J., Dyrynda, E.A., 2005. Comparison of polychlorinated biphenyl (PCB) induced effects on innate immune functions in harbour and grey seals. Aquatic Toxicology 74, 126–138.

Härkönen, T., Dietz, R., Reijnders, P., Teilmann, J., Harding, K., Hall, A., Brasseur, S., Siebert, U., Goodman, S.J., Jepson, P.D., Dau Rasmussen, T., Thompson, P., 2006. The 1988 and 2002 phocine distemper virus epidemics in European harbour seals. Diseases of Aquatic Organisms 68 (2), 115–130.

Harms, U., Drescher, H.E., Huschenbeth, E., 1978. Further data on heavy metals and organochlorines in marine mammals from German coastal waters. Meeresforschung - Reports on Marine Research 26 (3–4), 153–161.

Harrison, P.J., Heath, P.R., Eastwood, S.L., Burnet, P.W.J., McDonald, B., Pearson, R.C.A., 1995. The relative importance of premortem acidosis and postmortem interval for human brain gene expression studies: selective mRNA vulnerability and comparison with their encoded proteins. Neuroscience Letters 200 (3), 151–154.

Hasselmeier, I., Fonfara, S., Driver, J., Siebert, U., 2008. Differential hematology profiles of free-ranging, rehabilitated, and captive harbor seals (*Phoca vitulina*) of the German North Sea. Aquatic Mammals 34 (2), 149.

Helle, E., Olsson, M., Jensen, S., 1976. PCB levels correlated with pathological changes in seal uteri. Ambio 5, 261–262.

Hellwig, V., 2011. Effects of persistent polutants on marine mammals. In: Persistent Pollution–Past, Present and Future. Springer, Berlin, Heidelberg, pp. 337–362.

Hobbs, K., Lebeuf, M., Hammill, M., 2002. PCBs and OCPs in male harbour, grey, harp and hooded seals from the Estuary and Gulf of St Lawrence, Canada. Science of the Total Environment 296, 1–18.

Huber, S., Ahrens, L., Bårdsen, B.J., Siebert, U., Bustnes, J.O., Víkingsson, G.A., Ebinghaus, R., Herzke, D., 2012. Temporal trends and spatial differences of perfluoroalkylated substances in livers of harbor porpoise (*Phocoena phocoena*) populations from Northern Europe, 1991-2008. The Science of the Total Environment 419, 216–224.

Ishibashi, H., Iwata, H., Kim, E.Y., Tao, L., Kannan, K., Tanabe, S., et al., 2008. Contamination and effects of perfluorochemicals in baikal seal (*Pusa sibirica*). 2. Molecular characterization, expression level, and transcriptional activation of peroxisome proliferator-activated receptor α. Environmental Science and Technology 42 (7), 2302–2308.

Jauniaux, T., Petitjean, D., Brenez, C., Borrens, M., Brosens, L., Haelters, J., Tavernier, T., Coignoul, F., 2002. Post-mortem findings and causes of death of harbour porpoises (*Phocoena phocoena*) stranded from 1990 to 2000 along the coastlines of Belgium and Northern France. Journal of Comparative Pathology 126, 243–253.

Jepson, P.D., Bennett, P.M., Allchin, C.R., Law, R.J., Kuiken, T., Baker, J.R., et al., 1999. Investigating potential associations between chronic exposure to polychlorinated biphenyls and infectious disease mortality in harbour porpoises from England and Wales. Science of the Total Environment 7, 339–348.

Jepson, P.D., Baker, J.R., Kuiken, T., Simpson, V.R., Kennedy, S., Bennett, P.M., 2000. Pulmonary pathology of harbour porpoises stranded in England and Wales between 1990 and 1996. The Veterinary Record 146, 721–728.

Jepson, P.D., Bennett, P.M., Deaville, R., Allchin, C.R., Baker, J.R., Law, R.J., 2005. Relationships between polychlorinated biphenyls and health status in harbor porpoises (*Phocoena phocoena*) stranded in the United Kingdom. Environmental Toxicology and Chemistry 24, 238–248.

Jepson, P.D., Deaville, R., Barber, J.L., Aguilar, A., Borrell, A., Murphy, S., Barry, J., Brownlow, A., Barnett, J., Berrow, S., Cunningham, A.A., Davison, N.J., Ten Doeschate, M., Esteban, R., Ferreira, M., Foote, A.D., Genov, T., Gimenez, J., Loveridge, J., Llavona, A., Martin, V., Maxwell, D.L., Papachlimitzou, A., Penrose, R., Perkins, M.W., Smith, B., de Stephanis, R., Tragenza, N., Verborgh, P., Fernandez, A., Law, R.J., 2016. PCB pollution continues to impact populations of orcas and other dolphins in European waters. Scientific Report. https://doi.org/10.1038/srep18573.

Kakuschke, A., Valentine-Thon, E., Griesel, S., Fonfara, S., Siebert, U., Prange, A., 2005. Immunological impact of metals in harbour seals (*Phoca vitulina*) of the North Sea. Environmental Science and Technology 39, 7568–7575.

Kakuschke, A., Valentine-Thon, E., Fonfara, S., Griesel, S., Siebert, U., Prange, A., 2006. Metal sensitivity of marine mammals: a case study of a grey seal (*Halicoerus grypus*). Marine Mammal Science 22 (4), 985–996.

Kakuschke, A., Valentine-Thon, E., Fonfara, S., Griesel, S., Rosenberger, T., Siebert, U., Prange, A., 2008. Metal-induced impairment of the cellular immunity of newborn harbour seals (*Phoca vitulina*). Archives of Environmental Contamination and Toxicology 55, 129–136.

Kakuschke, A., Valentine-Thon, E., Fonfara, S., Kramer, K., Prange, A., 2009a. Effects of methyl-, phenyl-, ethylmercury and mercurychloride on immune cells of harbor seals (*Phoca vitulina*). Journal of Environmental Sciences 21, 1716–1721.

Kakuschke, A., Griesel, S., Fonfara, S., Rosenberger, T., Prange, A., 2009b. Concentrations of selected essential and non-essential elements in blood of harbour seal (*Phoca vitulina*) pups of the German North Sea. Biological Trace Element Research 127 (1), 28–36.

Kakuschke, A., Valentine-Thon, E., Griesel, S., Fonfara, S., Siebert, U., Prange, A., 2011. Are metal-induced hypersensitivities in harbor seals associated with liver function? Marine Pollution Bulletin 62, 1891–1894.

Kannan, K., Corsolini, S., Falandysz, J., Oehme, G., Focardi, S., Giesy, J.P., 2002. Perfluorooctanesulfonate and related fluorinated hydrocarbons in marine mammals, fishes, and birds from coasts of the baltic and the Mediterranean Seas. Environmental Science and Technology 36, 3210–3216.

Kersten, S., Desvergne, B., Wahli, W., 2000. Roles of PPARs in health and disease. Nature 405 (6785), 421–424.

Kim, E.Y., Hahn, M.E., 2002. cDNA cloning and characterization of an aryl hydrocarbon receptor from the harbor seal (*Phoca vitulina*): a biomarker of dioxin susceptibility? Aquatic Toxicology 58 (1), 57–73.

Kim, E.Y., Iwata, H., Suda, T., Tanabe, S., Amano, M., Miyazaki, N., Petrov, E.A., 2005. Aryl hydrocarbon receptor (AHR) and AHR nuclear translocator (ARNT) expression in Baikal seal (*Pusa sibirica*) and association with 2, 3, 7, 8–TCDD toxic equivalents and CYP1 expression levels. Comparative Biochemistry and Physiology - Part C: Toxicology and Pharmacology 141 (3), 281–291.

Kirkegaard, M., Sonne, C., Leifsson, P.S., Dietz, R., Born, E.W., Muir, D.C.G., Letcher, R.J., 2005. Histology of selected immunological organs in polar bear (*Ursus maritimus*) from East Greenland in relation to concentrations of organohalogen contaminants. The Science of the Total Environment 341 (1), 119–132.

Korff, V., Wargel, A., Lehnert, K., Siebert, U., Hellwig, V., 2009. Assessment of cell viability and specific metabolism in primary seal hepatocytes after exposure to PFOS and PCBs. In: Abstracts of the Society of Marine Mammalogy, 18th Biennual Conference, 12–16.10.2009, Quebec, Canada.

Korff, V., Behr, A., Wargel, A., Lehnert, K., Siebert, U., Hellwig, V., 2010. From seals to cells: protein biomarkers to reveal effects of persistent pollutants on primary hepatocytes of *Phoca vitulina*. In: ICES ASC Handbook of the Annual Science Conference, 20–24.9.2010, Nantes, France.

Kuenstl, L., Griesel, S., Prange, A., Goessler, W., 2009. Arsenic speciation in bodily fluids of harbour seals (*Phoca vitulina*) and harbour porpoises (*Phocoena phocoena*). Environmental Chemistry 6 (4), 319.

Lalancette, A., Morin, Y., Measures, L., Fournier, M., 2003. Contrasting changes of sensitivity by lymphocytes and neutrophils to mercury in developing grey seals. Developmental and Comparative Immunology 27, 735–747.

Law, R.J., Allchin, C.R., Harwood, J., 1989. Concentrations of organochlorine compounds in the blubber of seals from eastern and north-eastern England, 1988. Marine Pollution Bulletin 20 (3), 110–115 1989 Printed in Great Britain.

Law, R.J., Hopkin, A., Baker, J.R., Harwood, J., Jackson, D.B., Kennedy, S., Martin, A.R., Morris, R.J., 1991. Concentrations of trace metals in the livers of marine mammals (seals, porpoises and dolphins) from waters around the British Isles. Marine Pollution Bulletin 22 (4), 183–191.

Law, R.J., Jepson, P.D., Deaville, R., Reid, R.J., Patterson, I.A.P., Allchin, C.R., Jones, B.R., 2006. Collaborative UK Marine Mammals Strandings Project: summary of contaminant data for the period 1993–2001. Science Series Technical Report, Cefas Lowestoft 131, 72.

Lehnert, K., Müller, S., Weirup, L., Ronnenberg, K., Pawliczka, I., Rosenberger, T., Siebert, U., 2014a. Molecular biomarkers in grey seals (*Halichoerus grypus*) to evaluate pollutant exposure, health and immune status. Marine Pollution Bulletin.

Lehnert, K., Raga, J.A., Siebert, U., 2005. Macroparasites in stranded and bycaught harbour porpoises from German and Norwegian waters. Diseases of Aquatic Organisms 64 (3), 265–269.

Lehnert, K., Seibel, H., Hasselmeier, I., Wohlsein, P., Iversen, M., Nielsen, N.H., et al., 2014b. Increase in parasite burden and associated pathology in harbour porpoises (*Phocoena phocoena*) in West Greenland. Polar Biology 37 (3), 321–331.

Lehnert, K., Ronnenberg, K., Weijs, L., Covaci, A., Das, K., Hellwig, V., Siebert, U., 2016. Xenobiotic and immune-relevant molecular biomarkers in harbor seals as proxies for pollutant burden and effects. Archives of Environmental Contamination and Toxicology 70 (1), 106–120.

Lehnert, K., Weirup, L., Harding, K.C., Härkönen, T., Karlsson, O., Teilmann, J., 2017. Antarctic seals: molecular biomarkers as indicators for pollutant exposure, health effects and diet. The Science of the Total Environment 599, 1693–1704.

Levin, M., De Guise, S., Ross, P.S., 2005. Association between lymphocyte proliferation and polychlorinated biphenyls in free-ranging harbor seal (*Phoca vitulina*) pups from British Columbia, Canada. Environmental Toxicology and Chemistry 24, 1247–1252.

Levin, M., Gebhard, E., Jasperse, L., Desforges, J.P., Dietz, R., Sonne, C., et al., 2016. Immunomodulatory effects of exposure to polychlorinated biphenyls and perfluoroalkyl acids in East Greenland ringed seals (*Pusa hispida*). Environmental Research 151, 244–250.

Lind, Y., Darnerud, P.O., Atuma, S., Aune, M., Becker, W., Bjerselius, R., Cnattingius, S., Glynn, A., 2003. Polybrominated diphenyl ethers in breast milk from Uppsala County, Sweden. Environmental Research 2, 186–194.

Lucke, K., Lepper, P.A., Blanchet, M.A., Siebert, U., 2009. Temporary shift in masked hearing thresholds in a harbor porpoise (*Phocoena phocoena*) after exposure to seismic airgun stimuli. Journal of the Acoustical Society of America 125, 6.

Malek, T.R., 2003. The main function of IL-2 is to promote the development of T regulatory cells. Journal of Leukocyte Biology 74 (6), 961–965.

Mancia, A., Ryan, J.C., Van Dolah, F.M., Kucklick, J.R., Rowles, T.K., Wells, R.S., et al., 2014. Machine learning approaches to investigate the impact of PCBs on the transcriptome of the common bottlenose dolphin (*Tursiops truncatus*). Marine Environmental Research 100, 57–67.

Marcogliese, D.J., Pietrock, M., 2011. Combined effects of parasites and contaminants on animal health: parasites do matter. Trends in Parasitology 27 (3), 123–130.

Marsili, L., Fossi, M.C., Neri, G., Casini, S., Gardi, C., Palmeri, S., et al., 2000. Skin biopsies for cell cultures from Mediterranean free-ranging cetaceans. Marine Environmental Research 50 (1), 523–526.

Mayer, M.P., Bukau, B., 2005. Hsp70 chaperones: cellular functions and molecular mechanism. Cellular and Molecular Life Sciences 62, 670–684.

Mos, L., Morsey, B., Jeffries, S.J., Yunker, M.B., Raverty, S., De Guise, S., Ross, P.S., 2006. Chemical and biological pollution contribute to the immunological profiles of freeranging harbor seals. Environmental Toxicology and Chemistry 25, 3110–3117.

Morris, R.J., Law, R.J., Allchin, C.R., Kelly, C.A., Fileman, C.F. 1989. Metals and organochlorines in dolphins and porpoises of Cardigan Bay, West Wales. Marine Pollution Bulletin 20, 512.

Müller, S., Lehnert, K., Seibel, H., Driver, J., Ronnenberg, K., Teilmann, J., et al., 2013. Evaluation of immune and stress status in harbour porpoises (*Phocoena phocoena*): can hormones and mRNA expression levels serve as indicators to assess stress? BMC Veterinary Research 9 (1), 145.

Neale, J.C.C., Van deWater, J.A., Harvey, J.T., Tjeerdema, R.S., Gershwin, M.E., 2002. Proliferative responses of harbor seal (*Phoca vitulina*) T lymphocytes to model marine pollutants. Developmental Immunology 9, 215–221.

Neale, J.C.C., Kenny, T.P., Tjeerdema, R.S., Gershwin, M.E., 2005. PAH- and PCB-induced alterations of protein tyrosine kinase and cytokine gene transcription in harbor seal (*Phoca vitulina*) PBMC. Clinical and Developmental Immunology 12, 91–97.

Noël, M., Dangerfield, N., Jeffries, S., Lambourn, D., Lance, M., Helbing, C., et al., 2017. Polychlorinated biphenyl-related alterations of the expression of essential genes in harbour seals (*Phoca vitulina*) from coastal sites in Canada and the United States. Archives of Environmental Contamination and Toxicology 1–12.

Nyman, M., Bergknut, M., Fant, M.L., Raunio, H., Jestoi, M., Bengs, C., Murk, A., Koistinen, J., Bäckman, C., Pelkonen, O., Tysklind, M., Hirvi, T., Helle, E., 2003. Contaminant exposure and effects in Baltic ringed and grey seals as assessed by biomarkers. Marine Environmental Research 55, 73–99.

Olsson, M., Karlsson, B., Ahnland, E., 1994. Diseases and environmental contaminants in seals from the Baltic and the Swedish West Coast. The Science of the Total Environment 154, 217–227.

Pascual, S., Abollo, E., 2005. Whaleworms as a tag to map zones of heavy-metal pollution. Trends in Parasitology 21 (5), 204–206.

Pillet, S., Lesage, V., Hammill, M., Cyr, D.G., Bouquegneau, J., Fournier, M., 2000. In vitro exposure of seal peripheral blood leukocytes to different metals reveal a sex-dependent effect of zinc on phagocytic activity. Marine Pollution Bulletin 40, 921–927.

Reijnders, P.J.H., 1980. Organochlorine and heavy metal residues in harbour seals from the Wadden Sea and their possible effects on reproduction. Netherlands Journal of Sea Research 14, 30–65.

Reese, J.A., Byard, J.L., 1981. Isolation and culture of adult hepatocytes from liver biopsies. In Vitro 17, 935–940.

Reijnders, P.J.H., 1986. Reproductive failure in common seals feeding on fish from polluted coastal waters. Nature 324, 456–457.

Reijnders, P.J.H., 1988. Ecotoxicology perspectives in marine mammalogy: research principles and goals for a conservation policy. Marine Mammal Science 4, 91–102.

Richardson, W.J., Greene Jr., C.R., Malme, C.I., Thomson, D.I., 1995. Marine Mammals and Noise. Academic Press, San Diego. 576 p.

Roos, A.M., Bäcklin, B.M.V.M., Helander, B.O., Rigét, F.F., Eriksson, U.C., 2012. Improved reproductive success in otters (*Lutra lutra*), grey seals (*Halichoerus grypus*) and sea eagles (*Haliaeetus albicilla*) from Sweden in relation to concentrations of organochlorine contaminants. Environmental Pollution 170, 268–275.

Ross, P.S., Swart, R.L.D., Reijnders, P.J.H., van Loveren, H., Vos, J.G., Osterhaus, A.D.M.E., 1995. Contaminant-related suppression of delayed-type hypersensitivity and antibody responses in harbor seals fed herring from the Baltic Sea. Environmental Health Perspectives 103, 162–167.

Ross, P.S., de Swart, R.L., Addison, R.F., van Loveren, H., Vos, J.G., Osterhaus, A.D.M.E., 1996a. Contaminant-induced immunotoxicity in harbour seals: wildlife at risk? Toxicology 112, 157–169.

Ross, P.S., De Swart, R.L., Timmerman, H.H., Reijnders, P.J.H., Vos, J.G., Van Loveren, H., Osterhaus, A.D., 1996b. Suppression of natural killer cell activity in harbour seals (*Phoca vitulina*) fed Baltic Sea herring. Aquatic Toxicology 34, 71–84.

Ross, P.S., De Guise, S., 2007. Marine mammal immunotoxicology. In: Luebke, R., House, R., Kimber, I. (Eds.), Immunotoxicology and Immunopharmacology, third ed. CRC Press, Taylor & Francis Group, Boca Raton, FL.

Routti, H., Arukwe, A., Jenssen, B.M., Letcher, R.J., Nyman, M., Bäckman, C., Gabrielsen, G.W., 2010a. Comparative endocrine disruptive effects of contaminants in ringed seals (*Phoca hispida*) from Svalbard and the Baltic Sea. Comparative Biochemistry and Physiology - Part C: Toxicology and Pharmacology 152 (3), 306–312.

Routti, H., Jenssen, B.M., Lydersen, C., Bäckman, C., Arukwe, A., Nyman, M., et al., 2010b. Hormone, vitamin and contaminant status during the moulting/fasting period in ringed seals (*Pusa [Phoca] hispida*) from Svalbard. Comparative Biochemistry and Physiology Part A: Molecular and Integrative Physiology 155 (1), 70–76.

Sabat, R., Grütz, G., Warszawska, K., Kirsch, S., Witte, E., Wolk, K., Geginet, J., 2010. Biology of interleukin-10. Cytokine and Growth Factor Reviews 21 (5), 331–344.

Shaw, S.D., Brenner, D., Bourakovsky, A., Mahaffey, C.A., Perkins, C.R., 2005. Polychlorinated biphenyls and chlorinated pesticides in harbor seals (*Phoca vitulina* concolor) from the northwestern Atlantic coast. Marine Pollution Bulletin 50, 1069–1084.

Shaw, S., Berger, M.L., Brenner, D., Tao, L., Wu, Q., Kannan, K., 2009. Specific accumulation of perfluorochemicals in harbour seals (*Phoca vitulina concolor*) from the northwest Atlantic. Chemosphere 74 (8), 1037–1043.

Shizu, R., Benoki, S., Numakura, Y., Kodama, S., Miyata, M., Yamazoe, Y., Yoshinari, K., 2013. Xenobiotic-induced hepatocyte proliferation associated with constitutive active/androstane receptor (CAR) or peroxisome proliferator-activated receptor α (PPARα) is enhanced by pregnane X receptor (PXR) activation in mice. PLoS One 8 (4), e61802.

Siebert, U., Joiris, C., Holsbeek, L., Benke, H., Failing, K., Frese, K., Petzinger, E., 1999. Potential relation between mercury concentrations and necropsy findings in cetaceans from German waters of the North and Baltic Seas. Marine Pollution Bulletin 38 (4), 285–295.

Siebert, U., Wünschmann, A., Weiss, R., Frank, H., Benke, H., Frese, K., 2001. Post-mortem findings in harbour porpoises (*Phocoena phocoena*) from the German North and Baltic Seas. Journal of Comparative Pathology 124, 102–114.

Siebert, U., Vossen, A., Baumgärtner, W., Müller, G., Beineke, A., McLachlan, M., Bruhn, R., Thron, K., 2002. Untersuchungen zu Auswirkungen von Umweltchemikalien auf das Endokrinium und Immunsystem von Schweinswalen aus der deutschen Nord- und Ostsee: Umweltforschungsplan des Bundesministeriums für Umwelt, Naturschutz und Reaktorsicherheit. Chemikaliensicherheit, allgemeine Bewertungskriterien für chemische Produkte Forschungsbericht 29965221/01.

Siebert, U., Wünschmann, A., Tolley, K., Vikingsson, G., Olafsdottir, D., Lehnert, K., Weiss, R., Baumgärtner, W., 2006. Pathological findings in harbour porpoises (*Phocoena phocoena*) originating from Norwegian and Icelandic waters. Journal of Comparative Pathology 134 (2–3), 134–142.

Siebert, U., Wohlsein, P., Lehnert, K., Baumgärtner, W., 2007. Pathological findings in harbour seals (*Phoca vitulina*): 1996-2005. Journal of Comparative Pathology 137, 47–58.

Siebert, U., Prenger-Berrninghoff, E., Weiss, R., 2009. Regional differences in bacteria flora in harbour porpoises from the North Atlantic: environmental effects. Journal of Applied Microbiology 106, 329–337.

Siebert, U., Müller, S., Gilles, A., Sundermeyer, J., Narberhaus, I., 2012a. Chapter VII Species profiles marine mammals. In: Narberhaus, I., Krause, J., Bernitt, U. (Eds.), Threatened Biodiversity in the German North and Baltic Seas - Sensitivities towards Human Activities and the Effects of Climate Change. Naturschutz und Biologische Vielfalt, Heft 117, Bonn – Bad Godesberg, pp. 487–501.

Siebert, U., Heidmann, A., Friedhoff, N., Kruse, H., Rigét, F., Adler, S., Maser, E., 2012b. Organochlorine burdens in harbour seals from the German Wadden Sea collected during two phocine distemper epizootics and ringed seals from west Greenland waters. Environmental and Analytical Toxicology 2 (2).

Simms, W., Ross, P.S., 2000. Vitamin A physiology and its application as a biomarker of contaminant-related toxicity in marine mammals: a review. Toxicology and Industrial Health 16 (7–8), 291–302.

Skaare, J.U., Markussen, N.H., Norheim, G., Haugen, S., Holt, G., 1990. Levels of polychlorinated biphenyls, organochlorine pesticides, mercury, cadmium, copper, selenium, arsenic, and zinc in the harbour seal, *Phoca vitulina*, in Norwegian waters. Environmental Pollution 66 (4), 309–324.

Sørmo, E.G., Larsen, H.J.S., Johansen, G.M., Skaare, J.U., Jenssen, B.M., 2009. Immunotoxicity of polychlorinated biphenyls (PCB) in free-ranging gray seal pups with special emphasis on dioxin-like congeners. Journal of Toxicology and Environmental Health, Part A 72, 266–276.

Stevens, E.A., Mezrich, J.D., Bradfield, C.A., 2009. The aryl hydrocarbon receptor: a perspective on potential roles in the immune system. Immunology 127 (3), 299–311.

Sures, B., 2004. Environmental parasitology: relevancy of parasites in monitoring environmental pollution. Trends in Parasitology 20 (4), 170–177.

Tabuchi, M., Veldhoen, N., Dangerfield, N., Jeffries, S., Helbing, C.C., Ross, P.S., 2006. PCB-related alteration of thyroid hormones and thyroid hormone receptor gene expression in free-ranging harbor seals (*Phoca vitulina*). Environmental Health Perspectives 114 (7), 1024.

Thron, K.U., Bruhn, R., McLachlan, M.S., 2004. The influence of age, sex, body-condition and regions on the levels of PBDEs and toxaphene in harbour porpoises from European waters. Fresenius Environmental Bulletin 13, 146–155.

Tomita, H., Vawter, M.P., Walsh, D.M., Evans, S.J., Choudary, P.V., Li, J., et al., 2004. Effect of agonal and postmortem factors on gene expression profile: quality control in microarray analyses of postmortem human brain. Biological Psychiatry 55 (4), 346–352.

Troisi, G.M., Haraguchi, K., Kaydoo, D.S., Nyman, M., Aguilar, A., Borrell, A., Siebert, U., Mason, C.F., 2001. Bioaccumulation of polychlorinated biphenyls (PCBs) and dichlorodiphenylethane (DDE) methyl sulfones in tissues of seal and dolphin morbillivirus epizootic victims. Journal of Toxicology and Environmental Health, Part A 62, 1–8.

Ungherese, G., Ugolini, A., 2009. Sandhopper solar orientation as a behavioural biomarker of trace metals contamination. Environmental Pollution 157 (4), 1360–1364.

Van Bressem, M.-F., Raga, J.A., Di Guardo, G., Jepson, P.D., Duignan, P.J., Siebert, U., Barrett, T., Santos, M.C.D.O., Moreno, I.B., Siciliano, S., Aguilar, A., Van Waerebeek, K., 2009. Emerging infectious diseases in cetaceans worldwide and the possible role of environmental stressors. Diseases of Aquatic Organisms 86, 143–157.

Van de Viejver, K.I., Hoff, P.T., Das, K., Van Dongen, W., Esmans, E.L., Jauniaux, T., Bouquegneau, J.-M., Blust, R., De Coen, W., 2003. Perfluorinated chemicals infiltrate ocean waters: link between exposure levels and stable isotope ratios in marine mammals. Environmental Science and Technology 37 (24), 5545–5550.

Van de Viejver, K.I., Hoff, P.T., Das, K., Brasseur, S., Van Dongen, W., Esmans, E., Reijnders, P., Blust, R., De Coen, W., 2005. Tissue distribution of perfluorinate chemicals in harbour seals (*Phoca vitulina*) from the Dutch Wadden Sea. Environmental Science and Technology 39 (18), 6978–6984.

Weijs, L., Dirtu, C.A., Das, K., Gheorghe, A., Reijnders, J.H.P., Neels, H., Blust, R., Covaci, A., 2009. Inter-species differences for polychlorinated biphenyls and polybrominated diphenyl ethers in marine top predators from the Southern North Sea: part 1. Accumulation patterns in harbour seals and harbour porpoises. Environmental Pollution 437–444 Elsevier.

Weijs, L., Roach, A.C., Yang, R.S., McDougall, R., Lyons, M., Housand, C., et al., 2014. Lifetime PCB 153 bioaccumulation and pharmacokinetics in pilot whales: Bayesian population PBPK modeling and Markov chain Monte Carlo simulations. Chemosphere 94, 91–96.

Weijs, L., Zaccaroni, A., 2016. Toxicology of marine mammals: new developments and opportunities. Archives of Environmental Contamination and Toxicology 70 (1), 1–8.

Weirup, L., Müller, S., Ronnenberg, K., Rosenberger, T., Siebert, U., Lehnert, K., 2013. Immune-relevant and new xenobioticmolecular biomarkers to assess anthropogenic stress in seals. Marine Environmental Research 92, 43–51.

Wenzel, C., Adelung, D., Kruse, H., Wassermann, O., 1993. Trace metal accumulation in hair and skin of the harbour seal, *Phoca vitulina*. Marine Pollution Bulletin 26, 152–155.

Wolkers, H., Lydersen, C., Kovacs, K.M., 2004. Accumulation and lactational transfer of PCBs and pesticides in harbor seals (*Phoca vitulina*) from Svalbard. Norway. Science of the Total Environment 319, 137–146.

Wünschmann, A., Siebert, U., Frese, K., Weiss, R., Lockyer, C., Heide-Jørgensen, M.P., Müller, G., Baumgärtner, W., 2001. Evidence of infectious diseases in harbour porpoises (*Phocoena phocoena*) hunted in the water of Greenland and by-caught in the German North Sea and Baltic Sea. The Veterinary Record 148, 715–720.

Zelikoff, J.T., 1998. Biomarkers of immunotoxicity in fish and other non-mammalian sentinel species: predictive value for mammals? Toxicology 129 (1), 63–71.

Zenteno-Savin, T., Castellini, M., Rea, L., Fadely, B., 1997. Plasma haptoglobin levels in threatened Alaskan pinniped populations. Journal of Wildlife Diseases 33, 64–71.

Chapter 11

New Technologies for Monitoring Marine Mammal Health

Annalaura Mancia
University of Ferrara, Ferrara, Italy

INTRODUCTION

Traditionally, biology was a descriptive discipline, and that is how the biology of marine mammals has also been for a long time. In his work *Historia Animalium*, Aristotle was the first to make many pertinent observations about dolphins, including the fact that they would bear their young alive, suckle them, breathe air, and communicate by underwater sounds. From his descriptions and writings, we know that in the 4th century BC there were dolphins in the Mediterranean and porpoises in the Black Sea. Accurate natural history observation on the biology of marine mammals expanded throughout the time associated to morphologic descriptions in the pre-1900s and followed by description of behavior and distribution during the times of hunting and whaling activities. Studies of life history patterns, habitat use, and behavior in captivity or in nature started only in the second half of the last century, slowly combining aspects of mammalogy, ethology, ecology, conservation, evolutionary biology, and, finally, molecular biology.

Molecular biology is the field of biology that studies the composition, structure, and interactions of the most important macromolecules for each living organism, such as nucleic acids (DNA and RNA) and proteins that carry out the biologic processes essential for each cell's functions and maintenance. The DNA contains the genes that determine how the individual organism will be. The therapeutic control of a medical condition is generated from the basic concept that genes make proteins that catalyze a biochemical reaction and control the phenotype of the organism. Thus, the understanding of DNA and the identification of gene and protein defects responsible for a specific disease is very important in maintaining the health of an organism.

The method of dissecting biologic systems into their constituent parts, known as reductionist method, has been very useful in explaining the chemical basis of

Marine Mammal Ecotoxicology. https://doi.org/10.1016/B978-0-12-812144-3.00011-5

numerous living processes and has been largely responsible for the amazing progress seen in biology during the past five decades or so. However, biologic systems are extremely complex and have emergent properties that cannot be explained, or even predicted, by studying a gene at the time. New experimental techniques for investigating the unique complexity of biologic systems that results from the diversity of interactions and regulatory networks were, all of a sudden, a necessity.

Recent developments in high-throughput nanotechnologies and bioinformatics have enabled the examination of biologic systems in remarkable detail, providing the data that molecular biologists need to simulate the behavior of complex biologic networks and systems. We can now monitor thousands of molecules simultaneously and generate real-time pictures of any biologic system in any condition. The opportunity to assess the expression of hundreds to thousands of genes, proteins, or metabolites simultaneously has been made possible by the development of the "omics" technologies.

THE OMICS TECHNOLOGIES

Omics technologies aim at understanding a complex system, considered a whole. Therefore, in a specific biologic sample, the universal detection of genes is called *genomics*, of the totality of the mRNA is called *transcriptomics*, the study of the entire set of peptides and proteins is called *proteomics*, and that of all the intermediate products of metabolism is called *metabolomics*. Taking advantages of the new technologies, many other applications can be added to the list of these major omics technologies, generated by the necessity of deepening information on a specific mechanism or/and at a specific level.

The integration of all the omics technologies is characterized by the generation of an enormous amount of data that can be interpreted only by the involvement of many scientific disciplines (e.g., biology, computer science, engineering, bioinformatics, physics) (Chen et al., 2010). The result aims at deciphering systems in their complexity, at predicting how they can change over time and conditions, and at proposing solutions to health and environmental issues (Fig. 11.1).

Genomics

Genomics is an area that concerns the sequencing and analysis of an organism's genome. The cost of genome sequencing has gone down drastically from the 3 billion dollars used for the first human genome sequencing in the beginning of this century to about 3 thousand dollars in the year 2016 (and probably the half of that by the end of 2017), resulting in the establishment of the size and sequence of the genome of over a thousand species of organisms. Such knowledge of genome sequence has been useful in many ways. For example, we know that the minimum number of genes to sustain life by a bacterium is only 260. It also led to the surprising understanding that humans carry only 20,000 genes, against the previous guess that humans may have up to 100,000

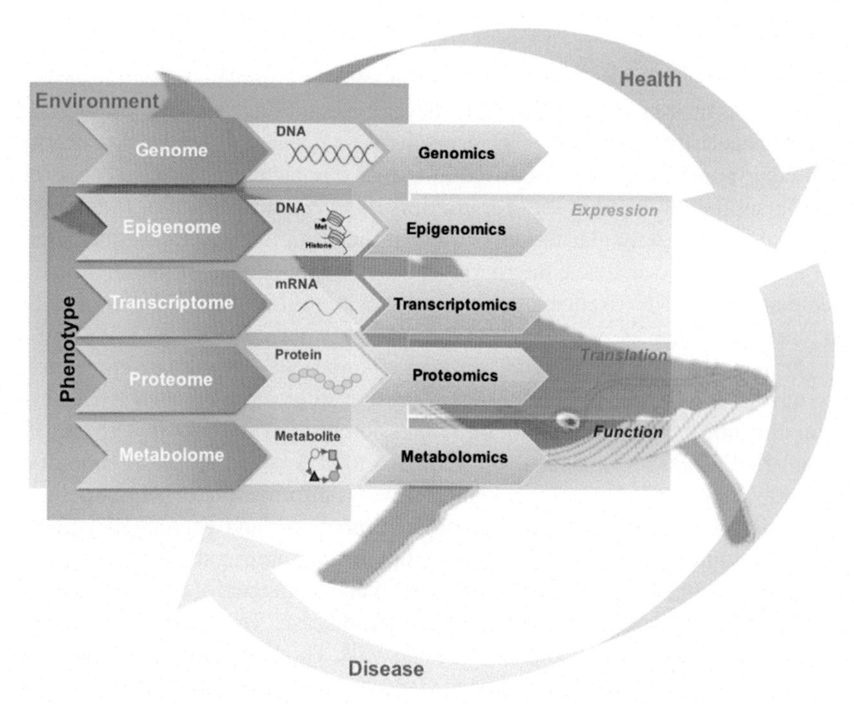

FIGURE 11.1 Outline of the omics sciences describing the complexity of systems. The omics profile of each individual is characterized by its biologic molecules that translate into the structure, function, and dynamics of the organism. Genome, epigenome, transcriptome, proteome, and metabolome are linked to the phenotype; they can change over time and conditions, in relation to health and environmental issues.

genes. The sequencing of the entire DNA of an organism is called a whole genome sequence (WGS). A WGS project involves the sequencing of DNA, the assembly of that sequence to create a representation of the original chromosome, and the annotation and analysis of that representation, with an emphasis on significance and function. It aims at the collective characterization of coding and noncoding DNA regions, their structure, function, and evolution. It reveals the complete DNA makeup of an organism, enabling the better understanding of the variations both within and between species, allowing an accurate differentiation between organisms. Comparisons of genome sequences from different individuals have led to the establishment of copy number variation with over 10 million single-nucleotide polymorphisms (SNP). SNPs can provide a genetic fingerprint for use in identity testing and are found to be involved in the etiology of many diseases. Several studies have demonstrated the diagnostic utility of WGS, for example, in mutation detection (Lupski et al., 2010; Herdewyn et al., 2012; Bae et al., 2014) or in the identification of clinically relevant variants in ~40% of pediatric populations with autism (Weedon et al., 2014) and ~60% of those with intellectual disability (Gilissen et al., 2014).

Transcriptomics

Transcriptomics is the study of the transcriptome: the complete set of transcripts in a cell and their quantity in a specific developmental stage or physiologic condition. Transcriptomics studies RNA in any of its forms: mRNA, rRNA, tRNA, and other noncoding RNA produced in one or a population of cells. The term can be applied to the total set of transcripts in a given organism, or to the specific subset in a particular tissue. Currently, there are two key techniques used for transcriptomic analysis: microarrays, which quantify a set of predetermined sequences, and RNA sequencing, which uses high-throughput sequencing to capture all sequences.

Differently from the genome, the transcriptome can vary with external environmental conditions. Generally transcriptomic studies are referred to expression profiling studies, examining the expression level of mRNAs in a given cell population. Transcriptome analysis allows us to understand the expression of a genome at the transcription level, which provides information on gene structure, regulation of gene expression, gene product functions, and genome dynamics. Understanding the transcriptome is essential for interpreting the functional elements of the genome and revealing the molecular constituents of cells and tissues, and also for understanding development and disease. Transcriptome analysis can be carried out at the resolution of single cells, a powerful strategy to connect gene expression networks, cell lineage, and phenotype of individual cells and to study complex disease such as cancer, as well as other biologic phenomena such as tissue regeneration, embryonic development, and immune response (Liu and Trapnell, 2016; Kanter and Kalisky, 2015). Comparison of transcriptomes allows the analysis of interspecies differences (Shay et al., 2013), the identification of genes that are differentially expressed in distinct cell populations, across tissues and individuals (Mele et al., 2015), in response to an injury (Khan et al., 2017) or to different treatments (Datta et al., 2016). Transcriptome sequencing can evaluate absolute transcript levels of sequenced and unsequenced organisms, detect novel transcripts and isoforms, and reveal sequence variations and splice variants. Affordable and fast, it is the most informative assay to start with, offering an overview of the expressed genes to guide subsequent analyses (carried through proteomics, metabolomics, and other methods).

The combination of genomics and transcriptomics is called functional genomics. Functional genomic studies aim at deciphering the connection between phenotypes and genotypes and have brought a revolution in the fields of medicine and modern biology.

Proteomics

Proteomics is the study of the proteome, the totality of the proteins in a cell, tissue, or organism and their identity, their biochemical properties, and functional roles. Differently from genomics and transcriptomics, propelled by advancement in sequencing technologies, proteomics has been driven by advances in mass spectrometry (MS) and other techniques that allow the analysis of a large number of protein samples at low cost (e.g., 2D gel electrophoresis). Proteomics is the study of how protein quantities, modifications, and structures change during development and in response to internal or external stimuli. The proteome of an organism is much larger and complex than its genome. Contrarily from the genome, the proteome changes constantly: DNA in organisms is essentially constant throughout their lives, while the kinds and amounts of proteins that are synthesized at any instant are subject to much variation. Proteins are continuously made, modified, and eliminated; many genes encode for more than one version of a protein, and a protein can be modified differently in response to cellular stimuli. The field of proteomics investigates which proteins are expressed at what stages in an organism's life and exactly how and why these proteins are expressed. Protein controls the structure and function of a cell by facilitating all biochemical reactions. Proteomics contributes to improve biomarker translation to modern medicine (Rifai et al., 2006; Veenstra, 2007). Biomarkers are measurable characteristics that reflect physiologic, pharmacologic, or disease processes and can be used to screen an individual for diagnostic and therapeutic purposes. Proteomic-based approaches for biomarker investigation can be employed in different aspects of medicine, to better understand pathways affected in a disease, to identify individuals at a high risk of developing the disease, and to identify individuals who are most likely to respond to specific treatment (Guest et al., 2013). The application goes beyond medicine: biomarkers may, in fact, be the best approach to identify an early response to contaminants (Broeg et al., 2005) and are very sensitive for identifying an organism's stress, although it is not always clear the origin of the stress, and there are often multiple stressors present (Smit et al., 2009). Biomarker responses can be measured in organisms collected from or deployed in field sites to integrate the effects of chemical and nonchemical stressors, reducing the need for complex laboratory exposure scenarios.

The integration of proteomics with genomics and transcriptomics is called proteogenomics. Posttranscription regulation, protein half-life, and posttranslational modification are examples of what defines the proteome dynamics and cannot be deduced from data of functional genomics; consequently, proteomics is a crucial, corresponding methodology to both transcriptomics and genomics.

Metabolomics

Metabolomics refers to the systematic identification and quantification of the metabolome of a biologic system (cell, tissue, organ, biologic fluid, or organism) at a specific point in time. Metabolome refers to low molecular weight organic molecules, intermediates, and products of metabolism, such as hormones, other signaling molecules, and secondary metabolites: like the transcriptome and the proteome, the metabolome is dynamic. Metabolomics is a powerful approach because metabolites and their concentrations directly reflect the underlying biochemical activity and state of the cell, thus representing best the molecular phenotype. Mass spectrometry (MS) and nuclear magnetic resonance (NMR) spectroscopy are the techniques most often used for metabolome profiling.

One of the applications of metabolic profiling studies is in the toxicology field. Metabolomics can detect the physiologic changes caused by toxic insult of a chemical (or a mixture of chemicals) especially of urine or blood plasma samples. The observed changes can be often related to specific lesion in liver or kidney.

A wider application of metabolomics is in the characterization of the interactions of organisms with their environment (environmental metabolomics) (Samuelsson and Larsson, 2008; Viant, 2008). A study can involve individuals to populations, and it can be related to the traditional fields of ecophysiology and ecology, and from an instantaneous effect to those over evolutionary time scales, enabling studies of genetic adaptation (Bundy et al., 2009). Metabolomics has a number of advantages over the other omics approaches. The metabolome is the final product of gene transcription, and therefore, changes in the metabolome are amplified relative to changes in the transcriptome and the proteome (Urbanczyk-Wochniak et al., 2003). The metabolome contains the smallest domain (made up of approximately 5000 metabolites vs. 100,000 proteins in the proteome and 20,000 expressed genes in the genome), but it also contains the most diverse biologic molecules, which can make it more physically and chemically complex than the other "omes."

Epigenomics

Different from the omics technologies listed so far describing the "ome" representing the totality of the principal macromolecules presents in all living organisms, epigenomic studies the modifications of the DNA that do not change its sequence, while changing the way the DNA sequence gives its instructions. The epigenome is made up of chemical compounds and proteins that can attach to DNA and can turn genes on or off, controlling the production of proteins in precise cells. These modifications are sometimes passed on from cell to cell as cells divide (mitosis) and can be passed down from one generation to the next (meiosis); hence, they are heritable. The epigenome can also change throughout a person's lifetime. Adjustments to next-generation sequencing protocols to enrich for the DNA regions carrying modifications (e.g., bisulfite treatment or antibody-based methods) are applied to the study of epigenetic processes. One type of modification is called DNA methylation, where methyl groups attached to the bases of the DNA molecule turn genes on or off. A second type is called histone modification, where chemical tags are attached to the histones (DNA-wrapping proteins that enable the DNA to be wound up into chromosomes), giving information about regions in the DNA that should be used or ignored.

Usually, the modifications occur as a natural process of development and tissue differentiation, but they can be altered in response to environmental exposures or disease. Epidemiological evidence increasingly suggests that early life exposures to dietary and environmental exposures can have a profound effect on the epigenome, resulting in birth defects and diseases developed later in life (Dolinoy et al., 2007; Jirtle and Skinner, 2007). The normal role in development and differentiation of epigenetic regulation can be misdirected, leading to a number of diseases including cancer (Jones and Baylin, 2007; Robertson, 2005). The epigenetic alterations are more readily reversible than genetic events, offering potential for the development of therapies (Reamon-Buettner and Borlak, 2007).

Toxicogenomics

Toxicogenomics is the application of genomic technologies (genome sequence analysis, gene expression profiling, proteomics, metabolomics) to study the adverse effects of environmental stressors and toxicants on the organisms and the environment health. The application of transcriptomics, proteomics, and metabolomics enables the study of adverse effects of xenobiotic substances in relation to structure and activity of the genome to understand the role of gene–environment interactions in disease (Heijne et al., 2005). This area promises to have a large impact on many other scientific and medical disciplines, as scientists can now generate complete descriptions of how components of biologic systems work together in response to various stresses, drugs, or toxicants (Gomase and Tagore, 2008). Due to the rapid advent in genomics technologies and attention to ecologic risk assessment, the term "ecotoxicogenomics" has recently emerged to describe integration of omics technologies (i.e., transcriptomics, proteomics, metabolomics, and epigenomics) into ecotoxicologic fields. Ecotoxicogenomics is defined as study of an entire set of genes or proteins expression in nontarget organisms that is important in responses to environmental toxicant exposures, offering benefit in ecologic risk assessment (Kim et al., 2015). While the availability of genomic information about nonmodel organisms is expanding, the application of ecotoxicogenomics to a variety of organisms becomes a powerful tool for evaluating the effects of chemicals on the entire ecosystems (Iguchi et al., 2006; Watanabe and Iguchi, 2003).

OMICS OF THE MARINE MAMMAL WORLD

Many marine mammal species and populations, considered to be most vulnerable to human activities, are endangered or threatened, or designated as depleted worldwide (Endangered Species Act; International Union for Conservation of Nature Species Programme). Marine mammals cannot escape the legacy of the global decline of the ocean health due to anthropogenic impacts such as overfishing, coastal habitat destruction, deep sea mining, oil and gas exploration, release of chemical contaminants, and pollutants from industrial applications (Desforges et al., 2016; Lane et al., 2015; Schwacke et al., 2012; Van Bressem et al., 2009). Noise can also critically impact marine mammal behavior and fitness (Blair et al., 2016; Ellison et al., 2012; Peng et al., 2015). Moreover, marine mammals are also constantly exposed to natural factors; biotoxins from harmful algae can produce mass mortalities that have devastating effects on population dynamics both in the short and long term (Van Dolah, 2000). Among the consequences, there is an increase in reports of diseases, such as metabolic disorder, opportunistic infections, and population changes in growth, reproduction, and survival (Di Guardo et al., 2011; Gulland and Hall, 2007). Current understanding of the long-term effects of these factors, alone and in combination, has been limited by the lack of comprehensive methodologies and by the protected status of the animals.

The potential of the application of omics technologies to marine environmental science and especially in environmental risk assessment is recognized by many researchers and official organizations (Bozinovic and Oleksiak, 2011; Kim et al., 2015; Kumar and Denslow, 2017; Veldhoen et al., 2012). The application of the omics technologies holds promises toward a significant progress in the understanding of marine mammal health and physiology challenged by a marine environment subject to continuous changes.

Marine Mammal Genome Projects

The pregenomic era was characterized by the Human Genome Project (HGP), an international research effort to determine the sequence of the human genome and to identify the genes that it contains. The first draft was released in 2001 (Lander et al., 2001), and the project was finally completed in 2003, 2.5 years ahead of time, and also significantly under budget, thanks to the advancement in sequencing technology and reduced cost. By the time the HGP began in the late 1990s, the highest estimates put the number of human protein-coding genes at 100,000, and since then the number has continued to shrink. However, at present, the final number of true protein-coding genes in the reference genome lies between 19,000 and 20,000, accounting for 1.5% of the 3 billion base pairs representing the entire genome. Of course, this information led to many unanswered questions that characterize what is presently called "the postgenomic era." The focus is on the discovery and explanation of all the functional elements encoded within the genome sequence, and the comparison of related genomes has emerged as a powerful instrument for genome interpretation. In an effort seeking the identification of functional elements that are conserved across mammals, a project funded by the National Human Genome Research Institute, the bottlenose dolphin has been chosen as one of 24 animals whose genome has been sequenced as part of the comparative genomic annotation (Lindblad-Toh et al., 2011). The first marine mammal genome to be sequenced was sequenced at 2× coverage, with its first version publicly available in 2010. In a 2× coverage, each genomic base is represented in roughly two sequence reads ("2×" redundancy), leaving many gaps in the final sequence due to statistical fluctuations in read placement, biases in preparative libraries, and difficulties leading to a low-quality assembly. The low coverage has significant effects on the subsequent analyses: the absence of a protein-coding gene, or a disruption of its open reading frame, may represent a deficiency of the assembly or may represent a real evolutionary gene loss. Moreover, low depth can introduce sequence errors that can be propagated, leading to wrong conclusions of a study (Green, 2007). However, a low-redundancy genome, such as the 2× dolphin genome sequenced, is useful to obtain biologic information: partial sequences of most genes and other evolutionarily conserved segments average estimates of mutation rates, as well as a comprehensive assessment of interspersed repeat content including the identification of lineage-specific families. A comparison of about 10,000 protein-coding sequences from the bottlenose dolphin genome with nine other amniotes genomes documented rates of synonymous substitution in the dolphin lineage that were significantly lower than other mammals and equivalent to that of humans and elephants. The dolphin lineage exhibited evidence of positive selection of multiple genes associated with the nervous system, metabolic processes, and glycemic regulation, and others possibly linked to cetacean specializations such as deep diving, blubber, and fat storage. In addition, the dolphin lineage showed a significant increase in selection on genes expressed in the mitochondrion in comparison with other mammalian genomes (McGowen et al., 2012). A deeper analysis using 11,838 high-quality orthologous gene alignments selected from the dolphin and four other terrestrial mammalian genomes identified genes that had undergone positive selection that are significantly enriched in the categories of lipid transport and localization, ATPase activity, sense perception of sound, and muscle contraction, all areas that are potentially related to cetacean adaptations (Nery et al., 2013; Sun et al., 2013). The dolphin genome has been used to obtain information on the independent evolution of echolocation in bats and cetaceans (Parker et al., 2013). More information has been obtained, increasing the coverage of the bottlenose dolphin genome; the analysis revealed that parallel substitutions are widespread in marine mammals, but also that while convergent phenotypic evolution can result from convergent molecular evolution, in the evolution process are more often used different molecular pathways to reach the same phenotypic outcome (Foote et al., 2015; Zhou et al., 2015). Among the whales, a comparative genomic analysis identified the expansion in the whale lineage of gene families associated with stress-responsive proteins and anaerobic metabolism, whereas gene families related to body hair and sensory receptors were contracted (Yim et al., 2014). Further insights into the genomic basis of aquatic adaptations in marine mammals that can be linked to their physiology and health may rely on functional or genomic analyses of noncoding regions, which will be soon achieved thanks to the latest release of novel genomes, first among all the bottlenose dolphin genome with 114.5× coverage. From the current annotation report, the bottlenose dolphin genome seems to contain about 17,000 protein-coding genes in 2.1 billion base pairs. The current coverage is about 95% versus the 86% of the previously implemented version. The advancement in sequencing technologies, together with the reduced cost, has made possible the sequencing of genomes of several other marine mammals in the last few years. To date, the genomes of several marine mammal species have been annotated and released by the National Center for Biotechnology Information (NCBI) Eukaryotic Genome Annotation Pipeline: five belonging to the order of Cetartiodactyla (*Balaenoptera acutorostrata scammoni*, minke whale; *Lipotes vexillifer*, Yangtze River dolphin; *Orcinus orca*, killer whale; *Physeter catodon*, sperm whale; *Tursiops truncatus*, bottlenose dolphin), three belonging to the order of Carnivora (*Leptonychotes weddellii*, Weddell seal; *Ursus maritimus*, polar bear; *Odobenus rosmarus divergens*, Pacific walrus), and one belonging to the order of Sirenia (*Trichechus manatus latirostris*, Florida manatee).

Omics Findings for Marine Mammal Health

Cetacea (Order: Cetartiodactyla)

Transcriptomics in marine mammal science is unquestionably the omics approach more represented in the literature. The reasons are several, the first being the cost of the experimental procedure: it is extremely less expensive to sequence only the small coding fraction of the genome instead of the whole with best coverage and resolution. Moreover, transcriptomic analysis does not necessarily need a reference genome. Therefore, the transcriptomic approach has been in use for many years now, refined by the advancement in sequencing technologies. Transcriptomic analyses aid discovery of novel gene functions and the connection of molecular and physiologic responses to a large scale of stimuli (ecologic, anthropogenic). The principal methods used are the gene expression microarrays and the more recent RNA-seq. With gene expression microarrays, thousands of genes are analyzed simultaneously in any given sample, thus obtaining a lot of information about the physiologic systems and the impact of environmental challenges. In 2007, the first microarray for marine mammal studies was developed. It was a species-specific cDNA microarray containing 1395 unigenes selected from targeted cloning and T and B cells cDNA libraries features (Mancia et al., 2007). The first microarray was useful for stress response and immune function studies in wild dolphins (Mancia et al., 2008, 2010). The blood transcriptome reveled that (1) the dolphin immune system mechanisms have high similarity to those of humans and other terrestrial mammals; (2) the immune system of the dolphins, resident inhabitants of coastal locations of temperate waters worldwide, reflects the environmental condition in which they live; and (3) the sampling method impact on downstream analysis underlines the need for species-specific baseline data. The same microarray was applied to screen a different tissue in a study focused on the study of the vitamin D3 pathway on cell cultures from dolphin skin; the analysis showed the importance of nonclassic functions of vitamin D3, such as its role in innate immunity, similar to what has been demonstrated in other mammals (Ellis et al., 2009).

In 2014, a much more comprehensive system, a species-specific oligo microarray, containing 24,418 unigene sequences from cDNA libraries of seven different tissues from bottlenose dolphin was generated (Mancia et al., 2014). The microarray was used to screen wild animal blood transcriptomes and was effective in the differentiation of populations of dolphins inhabiting different geographic locations and by the effects of environmental contaminants on dolphin health (Mancia et al., 2014, 2015). Dolphins inhabiting the coastal waters of Georgia in the United Statees, known to be heavily contaminated by Aroclor 1268 (Kucklick et al., 2011), an uncommon polychlorinated (PCB) mixture, displayed variation in expression of genes involved in xenobiotic metabolism, development/differentiation, and oncogenic pathways (Mancia et al., 2015). The same microarray was also applied to evaluate the skin tissue as a source of information, giving the minimal disruption inferred from the sampling methods (e.g., dart biopsy and/or stranding events). The skin was useful to evaluate the activation of an immune response to the exposure of contaminant of emerging concerns, such as bisphenol A and perfluorooctanoic acid (Lunardi et al., 2016) and also to establish baseline health parameters for investigations on contaminant exposure or health status. Gene expression was greatly impacted by season, with one-third of all the genes on the array varying between winter and summer, highlighting the need for creating a baseline for natural variability for a better investigation of the effects of a stressor (Van Dolah et al., 2015).

Despite its ability to interrogate the expression of thousands of genes, the microarray can still lack information, due to the incorrect abundance of some of the transcripts. The most recent RNA sequencing (RNA-seq) method, in which the cDNA made from the RNA sample is directly sequenced through high-throughput DNA sequencing, provides a less biased evaluation of the transcriptome. RNA-seq analyses are very sensitive and offer the advantages of detecting all the unique sequences and of quantifying levels of RNAs expressed at a very low level. Most importantly, the reduced cost of RNA-seq provides an efficient approach to generate sequences for functional genomics analyses in a nonmodel organism with unsequenced genomes using a de novo assembly procedure (Gui et al., 2013). Gui et al. (2013) characterized the leucocyte transcriptome of the Indo-Pacific humpbacked dolphin, now an endangered species because of the dramatic decline in population size of the past decades. The dataset provides a substantial genomic-level resource for the endangered species, while the identification of genetic markers and genes involved in immune system response and adaptive evolution can be useful in understanding the molecular mechanisms of various pathways in cetaceans.

Proteomic analyses have been demonstrated to support and validate studies on marine mammal health. One of the first proteomic studies on dolphins was a screening of the dolphin skin proteome by two-dimensional polyacrylamide gel electrophoresis (2D-PAGE), to compare the proteomic profile of skin tissue samples with that of skin-derived cultures. Results demonstrate that samples share distinct similarities sustaining the validity of the use of epidermal cell lines for the study of the interaction occurring between dolphins and the environment (Yu et al., 2005). Another example was a study investigating a dolphin condition similar to human metabolic syndrome using serum lipid profiles between two groups of dolphins, where high or low insulin were analyzed. The study found differences in phospholipid fatty acids in the two groups, which may play a role in the susceptibility to or development of an insulin-resistant-like state: the proteomic analysis of dolphin serum showed correlation of changes in the fatty acids with an insulin-sensitizing phenotype (Sobolesky et al., 2016).

An approach to direct, noninvasive, health assessments of wild marine mammals used metabolomics to profile exhaled breath metabolites, providing a first library of volatile and nonvolatile compounds in cetacean exhaled breath. Dolphin breath contains a large variety of low-abundance metabolites, many of which are common with those found in human breath and considered indicative of human health status. Providing a link between dolphin breath and certain health conditions or exposures, the monitoring of exhaled breath metabolome content provides useful diagnostic information that can be used by veterinary personnel and conservation managers in their decision-making processes (Aksenov et al., 2014; Zamuruyev et al., 2016).

Pinnipedia (Order: Carnivora)

Diseases of wild marine mammals can be difficult to diagnose because of their protected status and their unknown life history. A study using tools relying on cross-hybridization between closely related species presented the blood transcriptome of the California sea lion (*Zalophus californianus*) as a diagnostic classifier for rapid diagnosis and treatment of infection, disease, intoxication, or other causes of compromised health status in stranded animals (Mancia et al., 2012). The results achieved with this work indicate that cross-species microarray technologies, using a selected gene set for microarray design and analysis could be of use to wildlife managers and veterinarians in handling species with little genomic data. The California sea lion, a protected marine mammal inhabiting the western coast of North America, is a good sentinel species for coastal habitats, and the most common cause of stranding is reported to be domoic acid toxicosis, caused by the ingestion of domoic acid, a potent neurotoxin produced by diatom species of the genus *Pseudo-nizschia*. Human domoic acid intoxication cases present with abnormal behaviors, even seizures and epilepsy linked to hippocampal lesions, sharing the neuropathology profile observed in the California sea lions exposed to domoic acid, in particular in the recently discussed link between hippocampal lesions and epilepsy (Buckmaster et al., 2014; Ramsdell and Gulland, 2014). The blood transcriptomic analysis showed enrichment of pathways consistent with that neuropathology profile, confirmed by the analysis of the proteome of cerebrospinal fluid from intoxicated animals (Neely et al., 2015). Proteomic analysis of blood indicated that several apolipoprotein E charge forms decreased in domoic acid–intoxicated sea lions and may be important in the progression of domoic acid toxicosis (Neely et al., 2015).

The transcriptomic and the proteomic approaches taken, combined with machine learning approaches, performed as a robust and accurate tool to diagnose domoic acid toxicosis, demonstrating promise for future studies combining additional variables in multidimensional space to create robust classifiers (Mancia et al., 2012; Neely et al., 2015).

Omics studies on California sea lion (family: Otaridae) were the first attempt to combine a global-scale analysis to the health status of marine mammals belonging to the order of Carnivora, mostly related to the possibility of screening a high number of unhealthy animals. In fact, data analysis was guided and/or supported by evaluation obtained by clinical observations and health parameters measurements carried out on the high number of unhealthy and ill cases that seasonally were hospitalized and rehabilitated at the Marine Mammal Center, in Sausalito, CA. Lately, the attention seems to be focused on different species, pinnipeds belonging to the family of the Phocidae, and to the understanding of the mechanisms activated by environmental chemical or biologic stress. Transcriptomics was used to evaluate brain tissues from stranded harbor seals (*Phoca vitulina*) to understand the cause of death. The study identified pathways involved in innate and adaptive immunity in samples with a phocine herpesvirus (PhV-1) brain infection and found a strong upregulation of fatty acid metabolic genes in animals that did not die from viral infection. Although the cause of the dysregulation of fatty acid metabolism in the brains of these harbor seals is unknown, it may be correlated to exposure to toxins or nutrient depletion (Rosales and Vega Thurber, 2016). A de novo assembly of the blubber transcriptome in ringed seal (*Pusa hispida*) was used to identify molecular indicators of PCBs exposure. Transcript levels for gene targets were found to be correlated with increasing levels of blubber PCBs, linking an energy metabolism imbalance to the toxicity of the chemicals (Brown et al., 2017). Transcriptome analysis of muscle tissue of another member of the family Phocidae, the northern elephant seal (*Mirounga angustirostris*), provides a resource for a deeper investigation of the alteration of molecular pathways triggered by environmental stress, including modifications of metabolic and immune signaling as well as muscle tissue maintenance (Khudyakov et al., 2015a). An example comes from a study of the cellular responses to hypothalamic-pituitary-adrenal axis stimulation by measuring corticosteroid hormones, metabolites, and muscle gene expression before, during, and following administration of exogenous adrenocorticotropic hormone (ACTH); results suggests a compensatory, tissue-sparing mechanism used to maintain cortisol and aldosterone sensitivity while avoiding deleterious long-term consequences of stress (Khudyakov et al., 2015b). Reference sequences specific to elephant seals were also obtained from the blubber tissue and used to profile the transcriptomic response to hypothalamic-pituitary-adrenal stress axis activation and to identify tissue-specific molecular markers of stress in the pinniped fasting-adapted species (Khudyakov et al., 2017).

Sirenia (Superorder: Afroheria)

The completion of the genome draft of the Florida manatee (*Trichechus manatus latirostris*) in 2012 made immunogenetic exploration of the Sirenia order more feasible. The Florida manatee is one of the most endangered marine mammals in American coastal waters, continuously facing demographic challenges due to anthropogenic activities and stochastic factors. Brevetoxicosis and papillomavirus are disease-associated agents that have been described in their populations (Bossart et al., 2002; Walsh et al., 2015), in addition to several parasites including trematodes, nematodes, and coccidians (Bando et al., 2014). Discovering the genetic details of their immune system is an important step in the development of manatee-specific tools for monitoring health. The genomic scaffolds of the Florida manatee were used to characterize the organization and expression of the immunoglobulin heavy chain locus. The potential segmental diversity and constant region isotypic diversity described may be related to the mechanisms of defense against infectious disease in their environment.

New Technologies: Challenge and Progress

The number of advantages using the new omics technologies constantly increase, but there are also limitations and challenges that need to be considered. With the completion of the human genome, the relationship between one or more genes to a specific disease and the importance of the interaction with complex environmental factors became clear, making the understanding of the disease process more challenging. Moreover, knowledge of physiologic processes such as SNPs, epigenetic modifications, and posttranslational modifications increased the complexity of host–disease interactions. Combining the study of multiple genes, proteins, and metabolites, the multiomics approach proved to be the correct one to understand the global picture of how each disease affects its host. Each omics study provides information between disease and nondisease, describing alterations of genes (transcriptomics), proteins (proteomics), or metabolites (metabolomics). Thanks to the advent of the bioinformatics with the development of computational and statistical techniques, these large datasets can be combined to offer an integrated view of how an organism may react to a disease, leading to potential biomarkers that can be used in the future for therapeutic applications.

However, technical difficulties, expensive instruments, cost versus benefit uncertainty, and clinical stringency make the use of omics still limited. In marine mammal studies, these limitations are associated with limited sample availability and genetic heterogeneity. Beside these restrictions, the massive amount of data generated from each omics study with often small sample size complicates the data interpretation and restricts the clinical value of the results. While using experimental replication and validation methods as well as reference samples and baseline data will improve confidence in the reliability of the results, there are parameters that need to be well evaluated. The small sample size used in the experimental setting caused by the protected status of marine mammals and by the cost of the techniques/sample can be a source of false discovery. Solutions to maximize the value of an omics study rely on repeated observations over long period of times, the examination of multiple tissues or body fluids correlated to the disease, the use of more than one technique to validate results, and the integration of the results with the biology or pathology of disease. However, different types of high-throughput technologies, each simultaneously collecting a large set of molecular data, used to collect information on the molecular components of biologic systems need to be integrated and analyzed. Taken together, the complexity of biologic systems, the technological limits, the large number of biologic variables, and the relatively low number of biologic samples make data integration of multilayer datasets one of the most relevant problems computational scientists are addressing nowadays.

LINKING MARINE MAMMALS TO HUMANS: NEW RESOURCES FOR HEALTH

Mammals have returned to the water in at least seven separate lineages: Cetacea, Sirenia, Desmostylia (extinct order), Pinnipedia, *Ursus maritimus* (polar bear), *Enhydra lutris* (sea otter), and *Thalassocnus* spp. (aquatic sloths, extinct genus). Some of these lineages have retained most of their terrestrial form while spending most of their time in the water, while others have changed their morphology dramatically to spend the entire time in the water. Clades of extant marine mammals seem to have originated at two discrete times. Cetacea and Sirenia originated during the early Eocene (50 million years ago, MYA), a time of high productivity in aquatic environments, with warm, broad, shallow seas and abundant resources to exploit, whereas Pinnipedia originated during the Oligocene (35 MYA), when productivity of the oceans, like today, was more concentrated around areas of upwelling. Both Cetacea and Sirenia were fully aquatic by the end of the Eocene, while Pinnipedia are semiaquatic animals (Gingerich et al., 1983; Lipps and Mitchell, 1976; Uhen, 2007). Each of these evolutionary histories is different from the others. Despite the finding that these adaptations evolved in mosaic patterns, they all have aquatic characteristics in common, and different morphologic solutions to aquatic conditions were achieved separately in each clade. The transition from terrestrial animals to fully aquatic animals took about 12 MY, and eventually, they diversified into the species we know today: the genus *Tursiops*, which bottlenose dolphins belong to, first appeared in the fossil record about 5 MYA.

While some mammals were evolving in the oceans, others, on land, evolved into the first primates, humans ancestors. The human lineage split from the modern chimpanzees and bonobos around 7 MYA, with the early hominins (human-like primates) that were our direct ancestors. Current estimates of the similarity between the DNA sequences of both the human and chimpanzee genome range between 95% and 99% (Varki and Altheide, 2005).

The Earth is old and so is life: while the Earth formed 4.5 billion years ago, the oldest known fossils are around 3.5 billion years old. In this time scale, the evolution of marine mammals and primates are relatively recent events. Many studies have compared primates and cetaceans in the perspective to elucidate the social evolution of highly intellectual mammals in terrestrial and aquatic environments. Despite a deep evolutionary divergence, adaptation to physically dissimilar environments, and very different neuroanatomic organization, some primates and cetaceans show striking convergence in social behavior, artificial language comprehension, and self-recognition ability (Marino, 2002; Yamagiwa and Karczmarski, 2014). Thanks to next-generation sequencing and the availability of the genomes, we can now correlate the knowledge on the observations made on the ecology, social relationships, behavior, in the molecular mechanisms operated by genes, proteins, and metabolites. Using the current genomic available data, we can look briefly at the identity, at transcripts and protein level, between humans and marine mammals. In Table 11.1, coding regions of the Cetacea, the bottlenose dolphin (*Tursiops truncatus*), the minke whale (*Balaenoptera acutorostrata scammoni*), the killer whale (*Orcinus orca*), the sperm whale, (*Physeter catodon*), and the Yangtze River dolphin (*Lipotes vexillifer*) are compared to those of humans and to those of the closest relatives in the Cetartiodactyla order. Coding regions of the Pinnipeds, the Weddell seal (*Leptonychotes weddellii*), and the walrus (*Odobenus rosmarus divergens*) are compared to those of humans and to those of the closest relatives in the Carnivora order. Coding regions of the Florida manatee (*Trichechus manatus latirostris*) are compared to those of humans and to those of the closest relatives in the Afrotheria superorder. As expected, the similarity is higher between closely related species within the same order and superorder. But the results from alignment of sequences of human and marine mammals is still very high, at least in the coding regions analyzed. Unquestionably, more of the genome needs to be understood and analyzed in both humans and marine mammals, but the similarity that we can already observe is an important instrument for basic research and translational science. Marine mammals are used as sentinels for ocean and human health (Bossart, 2011), a choice driven by the characteristics shared with the human species (e.g., mammals with long life spans, long-term coastal residents, feed at a high trophic level). Marine sentinels allow the characterization and management of potentially negative impacts linked to the environment degradation that can affect animal and human health associated with the oceans. Marine mammals have also developed unique adaptations, some of which are species-specific, to live their entire life, or most of it, in the aquatic environment (e.g., deep long dives, swimming, thermoregulation, echolocations). These fascinating differences with land mammals can sometimes be turned into powerful tools to understand and treat complex pathologic processes in human medicine.

Applications of Modern Genomics Techniques: Metagenomics for Zoonoses

Marine mammals are top predators that are essential for the health and function of the oceans, too often affected by various factors that can be detrimental

TABLE 11.1 Transcript and Protein Alignment Within Marine Mammal Species of the Order of Cetartiodactyla (Infraorder: Cetacea), Carnivora (Clade: Pinnipedia), and Sirenia (Superorder: Afrotheria) and Between Marine Mammals and *Homo sapiens*

Cetacea

Tursiops truncatus	Transcript		Protein	
	% identity	% coverage	% identity	% coverage
Homo sapiens	89.52	82.84	76.30	78.73
Cetartiodactyla	91.99	93.52	79.73	81.75

Balaenoptera acutorostrata scammoni	Transcript		Protein	
	% identity	% coverage	% identity	% coverage
Homo sapiens			77.51	80.51
Cetartiodactyla	91.47	95.27	80.06	84.37

Lipotes vexillifer	Transcript		Protein	
	% identity	% coverage	% identity	% coverage
Homo sapiens	89.09	82.86	77.34	81.03
Cetartiodactyla	91.32	96.21	79.93	85.50

Orcinus orca	Transcript		Protein	
	% identity	% coverage	% identity	% coverage
Homo sapiens	89.39	84.93	76.97	83.03
Cetartiodactyla	91.88	96.98	80.17	87.05

Physeter catodon	Transcript		Protein	
	% identity	% coverage	% identity	% coverage
Homo sapiens			77.61	78.67
Cetartiodactyla	91.47	91.94	80.28	83.02

Pinnepedia

Leptonychotes weddellii	Transcript		Protein	
	% identity	% coverage	% identity	% coverage
Homo sapiens	89.18	76.02	77.85	75.98
Carnivora	92.67	89.45	78.05	80.65

Odobenus rosmarus divergens	Transcript		Protein	
	% identity	% coverage	% identity	% coverage
Homo sapiens	89.43	83.41	77.00	83.59
Carnivora	94.12	97.64	81.06	89.19

Sirenia

Trichechus manatus latirostris	Transcript		Protein	
	% identity	% coverage	% identity	% coverage
Homo sapiens	89.18	82.89	75.96	82.24
Afrotheria	90.64	96.85	78.42	88.42
Cetartiodactyla			77.29	84.75

Sets of transcripts and proteins were retrieved from Entrez, aligned to the genome by Splign (transcripts) or ProSplign (proteins) and passed to Gnomon, NCBI's gene prediction software. The sequences used for the alignments are reads from RefSeq or from the GenBank database. The number in the report is the sum of multiple GenBank queries. The query of GenBank should retrieve the list of organisms with txid91561 for the Cetartiodactyla, txid311790 for the Afrotheria, and txid33554.

to their populations. Forty-four percent of stranded marine mammals die from unknown causes (Gulland and Hall, 2007). Disease is a major cause of marine mammal population decline, and the role of the microbiome in disease has generated considerable interest. The etiology of stranding events still remains poorly characterized, but high-throughput sequencing technology can identify and yield new insights into the virome and microbiome for disease identification and surveillance. The microbiome acts strongly and significantly in maintaining host health with a vital role in disease manifestation and immune system function. Members of the microbial community can directly influence the progression of a disease and modulate the host's immune system regulation and response, making the host's microbial partners essential to immune system function (Maynard et al., 2012).

A promising approach for pathogen identification in stranded marine mammals is the use of metagenomics, the characterization of the collective genome of microorganisms isolated from an organism using high-throughput sequencing technologies. Metagenomic studies do not require prior information about the disease agents and allow detailed comparisons of health and disease, identifying new insights into the virome and microbiome of wildlife. Metagenomic studies of marine mammals focused on the viral and microbial community of many tissues and body niches have displayed a big diversity of the microbiota according to the organ type and may be used as a baseline survey for comparison with samples from stranded animals during unexplained disease outbreaks (Godoy-Vitorino et al., 2017). A viral metagenomic study to investigate potential viral pathogens associated with a mortality event of captive California sea lions identified a novel species-specific anellovirus (ZcAV) (Ng et al., 2009). A similar study on lung samples of the Pacific harbor seals (*Phoca vitulina richardsii*) revealed another novel seal anellovirus (SealAV), which clusters phylogenetically with anelloviruses from California sea lions and domestic cats (Ng et al., 2011). The description of ZcAV and SealAV in the lungs of pinnipeds suggests that anellovirus infections may be common and play a role in marine mammal health and disease. A different metagenomic study used next-generation sequencing to get a comprehensive view of the fecal viral populations from wild and temporarily captive California sea lions, reporting previously uncharacterized viruses, including astroviruses, picornaviruses, bocaviruses, and sapoviruses (Li et al., 2011). A metagenomic survey of viromes from feces of the Subantarctic fur seal (*Arctocephalus tropicalis*) and south American fur seals (*Arctocephalus australis*) used next-generation sequencing to explore the viral diversity of southern hemisphere marine mammals (Kluge et al., 2016). A study of the bacterial communities in hundreds of samples from different body sites in healthy dolphins and sea lions showed a highly diverse bacterial taxonomic composition, which varies according to body site and host species (Bik et al., 2016). These findings provide species-specific databases that can be compared to later virome surveys and microbiota to detect alterations associated with changes in marine mammal health or population size.

Metagenomic studies in cetaceans have also indicated how the cetacean microbiome is affected by human-related bacteria (Godoy-Vitorino et al., 2017). This is particularly relevant since many human infections have a zoonotic, i.e., wild or domestic animal, origin. The rise in zoonotic diseases is driven by a complex interplay of environmental, ecologic, and epidemiologic factors. Therefore, the identification of pathogens in marine mammals may also moderate disease outbreaks and prevent zoonotic transmission (Delwart, 2012). Currently, there are about 15 known zoonotic marine mammal pathogens (Waltzek et al., 2012). The bacterial pathogen that causes tuberculosis, *Mycobacterium tuberculosis*, was introduced to the Americas via pinnipeds (Bos et al., 2014). The influenza A virus, a global human threat, is present in cetaceans and pinnipeds and has been shown to be transmitted from seals to humans (Geraci et al., 1982; Reperant et al., 2009; Webster et al., 1981). Most recent emerging diseases have been associated with host switches, including severe acute respiratory syndrome coronavirus, H5N1 avian influenza, Hendra virus, Nipah virus, and acquired immunodeficiency syndrome (Woolhouse and Gowtage-Sequeria, 2005). The risk of being injured or acquiring zoonotic diseases is highest in marine mammal researchers, rehabilitators, trainers, veterinarians, and volunteers, but there also numerous recreational activities now permitting contact with these animals. Subsistence hunters such as whalers and sealers, as well as human rescuers during marine mammal stranding events, are at risk of disease acquisition through their direct physical contact with infected marine mammals or through the ingestion of marine mammal food products (Hunt et al., 2008; McLaughlin, 2004; Webster et al., 1981). Luckily, so far the majority of zoonotic marine mammal diseases have resulted in localized skin infections in man that resolved spontaneously or with appropriate medical therapy. However, other marine mammal zoonoses, if left untreated, could induce life-threatening systemic diseases that could pose public health risks.

As the number of zoonotic diseases rises, the identification of pathogens in marine mammals has become an indicator of environmental health. The list comprises bacterial, viral, and fungal infections producing seal finger, brucellosis, leptospirosis, mycobacteriosis, mycoplasmosis, influenza, lobomycosis, and blastomycosis (Waltzek et al., 2012). As the closest oceanic relatives of humans, marine mammals are sentinel species for both human and ocean health, and they are long-lived, top predators, inhabiting the same inshore ecosystems utilized by man (Bossart, 2011).

Applications of Marine Mammal Unique Adaptations: Marine Biomedicine

Medical research has necessitated the integration of the omics and computational biology data to diagnose, interpret, and prognosticate human disorders, even with the current more comprehensive knowledge of the human physiology and molecular mechanisms of diseases. Omics technologies, including genomics, transcriptomics, proteomics, metabolomics, and epigenomics have transformed human medical research in the last decade, but there are still unresolved underlying mechanisms in human disease. The exploitation of marine mammals' unique adaptations through omics technologies can give valuable insights into uncertain human biomedical conditions.

An example is the astonishing wound healing mechanisms observed in dolphins. In humans, the healing of tissue wounds is often associated with infection and results in scars. Zasloff, a professor of immunology at the Georgetown University, observed the clinical course of the recovery of two dolphins showing shark's bites about 30 cm in length and 3 cm in depth (including the blubber layer and the underlying muscle). He reported that, during the first day postinjury, blubber from surrounding tissues had already migrated over the open wound surface. On the second day, newly generated tissues were described, which would gradually fill the wound from its base, restoring the original volume (Zasloff, 2011). The wound healed completely in 4 weeks and without any infections, which is remarkable considering the analogies between their immune system and that of terrestrial mammals (Beineke et al., 2010; Mancia et al., 2007). Zasloff suggested a role of the components of the dolphin blubber during the healing process. Stem cells present in the blubber could have a role in this remarkable wound healing process. The composition of dolphin's blubber is in fact different from other marine mammals: the isovaleric acid is higher in concentration, being 2%–5% of total fatty acid in bottlenose dolphin, while there is no detection in several species of whales (Koopman et al., 2003). The isovaleric acid accumulates in the blubber and does not get burned for fuel during times of starvation; moreover, it seems to have antimicrobial activity (Hayashida-Soiza et al., 2008). A multilayered omics approach describing transcripts, proteins, and metabolites present in the blubber after a serious injury may unravel the mechanisms involved in the healing process and in the protection from infection. Advances in the comprehension of the mechanisms controlling the healing process could lead to the improvement in the control of wound healing in humans and terrestrial mammals, leading to new therapies: regenerative medicine is an emerging field, with numerous open trials in the veterinary and human fields.

Another example is given by the recent discoveries of the unique dolphin metabolism that can aid research for human metabolic syndrome and diabetes. People affected by metabolic syndrome can develop not only type II diabetes but also cardiovascular disease and the possibilities to have strokes. Like humans, dolphin can develop metabolic syndrome, characterized by elevated insulin, glucose, triglycerides, and ferritin (Venn-Watson et al., 2011, 2015). Fatty liver disease has been found both in wild dolphins and dolphins under human care, suggesting that dolphins are susceptible to metabolic disorders. A study compared a wild population of dolphins to one kept under human care, which have higher annual survival and lower mortality rates. A deep study into these animals' diets as a possible risk factor for longer life and metabolic disease highlighted the potential benefits of C17:0, the margaric or heptadecanoic acid, a saturated fatty acid also present in bovine milk fat. High levels of C17:0 on erythrocyte membranes or plasma phospholipids have been identified as protective factors against development of metabolic syndrome, type 2 diabetes, and associated inflammation (Venn-Watson et al., 2015). In a parallel study supported by proteomic analysis, differences in serum lipid profiles between two groups of dolphins, with high or low insulin, found differences in phospholipid fatty acids. Shifting the dolphin diet to fish rich in odd chain saturated fatty acids, such as C17:0, resulted in increased serum levels of the insulin-sensitizing hormone adiponectin and serum sphingosines consistent with an insulin-sensitizing phenotype (Sobolesky et al., 2016). A better understanding of the networks activated by the mechanisms regulating adiponectin in dolphins could aid research of metabolic syndrome and diabetes affecting human populations.

Each year, millions of individuals die or become ill because of conditions or diseases that reduce the oxygen supply to hypoxia-sensitive tissues such as the brain. Hypoxia is also involved in and is the cause of many neuronal disorders in humans, for example, Alzheimer disease (Peers et al., 2007), Parkinson disease (Speer et al., 2013), and cerebral ischemia (stroke). Acute metabolic insults like stroke have an especially devastating impact, which is mostly impossible to repair. By contrast, brains of diving mammals tolerate extended periods of systemic hypoxia without damage. There are several behavioral, anatomic, and physiologic adaptations that are associated with the remarkable dive capacity of many marine mammals (Davis, 2014) that may be partly explained on the genetic level by specific substitutions within the coding sequences (Foote et al., 2015) or by selective gene duplications and losses (Yim et al., 2014). Different studies used transcriptional analysis of the brain of the hooded seal (*Cystophora cristata*) and showed differential regulation of specific genes that may have a central role in the protection of the diving brain. In large parts, the seal brain responds to the hypoxic challenge in a similar way as the brain of other mammals, which includes the upregulation of typical stress proteins like cytokines and immediate early genes. However, the genes involved in the energy metabolism seem to be a specific response of the seal brain to hypoxia. These genes may also be suitable drug targets for human neuronal disorders associated with hypoxia (Fabrizius et al., 2016; Hoff et al., 2017).

CONCLUDING REMARKS

A new era of discoveries in the marine mammal field began in 2011, with the completion of the first marine mammal genome, 8 years after the first whole-genome assembly of the human genome. Since then, the bottlenose dolphin genome has been greatly implemented thanks to supplement of new high-throughput sequences data and to the newer high-coverage version of the genome. To date, genomes from nine species of marine mammals have been sequenced, annotated, and are publicly available at the NCBI database. Genome-guided transcriptome assembly as well as de novo transcriptome assembly are providing novel insights into the mechanisms underlying the variety and diversity of marine mammals, revealing novel genes, functions, and connections of molecular and physiologic mechanisms related to their adaptations, health, and disease.

But the study of the genome and of the information that is encoded within the full DNA sequence of an organism will not unlock the code of life. The genotype is not the only thing responsible for the final physical makeup of the organism. The study of the epigenome with the genome-wide mapping of DNA methylation, histone modifications, nucleosome positioning, and three-dimensional architecture and the integration of genome and epigenome, of the RNA information from coding (mRNA) and noncoding RNA (epigenetic-related RNAs, such as micro RNA, short interfering RNA, piwi-interacting RNA, and long noncoding RNA) are necessary to understand how environment and genetic inclinations can intertwine in the complexity of cell biology. The network and pathways observed can in predictive models identify potential risks and propose solutions, making the new technologies relevant as prognostic and diagnostic tools.

The exploitation and combination of the new advanced omics techniques and computational methods will finally allow the discovery of a constantly changing environment thanks to the understanding of the interaction between marine mammals, humans, and the oceans.

REFERENCES

Aksenov, A.A., Yeates, L., Pasamontes, A., Siebe, C., Zrodnikov, Y., Simmons, J., Mccartney, M.M., Deplanque, J.P., Wells, R.S., Davis, C.E., 2014. Metabolite content profiling of bottlenose dolphin exhaled breath. Analytical Chemistry 86, 10616–10624.

Bae, B.I., Tietjen, I., Atabay, K.D., Evrony, G.D., Johnson, M.B., Asare, E., Wang, P.P., Murayama, A.Y., Im, K., Lisgo, S.N., Overman, L., Sestan, N., Chang, B.S., Barkovich, A.J., Grant, P.E., Topcu, M., Politsky, J., Okano, H., Piao, X., Walsh, C.A., 2014. Evolutionarily dynamic alternative splicing of GPR56 regulates regional cerebral cortical patterning. Science 343, 764–768.

Bando, M., Larkin, I.V., Wright, S.D., Greiner, E.C., 2014. Diagnostic stages of the parasites of the Florida manatee, *Trichechus manatus* latirostris. The Journal of Parasitology 100, 133–138.

Beineke, A., Siebert, U., Wohlsein, P., Baumgartner, W., 2010. Immunology of whales and dolphins. Veterinary Immunology and Immunopathology 133, 81–94.

Bik, E.M., Costello, E.K., Switzer, A.D., Callahan, B.J., Holmes, S.P., Wells, R.S., Carlin, K.P., Jensen, E.D., Venn-Watson, S., Relman, D.A., 2016. Marine mammals harbor unique microbiotas shaped by and yet distinct from the sea. Nature Communications 7, 10516.

Blair, H.B., Merchant, N.D., Friedlaender, A.S., Wiley, D.N., Parks, S.E., 2016. Evidence for ship noise impacts on humpback whale foraging behaviour. Biology Letters 12.

Bos, K.I., Harkins, K.M., Herbig, A., Coscolla, M., Weber, N., Comas, I., Forrest, S.A., Bryant, J.M., Harris, S.R., Schuenemann, V.J., Campbell, T.J., Majander, K., Wilbur, A.K., Guichon, R.A., Wolfe Steadman, D.L., Cook, D.C., Niemann, S., Behr, M.A., Zumarraga, M., Bastida, R., Huson, D., Nieselt, K., Young, D., Parkhill, J., Buikstra, J.E., Gagneux, S., Stone, A.C., Krause, J., 2014. Pre-Columbian mycobacterial genomes reveal seals as a source of New World human tuberculosis. Nature 514, 494–497.

Bossart, G.D., 2011. Marine mammals as sentinel species for oceans and human health. Veterinary Pathology Online 48, 676–690.

Bossart, G.D., Ewing, R.Y., Lowe, M., Sweat, M., Decker, S.J., Walsh, C.J., Ghim, S.J., Jenson, A.B., 2002. Viral papillomatosis in Florida manatees (*Trichechus manatus* latirostris). Experimental and Molecular Pathology 72, 37–48.

Bozinovic, G., Oleksiak, M.F., 2011. Genomic approaches with natural fish populations from polluted environments. Environmental Toxicology and Chemistry 30, 283–289.

Broeg, K., Westernhagen, H.V., Zander, S., Korting, W., Koehler, A., 2005. The "bioeffect assessment index" (BAI). A concept for the quantification of effects of marine pollution by an integrated biomarker approach. Marine Pollution Bulletin 50, 495–503.

Brown, T.M., Hammond, S.A., Behsaz, B., Veldhoen, N., Birol, I., Helbing, C.C., 2017. De novo assembly of the ringed seal (*Pusa hispida*) blubber transcriptome: a tool that enables identification of molecular health indicators associated with PCB exposure. Aquatic Toxicology 185, 48–57.

Buckmaster, P.S., Wen, X., Toyoda, I., Gulland, F.M., Van Bonn, W., 2014. Hippocampal neuropathology of domoic acid-induced epilepsy in California sea lions (*Zalophus californianus*). The Journal of Comparative Neurology 522, 1691–1706.

Bundy, J.G., Davey, M.P., Viant, M.R., 2009. Environmental metabolomics: a critical review and future perspectives. Metabolomics 5, 3.

Chen, C., Mcgarvey, P.B., Huang, H., Wu, C.H., 2010. Protein bioinformatics infrastructure for the integration and analysis of multiple high-throughput "omics" data. Advances in Bioinformatics:423589.

Datta, A., Dey, S., Das, P., Alam, S.K., Roychoudhury, S., 2016. Transcriptome profiling identifies genes and pathways deregulated upon floxuridine treatment in colorectal cancer cells harboring GOF mutant p53. Genomics Data 8, 47–51.

Davis, R.W., 2014. A review of the multi-level adaptations for maximizing aerobic dive duration in marine mammals: from biochemistry to behavior. Journal of Comparative Physiology B 184, 23–53.

Delwart, E., 2012. Animal virus discovery: improving animal health, understanding zoonoses, and opportunities for vaccine development. Current Opinion in Virology 2, 344–352.

Desforges, J.P., Sonne, C., Levin, M., Siebert, U., De Guise, S., Dietz, R., 2016. Immunotoxic effects of environmental pollutants in marine mammals. Environment International 86, 126–139.

Di Guardo, G., Mazzariol, S., Fernandez, A., 2011. Biologically threatened dolphins and whales. Environmental Microbiology 13, 2833–2834.

Dolinoy, D.C., Weidman, J.R., Jirtle, R.L., 2007. Epigenetic gene regulation: linking early developmental environment to adult disease. Reproductive Toxicology 23, 297–307.

Ellis, B.C., Gattoni-Celli, S., Mancia, A., Kindy, M.S., 2009. The vitamin D3 transcriptomic response in skin cells derived from the Atlantic bottlenose dolphin. Developmental and Comparative Immunology 33, 901–912.

Ellison, W.T., Southall, B.L., Clark, C.W., Frankel, A.S., 2012. A new context-based approach to assess marine mammal behavioral responses to anthropogenic sounds. Conservation Biology 26, 21–28.

Fabrizius, A., Hoff, M.L., Engler, G., Folkow, L.P., Burmester, T., 2016. When the brain goes diving: transcriptome analysis reveals a reduced aerobic energy metabolism and increased stress proteins in the seal brain. BMC Genomics 17, 583.

Foote, A.D., Liu, Y., Thomas, G.W., Vinar, T., Alfoldi, J., Deng, J., Dugan, S., Van Elk, C.E., Hunter, M.E., Joshi, V., Khan, Z., Kovar, C., Lee, S.L., Lindblad-Toh, K., Mancia, A., Nielsen, R., Qin, X., Qu, J., Raney, B.J., Vijay, N., Wolf, J.B., Hahn, M.W., Muzny, D.M., Worley, K.C., Gilbert, M.T., Gibbs, R.A., 2015. Convergent evolution of the genomes of marine mammals. Nature Genetics 47, 272–275.

Geraci, J.R., St Aubin, D.J., Barker, I.K., Webster, R.G., Hinshaw, V.S., Bean, W.J., Ruhnke, H.L., Prescott, J.H., Early, G., Baker, A.S., Madoff, S., Schooley, R.T., 1982. Mass mortality of harbor seals: pneumonia associated with influenza A virus. Science 215, 1129–1131.

Gilissen, C., Hehir-Kwa, J.Y., Thung, D.T., Van De Vorst, M., Van Bon, B.W., Willemsen, M.H., Kwint, M., Janssen, I.M., Hoischen, A., Schenck, A., Leach, R., Klein, R., Tearle, R., Bo, T., Pfundt, R., Yntema, H.G., De Vries, B.B., Kleefstra, T., Brunner, H.G., Vissers, L.E., Veltman, J.A., 2014. Genome sequencing identifies major causes of severe intellectual disability. Nature 511, 344–347.

Gingerich, P.D., Wells, N.A., Russell, D.E., Shah, S.M., 1983. Origin of whales in epicontinental remnant seas: new evidence from the early eocene of Pakistan. Science 220, 403–406.

Godoy-Vitorino, F., Rodriguez-Hilario, A., Alves, A.L., Goncalves, F., Cabrera-Colon, B., Mesquita, C.S., Soares-Castro, P., Ferreira, M., Marcalo, A., Vingada, J., Eira, C., Santos, P.M., 2017. The microbiome of a striped dolphin (*Stenella coeruleoalba*) stranded in Portugal. Research in Microbiology 168, 85–93.

Gomase, V.S., Tagore, S., 2008. Toxicogenomics. Current Drug Metabolism 9, 250–254.

Green, P., 2007. 2× genomes–does depth matter? Genome Research 17, 1547–1549.

Guest, P.C., Gottschalk, M.G., Bahn, S., 2013. Proteomics: improving biomarker translation to modern medicine? Genome Medicine 5, 17.

Gui, D., Jia, K., Xia, J., Yang, L., Chen, J., Wu, Y., Yi, M., 2013. De novo assembly of the Indo-Pacific humpback dolphin leucocyte transcriptome to identify putative genes involved in the aquatic adaptation and immune response. PLoS One 8, e72417.

Gulland, F.M.D., Hall, A.J., 2007. Is marine mammal health deteriorating? Trends in the global reporting of marine mammal disease. EcoHealth 4, 135–150.

Hayashida-Soiza, G., Uchida, A., Mori, N., Kuwahara, Y., Ishida, Y., 2008. Purification and characterization of antibacterial substances produced by a marine bacterium *Pseudoalteromonas haloplanktis* strain. Journal of Applied Microbiology 105, 1672–1677.

Heijne, W.H., Kienhuis, A.S., Van Ommen, B., Stierum, R.H., Groten, J.P., 2005. Systems toxicology: applications of toxicogenomics, transcriptomics, proteomics and metabolomics in toxicology. Expert Review of Proteomics 2, 767–780.

Herdewyn, S., Zhao, H., Moisse, M., Race, V., Matthijs, G., Reumers, J., Kusters, B., Schelhaas, H.J., Van Den Berg, L.H., Goris, A., Robberecht, W., Lambrechts, D., Van Damme, P., 2012. Whole-genome sequencing reveals a coding non-pathogenic variant tagging a non-coding pathogenic hexanucleotide repeat expansion in C9orf72 as cause of amyotrophic lateral sclerosis. Human Molecular Genetics 21, 2412–2419.

Hoff, M.L., Fabrizius, A., Czech-Damal, N.U., Folkow, L.P., Burmester, T., 2017. Transcriptome analysis identifies key metabolic changes in the hooded seal (*Cystophora cristata*) brain in response to hypoxia and reoxygenation. PLoS One 12, e0169366.

Hunt, T.D., Ziccardi, M.H., Gulland, F.M., Yochem, P.K., Hird, D.W., Rowles, T., Mazet, J.A., 2008. Health risks for marine mammal workers. Diseases of Aquatic Organisms 81, 81–92.

Iguchi, T., Watanabe, H., Katsu, Y., 2006. Application of ecotoxicogenomics for studying endocrine disruption in vertebrates and invertebrates. Environmental Health Perspectives 114 (Suppl. 1), 101–105.

Jirtle, R.L., Skinner, M.K., 2007. Environmental epigenomics and disease susceptibility. Nature Reviews Genetics 8, 253–262.

Jones, P.A., Baylin, S.B., 2007. The epigenomics of cancer. Cell 128, 683–692.

Kanter, I., Kalisky, T., 2015. Single cell transcriptomics: methods and applications. Frontiers in Oncology 5, 53.

Khan, A., Ju, F., Xie, W., Tariq Hafeez, M., Cheng, X., Yang, Z., Zhu, L., Li, T., Zhang, S., 2017. Transcriptomic analysis reveals differential activation of microglial genes after ischemic stroke in mice. Neuroscience 348, 212–227.

Khudyakov, J.I., Champagne, C.D., Meneghetti, L.M., Crocker, D.E., 2017. Blubber transcriptome response to acute stress axis activation involves transient changes in adipogenesis and lipolysis in a fasting-adapted marine mammal. Scientific Reports 7, 42110.

Khudyakov, J.I., Champagne, C.D., Preeyanon, L., Ortiz, R.M., Crocker, D.E., 2015a. Muscle transcriptome response to ACTH administration in a free-ranging marine mammal. Physiological Genomics 47, 318–330.

Khudyakov, J.I., Preeyanon, L., Champagne, C.D., Ortiz, R.M., Crocker, D.E., 2015b. Transcriptome analysis of northern elephant seal (*Mirounga angustirostris*) muscle tissue provides a novel molecular resource and physiological insights. BMC Genomics 16, 64.

Kim, H.J., Koedrith, P., Seo, Y.R., 2015. Ecotoxicogenomic approaches for understanding molecular mechanisms of environmental chemical toxicity using aquatic invertebrate, Daphnia model organism. International Journal of Molecular Sciences 16, 12261–12287.

Kluge, M., Campos, F.S., Tavares, M., De Amorim, D.B., Valdez, F.P., Giongo, A., Roehe, P.M., Franco, A.C., 2016. Metagenomic survey of viral diversity obtained from feces of Subantarctic and South American fur seals. PLoS One 11, e0151921.

Koopman, H.N., Iverson, S.J., Read, A.J., 2003. High concentrations of isovaleric acid in the fats of odontocetes: variation and patterns of accumulation in blubber vs. stability in the melon. Journal of Comparative Physiology B 173, 247–261.

Kucklick, J., Schwacke, L., Wells, R., Hohn, A., Guichard, A., Yordy, J., Hansen, L., Zolman, E., Wilson, R., Litz, J., Nowacek, D., Rowles, T., Pugh, R., Balmer, B., Sinclair, C., Rosel, P., 2011. Bottlenose dolphins as indicators of persistent organic pollutants in the western North Atlantic Ocean and northern Gulf of Mexico. Environmental Science and Technology 45, 4270–4277.

Kumar, G., Denslow, N.D., 2017. Gene expression profiling in fish toxicology: a review. Reviews of Environmental Contamination and Toxicology 241, 1–38.

Lander, E.S., Linton, L.M., Birren, B., Nusbaum, C., Zody, M.C., Baldwin, J., Devon, K., Dewar, K., Doyle, M., Fitzhugh, W., Funke, R., Gage, D., Harris, K., Heaford, A., Howland, J., Kann, L., Lehoczky, J., Levine, R., Mcewan, P., Mckernan, K., Meldrim, J., Mesirov, J.P., Miranda, C., Morris, W., Naylor, J., Raymond, C., Rosetti, M., Santos, R., Sheridan, A., Sougnez, C., Stange-Thomann, Y., Stojanovic, N., Subramanian, A., Wyman, D., Rogers, J., Sulston, J., Ainscough, R., Beck, S., Bentley, D., Burton, J., Clee, C., Carter, N., Coulson, A., Deadman, R., Deloukas, P., Dunham, A., Dunham, I., Durbin, R., French, L., Grafham, D., Gregory, S., Hubbard, T., Humphray, S., Hunt, A., Jones, M., Lloyd, C., Mcmurray, A., Matthews, L., Mercer, S., Milne, S., Mullikin, J.C., Mungall, A., Plumb, R., Ross, M., Shownkeen, R., Sims, S., Waterston, R.H., Wilson, R.K., Hillier, L.W., Mcpherson, J.D., Marra, M.A., Mardis, E.R., Fulton, L.A., Chinwalla, A.T., Pepin, K.H., Gish, W.R., Chissoe, S.L., Wendl, M.C., Delehaunty, K.D., Miner, T.L., Delehaunty, A., Kramer, J.B., Cook, L.L., Fulton, R.S., Johnson, D.L., Minx, P.J., Clifton, S.W., Hawkins, T., Branscomb, E., Predki, P., Richardson, P., Wenning, S., Slezak, T., Doggett, N., Cheng, J.F., Olsen, A., Lucas, S., Elkin, C., Uberbacher, E., Frazier, M., Gibbs, R.A., Muzny, D.M., Scherer, S.E., Bouck, J.B., Sodergren, E.J., Worley, K.C., Rives, C.M., Gorrell, J.H., Metzker, M.L., Naylor, S.L., Kucherlapati, R.S., Nelson, D.L., Weinstock, G.M., Sakaki, Y., Fujiyama, A., Hattori, M., Yada, T., Toyoda, A., Itoh, T., Kawagoe, C., Watanabe, H., Totoki, Y., Taylor, T., Weissenbach, J., Heilig, R., Saurin, W., Artiguenave, F., Brottier, P., Bruls, T., Pelletier, E., Robert, C., Wincker, P., Smith, D.R., Doucette-Stamm, L., Rubenfield, M., Weinstock, K., Lee, H.M., Dubois, J., Rosenthal, A., Platzer, M., Nyakatura, G., Taudien, S., Rump, A., Yang, H., Yu, J., Wang, J., Huang, G., Gu, J., Hood, L., Rowen, L., Madan, A., Qin, S., Davis, R.W., Federspiel, N.A., Abola, A.P., Proctor, M.J., Myers, R.M., Schmutz, J., Dickson, M., Grimwood, J., Cox, D.R., Olson, M.V., Kaul, R., Raymond, C., Shimizu, N., Kawasaki, K., Minoshima, S., Evans, G.A., Athanasiou, M., Schultz, R., Roe, B.A.,

Chen, F., Pan, H., Ramser, J., Lehrach, H., Reinhardt, R., Mccombie, W.R., De La Bastide, M., Dedhia, N., Blocker, H., Hornischer, K., Nordsiek, G., Agarwala, R., Aravind, L., Bailey, J.A., Bateman, A., Batzoglou, S., Birney, E., Bork, P., Brown, D.G., Burge, C.B., Cerutti, L., Chen, H.C., Church, D., Clamp, M., Copley, R.R., Doerks, T., Eddy, S.R., Eichler, E.E., Furey, T.S., Galagan, J., Gilbert, J.G., Harmon, C., Hayashizaki, Y., Haussler, D., Hermjakob, H., Hokamp, K., Jang, W., Johnson, L.S., Jones, T.A., Kasif, S., Kaspryzk, A., Kennedy, S., Kent, W.J., Kitts, P., Koonin, E.V., Korf, I., Kulp, D., Lancet, D., Lowe, T.M., Mclysaght, A., Mikkelsen, T., Moran, J.V., Mulder, N., Pollara, V.J., Ponting, C.P., Schuler, G., Schultz, J., Slater, G., Smit, A.F., Stupka, E., Szustakowki, J., Thierry-Mieg, D., Thierry-Mieg, J., Wagner, L., Wallis, J., Wheeler, R., Williams, A., Wolf, Y.I., Wolfe, K.H., Yang, S.P., Yeh, R.F., Collins, F., Guyer, M.S., Peterson, J., Felsenfeld, A., Wetterstrand, K.A., Patrinos, A., Morgan, M.J., De Jong, P., Catanese, J.J., Osoegawa, K., Shizuya, H., Choi, S., Chen, Y.J., Szustakowki, J., International Human Genome Sequencing Consortium, 2001. Initial sequencing and analysis of the human genome. Nature 409, 860–921.

Lane, S.M., Smith, C.R., Mitchell, J., Balmer, B.C., Barry, K.P., Mcdonald, T., Mori, C.S., Rosel, P.E., Rowles, T.K., Speakman, T.R., Townsend, F.I., Tumlin, M.C., Wells, R.S., Zolman, E.S., Schwacke, L.H., 2015. Reproductive outcome and survival of common bottlenose dolphins sampled in Barataria Bay, Louisiana, USA, following the Deepwater Horizon oil spill. Proceedings of the Royal Society B: Biological Sciences 282, 20151944.

Li, L., Shan, T., Wang, C., Cote, C., Kolman, J., Onions, D., Gulland, F.M., Delwart, E., 2011. The fecal viral flora of California sea lions. Journal of Virology 85, 9909–9917.

Lindblad-Toh, K., Garber, M., Zuk, O., Lin, M.F., Parker, B.J., Washietl, S., Kheradpour, P., Ernst, J., Jordan, G., Mauceli, E., Ward, L.D., Lowe, C.B., Holloway, A.K., Clamp, M., Gnerre, S., Alfoldi, J., Beal, K., Chang, J., Clawson, H., Cuff, J., Di Palma, F., Fitzgerald, S., Flicek, P., Guttman, M., Hubisz, M.J., Jaffe, D.B., Jungreis, I., Kent, W.J., Kostka, D., Lara, M., Martins, A.L., Massingham, T., Moltke, I., Raney, B.J., Rasmussen, M.D., Robinson, J., Stark, A., Vilella, A.J., Wen, J., Xie, X., Zody, M.C., Broad Institute Sequencing Platform and Whole Genome Assembly, Baldwin, J., Bloom, T., Chin, C.W., Heiman, D., Nicol, R., Nusbaum, C., Young, S., Wilkinson, J., Worley, K.C., Kovar, C.L., Muzny, D.M., Gibbs, R.A., Baylor College of Medicine Human Genome Sequencing Center Sequencing Center Sequencing Team, Cree, A., Dihn, H.H., Fowler, G., Jhangiani, S., Joshi, V., Lee, S., Lewis, L.R., Nazareth, L.V., Okwuonu, G., Santibanez, J., Warren, W.C., Mardis, E.R., Weinstock, G.M., Wilson, R.K., Genome Institute at Washington University, Delehaunty, K., Dooling, D., Fronik, C., Fulton, L., Fulton, B., Graves, T., Minx, P., Sodergren, E., Birney, E., Margulies, E.H., Herrero, J., Green, E.D., Haussler, D., Siepel, A., Goldman, N., Pollard, K.S., Pedersen, J.S., Lander, E.S., Kellis, M., 2011. A high-resolution map of human evolutionary constraint using 29 mammals. Nature 478, 476–482.

Lipps, J.H., Mitchell, E., 1976. Trophic model for the adaptive radiations and extinctions of pelagic marine mammals. Paleobiology 2, 147–155.

Liu, S., Trapnell, C., 2016. Single-cell transcriptome sequencing: recent advances and remaining challenges. F1000Research 5.

Lunardi, D., Abelli, L., Panti, C., Marsili, L., Fossi, M.C., Mancia, A., 2016. Transcriptomic analysis of bottlenose dolphin (*Tursiops truncatus*) skin biopsies to assess the effects of emerging contaminants. Marine Environmental Research 114, 74–79.

Lupski, J.R., Reid, J.G., Gonzaga-Jauregui, C., Rio Deiros, D., Chen, D.C., Nazareth, L., Bainbridge, M., Dinh, H., Jing, C., Wheeler, D.A., Mcguire, A.L., Zhang, F., Stankiewicz, P., Halperin, J.J., Yang, C., Gehman, C., Guo, D., Irikat, R.K., Tom, W., Fantin, N.J., Muzny, D.M., Gibbs, R.A., 2010. Whole-genome sequencing in a patient with Charcot-Marie-Tooth neuropathy. New England Journal of Medicine 362, 1181–1191.

Mancia, A., Abelli, L., Kucklick, J.R., Rowles, T.K., Wells, R.S., Balmer, B.C., Hohn, A.A., Baatz, J.E., Ryan, J.C., 2015. Microarray applications to understand the impact of exposure to environmental contaminants in wild dolphins (*Tursiops truncatus*). Marine Genomics 19, 47–57.

Mancia, A., Lundqvist, M.L., Romano, T.A., Peden-Adams, M.M., Fair, P.A., Kindy, M.S., Ellis, B.C., Gattoni-Celli, S., Mckillen, D.J., Trent, H.F., Chen, Y.A., Almeida, J.S., Gross, P.S., Chapman, R.W., Warr, G.W., 2007. A dolphin peripheral blood leukocyte cDNA microarray for studies of immune function and stress reactions. Developmental and Comparative Immunology 31, 520–529.

Mancia, A., Ryan, J.C., Chapman, R.W., Wu, Q., Warr, G.W., Gulland, F.M., Van Dolah, F.M., 2012. Health status, infection and disease in California sea lions (*Zalophus californianus*) studied using a canine microarray platform and machine-learning approaches. Developmental and Comparative Immunology 36, 629–637.

Mancia, A., Ryan, J.C., Van Dolah, F.M., Kucklick, J.R., Rowles, T.K., Wells, R.S., Rosel, P.E., Hohn, A.A., Schwacke, L.H., 2014. Machine learning approaches to investigate the impact of PCBs on the transcriptome of the common bottlenose dolphin (*Tursiops truncatus*). Marine Environmental Research 100, 57–67.

Mancia, A., Warr, G.W., Almeida, J.S., Veloso, A., Wells, R.S., Chapman, R.W., 2010. Transcriptome profiles: diagnostic signature of dolphin populations. Estuaries and Coasts 33, 919–929.

Mancia, A., Warr, G.W., Chapman, R.W., 2008. A transcriptomic analysis of the stress induced by capture-release health assessment studies in wild dolphins (*Tursiops truncatus*). Molecular Ecology 17, 2581–2589.

Marino, L., 2002. Convergence of complex cognitive abilities in cetaceans and primates. Brain, Behavior and Evolution 59, 21–32.

Maynard, C.L., Elson, C.O., Hatton, R.D., Weaver, C.T., 2012. Reciprocal interactions of the intestinal microbiota and immune system. Nature 489, 231–241.

McGowen, M.R., Grossman, L.I., Wildman, D.E., 2012. Dolphin genome provides evidence for adaptive evolution of nervous system genes and a molecular rate slowdown. Proceedings of the Royal Society B: Biological Sciences 279, 3643–3651.

McLaughlin, J.B., 2004. Botulism type E outbreak associated with eating a beached whale, Alaska. Emerging Infectious Diseases 10, 1685–1687.

Mele, M., Ferreira, P.G., Reverter, F., Deluca, D.S., Monlong, J., Sammeth, M., Young, T.R., Goldmann, J.M., Pervouchine, D.D., Sullivan, T.J., Johnson, R., Segre, A.V., Djebali, S., Niarchou, A., GTEx Consortium, Wright, F.A., Lappalainen, T., Calvo, M., Getz, G., Dermitzakis, E.T., Ardlie, K.G., Guigo, R., 2015. Human genomics. The human transcriptome across tissues and individuals. Science 348, 660–665.

Neely, B.A., Soper, J.L., Gulland, F.M., Bell, P.D., Kindy, M., Arthur, J.M., Janech, M.G., 2015. Proteomic analysis of cerebrospinal fluid in California sea lions (*Zalophus californianus*) with domoic acid toxicosis identifies proteins associated with neurodegeneration. Proteomics 15, 4051–4063.

Nery, M.F., Gonzalez, D.J., Opazo, J.C., 2013. How to make a dolphin: molecular signature of positive selection in cetacean genome. PLoS One 8, e65491.

Ng, T.F., Suedmeyer, W.K., Wheeler, E., Gulland, F., Breitbart, M., 2009. Novel anellovirus discovered from a mortality event of captive California sea lions. Journal of General Virology 90, 1256–1261.

Ng, T.F., Wheeler, E., Greig, D., Waltzek, T.B., Gulland, F., Breitbart, M., 2011. Metagenomic identification of a novel anellovirus in Pacific harbor seal (*Phoca vitulina* richardsii) lung samples and its detection in samples from multiple years. Journal of General Virology 92, 1318–1323.

Parker, J., Tsagkogeorga, G., Cotton, J.A., Liu, Y., Provero, P., Stupka, E., Rossiter, S.J., 2013. Genome-wide signatures of convergent evolution in echolocating mammals. Nature 502, 228–231.

Peers, C., Pearson, H.A., Boyle, J.P., 2007. Hypoxia and Alzheimer's disease. Essays in Biochemistry 43, 153–164.

Peng, C., Zhao, X., Liu, G., 2015. Noise in the sea and its impacts on marine organisms. International Journal of Environmental Research and Public Health 12, 12304–12323.

Ramsdell, J.S., Gulland, F.M., 2014. Domoic acid epileptic disease. Marine Drugs 12, 1185–1207.

Reamon-Buettner, S.M., Borlak, J., 2007. A new paradigm in toxicology and teratology: altering gene activity in the absence of DNA sequence variation. Reproductive Toxicology 24, 20–30.

Reperant, L.A., Rimmelzwaan, G.F., Kuiken, T., 2009. Avian influenza viruses in mammals. Revue Scientifique et Technique 28, 137–159.

Rifai, N., Gillette, M.A., Carr, S.A., 2006. Protein biomarker discovery and validation: the long and uncertain path to clinical utility. Nature Biotechnology 24, 971–983.

Robertson, K.D., 2005. DNA methylation and human disease. Nature Reviews Genetics 6, 597–610.

Rosales, S.M., Vega Thurber, R.L., 2016. Brain transcriptomes of harbor seals demonstrate gene expression patterns of animals undergoing a metabolic disease and a viral infection. PeerJ 4, e2819.

Samuelsson, L.M., Larsson, D.G., 2008. Contributions from metabolomics to fish research. Molecular BioSystems 4, 974–979.

Schwacke, L.H., Zolman, E.S., Balmer, B.C., De Guise, S., George, R.C., Hoguet, J., Hohn, A.A., Kucklick, J.R., Lamb, S., Levin, M., Litz, J.A., Mcfee, W.E., Place, N.J., Townsend, F.I., Wells, R.S., Rowles, T.K., 2012. Anaemia, hypothyroidism and immune suppression associated with polychlorinated biphenyl exposure in bottlenose dolphins (*Tursiops truncatus*). Proceedings of the Royal Society B: Biological Sciences 279, 48–57.

Shay, T., Jojic, V., Zuk, O., Rothamel, K., Puyraimond-Zemmour, D., Feng, T., Wakamatsu, E., Benoist, C., Koller, D., Regev, A., Immgen, C., 2013. Conservation and divergence in the transcriptional programs of the human and mouse immune systems. Proceedings of the National Academy of Sciences of the United States of America 110, 2946–2951.

Smit, M.G., Bechmann, R.K., Hendriks, A.J., Skadsheim, A., Larsen, B.K., Baussant, T., Bamber, S., Sanni, S., 2009. Relating biomarkers to whole-organism effects using species sensitivity distributions: a pilot study for marine species exposed to oil. Environmental Toxicology and Chemistry 28, 1104–1109.

Sobolesky, P.M., Harrell, T.S., Parry, C., Venn-Watson, S., Janech, M.G., 2016. Feeding a modified fish diet to bottlenose dolphins leads to an increase in serum adiponectin and sphingolipids. Frontiers in Endocrinology (Lausanne) 7, 33.

Speer, R.E., Karuppagounder, S.S., Basso, M., Sleiman, S.F., Kumar, A., Brand, D., Smirnova, N., Gazaryan, I., Khim, S.J., Ratan, R.R., 2013. Hypoxia-inducible factor prolyl hydroxylases as targets for neuroprotection by "antioxidant" metal chelators: from ferroptosis to stroke. Free Radical Biology and Medicine 62, 26–36.

Sun, Y.B., Zhou, W.P., Liu, H.Q., Irwin, D.M., Shen, Y.Y., Zhang, Y.P., 2013. Genome-wide scans for candidate genes involved in the aquatic adaptation of dolphins. Genome Biology and Evolution 5, 130–139.

Uhen, M.D., 2007. Evolution of marine mammals: back to the sea after 300 million years. The Anatomical Record (Hoboken) 290, 514–522.

Urbanczyk-Wochniak, E., Luedemann, A., Kopka, J., Selbig, J., Roessner-Tunali, U., Willmitzer, L., Fernie, A.R., 2003. Parallel analysis of transcript and metabolic profiles: a new approach in systems biology. EMBO Reports 4, 989–993.

Van Bressem, M.F., Raga, J.A., Di Guardo, G., Jepson, P.D., Duignan, P.J., Siebert, U., Barrett, T., Santos, M.C., Moreno, I.B., Siciliano, S., Aguilar, A., Van Waerebeek, K., 2009. Emerging infectious diseases in cetaceans worldwide and the possible role of environmental stressors. Diseases of Aquatic Organisms 86, 143–157.

Van Dolah, F.M., 2000. Marine algal toxins: origins, health effects, and their increased occurrence. Environmental Health Perspectives 108 (Suppl. 1), 133–141.

Van Dolah, F.M., Neely, M.G., Mcgeorge, L.E., Balmer, B.C., Ylitalo, G.M., Zolman, E.S., Speakman, T., Sinclair, C., Kellar, N.M., Rosel, P.E., Mullin, K.D., Schwacke, L.H., 2015. Seasonal variation in the skin transcriptome of common bottlenose dolphins (*Tursiops truncatus*) from the northern Gulf of Mexico. PLoS One 10, e0130934.

Varki, A., Altheide, T.K., 2005. Comparing the human and chimpanzee genomes: searching for needles in a haystack. Genome Research 15, 1746–1758.

Veenstra, T.D., 2007. Global and targeted quantitative proteomics for biomarker discovery. Journal of Chromatography. B, Analytical Technologies in the Biomedical and Life Sciences 847, 3–11.

Veldhoen, N., Ikonomou, M.G., Helbing, C.C., 2012. Molecular profiling of marine fauna: integration of omics with environmental assessment of the world's oceans. Ecotoxicology and Environmental Safety 76, 23–38.

Venn-Watson, S., Carlin, K., Ridgway, S., 2011. Dolphins as animal models for type 2 diabetes: sustained, post-prandial hyperglycemia and hyperinsulinemia. General and Comparative Endocrinology 170, 193–199.

Venn-Watson, S.K., Parry, C., Baird, M., Stevenson, S., Carlin, K., Daniels, R., Smith, C.R., Jones, R., Wells, R.S., Ridgway, S., Jensen, E.D., 2015. Increased dietary intake of saturated fatty acid heptadecanoic acid (C17:0) associated with decreasing ferritin and alleviated metabolic syndrome in dolphins. PLoS One 10, e0132117.

Viant, M.R., 2008. Recent developments in environmental metabolomics. Molecular BioSystems 4, 980–986.

Walsh, C.J., Butawan, M., Yordy, J., Ball, R., Flewelling, L., De Wit, M., Bonde, R.K., 2015. Sublethal red tide toxin exposure in free-ranging manatees (*Trichechus manatus*) affects the immune system through reduced lymphocyte proliferation responses, inflammation, and oxidative stress. Aquatic Toxicology 161, 73–84.

Waltzek, T.B., Cortes-Hinojosa, G., Wellehan Jr., J.F., Gray, G.C., 2012. Marine mammal zoonoses: a review of disease manifestations. Zoonoses and Public Health 59, 521–535.

Watanabe, H., Iguchi, T., 2003. Evaluation of endocrine disruptors based on gene expression using a micorarray. Environmental Science 10, 61–67.

Webster, R.G., Geraci, J., Petursson, G., Skirnisson, K., 1981. Conjunctivitis in human beings caused by influenza A virus of seals. New England Journal of Medicine 304, 911.

Weedon, M.N., Cebola, I., Patch, A.M., Flanagan, S.E., De Franco, E., Caswell, R., Rodriguez-Segui, S.A., Shaw-Smith, C., Cho, C.H., Lango Allen, H., Houghton, J.A., Roth, C.L., Chen, R., Hussain, K., Marsh, P., Vallier, L., Murray, A., International Pancreatic Agenesis Consortium, Ellard, S., Ferrer, J., Hattersley, A.T., 2014. Recessive mutations in a distal PTF1A enhancer cause isolated pancreatic agenesis. Nature Genetics 46, 61–64.

Woolhouse, M.E., Gowtage-Sequeria, S., 2005. Host range and emerging and reemerging pathogens. Emerging Infectious Diseases 11, 1842–1847.

Yamagiwa, J., Karczmarski, L., 2014. Primates and Cetaceans: Field Research and Conservation of Complex Mammalian Societies. Springer, New York.

Yim, H.S., Cho, Y.S., Guang, X., Kang, S.G., Jeong, J.Y., Cha, S.S., Oh, H.M., Lee, J.H., Yang, E.C., Kwon, K.K., Kim, Y.J., Kim, T.W., Kim, W., Jeon, J.H., Kim, S.J., Choi, D.H., Jho, S., Kim, H.M., Ko, J., Kim, H., Shin, Y.A., Jung, H.J., Zheng, Y., Wang, Z., Chen, Y., Chen, M.,

Jiang, A., Li, E., Zhang, S., Hou, H., Kim, T.H., Yu, L., Liu, S., Ahn, K., Cooper, J., Park, S.G., Hong, C.P., Jin, W., Kim, H.S., Park, C., Lee, K., Chun, S., Morin, P.A., O'brien, S.J., Lee, H., Kimura, J., Moon, D.Y., Manica, A., Edwards, J., Kim, B.C., Kim, S., Wang, J., Bhak, J., Lee, H.S., Lee, J.H., 2014. Minke whale genome and aquatic adaptation in cetaceans. Nature Genetics 46, 88–92.

Yu, J., Kindy, M.S., Ellis, B.C., Baatz, J.E., Peden-Adams, M., Ellingham, T.J., Wolff, D.J., Fair, P.A., Gattoni-Celli, S., 2005. Establishment of epidermal cell lines derived from the skin of the Atlantic bottlenose dolphin (*Tursiops truncatus*). The Anatomical Record. Part A, Discoveries in Molecular, Cellular, and Evolutionary Biology 287, 1246–1255.

Zamuruyev, K.O., Aksenov, A.A., Baird, M., Pasamontes, A., Parry, C., Foutouhi, S., Venn-Watson, S., Weimer, B.C., Delplanque, J.P., Davis, C.E., 2016. Enhanced non-invasive respiratory sampling from bottlenose dolphins for breath metabolomics measurements. Journal of Breath Research 10, 046005.

Zasloff, M., 2011. Observations on the remarkable (and mysterious) wound-healing process of the bottlenose dolphin. Journal of Investigative Dermatology 131, 2503–2505.

Zhou, X., Seim, I., Gladyshev, V.N., 2015. Convergent evolution of marine mammals is associated with distinct substitutions in common genes. Scientific Reports 5, 16550.

Chapter 12

Immunotoxic Effects of Environmental Pollutants in Marine Mammals

Jean-Pierre Desforges[1], Christian Sonne[1], Rune Dietz[1], Milton Levin[2]
[1]Aarhus University, Roskilde, Denmark; [2]University of Connecticut, Storrs, CT, United States

INTRODUCTION

The health of marine ecosystems worldwide is considered to be deteriorating, as evidenced by increased reporting of diseases in marine organisms and the ever-rising threat of anthropogenic pollution to marine mammal populations (Ross et al., 1996a; Duignan et al., 2014; Gulland and Hall, 2007). Marine mammals are particularly vulnerable to the accumulation of environmental pollutants since they often inhabit polluted coastal waters, have long lifespans to accumulate contaminants, eat at the top of food webs and thus biomagnify contaminants, and cannot metabolically eliminate persistent chemicals (Ross et al., 2000). There have been many instances over the past decades in which high body burdens of environmental contaminants, particularly persistent organic pollutants (POPs), have been linked to severe health and population-level effects in marine mammals. Elevated POP levels in Baltic seal populations in the 1970s–80s were linked to widespread reproductive failure and high prevalence of tissue lesions causing the so called "Baltic Seal Disease Complex" (Helle, 1980; Helle et al., 1976; Olsson et al., 1994; Roos et al., 2012). Beluga whales (*Delphinapterus leucas*) in the St. Lawrence Estuary had similarly high contaminant exposure in the 1980s, which was linked to the high prevalence of degenerative, infectious, neoplastic, hyperplastic, and necrotic lesions (De Guise et al., 1995b; Martineau et al., 1994). Lastly, epizootic outbreaks of morbillivirus in 1988/89, 1990/91, and 2002 in harbor seals (*Phoca vitulina*) and stripped dolphins (*Stenella coeruleoalba*) in European and Mediterranean waters resulted in the death of thousands of animals, and there were suspected links to contaminant-induced immunosuppression as a trigger for disease susceptibility (Aguilar and Borrell, 1994; Dietz et al., 1989; Duignan et al., 2014; Hall et al., 1992; Härkönen et al., 2006). These instances have highlighted the possible hazards associated with exposure to environmental contaminants and have driven the field of marine mammal immunotoxicology.

Marine Mammal Ecotoxicology. https://doi.org/10.1016/B978-0-12-812144-3.00012-7

Marine mammal immunotoxicology aims to study the toxic effects of compounds, both naturally occurring and anthropogenic, on the immune system of marine mammals. Many environmental contaminants are known to be immunotoxic in laboratory animals, and accumulating evidence suggests the same is true for marine mammals. Though cause and effect relationships between contaminants and health effects in marine mammals have yet to be fully defined, evidence of effects such as reproductive impairment, endocrine disruption, and immune modulation support the hypothesis that POPs are detrimental to the overall health of marine mammals around the world. A better understanding of the immunomodulatory effects of environmental contaminants will be useful for future assessment of the possible impact of those pollutants on the health of marine mammals, as well as other aquatic species, and managing the risks associated with exposure to the ubiquitous contaminants in general.

MARINE MAMMAL IMMUNOLOGY

The Immune System: Innate Versus Adaptive

Marine mammal immunology is a rapidly evolving discipline. Though the immune system of marine mammals is not fully understood, it is, however, comparable to that of other mammalian species in terms of anatomy and function. Similar to terrestrial species, the marine mammal immune system is composed of two branches, the innate (natural) and acquired (adaptive). The innate arm provides the first line of defense against pathogens and consists of anatomic (skin), physiologic (change in pH and temperature), phagocytic (neutrophils and macrophages), and inflammatory (serum proteins) barriers, which block the entrance and establishment of infectious agents (Kuby, 2012). These responses are rapid, ranging from minutes to a few days, nonspecific, and require no previous encounter with the agent. If the innate immune system is unable to effectively combat the pathogen, an acquired immune response will be mounted. This arm of the immune system is specific, possesses memory, and can distinguish between self and nonself. Acquired immunity can be further divided into humoral (antibody) and cell-mediated (T and B lymphocytes) responses. Cell-mediated responses involve different T cell subsets (helper, cytotoxic, and regulatory cells) and mediate cellular immunity via interactions with antigen-presenting cells (e.g., macrophages, dendritic cells). Humoral immune responses are mediated by B cells (plasma cells) and their secreted antibodies, which act against extracellular pathogens (Abbas et al., 2012). Taken together, both arms of the immune system can mount an effective battle against infectious agents, central to the health and well-being of an individual.

Immune Organs, Cells, and Proteins

Marine mammals possess primary immune organs (e.g., bone marrow, thymus) and secondary immune organs (e.g., spleen, lymph nodes, and mucosal-associated lymphoid tissue [MALT]). Macroscopic and microscopic investigation of immune organs from cetaceans, pinnipeds, and polar bears have revealed similar tissue distribution and organization to that of terrestrial mammals, as previously summarized (Beineke et al., 2010). Peripheral blood immune cells (e.g., neutrophils, basophils, eosinophils, lymphocytes, and monocytes) and tissue resident immune cells (e.g., dendritic, lymphocytes, macrophages) are the key cells involved in protecting the body against both infectious disease and foreign invaders.

Immunoglobulins (Ig), also known as antibodies (Ab), are glycoprotein molecules produced by plasma cells, which are differentiated B lymphocytes that produce a single type of antibody, and they are important in the humoral immune response. Immunoglobulins are a critical part of the immune response by specifically recognizing and binding to particular antigens, such as bacteria or viruses, and aiding in their destruction. Antibodies can exist as different classes, or isotypes, and are defined by the different types of heavy chains they contain, such as alpha (IgA), gamma (IgG), and mu (IgM).

Cytokines are soluble messengers of the immune system and have the capacity to regulate many different cells. These include interferons, interleukins, and growth factors that are secreted by cells (e.g., lymphocytes, macrophages) and have an effect on other cells (Kuby, 2012). Examples of proinflammatory cytokines, secreted in the beginning of an immune response, include interleukin (IL)-1, IL-6, IL-8, and tumour necrosis factor (TNF), and they are produced predominantly by macrophages, monocytes, and T helper 1 (Th1) lymphocytes. Antiinflammatory cytokines, secreted to dampen an inflammatory response, include IL-4 and IL-10, and they are secreted predominately by T helper 2 (Th2) lymphocytes. T helper cells can also direct an immune response based on the release of one or more cytokines. Th1 cells secrete interferon gamma (INFγ), IL-2, and TNF, which stimulate cell-mediated immunity to help combat intracellular pathogens (e.g., intracellular viruses), whereas Th2 cells produce IL-4, IL-10, IL-6, and IL-13, which inhibit cell-mediated (Th1) immunity and promote humoral immune responses to help combat extracellular pathogens (Scheerlinck and Yen, 2005).

Acute phase proteins (APP) are a class of proteins, predominantly produced by the liver, whose plasma concentrations increase or decrease in response to inflammation. APPs are considered to be nonspecific innate immune components involved in the restoration of homeostasis and the restriction of microbial growth before animals develop acquired immunity to a challenge (Murata et al., 2004). Examples of APPs include C-reactive protein, haptoglobin (Hp), serum amyloid A (SAA), and serum amyloid P (SAP).

The complement system, also referred to as the complement cascade, is a part of the innate immune system that enhances (i.e., complements) the ability of phagocytic cells to clear microbes and damaged cells. This system consists of a number of small proteins found in the blood and normally circulating as inactive precursors (pro-proteins). The complement system can be activated by the classical pathway, the alternative pathway, or by the lectin pathway (Kuby, 2012).

Characterizing Immune Function

In vitro functional immune assays for laboratory animals have been modified and/or optimized to assess and quantify several key innate and adaptive immune functions in marine mammals. The in vitro assays, including mitogen-induced lymphocyte proliferation, phagocytosis, and natural killer (NK) cell activity, are part of Tiers I and II of the National Toxicology Program, and considered sensitive and reproducible to predict immunotoxicity (Luster et al., 1993, 1992). Importantly, assays for use in marine mammals can be performed with whole blood, which can be obtained with nonlethal, routine, and minimally invasive techniques, from both managed-care and wild (restrained) individuals.

Innate Immunity: Phagocytosis

Phagocytosis is the principal effector mechanism for the disposal of invading, foreign, or otherwise unwanted cells or particles (van Oss, 1986). Upon interaction of the particle and surface receptors on the phagocyte, phagocytosis is initiated, leading to internalization of the particle in a phagolysosome, the site of particle destruction involving the respiratory burst (described subsequently) (Allen and Aderem, 1996). Neutrophils are the most important circulating phagocyte, providing the first line of defense against invading particles, especially bacteria. Circulating monocytes, precursors to tissue macrophages, also have the ability to phagocytize. Assays have been developed and optimized to measure phagocytosis in vitro for beluga whales (De Guise et al., 1995a) and bottlenose dolphins (*Tursiops truncatus*) (Keogh et al., 2011; Noda et al., 2003) and have been adapted to measure phagocytosis in several other cetacean and pinniped species (Levin et al., 2010, 2005b; Mos et al., 2006).

Innate Immunity: Respiratory Burst

The respiratory burst is the principal effector mechanism for the production of reactive oxygen species (ROS) used to kill internalized pathogens following phagocytosis. The respiratory burst generates ROS by an oxygen-dependent process in which membrane-bound NADPH oxidases catalyze the reduction of molecular oxygen to the reactive oxygen intermediate, superoxide (O_2^-) (Roos et al., 2003; Dahlgren and Karlsson, 1999). Superoxide dismutase (SOD) can convert the superoxide anion into hydrogen peroxide (H_2O_2), followed by either production of hypochlorous acid in a reaction catalyzed by myeloperoxidase (MPO) or production of water catalyzed by catalase. These reactive species are very effective in killing phagocytized microorganisms. Professional phagocytes (neutrophils, monocytes, and macrophages) are the most effective cells to generate the respiratory burst. Assays have been developed to measure respiratory burst in cells from beluga whales (De Guise et al., 1995a) and bottlenose dolphins (Keogh et al., 2011; Shiraishi et al., 2002), and they have been adapted to measure respiratory burst in vitro in several cetacean and pinniped species (Frouin et al., 2013; Levin et al., 2007).

Innate Immunity: Natural Killer Cell Activity

NK cell activity is the principle effector mechanism used to kill tumor cells and virus-infected cells in the early phase of an infection. NK cells represent a heterogeneous population of CD3 negative, T cell receptor negative, large granular lymphocytes that commonly express surface markers such as CD16 and CD56 in humans (O'Shea and Ortaldo, 1992). NK cell cytotoxicity is regulated by a complex balance between activating and inhibitory signals (Moretta and Moretta, 2004; Moretta et al., 2003). NK cells can kill target cells by releasing small granzymes containing perforins and proteases into the cytoplasm of a target cell in close proximity. Perforins form pores in the cell membrane of the target cell, creating an aqueous channel through which the granzymes and associated molecules can enter, inducing either apoptosis or osmotic cell lysis. Assays have been optimized to measure NK cell activity in beluga whales (De Guise et al., 1997) and harbor seals (*Phoca vitulina*) (Ross et al., 1996c).

Adaptive Immunity: Lymphocyte Proliferation

Lymphocyte proliferation is the first step in a proper immune response to create effector lymphocytes, necessary to eliminate a current antigen, or memory lymphocytes, necessary to eliminate the same antigen the host may encounter in the future; this memory function insures that future responses to an antigen are faster and stronger compared to the first encounter. T lymphocytes, including T helper (Th) cells, which help other white blood cells in the immune response, and cytotoxic T lymphocytes, which destroy virus-infected cells and tumor cells, help orchestrate a proper cell-mediated immune response. B lymphocytes, including antibody-producing plasma cells, help orchestrate a proper humoral immune response. Assays have been developed to measure lymphocyte proliferation in vitro for cells from harbor seals (de Swart et al., 1993; DiMolfetto-Landon et al., 1995), beluga whales (De Guise et al., 1996a), bottlenose dolphins, pilot whales (*Globicephala melas*), and killer whales (*Orcinus orca*) (Colgrove, 1978; Mumford et al., 1975).

IMMUNOTOXICOLOGY IN MARINE MAMMALS

Study Designs and Methodologies for Marine Mammal Immunotoxicology

The immunotoxicity of many environmental and industrial contaminants are well documented in laboratory rodents; however, similar characterization of risk for marine mammals represents a considerable challenge given the many logistical and ethical constraints. Nonetheless, marine mammologists and immunotoxicologists have developed many different approaches to evaluate if environmental contaminants are modulating immunity in marine mammals. These approaches—natural exposure (in vivo), captive studies (in vivo), laboratory exposure (in vitro), and effect modeling (in silico)—sweep a broad spectrum of specificity and complexity in terms of effects and mechanisms and come with inherent levels of certainty regarding interpretation of results for free-ranging animals.

Field studies of naturally exposed populations represent the most realistic and comprehensive approach when evaluating the effects of acute and chronic exposure to environmental contaminants. These types of studies are, however, complicated by logistical, financial, and ethical challenges of sampling free-ranging marine animals, as well as confounding factors including age, sex, diet, habitat, and temporal offsets between potential exposure and measured immunological changes. Still, in vivo exposures capture the toxicokinetics of contaminant exposure as well as the complex interplay between cells and tissues of the body. These studies entail the live capture of wild marine mammals and sampling of tissues and/or blood for both contaminant analysis and immunological assessments, followed by correlative analysis of the two (Beckmen et al., 2003; Lahvis et al., 1995; Levin et al., 2005a; Lie et al., 2004; Schwacke et al., 2012; Sørmo et al., 2009).

Captive studies retain cost and logistical constraints, as well as different ethical challenges, but by carefully selecting individuals of the same age and sex and keeping habitat and feeding conditions constant, confounding are factors kept to a minimum, providing more definitive causation of observed contaminant-related effects. The landmark study of this type involved keeping harbor seals in semifield conditions and comparing the immunological changes between two groups of seals fed different diets; the exposed group had a diet of fish from the contaminated Baltic Sea, while the "control" group was fed fish from the much less contaminated North Atlantic Ocean (de Swart et al., 1996; Ross et al., 1996a). The only other similar study used sledge dogs in West Greenland as surrogates for polar bears; here, dogs were divided into two feeding groups, one fed contaminated marine mammal blubber and the other fed much less contaminated pork fat (Sonne, 2010). These studies generated compelling evidence that natural and realistic exposure to environmental contaminants can significantly modulate almost all measurable markers of immunity.

In vitro exposure experiments provide the unique advantage of evaluating the effects of contaminants on species such as marine mammals where in vivo studies are often impossible. Furthermore, in vitro exposure provides rapid detection of direct effects of a contaminant of interest on the immune system at relatively low cost. It also easily allows comparison of immune effects among species and at broad ranges of environmentally relevant concentrations. In vitro experiments typically utilize live immune cells harvested from captive or wild animals to conduct controlled concentration-response studies on particular immune functions (see Immune Organs, Cells, and Proteins section) after exposure to individual or mixtures of contaminants (e.g., Das et al., 2008; De Guise et al., 1998, 1996b; Dupont et al., 2015; Frouin et al., 2010b; Levin et al., 2016; Mori et al., 2008).

In silico modeling approaches are the farthest removed from live animal studies; however, they can provide useful information on the risks of immunotoxicity for whole populations of marine mammals. These approaches can make use of the results from controlled laboratory studies to extrapolate effects in species that have not been studied, as is often the case for various marine mammals. One such approach involves physiologically based pharmacokinetic (PBPK) modeling to calculate risk quotients based on actual tissue contaminant levels in the species of interest and critical body residues determined from laboratory rodent studies. In this way, risk quotients provide a quantitative measure of immunotoxicity risk. This approach has been used to evaluate and compare the current risk of immunotoxicity in 11 circumpolar populations of polar bears (Dietz et al., 2015) and to corroborate immune effects observed in sledge dogs from the controlled feeding study described previously (Sonne et al., 2016).

Synthesis of Immunotoxicity in Marine Mammals

The weight of evidence from field studies, captive feeding experiments, and in vitro investigations strongly suggest that contaminants in the marine environment can modulate the immune response of marine mammals, leading to potential conservation-level concerns in some highly exposed populations (Desforges et al., 2016; Letcher et al., 2010; Ross et al., 1996a). Bioaccumulative organic contaminants, such as legacy POPs and emerging organohalogen flame-retardants, are of great concern to marine mammals, as these resist metabolic breakdown and reach elevated levels in adipose tissue (Ross et al., 2000). Metals, such as mercury, cadmium, lead, and tin compounds can be of concern particularly for local populations of marine mammals near point sources or in ultimate sink regions within the Arctic (Dietz et al., 2013; Kakuschke et al., 2005; Nakata et al., 2002). There is also evidence that nonbioaccumulative and/or natural compounds, such as algal biotoxins or oil compounds, are immunotoxic to marine mammals. Recent work by Desforges et al. (2016) provided an extensive review of immunotoxicity studies in marine mammals and reported observed effects on immune endpoints into the following specific categories: immune tissue histopathology, circulating immune cell populations, lymphocyte proliferation, phagocytosis, respiratory burst, NK cell activity, immunoglobulin production, and cytokine gene expression. A summary of effects on all endpoints is presented in Fig. 12.1 and discussed subsequently in four broad immune effect classifications: tissue pathology, functional immune assays, gene expression, and host-resistance.

Pathology of Lymphoid Tissues

Any effect on primary or secondary lymphoid tissues can lead to downstream consequences on many aspects of the immune system, as these are crucial sites

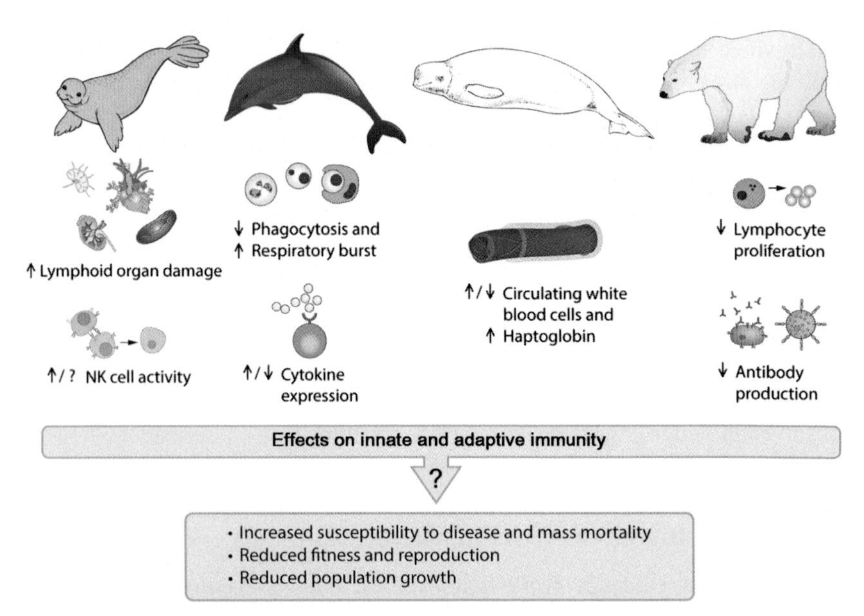

FIGURE 12.1 Summary of documented contaminant-mediated effects on the immune system of marine mammals. The weight of evidence in the marine mammal immunotoxicology literature suggests that environmental pollutants modulate both innate and adaptive immunity, including aspects of cellular and humoral responses. Although still unclear, effects on the immune system may have consequences on intimately tied reproductive and endocrine systems, potentially leading to reduced fitness and population growth. *(Taken with permission from Desforges, J.-P.W., Sonne, C., Levin, M., Siebert, U., De Guise, S., Dietz, R., 2016. Immunotoxic effects of environmental pollutants in marine mammals. Environment International 86, 126–139.)*

for immune cell generation, maturation, and activation (Moser and Leo, 2010). Thus, pathology in the thymus or bone marrow can induce changes in downstream cellular and humoral immune responses (Luebke et al., 2007; Silkworth and Antrim, 1985; Suh et al., 2002; Thurmond et al., 2000). It is possible that effects on other immune endpoints are directly or indirectly related to an initial lymphoid tissue pathology.

As high trophic predators and ecosystem sentinels, polar bears (*Ursus maritimus*) and harbor porpoises (*Phocoena phocoena*) have been used to examine the link between marine contaminant exposure and lymphoid tissue pathology. Tissue levels of POPs, particularly PCBs, have been correlated with specific pathologies of the thymus, spleen, and lymph nodes in these two species. Thymic and spleen dysfunction in porpoises inhabiting northern European waters was characterized by interfollicular fibrosis, cortical lymphoid depletion, loss of the corticomedullary junction, and splenic lymphocyte depletion (Beineke et al., 2005). Similarly, East Greenland polar bears were observed to have secondary follicles in spleen and lymph nodes correlating with blubber PCBs (Kirkegaard et al., 2005). Laboratory rodent studies have long confirmed that pathology of lymphoid tissues, particularly thymic atrophy (Vos et al., 1989; Vos and Van Loveren, 1994), is a common and sensitive effect of POP exposure, supporting the limited findings specific to marine mammals.

Altered Immune Function or Cell Populations

Blood collection is relatively common in captive and field studies of marine mammals, allowing for various types of analyses. Hematology—quantifying and characterizing populations of cells and proteins in blood—is used in clinical health assessments, and a shift in white blood cell populations from a known healthy reference or background can be indicative of immune perturbations. Live cells can also be harvested from blood samples and used for functional immune assays and in vitro experimentation. These types of immune characterizations are the most commonly reported in immunotoxicology assessments for marine mammal populations, and a detailed review of these is provided in Desforges et al. (2016). The following provides an overview of the types of studies that exist and the classes of contaminants that have been assessed thus far.

Organochlorines (OCs) have been the predominant focus of immunotoxicology assessments in marine mammals. This includes field campaigns and feeding studies in which exposure is to the "natural" complex mixture of hundreds of different chemicals (e.g., Beckmen et al., 2003; Dupont et al., 2013; Lahvis et al., 1995; Levin et al., 2005a; Lie et al., 2005; Ross et al., 1996b) as well as in vitro investigations in which specific compounds are used for laboratory exposure experiments (e.g., De Guise et al., 1998; Frouin et al., 2010a,b; Hammond et al., 2005; Levin et al., 2005b). PCB is most often used as the reference contaminant of interest in these studies because it remains the contaminant with the highest burden and greatest risk for immunotoxicity in most marine mammal populations despite having been banned decades ago (Dietz et al., 2015). For instance, tissue levels of PCBs were positively correlated with blood levels of granulocytes, neutrophils, and basophils in free-ranging harbor seals and northern fur seals (*Callorhinus ursinus*) (Beckmen et al., 2003; de Swart et al., 1994). While some studies have linked blood cell counts to contaminant exposure, many have found mixed results or no link at all (Hall et al., 1997; Nyman et al., 2003; Schumacher et al., 1995), revealing the difficulty of using naturally fluctuating blood parameters as markers of immunotoxicity. Functional immune assays have more consistently demonstrated that OCs modulate immune responses. The weight of evidence suggests that PCBs have a negative and concentration-dependent effect on the ability of lymphocytes to proliferate in response to a mitogen as well as the ability of phagocytic cells to engulf invading particles (Desforges et al., 2016). Several studies have also suggested that PCBs stimulate respiratory burst in phagocytes from various marine mammal species, increasing the risk of oxidative stress and tissue damage (Levin et al., 2004; Mos et al., 2006). Few data are available on the cytotoxic activity of NK cells, and these suggest a possible negative effect from POP exposure (Ross et al., 1996c) or none at all (Hammond et al., 2005). In vivo delayed-type hypersensitivity response, measured as the level of skin swelling after intradermal injection of particular antigens, requires often impossible serial animal assessments, but it has been shown to be suppressed by POPs in harbor seals (Ross et al., 1995).

Mercury has been shown to be a potent immunosuppressant in seals and cetaceans. In vitro exposures have consistently shown a strong negative effect of mercury on lymphocyte proliferation and phagocytosis (e.g., Das et al., 2008; De Guise et al., 1996b; Dupont et al., 2015; Frouin et al., 2012; Kakuschke et al., 2009; Pellissó et al., 2008). Immune modulation and sensitization have also been documented for various essential and nonessential metals (e.g., Kakuschke et al., 2008, 2006). Butyltins, organotin compounds used as antibiofouling agents for marine vessels, have also been shown to suppress the lymphocyte proliferation response in vitro using PBMCs from several marine mammal species (Nakata et al., 2002).

Polybrominated diphenyl ethers (PBDEs) and perfluorinated compounds (PFCs) are two more recent and emerging contaminants that have been assessed, albeit in very few studies. Frouin et al. (2010b) tested various PBDE congeners on harbor seal immune cells and found that BDE-47, -99 and, -153 caused a concentration-dependent decrease in phagocytic activity and efficiency. Contrarily, a penta-PBDE mixture (DE-71) did not significantly modulate lymphocyte proliferation or NK cell activity in bottlenose dolphin immune cells at what the authors considered environmentally relevant exposure concentrations. The authors concluded that dolphins are likely not at risk of immune toxicity from current levels of PBDE exposure. PFCs, particularly perfluorooctane sulfonate, can accumulate to high levels in marine mammals, and in vivo studies of highly exposed Atlantic bottlenose dolphins suggested a positive correlation between exposure levels and several blood clinical parameters as well as B cell proliferation (Fair et al., 2013). A follow-up in vitro study using dolphin PBMCs supported the effect of PFCs on B cell proliferation (Wirth et al., 2014b). In contrast, no effect of PFCs on B or T cell proliferation was found in East Greenland ringed seals (*Pusa hispida*) upon in vitro exposure, though levels were lower than tested in the dolphin study (Levin et al., 2016).

Large blooms of algae, i.e., harmful algal blooms, are known to produce natural toxins that can cause severe deleterious health effects in marine organisms, including mortality events (Flewelling et al., 2005). These naturally produced compounds, including brevetoxin, domoic acid, and saxitoxin, are potent neurotoxins and potential contributors to increased susceptibility to disease. In vitro investigations of these compounds found positive concentration-response relationships with different innate and adaptive immune functions in bottlenose dolphins, California sea lions (*Zalophus californianus*), and harbor seals (Gebhard et al., 2015; Levin et al., 2010; Bogomolni et al., 2016), confirming that these compounds can modulate the immune system of marine mammals. Saxitoxin was also found to increase the amount of phocine distemper virus present in lymphocytes during in vitro exposure studies with harbor seal cells, suggesting a role for this compound in the susceptibility of disease (Bogomolni et al., 2016).

Pharmaceuticals represent another important class of marine contaminants that have received little attention in marine mammals. Only one recent study utilizing a harbor seal B lymphoma cell line has evaluated the immunotoxicity of 10 pharmaceuticals. The authors reported that 17α-ethinyl estradiol and naproxen suppressed lymphocyte proliferation individually and that binary mixtures of several pharmaceuticals had synergistic effects on the suppression of immune activity (Kleinert et al., 2018). This represents important findings for marine mammals inhabiting coastal regions receiving large inputs of pharmaceuticals through wastewater treatment effluent.

Altered Gene Expression

New genetic and genomic tools have permitted immunotoxicologists to evaluate contaminant effects at the molecular level to complement standard cellular, tissue, and organism markers of immunotoxicity. Gene expression in marine mammal immunotoxicology is a relatively new field with few publications to date, and these have focused on cytokines. A brief summary of these studies is presented in Table 12.1, which reveals that several classes of contaminants as well as different cytokines have been investigated through in vitro exposure experiments as well as in vivo field studies of free-ranging animals.

Hepatic expression of the proinflammatory cytokine IL-1 in ringed seals from Labrador, the Baltic Sea, and Svalbard positively correlated with blubber levels of PCBs (Brown et al., 2014; Routti et al., 2010). In vitro exposure of harbor seal leukocytes to PCB-169 and benzo-a-pyrene (BaP) resulted in reduced expression of IL-1 (Neale et al., 2005); this contrast in directionality of effects may be linked to tissue type or differences in responses between in vivo and in vitro systems for cytokine expression. Because of the few number of studies examining cytokines, it is too early to clearly associate contaminant exposure to a certain pattern of cytokine expression. Nonetheless, cytokine profiling is an exciting new field in marine mammal immunology and toxicology, and further studies will shed light into the mechanisms of cytokine disruption by contaminants in marine mammals.

Altered Host Resistance to Disease

All previously discussed, indicators of immunological change contribute to the overall immunological health of animals, but the ultimate descriptor of immunity is susceptibility to disease. Controlled host-resistance experiments, while commonly done in laboratory rodents, are ethically challenging for marine mammals. One clever strategy used by marine mammal toxicologists involves comparing contaminant levels within a population of marine mammals that have died from either infectious disease or from trauma. After considering the impact of confounding factors, such as condition of the animals in relation to contaminant levels, this exercise can provide valuable insight to the link between contaminant exposure and vulnerability to disease. As mentioned in the introduction, there have been several large outbreaks of morbillivirus in marine mammals since the 1980s, and researchers later reported that animals

TABLE 12.1 Altered Cytokine Gene Expression in Marine Mammals Exposed to Environmental Contaminants

Species	Location	Cytokine Analyzed	Pollutant	Study Type	Effect on Gene Expression	References
Harbor seal	North Sea	IL-2, IL-4, TGF-β	MeHg	In vitro	↓ IL-2 & TGF-β	Das et al. (2008)
					↑ IL-4	
	North Sea	IL-2, IL-4, TGF-β	MeHg	In vitro	No effect	Kakuschke et al. (2009)
	California	IL-1β, IL-2	PCB-169	In vitro	↓ IL-1 and IL-2	Neale et al. (2005)
			BaP			
Grey seal	North Sea	IL-2, IL-4	Hg, Cd, Be, Ni	In vitro	↑ IL-2 and IL-4 (Hg)	Kakuschke et al. (2006)
					↓ IL-2 and IL-4 (Be, Ni)	
					No effect (Cd)	
Ringed seal	Baltic Sea	IL-1β	∑PCB	In vivo	↑ IL-1β	Routti et al. (2010)
	Labrador	IL-1β	∑PCB	In vivo	↑ IL-1β	Brown et al. (2014)

Modified from Desforges, J.-P.W., Sonne, C., Levin, M., Siebert, U., De Guise, S., Dietz, R., 2016. Immunotoxic effects of environmental pollutants in marine mammals. Environment International 86, 126–139.

that had died from the disease had higher concentrations of PCBs than survivors or healthy individuals before the event (Aguilar and Borrell, 1994; Hall et al., 1992). Extensive studies of harbor porpoises in the United Kingdom have compared diseased animals with trauma-killed animals and found PCB levels to be higher in diseased animals; statistical approaches estimated that each increase in blubber PCBs of 1 mg/kg created an additional risk of infectious disease mortality of 2% (Hall et al., 2006; Jepson et al., 1999). Altogether, there appears to be a strong case showing that environmental contaminants, especially PCBs, are modulating the immune system of marine mammals and increasing the susceptibility and risk of mortality to infectious diseases.

CONSIDERATIONS, CHALLENGES, AND FUTURE NEEDS

Reference Intervals

There is a knowledge gap in understanding the normal range, or reference intervals (RIs), for important physiologic endpoints in wildlife. Examples of endpoints include serum biochemistry, hematology, hormones and cytokines values, functional immune assays, and body condition. Statistically derived RIs provide a baseline to which individual test results can be compared, thus allowing the evaluation of individual health relative to a "normal" healthy population. To date, RIs have been established for serum chemistry values for free-ranging beluga whales, harp seals, and ringed seals (Nordoy and Thoresen, 2002; Tryland et al., 2006a,b), as well as for managed-care beluga whales (Norman et al., 2013).

There are several challenges in developing RIs for free-ranging marine mammals. The main challenge is obtaining an adequate sample size to establish an RI for a particular biomarker. Procedures for establishing RIs based on reference sample size and distribution have been recommended by the American Society for Veterinary Clinical Pathology for wildlife species, with sample sizes ranging from $n=20$ up to $n>120$ (Friedrichs et al., 2013). In addition to sample size, other considerations include the proper tissue/sample to collect (e.g., blood/serum/plasma, urine), time to collection (e.g., fresh/live capture vs. stranded vs. fresh dead), proper tissue storage (e.g., room temperature, −20°C, −80°C, liquid nitrogen), central tissue bank/repository, and proper permits to ship samples internationally (i.e., endangered/protected species) to a laboratory with the proper equipment to analyze the sample. Most importantly is the validation and usefulness of the biomarker and RIs as tools to help make the link between contaminant exposure and adverse health outcomes in individuals and populations.

Marine Mammal Cell Lines

Because live immune cells are difficult to acquire for many marine mammal species, relevant immortalized cell lines can be invaluable for in vitro risk assessment. Immortalized cell lines that retain important immunological properties have the potential to help further characterize the immune system in controlled laboratory settings. Although valuable, several established marine mammal cell lines are nonimmune, skin/fibroblast cell lines (Annalaura et al., 2012; Burkard et al., 2015; Wise et al., 2015; Wang et al., 2011). One marine mammal specific immune cell line does exist, the harbor seal 11B7501 lymphoma B cell line, and it has been used to characterize the immunotoxic effects of heavy metals and pharmaceuticals (Frouin et al., 2010a; Kleinert et al., 2018).

Marine Mammal Immune Cell Tissue Bank

Managed-care facilities routinely internally bank serum and/or plasma collected during medical examinations of housed or wild animals undergoing rehabilitation. Although not routine, live blood lymphocytes can also be isolated and cryopreserved. The inclusion of immune cells and serum/plasma to a controlled tissue bank should be considered and may help to conduct retrospective analyses. The National Marine Mammal Tissue Bank, jointly operated by the National Oceanic and Atmospheric Administration and the National Institute of Standards and Technology, currently maintains cryopreserved tissue samples (blubber, kidney, liver) from a variety of marine mammals, and it could be the appropriate facility to bank immune cells, serum, and plasma.

In Vitro Extrapolation

The many benefits of in vitro immunotoxicology have been discussed, and of these, none are more important than facilitating controlled studies for difficult to sample animal species. There are, however, important drawbacks of current in vitro experimentation techniques, particularly the lack of pharmacokinetic considerations (absorption, excretion, and biotransformation) and biologic realism/complexity. Advanced multitissue cell culture systems (Materne et al., 2015) may provide exciting future avenues to add complexity and cellular crosstalk for in vitro immunotoxicology assessments. Biokinetics are more difficult to mimic in the laboratory since culturing systems are different from in vivo conditions; thus, directly extrapolating in vitro toxicity data to in vivo animal exposure may not be accurate. This is because blood has many more proteins and lipids available for contaminants to bind to (relative to culture media), lowering the bioavailability of the contaminant(s) for target cells; in this sense, culture concentrations may not directly compare to serum or blood concentrations. Many approaches have been developed to generate biokinetic models of in vitro systems to extrapolate contaminant concentrations to in vivo equivalents (Armitage et al., 2014; Gülden et al., 2006; Hamon et al., 2015). This approach for in vitro immunotoxicology is needed to understand and improve its real-world applicability. That being said, these drawbacks of in vitro immunotoxicology may not be relevant to certain study designs, for example, when evaluating direct effects on specific cells, mechanisms of action, or when comparing effects between multiple species. Careful consideration is needed when designing a study to answer the questions of interest.

Multiple Stressors and Population-Level Effects

One of the great challenges for modern ecotoxicology is to understand and predict the biologic implications of exposure to multiple stressors. Contaminants are one of many potential natural and anthropogenic threats to an animal's well-being, and these include hundreds of individual compounds interacting in often complex ways. Immune assessments in naturally exposed marine mammals will inherently report the effects of the complex contaminant mixture, while in vitro experiments are more limited in this context. A few studies have assessed the impact of individual PCB congener mixtures (e.g., Levin et al., 2007; Mori et al., 2008) or commercial mixtures of PCBs or PBDEs (Hammond et al., 2005; Wirth et al., 2014a), but realistic contaminant cocktails are difficult to synthesize. A new method to rapidly extract contaminants from large samples of marine mammal blubber using a freeze-filtration method to retain contaminants but remove lipids has been recently reported and should facilitate the generation of realistic animal-derived cocktails that can be used for in vitro experiments (Desforges et al., 2017). Although this approach cannot provide information of which individual compound is causing effects, it does provide a unique, species-specific method to assess immunotoxicity in marine mammal species that are difficult to study in vivo.

Assessing the risk of immunotoxicity in a broader framework of multiple stressors is a critical challenge for future toxicologists, ecologists, and immunologists. Marine mammals in the wild are exposed to a wide range of possible stressors like anthropogenic noise, climate change, oil spills, ship strikes, decreasing food availability or quality, harmful algal blooms, marine debris, and infectious diseases. How these individually impact animal health is a difficult task on its own, but researchers must develop the tools necessary to evaluate how different stressors interact and influence animal physiology and fitness. Ethical constraints will limit experimental studies with captive marine mammals, so most work in this field will need to utilize model laboratory species, innovative in vitro approaches, and creative modeling methodologies. Furthermore, greater effort is needed to link results measured in the laboratory to conservation-relevant metrics, including population-level consequences. Population-effect modeling will ultimately provide the best information as to the real threat posed by immune and other stressors (e.g., reproductive and carcinogenic effects) to populations of marine mammals.

CONCLUSIONS

The weight of evidence demonstrating that environmental contaminants can modulate the immune system of marine mammals continues to grow and has become extremely convincing. Captive feeding studies, in vitro experiments, and modeling simulations have corroborated findings in free-ranging marine mammals, suggesting environmental pollutants pose a serious threat to the immunological health and consequent susceptibility to disease in highly exposed populations of marine mammals. Additional work is still required to understand the mechanisms behind immunotoxic effects as well as the complex interplay in chemical mixtures and multiple-stressor environments. The availability of species-specific reagents and commercial immune assays suitable for marine mammal species continues to limit further innovative developments in marine mammal immunology; however, the increasing use of genomics and other advanced molecular tools in ecotoxicology and marine mammal science will surely provide exciting new avenues for marine mammal immunotoxicology for years to come.

REFERENCES

Abbas, A.K., Lichtman, A.H., Pillai, S., 2012. Basic Immunology: Functions and Disorders on the Immune System, fourth ed. Elsevier Inc., Philadelphia.

Aguilar, A., Borrell, A., 1994. Abnormally high polychlorinated biphenyl levels in striped dolphins (*Stenella coeruleoalba*) affected by the 1990–1992 Mediterranean epizootic. The Science of the Total Environment 154, 237–247.

Allen, L.A., Aderem, A., 1996. Mechanisms of phagocytosis. Current Opinion in Immunology 1996 (8), 36–40.

Annalaura, M., Spyropoulos, D.D., McFee, W.E., Newton, D.A., Baatz, J.E., 2012. Cryopreservation and in vitro culture of primary cell types from lung tissue of a stranded pygmy sperm whale (*Kogia breviceps*). Comparative Biochemistry and Physiology Part C: Toxicology and Pharmacology 155, 136–142.

Armitage, J.M., Wania, F., Arnot, J.A., 2014. Application of mass balance models and the chemical activity concept to facilitate the use of in vitro toxicity data for risk assessment. Environmental Science and Technology 48, 9770–9779.

Beckmen, K.B., Blake, J.E., Ylitalo, G.M., Stott, J.L., O'Hara, T.M., 2003. Organochlorine contaminant exposure and associations with hematological and humoral immune functional assays with dam age as a factor in free-ranging northern fur seal pups (*Callorhinus ursinus*). Marine Pollution Bulletin 46, 594–606.

Beineke, A., Siebert, U., Mclachlan, M., Bruhn, R., Thron, K., Failing, K., Muller, G., Baumgartner, W., 2005. Investigations of the potential influence of environmental contaminants on the thymus and spleen of harbor porpoises (*Phocoena phocoena*). Environmental Science and Technology 39, 3933–3938.

Beineke, A., Siebert, U., Wohlsein, P., Baumgärtner, W., 2010. Immunology of whales and dolphins. Veterinary Immunology and Immunopathology 133, 81–94.

Bogomolni, A.L., Bass, A.L., Fire, S., Jasperse, L., Levin, M., Nielsen, O., Waring, G., De Guise, S., 2016. Saxitoxin increases phocine distemper virus replication upon in-vitro infection in harbor seal immune cells. Harmful Algae 51, 89–96.

Brown, T.M., Ross, P.S., Reimer, K.J., Veldhoen, N., Danger, N.J., Fisk, A.T., Helbing, C.C., 2014. PCB related effects thresholds as derived through gene transcript profiles in locally contaminated ringed seals (*Pusa hispida*). Environmental Science and Technology 48, 12952–12961.

Burkard, M., Whitworth, D., Schirmer, K., Nash, S.B., 2015. Establishment of the first humpback whale fibroblast cell lines and their application in chemical risk assessment. Aquatic Toxicology 167, 240–247.

Colgrove, G.S., 1978. Stimulation of lymphocytes from a dolphin (*Tursiops truncatus*) by phytomitogens. American Journal of Veterinary Research 39, 141–144.

Dahlgren, C., Karlsson, A., 1999. Respiratory burst in human neutrophils. Journal of Immunological Methods 232, 3–14.

Das, K., Siebert, U., Gillet, A., Dupont, A., Di-Poï, C., Fonfara, S., Mazzucchelli, G., De Pauw, E., De Pauw-Gillet, M.-C., 2008. Mercury immune toxicity in harbour seals: links to in vitro toxicity. Environmental Health 7, 52.

De Guise, S., Flipo, D., Boehm, J.R., Martineau, D., Béland, P., Fournier, M., 1995a. Immune functions in beluga whales (*Delphinapterus leucas*): evaluation of phagocytosis and respiratory burst with peripheral blood leukocytes using flow cytometry. Veterinary Immunology and Immunopathology 47, 351–362.

De Guise, S., Martineau, D., Beland, P., Fournier, M., 1995b. Possible mechanisms of action of environmental contaminants on St. Lawrence beluga whales (*Deiphinapterus leucas*). Environmental Health Perspectives 103, 73–77.

De Guise, S., Bernier, J., Dufresne, M.M., Martineau, D., Pierre, B., Fournier, M., 1996a. Immune functions in beluga whales (*Delphinapterus leucas*): evaluation of mitogen-induced blastic transformation of lymphocytes from peripheral blood, spleen and thymus. Veterinary Immunology and Immunopathology 50, 117–126.

De Guise, S., Bernier, J., Martineau, D., Beland, P., Fournier, M., 1996b. Effects on in vitro exposure of beluga whale splenocytes and thymocytes to heavy metals. Environmental Toxicology and Chemistry 15, 1357–1364.

De Guise, S., Ross, P.S., Osterhaus, A.D.M.E., Martineau, D., Beland, P., Fournier, M., 1997. Immune functions in beluga whales Delphinapterus leucas: evaluation of natural killer cell activity. Veterinary Immunology and Immunopathology 58, 345–354.

De Guise, S., Martineau, D., Beland, P., Fournier, M., 1998. Effects of in vitro exposure of beluga whale leukocytes to selected organochlorines. Journal of Toxicology and Environmental Health, Part A 55, 479–493.

de Swart, R.L., Kluten, R.M., Huizing, C.J., Vedder, L.J., Reijnders, P.J., Visser, I.K., UytdeHaag, F.G., Osterhaus, A.D., 1993. Mitogen and antigen induced B and T cell responses of peripheral blood mononuclear cells from the harbour seal (*Phoca vitulina*). Veterinary Immunology and Immunopathology 37, 217–230.

de Swart, R.L., Ross, P.S., Vedder, L.J., Timmerman, H.H., Van Loveren, H., Vos, J.G., Reijnders, P.J.H., Osterhaus, A.D.M.E., 1994. Impairment of immune function in harbor seals (*Phoca vitulina*) feeding on fish from polluted waters. Ambio 23, 155–159.

de Swart, R.L., Ross, P.S., Vos, J.G., Osterhaus, A.D., 1996. Impaired immunity in harbour seals (*Phoca vitulina*) exposed to bioaccumulated environmental contaminants: review of a long-term feeding study. Environmental Health Perspectives 104 (Suppl.), 823–828.

Desforges, J.-P.W., Sonne, C., Levin, M., Siebert, U., De Guise, S., Dietz, R., 2016. Immunotoxic effects of environmental pollutants in marine mammals. Environment International 86, 126–139.

Desforges, J.-P., Eulaers, I., Periard, L., Sonne, C., Dietz, R., Letcher, R.J., 2017. A rapid analytical method to quantify complex organohalogen contaminant mixtures in large samples of high lipid mammalian tissues. Chemosphere 176, 243–248.

Dietz, R., Heide-jorgensen, M.P., Harkonen, T., 1989. Mass deaths of harbor seals (*Phoca vitulina*) in Europe. Ambio 18, 258–264.

Dietz, R., Sonne, C., Basu, N., Braune, B., Hara, T.O., Letcher, R.J., Scheuhammer, T., Andersen, M., Andreasen, C., Andriashek, D., Asmund, G., Aubail, A., Baagøe, H., Born, E.W., Chan, H.M., Derocher, A.E., Grandjean, P., Knott, K., Kirkegaard, M., Krey, A., Lunn, N., Messier, F., Obbard, M., Olsen, M.T., Ostertag, S., Peacock, E., Renzoni, A., Rigét, F.F., Utne, J., Stern, G., Stirling, I., Taylor, M., Wiig, Ø., Wilson, S., Aars, J., 2013. What are the toxicological effects of mercury in Arctic biota? The Science of the Total Environment 443, 775–790.

Dietz, R., Gustavson, K., Sonne, C., Desforges, J.-P., Rigét, F.F., Pavlova, V., McKinney, M. a., Letcher, R.J., 2015. Physiologically-based pharmacokinetic modelling of immune, reproductive and carcinogenic effects from contaminant exposure in polar bears (*Ursus maritimus*) across the Arctic. Environmental Research 140, 45–55.

DiMolfetto-Landon, L., Erickson, K.L., Blanchard-Channell, M., Jeffries, S.J., Harvey, J.T., Jessup, D.A., Ferrick, D.A., Stott, J.L., 1995. Blastogenesis and interleukin-2 receptor expression assays in the harbor seal (*Phoca vitulina*). Journal of Wildlife Diseases 31, 150–158.

Duignan, P., Van Bressem, M.-F., Baker, J., Barbieri, M., Colegrove, K., De Guise, S., de Swart, R., Di Guardo, G., Dobson, A., Duprex, W., Early, G., Fauquier, D., Goldstein, T., Goodman, S., Grenfell, B., Groch, K., Gulland, F., Hall, A., Jensen, B., Lamy, K., Matassa, K., Mazzariol, S., Morris, S., Nielsen, O., Rotstein, D., Rowles, T., Saliki, J., Siebert, U., Waltzek, T., Wellehan, J., 2014. Phocine distemper virus: current knowledge and future directions. Viruses 6, 5093–5134.

Dupont, A., Siebert, U., Covaci, A., Weijs, L., Eppe, G., Debier, C., De Pauw-Gillet, M.-C., Das, K., 2013. Relationships between in vitro lymphoproliferative responses and levels of contaminants in blood of free-ranging adult harbour seals (*Phoca vitulina*) from the North Sea. Aquatic Toxicology 142–143, 210–220.

Dupont, A., De Pauw-Gillet, M.-C., Schnitzler, J., Siebert, U., Das, K., 2015. Effects of methylmercury on harbour seal peripheral blood leucocytes in vitro studied by electron microscopy. Archives of Environmental Contamination and Toxicology 70, 133–142.

Fair, P.A., Romano, T., Schaefer, A.M., Reif, J.S., Bossart, G.D., Houde, M., Muir, D., Adams, J., Rice, C., Hulsey, T.C., Peden-Adams, M., 2013. Associations between perfluoroalkyl compounds and immune and clinical chemistry parameters in highly exposed bottlenose dolphins (*Tursiops truncatus*). Environmental Toxicology and Chemistry 32, 736–746.

Flewelling, L.J., Naar, J.P., Abbott, J.P., Baden, D.G., Barros, N.B., Bossart, G.D., Bottein, M.D., Hammond, D.G., Haubold, E.M., Heil, C.A., Henry, M.S., Jacocks, H.M., Leighfield, T.A., Pierce, R.H., Pitchford, T.D., Rommel, S.A., Scott, P.S., Steidinger, K.A., Truby, E.W., Van Dolah, F.M., Landsberg, J.H., 2005. Brevetoxicosis: red tides and marine mammal mortalities. Nature 435, 755–756.

Friedrichs, K.R., Harr, K.E., Freeman, K.P., Szladovits, B., Walton, R.M., Barnhart, K.F., Blanco-Chavez, J., 2013. ASVCP reference interval guidelines: determination of de novo reference intervals in veterinary species and other related topics. Veterinary Clinical Pathology 41, 441–453.

Frouin, H., Fortier, M., Fournier, M., 2010a. Toxic effects of various pollutants in 11B7501 lymphoma B cell line from harbour seal (*Phoca vitulina*). Toxicology 270, 66–76.

Frouin, H., Lebeuf, M., Hammill, M., Masson, S., Fournier, M., 2010b. Effects of individual polybrominated diphenyl ether (PBDE) congeners on harbour seal immune cells in vitro. Marine Pollution Bulletin 60, 291–298.

Frouin, H., Loseto, L.L., Stern, G.A., Haulena, M., Ross, P.S., 2012. Mercury toxicity in beluga whale lymphocytes: limited effects of selenium protection. Aquatic Toxicology 109, 185–193.

Frouin, H., Haulena, M., Akhurst, L.M.F., Raverty, S., Ross, P.S., 2013. Immune status and function in harbor seal pups during the course of rehabilitation. Veterinary Immunology and Immunopathology 155, 98–109.

Gebhard, E., Levin, M., Bogomolni, A., De Guise, S., 2015. Immunomodulatory effects of brevetoxin (PbTx-3) upon in vitro exposure in bottlenose dolphins (*Tursiops truncatus*). Harmful Algae 44, 54–62.

Gülden, M., Dierickx, P., Seibert, H., 2006. Validation of a prediction model for estimating serum concentrations of chemicals which are equivalent to toxic concentrations in vitro. Toxicology in Vitro 20, 1114–1124.

Gulland, F.M.D., Hall, A.J., 2007. Is marine mammal health deteriorating? Trends in the global reporting of marine mammal disease. EcoHealth 4, 135–150.

Hall, A.J., Law, R.J., Wells, D.E., Harwood, J., Ross, H.M., Kennedy, S., Allchin, C.R., Campbell, L.A., Pomeroy, P.P., 1992. Organochlorine levels in common seals (*Phoca vitulina*) which were victims and survivors of the 1988 phocine distemper epizootic. The Science of the Total Environment 115, 145–162.

Hall, A., Pomeroy, P., Green, N., Jones, K., Harwood, J., 1997. Infection, haematology and biochemistry in grey seal pups exposed to chlorinated biphenyls. Marine Environmental Research 43, 81–98.

Hall, A.J., Hugunin, K., Deaville, R., Law, R.J., Allchin, C.R., Jepson, P.D., 2006. The risk of infection from polychlorinated biphenyl exposure in the harbor porpoise (*Phocoena phocoena*): a case–control approach. Environmental Health Perspectives 114, 704–711.

Hammond, J.A., Hall, A.J., Dyrynda, E.A., 2005. Comparison of polychlorinated biphenyl (PCB) induced effects on innate immune functions in harbour and grey seals. Aquatic Toxicology 74, 126–138.

Hamon, J., Renner, M., Jamei, M., Lukas, A., Kopp-Schneider, A., Bois, F.Y., 2015. Quantitative in vitro to in vivo extrapolation of tissues toxicity. Toxicology in Vitro 1–14.

Härkönen, T., Dietz, R., Reijnders, P., Teilmann, J., Harding, K., Hall, A., Brasseur, S., Siebert, U., Goodman, S.J., Jepson, P.D., Rasmussen, T.D., Thompson, P., 2006. A review of the 1988 and 2002 phocine distemper virus epidemics in European harbour seals. Diseases of Aquatic Organisms 68, 115–130.

Helle, E., 1980. Lowered reproductive capacity in female ringed seals (*Pusa hispida*) in the Bothnian Bay, northern Baltic Sea, with special reference to uterine occlusions. Annales Zoologici Fennici 17, 147–158.

Helle, E., Olsson, M., Jensen, S., 1976. PCB Levels Correlated with Pathological changes in seal uteri. Ambio 5, 261–262.

Jepson, P.D., Bennett, P.M., Allchin, C.R., Law, R.J., Kuiken, T., Baker, J.R., Rogand, E., James, K.K., 1999. Investigating potential associations between chronic exposure to polychlorinated biphenyls and infectious disease mortality in harbour porpoises from England and Wales. The Science of the Total Environment 7, 339–348.

Kakuschke, A., Valentine-Thon, E., Griesel, S., Fonfara, S., Siebert, U., Prange, A., 2005. Immunological impact of metals in harbor seals (*Phoca vitulina*) of the North Sea. Environmental Science and Technology 39, 7568–7575.

Kakuschke, A., Valentine-Thon, E., Fonfara, S., Griesel, S., Siebert, U., Prange, A., 2006. Metal sensitivity of marine mammals: a case study of a gray seal (*Halichoerus grypus*). Marine Mammal Science 22, 985–996.

Kakuschke, A., Valentine-Thon, E., Griesel, S., Rosenberger, T., Mundry, R., Siebert, U., Prange, A., 2008. Blood metal levels and metal-influenced immune functions of harbour seals in captivity. Marine Pollution Bulletin 56, 764–769.

Kakuschke, A., Valentine-Thon, E., Fonfara, S., Kramer, K., Prange, A., 2009. Effects of methyl-, phenyl-, ethylmercury and mercurychloride on immune cells of harbor seals (*Phoca vitulina*). Journal of Environmental Sciences 21, 1716–1721.

Keogh, M.J., Spoon, T., Ridgway, S.H., Jensen, E., Van Bonn, W., Romano, T.A., 2011. Simultaneous measurement of phagocytosis and respiratory burst of leukocytes in whole blood from bottlenose dolphins (*Tursiops truncatus*) utilizing flow cytometry. Veterinary Immunology and Immunopathology 144, 468–475.

Kirkegaard, M., Sonne, C., Leifsson, P.S., Dietz, R., Born, E.W., Muir, D.C.G., Letcher, R.J., 2005. Histology of selected immunological organs in polar bear (*Ursus maritimus*) from East Greenland in relation to concentrations of organohalogen contaminants. The Science of the Total Environment 341, 119–132.

Kleinert, C., Lacaze, E., Mounier, M., De Guise, S., Fournier, M., **118**, 2018. Immunotoxic effects of single and combined pharmaceuticals exposure on harbor seal (*Phoca vitulina*) B lymphoma cell line. Marine Pollution Bulletin 237–247.

Kuby, J., 2012. Immunology, seventh ed. W.H. Freeman and Company, New York, NY.

Lahvis, G.P., Wells, R.S., Kuehl, D.W., Stewart, J.L., Rhinehart, H.L., Via, C.S., 1995. Decreased lymphocyte responses in free-ranging bottlenose dolphins (*Tursiops truncatus*) are associated with increased concentrations of PCBs and DDT in peripheral blood. Environmental Health Perspectives 103, 67–72.

Letcher, R.J., Bustnes, J.O., Dietz, R., Jenssen, B.M., Jørgensen, E.H., Sonne, C., Verreault, J., Vijayan, M.M., Gabrielsen, G.W., 2010. Exposure and effects assessment of persistent organohalogen contaminants in arctic wildlife and fish. The Science of the Total Environment 408, 2995–3043.

Levin, M., Morsey, B., Mori, C., Guise, S., 2004. Specific non-coplanar PCB-mediated modulation of bottlenose dolphin and beluga whale phagocytosis upon in vitro exposure. Journal of Toxicology and Environmental Health, Part A 67, 1517–1535.

Levin, M., De Guise, S., Ross, P.S., 2005a. Association between lymphocyte proliferation and polychlorinated biphenyls in free-ranging harbor seal (*Phoca vitulina*) pups from British Columbia, Canada. Environmental Toxicology and Chemistry 24, 1247–1252.

Levin, M., Morsey, B., Mori, C., Nambiar, P.R., de Guise, S., 2005b. PCBs and TCDD, alone and in mixtures, modulate marine mammal but not B6C3F1 mouse leukocyte phagocytosis. Journal of Toxicology and Environmental Health, Part A 68, 635–656.

Levin, M., Morsey, B., De Guise, S., 2007. Modulation of the respiratory burst by organochlorine mixtures in marine mammals, humans, and mice. Journal of Toxicology and Environmental Health, Part A 70, 73–83.

Levin, M., Joshi, D., Draghi, A., Gulland, F.M., Jessup, D., De Guise, S., 2010. Immunomodulatory effects upon in vitro exposure of California sea lion and southern sea otter peripheral blood leukocytes to domoic acid. Journal of Wildlife Diseases 46, 541–550.

Levin, M., Gebhard, E., Jasperse, L., Desforges, J.-P., Dietz, R., Sonne, C., Eulaers, I., Covaci, A., Bossi, R., De Guise, S., 2016. Immunomodulatory effects of exposure to polychlorinated biphenyls and perfluoroalkyl acids in East Greenland ringed seals (*Pusa hispida*). Environmental Research 151, 244–250.

Lie, E., Larsen, H.J.S., Larsen, S., Johansen, G.M., Derocher, A.E., Lunn, N.J., Norstrom, R.J., Wiig, Ø., Skaare, J.U., 2004. Does high organochlorine (OC) exposure impair the resistance to infection in polar bears (*Ursus maritimus*)? Part I: Effect of OCs on the humoral immunity. Journal of Toxicology and Environmental Health, Part A 67, 555–582.

Lie, E., Larsen, H.J.S., Larsen, S., Johansen, G.M., Derocher, A.E., Lunn, N.J., Norstrom, R.J., Wiig, O., Skaare, J.U., 2005. Does high organochlorine (OC) exposure impair the resistance to infection in polar bears (*Ursus maritimus*)? Part II: possible effect of OCs on mitogen- and antigen-induced lymphocyte proliferation. Journal of Toxicology and Environmental Health, Part A 68, 457–484.

Luebke, R., House, R., Kimber, I., 2007. Immunotoxicology and Immunopharmacology, third ed. CRC Press, Florida.

Luster, M.I., Portier, C., Pait, D., White, K., Gennings, C., Munson, A.E., Rosenthal, G.J., 1992. Risk Assessment in Immunotoxicology: I. Sensitivity and Predictability of immune tests. Fundamental and Applied Toxicology 18, 200–210.

Luster, M., Portier, C., Pait, D.G., Rosenthal, G., Germolec, D., Corsini, E., Blaylock, B., Pollock, P., Kouchi, Y., Craig, W., White, K., Munson, A., Comment, C., 1993. Risk assessment in immunotoxicology: II. Relationships between immune and host resistance tests. Fundamental and Applied Toxicology 21, 71–82.

Martineau, D., De Guise, S., Fournier, M., Shugart, L., Girard, C., Lagacé, A., Béland, P., 1994. Pathology and toxicology of beluga whales from the St. Lawrence Estuary, Quebec, Canada. Past, present and future. The Science of the Total Environment 154, 201–215.

Materne, E.M., Ramme, A.P., Terrasso, A.P., Serra, M., Alves, P.M., Brito, C., Sakharov, D.A., Tonevitsky, A.G., Lauster, R., Marx, U., 2015. A multi-organ chip co-culture of neurospheres and liver equivalents for long-term substance testing. Journal of Biotechnology 205, 36–46.

Moretta, L., Moretta, A., 2004. Unravelling natural killer cell function: triggering and inhibitory human NK receptors. The EMBO Journal 23, 255–259.

Moretta, L., Mingari, M.C., Bottino, C., Pende, D., Biassoni, R., Moretta, A., 2003. Cellular and molecular basis of natural killer and natural killer-like activity. Immunology Letters 88, 89–93.

Mori, C., Morsey, B., Levin, M., Gorton, T.S., De Guise, S., 2008. Effects of organochlorines, individually and in mixtures, on B-cell proliferation in marine mammals and mice. Journal of Toxicology and Environmental Health, Part A 71, 266–275.

Mos, L., Morsey, B., Jeffries, S.J., Yunker, M.B., Raverty, S., De Guise, S., Ross, P.S., 2006. Chemical and biological pollution contribute to the immunological profiles of free-ranging harbor seals. Environmental Toxicology and Chemistry 25, 3110–3117.

Moser, M., Leo, O., 2010. Key concepts in immunology. Vaccine 28, 2–13.

Mumford, D.M., Stockman, G.D., Barsales, P.B., Whitman, T., Wilbur, J.R., 1975. Lymphocyte transformation studies of sea mammal blood. Experientia 31, 498–500.

Murata, H., Shimada, N., Yoshioka, M., 2004. Current research on acute phase proteins in veterinary diagnosis: an overview. The Veterinary Journal 168, 28–40.

Nakata, H., Sakakibara, A., Kanoh, M., Kudo, S., Watanabe, H., Nagai, N., Miyazaki, N., Asano, Y., Tanabe, S., 2002. Evaluation of mitogen-induced responses in marine mammal and human lymphocytes by in-vitro exposure of butyltins and non-ortho coplanar PCBs. Environmental Pollution 120, 245–253.

Neale, J.C.C., Kenny, T.P., Tjeerdema, R.S., Gershwin, M.E., 2005. PAH- and PCB-induced alterations of protein tyrosine kinase and cytokine gene transcription in harbor seal (*Phoca Vitulina*) PBMC. Clinical and Developmental Immunology 12, 91–97.

Noda, K., Aoki, M., Asaki, H., Shimada, T., Ohashi, F., 2003. Evaluation of the polymorphonuclear cell functions of bottlenose dolphins. Journal of Veterinary Medical Science 65, 727–729.

Nordoy, E.S., Thoresen, S.I., 2002. Reference values for serum biochemical parameters in free-ranging harp seals. Veterinary Clinical Pathology 31, 98–105.

Norman, S.A., Beckett, L.A., Miller, W.A., St Leger, J., Hobbs, R.C., 2013. Variation in hematologic and serum biochemical values of belugas (*Delphinapterus leucas*) under managed care. Journal of Zoo and Wildlife Medicine 44, 376–388.

Nyman, M., Bergknut, M., Fant, M.L., Raunio, H., Jestoi, M., Bengs, C., Murk, A., Koistinen, J., Bäckman, C., Pelkonen, O., Tysklind, M., Hirvi, T., Helle, E., 2003. Contaminant exposure and effects in Baltic ringed and grey seals as assessed by biomarkers. Marine Environmental Research 55, 73–99.

Olsson, M., Karlsson, B., Ahnland, E., 1994. Diseases and environmental contaminants in seals from the Baltic and the Swedish west coast. The Science of the Total Environment 154, 217–227.

O'Shea, J., Ortaldo, J.R., 1992. The Natural Killer Cell. IRL Press Oxford.

Pellissó, S.C., Muñoz, M.J., Carballo, M., Sánchez-Vizcaíno, J.M., 2008. Determination of the immunotoxic potential of heavy metals on the functional activity of bottlenose dolphin leukocytes in vitro. Veterinary Immunology and Immunopathology 121, 189–198.

Roos, D., van Bruggen, R., Meischl, C., 2003. Oxidative killing of microbes by neutrophils. Microbes and Infection 5, 1307–1315.

Roos, A.M., Bäcklin, B.M.V.M., Helander, B.O., Rigét, F.F., Eriksson, U.C., 2012. Improved reproductive success in otters (*Lutra lutra*), grey seals (*Halichoerus grypus*) and sea eagles (*Haliaeetus albicilla*) from Sweden in relation to concentrations of organochlorine contaminants. Environmental Pollution 170, 268–275.

Ross, P.S., De Swart, R.L., Reijnders, P.J.H., van Loveren, H., Vos, J.G., Osterhaus, A.D.M.E., 1995. Contaminant-related suppression of delayed-type hypersensitivity and antibody responses in harbor seals fed herring from the Baltic Sea. Environmental Health Perspectives 103, 162–167.

Ross, P.S., De Swart, R.L., Van Loveren, H., Osterhaus, A.D.M.E., Vos, J.G., 1996a. The immunotoxicity of environmental contaminants to marine wildlife: a review. Annual Review of Fish Diseases 6, 151–165.

Ross, P.S., DeSwart, R., Addison, R., Van Loveren, H., Vos, J., Osterhaus, A., 1996b. Contaminant-induced immunotoxicity in harbour seals: wildlife at risk? Toxicology 112, 157–169.

Ross, P.S., DeSwart, R.L., Timmerman, H.H., Reijnders, P.J.H., Vos, J.G., Van Loveren, H., Osterhaus, A.D., 1996c. Suppression of natural killer cell activity in harbour seals (*Phoca vitulina*) fed Baltic Sea herring. Aquatic Toxicology 34, 71–84.

Ross, P., Ellis, G.M., Ikonomou, M.G., Barrett-Lennard, L.G., Addison, R.F., 2000. High PCB concentrations in free-ranging pacific killer whales, *Orcinus orca*: effects of age, sex and dietary preference. Marine Pollution Bulletin 40, 504–515.

Routti, H., Arukwe, A., Jenssen, B.M., Letcher, R.J., Nyman, M., Bäckman, C., Gabrielsen, G.W., 2010. Comparative endocrine disruptive effects of contaminants in ringed seals (*Phoca hispida*) from Svalbard and the Baltic Sea. Comparative Biochemistry and Physiology Part C: Toxicology and Pharmacology 152, 306–312.

Scheerlinck, J.P., Yen, H.H., 2005. Veterinary applications of cytokines. Veterinary Immunology and Immunopathology 108, 17–22.

Schumacher, U., Heidemann, G., Skírnisson, K., Schumacher, W., Pickering, R.M., 1995. Impact of captivity and contamination level on blood parameters of harbour seals (*Phoca vitulina*). Comparative Biochemistry and Physiology Part A: Physiology 112, 455–462.

Schwacke, L.H., Zolman, E.S., Balmer, B.C., De Guise, S., George, R.C., Hoguet, J., Hohn, A.A., Kucklick, J.R., Lamb, S., Levin, M., Litz, J.A., McFee, W.E., Place, N.J., Townsend, F.I., Wells, R.S., Rowles, T.K., 2012. Anaemia, hypothyroidism and immune suppression associated with polychlorinated biphenyl exposure in bottlenose dolphins (*Tursiops truncatus*). Proceedings of the Royal Society of London B: Biological Sciences 279, 48–57.

Shiraishi, R., Itou, T., Sugisawa, H., Shoji, Y., Endo, T., Sakai, T., 2002. The respiratory burst activity of bottlenose dolphin neutrophils elicited by several stimulants. Journal of Veterinary Medical Science 64, 711–714.

Silkworth, J., Antrim, L., 1985. Relationship between Ah receptor-mediated polychiorinated biphenyl (PCB)-induced humoral immunosuppression and thymic atrophy. Journal of Pharmacology and Experimental Therapeutics 235, 606–611.

Sonne, C., 2010. Health effects from long-range transported contaminants in Arctic top predators: an integrated review based on studies of polar bears and relevant model species. Environment International 36, 461–491.

Sonne, C., Gustavson, K., Eulaers, I., Desforges, J.-P., Letcher, R.J., Rigét, F.F., Styrishave, B., Dietz, R., 2016. Risk evaluation of the Arctic environmental POP exposure based on critical body residue and critical daily dose using captive Greenland sledge dogs (*Canis familiaris*) as surrogate species. Environment International 88, 221–227.

Sørmo, E.G., Larsen, H.J.S., Johansen, G.M., Skaare, J.U., Jenssen, B.M., 2009. Immunotoxicity of polychlorinated biphenyls (PCB) in free-ranging gray seal pups with special emphasis on dioxin-like congeners. Journal of Toxicology and Environmental Health, Part A 72, 266–276.

Suh, J., Jeon, Y.J., Kim, H.M., Kang, J.S., Kaminski, N.E., Yang, K.-H., 2002. Aryl hydrocarbon receptor-dependent inhibition of AP-1 activity by 2,3,7,8-tetrachlorodibenzo-p-dioxin in activated B cells. Toxicology and Applied Pharmacology 181, 116–123.

Thurmond, T.S., Staples, J.E., Silverstone, A.E., Gasiewicz, T.A., 2000. The aryl hydrocarbon receptor has a role in the in vivo maturation of murine bone marrow B lymphocytes and their response to 2,3,7,8-tetrachlorodibenzo-p-dioxin. Toxicology and Applied Pharmacology 165, 227–236.

Tryland, M., Krafft, B.A., Lydersen, C., Kovacs, K.M., Thoresen, S.I., 2006a. Serum chemistry values for free-ranging ringed seals (*Pusa hispida*) in Svalbard. Veterinary Clinical Pathology 35, 405–412.

Tryland, M., Thoresen, S.I., Kovacs, K.M., Lydersen, C., 2006b. Serum chemistry of free-ranging white whales (*Delphinapterus leucas*) in Svalbard. Veterinary Clinical Pathology 35, 199–203.

van Oss, C.J., 1986. Phagocytosis: an overview. Methods in Enzymology 132, 3–15.

Vos, J.G., Van Loveren, H., 1994. Developments of immunotoxicology methods in the rat and applications to the study of environmental pollutants. Toxicology in Vitro 8, 951–956.

Vos, J., Van Loveren, H., Wester, P., Vethaak, D., 1989. Toxic effects of environmental chemicals on the immune system. Trends in Pharmacological Sciences 10, 289–292.

Wang, J., Su, W., Wang, J., Xiao, W., Wang, D., 2011. Establishment and characterization of fibroblast cell lines from the skin of the Yangtze finless porpoise. In Vitro Cellular and Developmental Biology Animal 47, 618–630.

Wirth, J.R., Peden-Adams, M.M., White, N.D., Bossart, G.D., Fair, P., 2014a. In vitro exposure of DE-71, a penta-PBDE mixture, on immune endpoints in bottlenose dolphins (*Tursiops truncatus*) and B6C3F1 mice. Journal of AppliedToxicology 191–198.

Wirth, J.R., Peden-Adams, M.M., White, N.D., Bossart, G.D., Fair, P.A., 2014b. In vitro PFOS exposure on immune endpoints in bottlenose dolphins (*Tursiops truncatus*) and mice. Journal of Applied Toxicology 34, 658–666.

Wise, C.F., Wise, S.S., Thompson, W.D., Perkins, C., Wise, J.P., 2015. Chromium is elevated in fin whale (*Balaenoptera physalus*) skin tissue and is genotoxic to fin whale skin cells. Biological Trace Element Research 166, 108–117.

Chapter 13

Ecotoxicologic Stress in Arctic Marine Mammals, With Particular Focus on Polar Bears

Heli Routti[1], Bjørn M. Jenssen[2,3], Sabrina Tartu[1]
[1]Norwegian Polar Institute, Fram Centre, Tromsø, Norway; [2]Norwegian University of Science and Technology, Trondheim, Norway; [3]The University Centre in Svalbard, Longyearbyen, Norway

CONTAMINANTS IN ARCTIC MARINE MAMMALS

Transport of Contaminants to the Arctic

A variety of environmental contaminants are present in Arctic marine mammals, although most Arctic areas are far away from industrial sources. The presence of contaminants in the pristine Arctic is mostly due to long-range transport. Both air and ocean currents transport contaminants to the Arctic from southern industrialized areas. Polychlorinated biphenyls (PCBs), most chlorinated pesticides, and brominated flame retardants (BFRs) are mainly subject to long-range transport by air currents (Frank and Dugani, 2003; Wania and Mackay, 1993; Macdonald et al., 2005). Cold condensation in Arctic areas facilitates the deposition of atmospherically transported compounds to the surface, meaning that with reducing temperature, contaminants in the atmosphere condense onto surfaces such as water, snow, and ice (Ma, 2010; Wania and Mackay, 1993). Volatile precursors of perfluoroalkyl substances (PFAS) may also undergo long-range atmospheric transport and degrade to perfluoroalkyl sulfonates or carboxylates (PFSAs and PFCAs) in the atmosphere and snow (Ellis et al., 2004; Martin et al., 2006; D'eon et al., 2006; Taniyasu et al., 2013). The main transport routes for beta-hexachlorocyclohexane (β-HCH) and PFASs, which are directly released into the aquatic system, are ocean currents (Li and Macdonald, 2005) (Zhao et al., 2012; Wania, 2007). Mercury (Hg) is mainly transported by air currents, followed by ocean currents, coastal erosion, and riverine inputs (Outridge et al., 2008). Local pollution may also be a source for contaminants in the Arctic marine environment. For example, poor waste handling at a remote military station at Saglek in Labrador, Canada, has polluted the surrounding

Marine Mammal Ecotoxicology. https://doi.org/10.1016/B978-0-12-812144-3.00013-9

areas with PCBs (Kuzyk et al., 2005), which has led to higher PCB exposure at all trophic levels, including marine mammals (Brown et al., 2015).

Biomagnification and Biotransformation

Environmental pollutants, in particular persistent organic pollutants (POPs) and Hg, are subject to biomagnification in Arctic marine food webs (Fisk et al., 2001; Atwell et al., 1998; Dehn et al., 2006). Biomagnification of POPs and their chemical profile in the Arctic marine food web depends on the physicochemical properties of the compounds, food web structure, and the biotransformation capacity of an organism (Fisk et al., 2001; Hallanger et al., 2011; McKinney et al., 2012). Compounds that easily biomagnify in marine mammal food webs are those that have a high octanol-water partitioning coefficient (lipid-soluble) (Fisk et al., 2001), or those that have a moderate octanol-water partitioning coefficient and high octanol-air partitioning coefficient (Kelly et al., 2007). Variation in food web structures also explains contaminant accumulation to Arctic marine mammals (McKinney et al., 2011c; Routti et al., 2012; Hobson et al., 2002). In general, biomagnification of, in particular, lipophilic pollutants is efficient in Arctic marine food webs, as Arctic marine food webs are relatively long, and lipids are rapidly transferred from primary producers to top predators (Welch et al., 1992; Falk-Petersen et al., 2009).

Biotransformation capacity, which ultimately defines the contaminant concentration in animals, varies largely between species. Several lipophilic POPs including PCBs, dichlorodiphenyltrichloroethane (DDT), and chlordanes may be biotransformed at least to some extent in Arctic marine mammals (Letcher et al., 2000). POPs induce xenobiotic-metabolizing phase I (cytochrome P450, i.e., CYP) and conjugating phase II enzymes (Letcher et al., 2000). CYPs belong to a large superfamily of enzymes that are responsible for catalyzing the synthesis/metabolism of endogenous compounds and oxidation or dehalogenation of halogenated exogenous compounds. The role of phase II enzymes is to catalyze the conjugation of, for example, glutathione, sulphate, and glucuronic acid to endo- and exogenous compounds. In Arctic marine mammals, the main biotransformation products of PCBs are hydroxylated (OH) and methylsulfonyl ($MeSO_2$) PCBs, whereas DDT is biotransformed to DDE and further to $MeSO_2$-DDE, and chlordane to oxychlordane (Letcher et al., 1998, 2009; McKinney et al., 2006b; Routti et al., 2008, 2009a; Wolkers et al., 2000). Both debromination and oxidation are biotransformation pathways for PBDEs (Krieger et al., 2016; Hakk and Letcher, 2003), although the tissue metabolite concentrations are very low in Arctic marine mammals (McKinney et al., 2006b; Routti et al., 2009b; Verreault et al., 2005a). Biotransformation capacity has been studied in Arctic marine mammals using different approaches: characterization of CYP enzymes (McKinney et al., 2004; White et al., 1994; Bandiera et al., 1995; Letcher et al., 1996; Nyman et al., 2001; Wolkers et al., 1998b), in vitro microsomal assays (Letcher et al., 2014; McKinney et al., 2006a, 2011a; Krieger et al., 2016), PCB congener profiles in predator/prey (Letcher et al., 1998, 2009; Wolkers et al., 1998a, 2000, 2004b, 2006), and ratios of parent compounds and their biotransformation products in tissues (e.g., OH-PCB/PCBs or $MeSO_2$-PCBs/PCBs) (Routti et al., 2008, 2009a; Gebbink et al., 2008; Hoekstra et al., 2003). These studies indicate that the biotransformation capacity is generally high in polar bears (*Ursus maritimus*) compared to seals and is lowest in whales (Fig. 13.1). However, not all compounds are subject to biotransformation. PFSAs and PFCAs are terminal products, whereas, for example, a highly produced precursor of perfluorooctane sulfonate (PFOS) may be biotransformed to an intermediate product (i.e., perfluorooctane sulfonamide, FOSA), and further to PFOS in Arctic mammals (Letcher et al., 2014; Gebbink et al., 2016).

Species, which feed at the top of the Arctic marine food webs, such as polar bears and killer whales (*Orcinus orca*), are among Arctic marine mammals with the highest concentrations of lipophilic POPs (Fig. 13.1) (Norstrom and Muir, 1994; Wolkers et al., 2007; Letcher et al., 2010). However, also species that feed at relatively high trophic level and have a low capacity to biotransform xenobiotics, such as white whales (*Delphinapterus leucas*) and narwhals (*Monodon monoceros*), show very high concentrations of lipophilic POPs (Wolkers et al., 2006; Andersen et al., 2001; Stern et al., 2005). On the other hand, concentrations of PFSAs and PFCAs, which are not subject to biotransformation, are clearly higher in polar bears compared to other Arctic marine mammals (Fig. 13.1) (Gebbink et al., 2016; Smithwick et al., 2005; Kelly et al., 2009; Tomy et al., 2009).

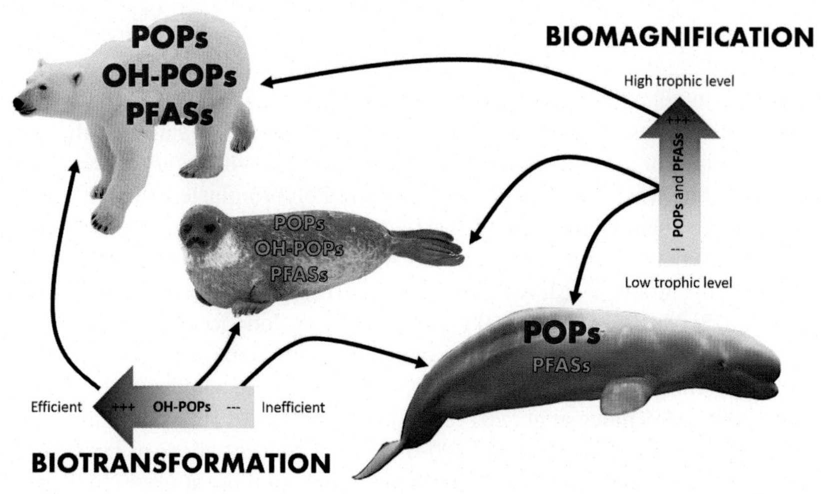

FIGURE 13.1 Biomagnification and biotransformation interplay to predict pollutant concentrations in marine mammals. Lipophilic persistent organic pollutants (POPs) and perfluoroalkyl sulfonates and carboxylates (here PFASs) are more concentrated at higher trophic levels than at lower trophic levels. Species efficient in biotransforming POPs have higher concentrations of metabolites, whereas perfluoroalkyl sulfonates and carboxylates are not biotransformed. *Yellow to red color scale* represents increasing pollutant concentrations; *black arrows* indicate relative trophic position and biotransformation efficiency in polar bears, white whales, and ringed seals.

Monitoring Contaminants in Arctic Marine Mammals

Contaminants are regularly monitored in Arctic marine mammals as part of national and international monitoring programs (e.g., Arctic Monitoring and Assessment Program, Northern Contaminants Program [Canada], Environmental Monitoring of Svalbard and Jan Mayen). In addition, samples from marine mammals are used to screen the presence of current-use chemicals in the Arctic environment. Detection of a compound in Arctic marine mammals gives input to international conventions, such as Stockholm Convention (www.pops.int), a global treaty to protect humans and the environment from POPs. The presence of an anthropogenic chemicals in Arctic marine mammals indicates that a compound is persistent and subject for long-range transport, bioaccumulation, and biomagnification. These, in addition to toxicity, are criteria to list compounds under the Stockholm Convention.

Compounds, which have been regulated at the national level since the 1970s and globally by Stockholm Convention since 2004 (Stockholm Convention, 2001), still make up a major portion of lipophilic contaminants found in Arctic marine mammals. For example, concentrations of ∑PCBs (1800–10,500 ng/g lipid weight [lw]), followed by chlordanes, are highest among legacy pollutants in adipose tissue of polar bears across subpopulations from Alaska to Svalbard (McKinney et al., 2011b). PCBs, followed by DDTs, are the main compounds found in ringed seal (*Pusa hispida*) blubber from most localities across the Arctic (Muir et al., 2013; Wolkers et al., 2008; Vorkamp et al., 2008; Savinov et al., 2011), and the PCB concentrations are ~10 times higher in polar bear versus ringed seal tissues (Letcher et al., 2009). The relative contribution of chlorinated pesticides is much higher in white whales and narwhals compared to polar bears and ringed seals (Wolkers et al., 2006; Villanger et al., 2011b; Stern et al., 2005). The most recent study from Svalbard reported that concentrations of the main groups of POPs were ~1000–4000 ng/g lw in adipose tissue of white whales (Villanger et al., 2011b). Polybrominated diphenyl ethers (PBDEs) are quantitatively the most abundant BFRs in polar bears, ringed seals, and white whales, but the concentrations are orders of magnitude lower than ∑PCB concentrations (McKinney et al., 2011b; Villanger et al., 2011b; Muir et al., 2013; Vorkamp et al., 2011). Hexabromocyclododecane (HBCDD) has also been reported in Arctic marine mammals (e.g., polar bears, ringed seals, and white whales) with concentrations up to 50 ng/g lw in adipose tissue (Sørmo et al., 2006; McKinney et al., 2011b; Villanger et al., 2011b; Muir et al., 2013; Vorkamp et al., 2011; Tomy et al., 2009), whereas concentrations of novel BFRs are considerably lower or not detected according to most studies (McKinney et al., 2011b; Vorkamp et al., 2015; Muir et al., 2013; Vorkamp and Riget, 2014; Villanger et al., 2013; Tartu et al., 2017a). However, blubber concentrations of tetrabromoethylcyclohexane (TBECH) have been reported to exceed HBCDD concentrations in white whales from the Canadian Arctic (Tomy et al., 2008). Concentrations of dechlorane plus and other dechloranes have been detected at low concentrations (<2 ng/g ww) in adipose tissue of ringed seals, polar bears, and/or white whales (Shen et al., 2012; Vorkamp et al., 2015). Short- and medium-chain chlorinated paraffins have also been reported in marine mammal blubber samples from the Canadian Arctic and Greenland at relatively high concentrations (~200–500 ng/g ww and 50–150 ng/g lw) (Tomy et al., 2000; Muir et al., 2013), but analytical challenges have prevented analyses of more samples (Muir et al., 2013). Current-use pesticides (CUPs) have been little studied in Arctic marine mammals (Vorkamp and Riget, 2014). A recent study from the Canadian Arctic suggests that most CUPs do not biomagnify in marine mammal food webs (Morris et al., 2016). The only exception was endosulfan (listed under Stockholm Convention in 2011), which biomagnified from ringed seal to polar bear but not from lower trophic level to ringed seal (Morris et al., 2016). However, the concentrations in polar bear adipose tissue were still relatively low (<1 ng/g lw) (Morris et al., 2016), which is in accordance with other recent studies on Arctic marine mammals (Muir et al., 2013; Weber et al., 2010). Recent studies from the Norwegian Arctic have reported the presence of nonylphenol in polar bears (Simon et al., 2013; Routti et al., 2016b), whereas organophosphorous flame retardants were only detected in a few marine mammal samples (Hallanger et al., 2015).

PFASs are detected at high concentrations in protein-rich tissues of Arctic marine mammals. Studies from Svalbard indicate that PFAS concentrations are higher than concentrations of PCBs and OH-PCBs in polar bears, whereas plasma concentrations of PFASs and PCBs are similar in ringed seals (Routti et al., 2008, 2016a; Bytingsvik et al., 2012a,b). As polar bears are efficient to biotransform xenobiotics, circulating concentrations of OH-PCBs exceed those of PCBs (Bytingsvik et al., 2012a), whereas PCBs dominate over OH-PCBs in ringed seal plasma (Routti et al., 2008).

Individual Variation

Concentrations of contaminants in Arctic marine mammals show high individual variation. In addition to position in the food web (as discussed earlier), the variation may be related to age, breeding status, seasonal changes in body mass, and energy needs. Lipophilic pollutants accumulate with age in males but not in females (Muir et al., 2000). Arctic marine mammal females transfer lipophilic pollutants to their offspring through lipid-rich milk, leading to high levels of pollutants in the offspring (Bytingsvik et al., 2012a; Villanger et al., 2013; Bernhoft et al., 1997; Wolkers et al., 2002, 2004a; Polischuk et al., 2002). For example, circulating levels of PCBs are twice as high in a few months old polar bear cubs than in their mothers (Bytingsvik et al., 2012a; Polischuk et al., 2002). Biologic factors influence generally less concentrations of PFASs than lipophilic POPs in Arctic marine mammals. PFAS levels in ringed seals, white whales, and polar bears have not been related to age, sex, or body condition (Routti et al., 2014, 2016a; Butt et al., 2008; Kelly et al., 2009; Tartu et al.,

2017b; Smithwick et al., 2005). Although maternal transfer of compounds that bind to proteins, OH-PCBs and PFASs, is relatively low in polar bears (Bytingsvik et al., 2012a,b), circulating concentrations of most PFASs are higher in hooded seal pups than their mothers (Grønnestad et al., 2017). Arctic marine mammals go through seasonal changes in body fatness due to limited access to food, molting, reproduction, or migration. Seasonal loss of body mass leads to higher levels of lipophilic pollutants (Lydersen et al., 2002; Polischuk et al., 2002; Routti et al., 2010b; Tartu et al., 2017a). Furthermore, higher energetic need has been suggested to lead to increased concentrations of PCBs in polar bears that take long migrations (Olsen et al., 2003).

Spatial Trends

Spatial trends have been mainly studied in two circumpolar species, namely ringed seals and polar bears, which both are key species in Arctic marine ecosystems. Data are mostly available from Alaska, Canada, and the European Arctic, whereas reports from the Russian Arctic are scarcer. Studies on polar bears spanning from Alaska to Svalbard indicate that concentrations of PCBs, PBDEs, HBCDD, and PFASs increase from west to east, whereas chlordane and DDT levels do not show as clear a spatial variation (Verreault et al., 2005b; McKinney et al., 2011b; Muir et al., 2006; Smithwick et al., 2005). In contrast, the pattern for β-HCH was the opposite (McKinney et al., 2011b; Verreault et al., 2005b). Similar trends were reported for ringed seals (Muir et al., 2000). Studies on polar bears and ringed seals, including the Russian Arctic, indicate that concentrations of PCBs and chlorinated pesticides are highest in Kara Sea and Franz Josef Land area (Andersen et al., 2001; Lie et al., 2003; Muir et al., 2000). However, the most recent study on ringed seals points out that only PCB and DDT concentrations are higher in the Russian Arctic, whereas concentrations of other organochlorines and PBDEs are similar or lower compared to other Arctic locations (Savinov et al., 2011). Several studies on polar bears indicate that concentrations of Hg are highest in the Beaufort Sea and lowest in the western Hudson Bay (Norstrom et al., 1986; Rush et al., 2008; Braune et al., 1991; Routti et al., 2011). Hg concentrations are also higher in ringed seals from the western Canadian Arctic compared to the eastern Canadian Arctic, Alaska, and Greenland (Wagemann et al., 1996; Brown et al., 2016), whereas spatial differences in white whales are less prominent within the Canadian Arctic (Lockhart et al., 2005). The observed spatial trends on lipophilic POPs and Hg are explained by both variations in food web structure as well as regional differences in contamination (McKinney et al., 2011c; Routti et al., 2012; Brown et al., 2016).

Although levels of some pollutants are relatively high in Arctic marine mammals, comparisons with conspecifics from industrialized areas indicate that most pollutant concentrations are considerably higher in more southerly populations. For example, PCB, DDT, and chlordane concentrations are 5–10 times higher in ringed seals from the Baltic Sea compared to those from Svalbard

(Routti et al., 2008, 2009a,b), and this spatial difference was even larger in the late 1990s (Nyman et al., 2002). Also, most legacy POPs, including PBDEs and HBCDD, and PFASs are considerably higher in southern harbor seal (*Phoca vitulina*) populations close to industrialized areas compared to the small population living in Svalbard, whereas HCB and PFCA concentrations were roughly similar (Jenssen et al., 2007; Routti et al., 2014). Moreover, contaminant levels in white whales from the Canadian Arctic are substantially lower compared to that of the St. Lawrence population (McKinney et al., 2006b).

Temporal Trends

Most studies from the Arctic mammals show that the old, classic pollutants such as PCBs, DDT, and chlordanes are decreasing, whereas the trends for PBDEs, HCHs, PFASs, and Hg are more variable (Rigét et al., 2010, 2011; Muir et al., 2013). Among the increasing trends, β-HCH is increasing in some ringed seal and polar bear populations from the Canadian Arctic (McKinney et al., 2010; Muir et al., 2013). PBDEs are increasing in particular in some marine mammal populations from the Canadian Arctic (Muir et al., 2013; McKinney et al., 2010). In addition, both PCB and PBDE concentrations have increased in East Greenland polar bears during recent years, likely due to changes in the food web (Rigét et al., 2016, McKinney et al., 2013). Also, concentrations of HBCDD and Hg have been increasing in several marine mammal populations from East Greenland and the Canadian Arctic (Muir et al., 2013; Dietz et al., 2013a; Rigét et al., 2016; Vorkamp et al., 2011). Although PFSAs are decreasing in most Arctic marine mammal populations, PFCAs show increasing trends in many species and locations (Rigét et al., 2013; Routti et al., 2016a, 2017; Muir et al., 2013).

EFFECTS OF CONTAMINANTS IN ARCTIC MARINE MAMMALS

Contaminant effects have mostly been investigated using correlative approaches based on field samples, whereas an increasing number of modeling and in vitro studies have been published during recent years.

Seals and Whales

Effects of contaminants in seal and whale species inhabiting Arctic areas have mostly been investigated in conspecifics living in more southerly and more polluted areas. For example, health effects, including reproduction impairment, pathologic changes, and endocrine disruption, have been associated with high contaminant exposure in the Baltic ringed seal (*Pusa hispida* ssp. *botnica*) (Helle et al., 1976; Bergman and Olsson, 1985; Olsson et al., 1994; Routti et al., 2010a; Nyman et al., 2003; Castelli et al., 2014). Furthermore, prevalence of cancer and immunosuppression has been related to pollutant exposure in white whales from St. Lawrence, Quebec, Canada (Deguise et al., 1995; Martineau et al., 1994).

Several studies focusing on the Baltic ringed seals have used ringed seal population from Svalbard as a less polluted reference area (Routti et al., 2010a; Nyman et al., 2003; Castelli et al., 2014; Kanerva et al., 2012), but few molecular-level parameters have been associated with contaminant exposure in the Arctic seals. Wolkers et al. (1998a, 2000, 2009) reported significant associations between activities of CYP enzymes and concentrations of organochlorine compounds in several seal species. Brown et al. (2014) studied mRNA transcript levels in locally contaminated ringed seals from the Saglek Bay, Canada. Gene transcript levels of the *aryl hydrocarbon receptor* (*Ahr*), *interleukin-1β, estrogen receptor α, insulin like growth factor receptor 1*, and *glucocorticoid receptor α* were positively related to blubber PCB concentrations; PCB threshold values calculated for these five genes averaged 1680±206 ng/g lw. However, a risk assessment study concluded that the health risk of contaminant mixtures is very low in Arctic seals (Villa et al., 2017). A similar study conducted on white whales ($n=43$) from the Canadian Arctic reported that *Ahr* and *Cyp1a1* transcripts were positively correlated with PCB concentrations, whereas the other studied genes involved in growth, metabolism, and development were not related to blubber contaminant concentrations (Noel et al., 2014). Circulating thyroid hormone concentrations have also been associated with concentrations of lipophilic pollutants in white whales and hooded seals (Villanger et al., 2011b, 2013).

The risk of Hg-associated neurotoxic effects in ringed seals and white whales was assessed by comparing brain Hg concentrations in these species with threshold concentrations based on an extensive literature research on other mammalian species (Krey et al., 2015). The study concluded that brain Hg concentrations in white whales exceeded the neurotoxicity thresholds, and those in ringed seal brain were within the range that may cause neurobehavioral effects. Similarly, Dietz et al. (2013b) concluded that among Arctic marine mammals, toothed whales show the highest Hg concentrations in brain tissue, and values exceed thresholds for neurochemical effects in fish-eating mammals. Furthermore, a correlative field study on white whales reports that components of the γ-aminobutyric acid and glutamatergic signaling pathways were associated with brain methyl-Hg (MeHg) concentrations (Ostertag et al., 2014). High ratios of selenium (Se):Hg have been suggested to protect Arctic marine mammals from Hg-associated toxicity by formation of HgSe complex (Ikemoto et al., 2004; Ostertag et al., 2014; Krey et al., 2015; Dietz et al., 2000; Yoneda and Suzuki, 1997). However, due to biologic formation of HgSe, depletion of the selenium pool may be a driving mechanism of neurotoxic effects in marine mammals (Gajdosechova et al., 2016). Furthermore, it was previously thought that the HgSe complex was inert, but a recent study on its nanoparticular form suggests that the biologically formed HgSe may be highly reactive (Gajdosechova et al., 2016).

Immunotoxic effects of contaminants on marine mammals, including Arctic marine mammals, have been recently reviewed (Desforges et al., 2016). As the topic is also part of this book, we will not conduct a thorough review on immunotoxic effects in Arctic marine mammals.

Polar Bears

Recent scientific publications indicate that POPs, and in particular PCBs, represent a significant threat at the population level in polar bears. Using physiologically based pharmacokinetic modeling, it has been estimated that the concentrations of the sum of anthropogenic organohalogenated compounds in 11 polar bear subpopulations (from Alaska to Svalbard) exceeded the threshold levels for affecting reproduction and immune health and causing carcinogenicity by 4.3–28.6 times (Dietz et al., 2015). Although the toxic effect threshold levels, termed risk quotients, applied were based on studies on laboratory rats (*Rattus rattus*), and species differences in toxic sensitivity therefore may be biased, this modeling study clearly indicates that it is not unlikely that the internal concentrations of organohalogenated compounds affect ecologically significant fitness traits in polar bears, and may thus be responsible for adverse population effects. Interestingly, PCBs were the main contributors for all these effect categories, and the PCBs alone contributed to fact that the toxic effect threshold levels were exceeded by 3.8–22.5 times (Dietz et al., 2015). Thus, even though the concentrations of PCBs generally have declined in polar bears during the decades, PCBs still appear to pose a significant risk for the reproduction and health of individual polar bears. Similarly, a risk assessment study based on calculations of daily intake of pollutants by polar bears in comparison to acceptable daily intake for humans concludes that the risk of pollutant mixtures for adult polar bears is very high and even higher for polar bears cubs (Villa et al., 2017).

Krey et al. (2015) assessed the risk for Hg-associated neurotoxic effects in polar bears from Nunavik, Canada, by comparing polar bear brain Hg concentrations with threshold concentrations in other mammalian species. The study concluded that brain Hg concentrations in polar bears were below the threshold levels for neurobehavioral effects. Despite low Hg concentrations in polar bear brain, *N*-methyl-D-aspartate receptor levels in brain were negatively correlated with both total Hg and MeHg, whereas other neurochemical biomarkers or global DNA methylation were not significantly related to Hg exposure (Basu et al., 2009; Pilsner et al., 2010). Another study showed negative relationships between monoamine oxidase activity and total Hg concentrations in polar bear brain, whereas cholinesterase activity or muscarinic acetylcholine receptor binding were not related to Hg exposure (Krey et al., 2014). In vitro experiments using isolates from polar bear brain tissue, however, demonstrated that Hg can inhibit activity/binding of all the three studied molecules (Krey et al., 2014). Pedersen et al. (2015) studied correlations between PFAS exposure and neurochemical transmitter systems in different brain regions of East Greenland polar bears. The results based on measurements from five to nine individuals support the hypothesis that PFAS concentrations in polar bears from East Greenland are above the threshold limits for neurochemical alterations (Pedersen et al., 2015).

Correlative field studies indicate adverse effects on fitness-related physiologic variables in polar bears, such as reproductive hormone concentrations (Ciesielski et al., 2018; Gustavson et al., 2015; Oskam et al., 2003; Pedersen et al., 2016), thyroid hormone concentrations (Skaare et al., 2001; Braathen et al., 2004; Gabrielsen et al., 2015; Villanger et al., 2011a; Bourgeon et al., 2017), and immune function (Lie et al., 2004, 2005). Gustavson et al. (2015) showed inverse relationships between plasma levels of pregnenolone (PRE) and androstenedione (AN) and OH-PCBs in female polar bears and suggested that CYP17 may be a target for OH-PCBs. Steroid hormone disruption in polar bears may also be linked to the activation of pregnane X receptor (PXR; also known as the steroid xenobiotic receptor), by environmental pollutants. Lille-Langøy et al. (2015) studied the activation of polar bear PXR using an in vitro luciferase reporter assay. A wide range of compounds polar bears are exposed to, including PCBs, organochlorine pesticides, and PBDEs, activated the receptor. As PXR induces CYP enzymes, which catalyze metabolism of various steroid hormones (Niwa et al., 1998; Kliewer et al., 1998), its activation may also be related to steroid disruption in polar bears.

A series of in vitro studies has focused on interactions between pollutants and polar bear plasma proteins involved in thyroid hormone transport. Gutleb et al. (2010) extracted thyroid transporting proteins from polar bear plasma and found that one of the major OH-PCBs in polar bear plasma has much higher affinity toward transthyretin (TTR) than its natural ligand, thyroxine. The study suggested that TTRs in free-ranging polar bears are completely saturated by contaminants. Simon et al. (2011) developed a sample preparation method to extract a broad range of thyroid hormone disrupting compounds in plasma and test the binding potency toward human TTR. The method was further used by Bytingsvik et al. (2013), who showed that contaminant-related TTR-binding activity was twice as high in polar bear cubs sampled in the late 1990s compared to those sampled in 2008, reflecting the decline in pollutant exposure. Furthermore, OH-PCBs explained ~60% of the TTR-binding activity. A follow-up study using effect-directed analyses revealed that the remaining TTR-binding activity (40%) in plasma from polar bear cubs was explained by nonylphenol and higher chlorinated OH-PCBs (Simon et al., 2013). In conclusion, the studies suggest that several compounds present in plasma of polar bear cubs have very high affinity to TTR. As these compounds are found at high concentrations, they can occupy all circulating TTRs, leaving no place for the natural hormone thyroxine. The exact consequences of fully saturated TTR are not known. However, as thyroid hormones are essential for growth and development, thyroid disrupting properties of pollutants in young polar bear cubs raise concern for potential effects on neurological development (Jenssen et al., 2015).

Pathologic changes have been examined principally in the East Greenland polar bear subpopulation, which is among the most polluted wildlife populations within the Arctic, and globally (Sonne et al., 2011). Size of the sexual organs in both males and females, as well as penile bone mineral density, were negatively related to contaminant exposure in East Greenland polar bears (Sonne et al., 2006). Furthermore, penile bone density was lower in polar bear subpopulations with high levels of contaminant compared to less contaminated subpopulations

(Sonne et al., 2015). Exposure to organohalogen compounds may also have contributed to reduce skull bone mineral density in East Greenland polar bears over the past 120 years (Sonne et al., 2004, 2013). A study examining hepatic and renal lesions in over 100 polar bears from East Greenland concluded that contaminants, including POPs and Hg, are likely to contribute in the development of the lesions (Sonne et al., 2012b). No histologic lesions in thyroid or adrenal were related to contaminant exposure (Sonne et al., 2012a).

Despite very high concentrations of organohalogenated compounds in polar bears, only one study has investigated population-level effects. In a meta-study, Nuijten et al. (2016) examined the effects of several groups of organohalogenated compounds and other anthropogenic stressors (ice cover [as a proxy of climate change], human population density, and harvest rate) on population-level characteristic (polar bear subpopulation density, mean litter size, and adult female survival rate) on circumpolar polar bear subpopulations. The results suggested that a model containing only PCBs or only DDTs as a predictor were best in explaining the variance in polar bear density between subpopulations. However, dieldrin, PBDEs, and human population density were also important predictor variables for polar bear densities in the subpopulations. Surprisingly, sea ice cover, being a proxy for global warming, and harvest rate were not included in the models that described the subpopulation densities. Thus, the study by Nuijten et al. (2016) indicates that levels of organohalogenated compounds in polar bears may actually have a more negative impact on polar bear populations than the effects associated with decreasing ice cover due to global warming. Nevertheless, in this study, adult female survival was negatively affected by the September sea ice cover, and together with reports of sea ice–related demographic declines in two subpopulations of polar bears (Bromaghin et al., 2015; Lunn et al., 2016), it is likely that the predicted losses of polar bear ice habitats in the polar basin (Durner et al., 2009) will have population-level effects in polar bears.

ARCTIC MARINE MAMMALS FACED BY THE GLOBAL CLIMATE CHANGE

A number of physical factors, such as dramatic seasonal change, extensive permanent and seasonal ice cover, and a large supply of freshwater from rivers and melting ice, make Arctic marine ecosystems unique (Ellingsen et al., 2008). As these conditions are challenging for marine biota, Arctic marine ecosystems have a large number of endemic specialists (Callaghan et al., 2004). In the Arctic, the amount of sea ice is steadily decreasing (Hoegh-Guldberg and Bruno, 2010), and since the last decade, this decrease seems to have accelerated (Stroeve and Notz, 2015).

This sea ice cover decline is likely to lead to qualitative and/or quantitative diet changes for a number of species, as a consequence of phenological mismatches between predator/prey distribution (Descamps et al., 2016; Grebmeier et al., 2006; Post, 2018). Sea ice is used as a platform for nearly all aspects of the life history of Arctic marine mammals (Kovacs et al., 2010; Moore and Huntington, 2008). For ice-associated whales (white whales, narwhals, and bowhead whales [*Balaena mysticetes*]), sea ice acts alternatively as a substrate or a barrier for breathing, feeding, or as a shelter for predator avoidance (Kovacs et al., 2011). Ice-obligate seals (e.g. walruses [*Odobenus rosmarus*], bearded seals [*Erignathus barbatus*], and ringed seals) rely on sea ice as a platform for hauling out, reproducing, and molting. Polar bears rely on seasonal sea ice to hunt seals but also to mate and for pregnant females to den (Kovacs et al., 2011; Laidre et al., 2008; Moore and Huntington, 2008). Therefore, a reduction and shrinking of the Arctic sea ice cover is likely to affect a number of resident species.

The polar bear appears to be among the most sensitive Arctic marine mammal species to climate change, primarily due to its reliance on sea ice and specialized feeding (Laidre et al., 2008). Within the past few years, a number of studies have brought the attention on the threat of declining sea ice conditions on the maintenance of polar bear populations (Lunn et al., 2016; Obbard et al., 2016; Regehr et al., 2016; Wilson et al., 2016). With an estimated population size of 26,000 individuals, and a generation length of 11.5 years (Regehr et al., 2016), one study estimated a 30% decline of polar bear populations over three generations (35–41 years) due to summer sea ice loss (Regehr et al., 2016). Behavioral adaptations to a changing habitat have led to preying on nonpreferred prey (Galicia et al., 2016; McKinney et al., 2017; Prop et al., 2015; Stempniewicz et al., 2014; Tartu et al., 2016), increased frequency of long-distance swims (Pilfold et al., 2017), and modification of summer activity (Ware et al., 2017). All of these behavioral modifications could lead to a negative energy balance; for example, a bear would spend more energy chasing than the actual energy value of the prey it will consume.

COMBINED EFFECTS OF CONTAMINANT EXPOSURE AND CLIMATE CHANGE IN POLAR BEAR POPULATION

It is possible that the effects of global warming may enhance the effects of anthropogenic pollutants on the individual health of polar bears, and thus also the global population of polar bears (Jenssen et al., 2015; McKinney et al., 2015). Climate warming results in earlier break-up of sea ice and is thus likely to prolong the fasting period in polar bears. Since many lipid-soluble POPs are not excreted, fasting and especially emaciation will result in large increases of the concentrations of these compounds in blood and target tissues (Helgason et al., 2013; Polischuk et al., 2002; Christensen et al., 2007). In some regions, the decline of sea ice extent has resulted in changes in the presence of seal species that polar bears predate, and this has been shown to cause increased bioaccumulation of certain POPs (McKinney et al., 2009, 2013, 2015; Thiemann et al., 2008). Thus, prolonged fasting due to climate-induced ice loss and resultant lowered prey availability that will result in fasting, or dietary changes will cause levels of POPs to exceed threshold levels for effects by even more than modeled by Dietz et al. (2015). It is also likely that such increases of body burdens of POPs that are reproductive endocrine disrupting chemicals (EDCs) will add to the predicted negative reproductive effects that these compounds may have on the reproductive success of polar bears. Recently, it has been modeled that contaminant-related male infertility may lead to a reduction in population growth via an Allee effect (Pavlova et al., 2016). Contamination-reduced pregnancy rates, caused by decreased availability of fertile males, could trigger a mate-finding Allee effect, particularly when male–male competition for females is limited or when infertile males are able to compete with fertile males for females. Reported negative associations between POPs and testosterone and dihydrotestosterone in male polar bears (Ciesielski et al., 2018; Oskam et al., 2003) may indicate that increased body burdens of POPs due to climate change may contribute to such an Allee effect.

Several correlative studies have added to the weight of evidence that POPs interfere with endocrine molecular and physiologic variables that are important for health and individual fitness in polar bears (Jenssen et al., 2015). Examples are the previously mentioned sex steroid hormones, thyroid hormones, and immune-responses. In polar bears, climate change will likely result in increased internal concentrations of lipophilic POPs due to ice-related loss of prey and an associated poorer body condition, and changes in food web structure (McKinney et al., 2009) (see also Case Study: Synergistic Effects of Sea Ice Decline and Contaminant Exposure in Polar Bears From the Barents Sea section). Thus, the climate-associated increases in internal concentrations of POPs will have a negative effect on variables related to the individual health and reproductive capacity of polar bears. In polar bears, habitat loss associated to climate change will have an important reinforcing effect on the impact of POPs at the population level. As previously mentioned, a meta-study has indicated that internal body concentrations of certain POPs were better descriptors of the subpopulation sizes of polar bears than ice habitat variables (Nuijten et al., 2016). Although the habitat loss caused by climate change will affect polar bear populations, it is possible that climate-associated increases of POPs will contribute to population effects beyond the effects caused by habitat loss alone (Jenssen et al., 2015).

CASE STUDY: SYNERGISTIC EFFECTS OF SEA ICE DECLINE AND CONTAMINANT EXPOSURE IN POLAR BEARS FROM THE BARENTS SEA

Background

The Barents Sea area is experiencing the fastest loss of sea ice recorded throughout the Arctic, with a reduction of 42 ± 7 ice-free days per decade (Laidre et al., 2015) because of intense air–sea interaction (Smedsrud et al., 2013). The length of the summer season (i.e., time with reduced ice) in the Barents Sea has increased by over 20 weeks between 1979 and 2013 (Laidre et al., 2015). These changes are also accompanied by winter sea ice retreat that has been ~50% since the late 1990s (Årthun et al., 2012) and is especially pronounced in the Barents Sea compared to other Arctic areas (Dobricic et al., 2016). In addition to the current climatic state of the Barents Sea, climate models predict that this region will experience the most significant sea ice retreat within the entire Arctic in the near future (Koenigk et al., 2013). This situation is alarming, considering that the Barents Sea is one of the most productive seas adjacent to the Arctic Ocean (Helland-Hansen and Nansen, 1909).

Because of the rapidly declining sea ice in the Barents Sea, the polar bear subpopulation from this area is subject to pronounced loss of habitat compared to other polar bear subpopulations (Durner et al., 2009). Since early 1990s, the optimal polar bear habitat declined by 11%–12% per decade, and projected changes for the 21st century suggest 50% loss for this subpopulation (Lone et al., 2017; Durner et al., 2009). In addition, this subpopulation is among the most contaminated polar bear subpopulations in the circumpolar Arctic, defined as a "hotspot" population in terms of contaminant exposure in Arctic wildlife (Letcher et al., 2010; McKinney et al., 2011b; Verreault et al., 2005b; Smithwick et al., 2005).

Recent knowledge suggests that contaminants may interfere with energy metabolism (Casals-Casas and Desvergne, 2011; Grün and Blumberg, 2009; Heindel et al., 2015). Exposure to POPs among other EDCs has also been related to common metabolic problems in humans, such as obesity and type 2 diabetes (Heindel et al., 2015; Lee et al., 2014; Legler et al., 2015). Based on the indications that contaminant exposure may interact with energy metabolism, we hypothesize that exposure to contaminants leads to suboptimal energy metabolism in polar bears.

Study Design

To acquire knowledge on combined effects of pollutants and sea ice decline on polar bears, we captured adult female polar bears during two contrasted seasons: spring, when sea ice is extended, and autumn, when most sea ice has retreated. The bears belonging to the Barents Sea subpopulation were sampled at Svalbard, Norway. Although the bears were sampled opportunistically, sampling areas differed in terms of sea ice with West Svalbard having less sea ice yearlong than East Svalbard. We sampled only adult (4–28 years, age estimated from tooth) females to avoid gender-specific differences in physiology. The 112 samples that we collected (April 2012, $n=33$, September 2012, $n=24$, April 2013, $n=29$, and September 2013, $n=26$) represented 78 females. Twenty-six females were captured either twice ($n=19$), three times ($n=6$), or four times ($n=1$). The bears were immobilized from a helicopter by a remote injection. We obtained morphometric measurements to estimate their body condition using a body condition index (Cattet et al., 2002). Female polar bear were classified in three groups according to their breeding status: solitary (i.e., alone or together with a male in spring), with one or two cubs of the year, or with one or two yearlings. We collected different biologic samples (blood, fat biopsy, hair); part of them were frozen in liquid nitrogen immediately.

We measured a number of the notorious legacy and emerging compounds: lipid-soluble PCBs, pesticides, and BFRs in plasma and fat, and PFASs and OH-PCBs in plasma (Tartu et al., 2017a,b). To get insight from polar bears' diet (i.e., trophic level) and lipid and carbon sources, we analyzed red blood cell and plasma samples for nitrogen and carbon stable isotope ratios $\delta^{15}N$ and $\delta^{13}C$, and fat samples for fatty acid composition (Tartu et al., 2016). We also measured plasma urea and creatinine concentrations as a proxy for a metabolic state (feeding/fasting) (Tartu et al., 2017a,b). Furthermore, we investigated a variety of biomarkers related to lipid metabolism from gene to metabolome level (Bourgeon et al., 2017; Tartu et al., 2017c). Specifically in fat samples, we measured the abundance profiles of nine lipid-related genes and fatty acid synthesis and elongation indices, in addition to plasma concentrations of thyroid hormones and lipid-related variables (cholesterol, triglycerides, high-density lipoprotein). Furthermore, we analyzed the plasma metabolome and lipidome.

Results

Sea Ice Conditions Around Svalbard Coasts

In the Barents Sea, sea ice concentration and the timing and duration of sea ice availability are irregular temporally and spatially (Falk-Petersen et al., 2000; Hop et al., 2000; Pavlova et al., 2014; Vinje and Kvambekk, 1991). Seasonal sea ice generally reaches its maximal extent in March and is minimal in September. Within the study period of our project, the sampling years, 2012–13, were also contrasted with more sea ice in winter 2011–12 than in 2012–13. In addition, sea ice is less extended and less dense along the west coast of Svalbard compared to the eastern coast (Falk-Petersen et al., 2000; Hop et al., 2000; Prop et al., 2015; Vinje and Kvambekk, 1991). In contrast, the southeast area of Svalbard, including the islands Barentsøya and Edgeøya, experiences the largest amplitude of sea ice retreat in summer (Falk-Petersen et al., 2000; Prop et al., 2015).

Polar Bears' Feeding Strategy

In response to highly variable environmental conditions, polar bears are metabolically unique and can shift from a feeding to a fasting state at any time (Cattet, 2000). Generally, they prepare themselves for periods of food scarcity or reproductive fasting by deposition of large amounts of body fat from April to June, during the ringed seals pupping and molting periods (Ramsay and Stirling, 1988). During that period, polar bears almost double their body mass (Atkinson and Ramsay, 1995; Derocher and Stirling, 1994). For much of the remaining year, polar bears feed little and undergo long fasting periods (Messier et al., 1992). For pregnant females, fasting can last for about 4–8 months, varying by subpopulations (Amstrup, 2003). When fasting, polar bears lose approximately 0.5% of their mass per day, and adult males could sustain a 7.4 months fasting period before dying of starvation (Pilfold et al., 2016). Polar bears have evolved a unique set of genes that enables them to efficiently develop adipose tissue and deal with a lipid-rich diet (Liu et al., 2014). To better understand the impact of sea ice decline on polar bear reproduction and health, recent studies (e.g., Atwood et al., 2016) have mentioned the urge to refine understanding of polar bear physiology, particularly processes that influence the allocation of nutrients and that determine the limits of fasting.

Influence of Sea Ice on Polar Bears' Diet and Body Condition

During the four sampling periods, April and September in 2012 and 2013, sea ice conditions were highly variable (see Fig. S5 in Tartu et al., 2017a), and the foraging ecology of the captured female polar bears varied according to habitat use and sampling period. For instance, in areas with extended sea ice, female polar bears foraged on higher trophic level prey and generally consumed fewer terrestrial items compared to females using a habitat with poor sea ice conditions (Tartu et al., 2016). Results based on diet proxies suggest that polar bears feed more on seabirds in areas with less sea ice, whereas polar bears inhabiting areas with more stable sea ice conditions are more selective toward seals (Tartu et al., 2016). Waterfowls and reindeers also make a part of autumn diet in areas where the sea ice retreats (Tartu et al., 2016). Polar bears use sea ice as a hunting platform and feed on bearded seals, harp seals, or hooded seals (Derocher et al., 2002). In contrast, in areas with poor sea ice conditions, polar bears would rather follow the coastline to prey on ringed seal pups born in fast ice habitats and probably on seabirds and terrestrial prey the remaining time (Prop et al., 2015; Smith and Lydersen, 1991; Stempniewicz, 2006; Stempniewicz et al., 2014). Therefore, females using habitats with poor sea ice conditions would rather have access to small (ringed seal pups) or low-energy terrestrial prey, and they may have to cover large areas to meet their energy requirements.

The spatial and temporal differences in foraging ecology in female polar bears captured in 2012–13 in Svalbard translated into body condition variations (Bourgeon et al., 2017; Tartu et al., 2017a). Female polar bears using the habitat with poor sea ice conditions were leaner than females using a habitat with extended and/or stable sea ice conditions (Bourgeon et al., 2017). This pattern is likely the result of ice-related prey distribution. In areas and periods with good sea ice conditions, polar bear prey are likely more abundant and available over an extended period in time.

Sea Ice–Related Influence of Body Condition and Diet on POP Concentrations

In Svalbard female polar bears, body condition was the most important predictor for concentrations of lipophilic POPs (Fig. 13.2) (Tartu et al., 2017a).

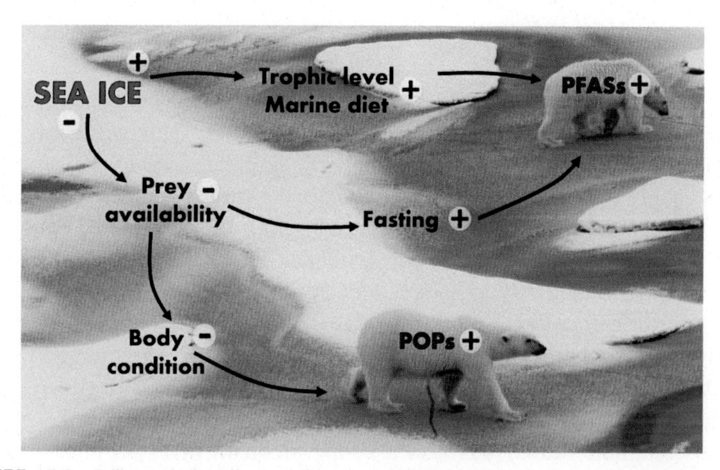

FIGURE 13.2 Influence of sea ice conditions on pollutant concentrations in female polar bears from Svalbard. Perfluoroalkyl substances (PFASs) were measured in plasma and persistent organic pollutants (POPs) were measured in both plasma and fat biopsies. The *arrows* indicate potential causative effects and the ± signs the direction of the relationships for the measured variables. Fasting state was inferred using plasma urea to creatinine ratios, trophic level, and diet by using nitrogen stable isotope ratios in red blood cells and body condition by calculating a body condition index developed for ursids.

Lean females using a habitat with poor sea ice conditions had higher concentrations of lipophilic POPs in plasma and adipose tissue than females using a good quality habitat. Although female polar bears using a poor condition habitat fed at a lower trophic level than females using the high-quality habitats (Tartu et al., 2016), they still had higher concentrations of lipophilic POPs. Body condition therefore prevailed on diet to predict lipophilic POP concentrations in plasma and adipose tissue (Tartu et al., 2017a). Season was also a strong predictor of lipophilic POPs; interestingly, in spring when sea ice was extended, female polar bears had higher concentrations of lipophilic POPs compared to females captured in September when sea ice had retreated (Tartu et al., 2017a). Among the four studied seasons (spring and autumn 2012–13), concentrations of POPs were highest in spring 2013, following a winter with poor sea ice condition, when the bears were also leanest. The relationship between sea ice and body condition is likely a result from the interval between the peak availability of prey and the lag for adipose tissue to build up. In April, polar bears are on average thinner after a winter period with relatively low prey availability; whereas in autumn, polar bears have built fat reserves after an intensive feeding period in spring/early summer (Cattet, 2000; Derocher and Stirling, 1994; Lønø, 1970). However, if sea ice conditions have been poor during winter, polar bears have relied even more on their fat reserves compared to a winter with more stable sea ice conditions. A winter with less sea ice will thus result in higher tissue POP concentrations in spring.

Sea Ice–Related Influence of Diet and Metabolic State on PFAS Concentrations

PFAS concentrations in plasma were related to diet and metabolic state, but not to body condition (Fig. 13.2) (Tartu et al., 2017b). This was expected, as in contrast to POPs, PFASs are proteinophilic pollutants. PFAS concentrations were strongly related to trophic level and marine carbon sources inferred from stable isotope values. Furthermore, concentrations of PFOS and PFCAs were higher in females using a habitat with extended sea ice conditions compared to females using a habitat with poor sea ice conditions, which was likely related to dietary differences (Tartu et al., 2017b). This suggest that preying on preferred marine higher trophic level prey, as found in a habitat with good sea ice conditions (Tartu et al., 2016), results in high PFAS exposure. Interestingly, metabolic state, defined as the individual feeding/fasting state (based on plasma urea to creatinine ratios) was also a strong predictor of PFAS concentrations (Tartu et al., 2017b). Fasting female polar bears had higher plasma concentrations of PFSAs, PFOA, PFNA, and PFDA than non-fasting females (Tartu et al., 2017b). When fasting, bears can conserve their protein pool by recycling urea nitrogen into plasma proteins (Nelson et al., 1975), for instance, concentrations of β-globulins in plasma are increased in fasting polar bears (Cattet, 2000). Considering that PFASs bind to proteins, an increased proportion of proteins in blood could explain the high PFAS concentrations in fasting female polar bears. Yet, another explanation could be that metabolic rate and contaminant excretion is reduced in fasting animals (Aas et al., 2014).

Effect of Pollutants on Polar Bear Energy Homeostasis

Concentrations of thyroid hormones were negatively related to organochlorine contaminant and PFAS concentrations in polar bears sampled in spring but not in autumn (Bourgeon et al., 2017). A further study by Tartu et al. (2017c) indicates that pollutants affect lipid biosynthesis and catabolism in female polar bears. PCBs, BDE-153, and oxychlordane measured in polar bear adipose tissue were positively related to transcript levels of genes involved in the accumulation of triglycerides via glucose and utilization of free fatty acids. *Trans*-nonachlor and the dioxin-like PCB-118 predicted an increase of transcript levels of genes related to fatty acid transport, oxidation, and insulin utilization. Furthermore, relationships between PFAS concentrations, thyroid hormones, lactate and lipid-related parameters in plasma, and transcript levels of a gene related to lipid synthesis and accumulation suggest that a PFAS-mediated disruption of the hypothalamic-pituitary-thyroid axis could lead to higher levels of plasma lipids and lower lactate concentrations. De novo synthesis of fatty acids was negatively related to BDE-153 concentrations, whereas fatty acid elongation index showed an opposite relationship with the same compound. Finally, although the polar bear metabolome and lipidome rather clustered according to sampling season than pollutant concentrations, we observed associations between the lipidome and pollutants. The aforementioned relationships between lipid metabolism and pollutant concentrations were more pronounced during periods with little sea ice. This suggests that sea ice decline and pollutant exposure may have synergistic negative effects on polar bear health.

Effect of Pollutants in Polar Bear Energy Homeostasis Using In Vitro and In Silico Approach

Parallel to the field study, pollutant effects on polar bear lipid homeostasis were studied using in vitro and in silico methods (Fig. 13.3) (Routti et al., 2016b). The authors focused on peroxisome proliferator-activated receptor gamma (PPARG), which is the major regulator of adipogenesis and promoter of lipid stores (Lefterova et al., 2014). A luciferase reporter assay was constructed to study activation (agonistic and antagonistic) of polar bear PPARG by environmental pollutants and their mixtures. Agonistic effects by BFRs, PCBs, and chlorinated pesticides were tested as single compounds and mixtures, whereas antagonistic effects were tested on the mixtures and a smaller range of single compounds. Single compounds and the mixtures were applied at similar concentrations as found in polar bear adipose tissue. Furthermore, PPARG–pollutant interactions were modeled using docking and scoring methods. In addition, the authors investigated the first and second wave adipogenesis in 3T3-L1 preadipocytes originating from mouse. The 3T3-L1 cells were exposed to polar bear adipose tissue and liver extracts and synthetic mixtures that corresponded to POP concentrations in the extracts. Exogenous compounds in the extracts were investigated using target and nontarget/suspect methods. Finally, stem cells from polar bear fat tissue were isolated and a method was developed to

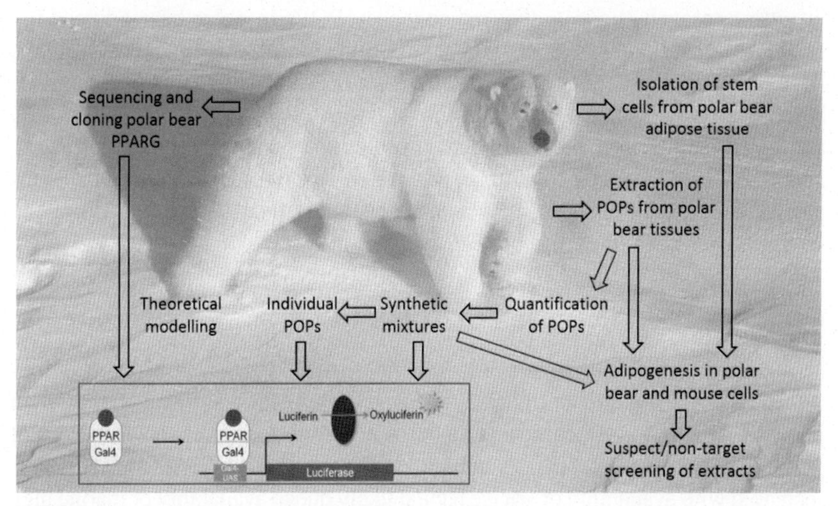

FIGURE 13.3 Study design to investigate effects of contaminants on polar bear energy homeostasis in vitro and in silico.

study the differentiation of the stem cells to mature fat cells (adipogenesis). Adipogenesis by contaminant extracts and synthetic mixtures was studied in the polar bear cells. The measured endpoints were lipid accumulation and transcript levels of *PPARG* and its target gene *fatty acid binding protein 4* (*FABP4*).

The results of the study by Routti et al. (2016b) showed that contaminant mixtures suppressed polar bear PPARG-mediated luciferase activity as well as the levels of *PPARG* and *FABP4* transcripts in polar bear adipose tissue–derived stem cells. IC50 (a response midway between the estimates of the upper and lower plateau) for the synthetic mixture was ~10 μM (concentration of summed lipophilic POPs), which is within the range of average POP concentrations in adipose tissue of polar bears from circumpolar subpopulations (McKinney et al., 2011b). Routti et al. proposed that PCB153 contributes the most to the antagonistic effects by the synthetic POP mixtures, since it is the major PCB in the mixture and a full antagonist of polar bear PPARG. Modeling pollutant–polar bear PPARG interactions suggested that the binding affinity was weak for chlorinated compounds that showed antagonistic activity toward polar bear PPARG. PPARG antagonism may, however, be acquired through weaker noncovalent binding (Leesnitzer et al., 2002; Fratev et al., 2015; Ohashi et al., 2015). Furthermore, contaminant extracts from polar bear liver and adipose tissue enhanced adipogenesis in both 3T3-L1 preadipocytes and polar bear adipose tissue-derived stem cells, whereas lipid accumulation was not affected by the synthetic mixtures (Routti et al., 2016b). The reason why the extracts enhanced adipogenesis, and synthetic mixture did not, is likely due to the presence of emerging chemicals in the extracts. Suspect/nontarget screening revealed the presence of emerging chemicals belonging to compound groups that have been previously reported to induce adipogenesis: phthalates, synthetic musks, and nonylphenols (Pereira-Fernandes et al., 2013). In conclusion, the in vitro approach suggests that major legacy POPs in polar bear adipose tissue antagonized polar bear PPARG at environmentally relevant concentrations, whereas emerging chemicals present in polar bear tissue extracts may enhance adipogenesis. In contrast to the in vitro approach, transcript levels of *PPARG* were positively related to PCBs and oxychlordane in the field samples (Tartu et al., 2017c). This may explained by activation of retinoic X receptor that forms a heterodimer with PPARG in vivo, and may thus activate PPARG as discussed by Tartu et al., 2017c.

CONCLUSION AND PERSPECTIVES

The main conclusions of the polar bear field study are that sea ice conditions influence concentrations of lipophilic POPs and PFAS in polar bears by different pathways (Fig. 13.2). Decline in sea ice was related to increased concentrations of lipophilic POPs through changes in body condition, whereas PFAS concentrations increased with availability of sea ice habitat likely due to availability of marine high trophic level prey. In addition, correlative field studies and in vitro studies on polar bear receptors and cells indicate that contaminants can disrupt lipid metabolism in polar bears (Routti et al., 2016b; Tartu et al., 2017c). Furthermore, effects of contaminants may disturb polar bear lipid metabolism through multiple pathways, and additive or synergistic effects of sea ice decline and contaminant exposure on polar bear lipid metabolism are likely to occur. There is a lack of knowledge on how much polar bears can adapt to pollutant exposure coupled with less predictable feeding opportunities; there is therefore a need for further research in this field, especially in the light of the ongoing accelerated climate change.

REFERENCES

Aas, C.B., Fuglei, E., Herzke, D., Yoccoz, N.G., Routti, H., 2014. Effect of body condition on tissue distribution of perfluoroalkyl substances (PFASs) in arctic fox (*Vulpes lagopus*). Environmental Science and Technology 48, 11654–11661.

Amstrup, S.C., 2003. Polar bear, *Ursus maritimus*. In: Wild Mamm. N. Am. Biol. Manag. Conserv.

Andersen, G., Kovacs, K.M., Lydersen, C., Skaare, J.U., Gjertz, I., Jenssen, B.M., 2001. Concentrations and patterns of organochlorine contaminants in white whales (*Delphinapterus leucas*) from Svalbard, Norway. The Science of the Total Environment 264, 267–281.

Årthun, M., Eldevik, T., Smedsrud, L.H., Skagseth, Ø., Ingvaldsen, R.B., 2012. Quantifying the influence of Atlantic heat on Barents Sea ice variability and retreat. Journal of Climate 25, 4736–4743.

Atkinson, S., Ramsay, M., 1995. The effects of prolonged fasting of the body composition and reproductive success of female polar bears (*Ursus maritimus*). Functional Ecology 559–567.

Atwell, L., Hobson, K.A., Welch, H.E., 1998. Biomagnification and bioaccumulation of mercury in an arctic marine food web: insights from stable nitrogen isotope analysis. Canadian Journal of Fisheries and Aquatic Sciences 55, 1114–1121.

Atwood, T.C., Marcot, B.G., Douglas, D.C., Amstrup, S.C., Rode, K.D., Durner, G.M., Bromaghin, J.F., 2016. Forecasting the relative influence of environmental and anthropogenic stressors on polar bears. Ecosphere 7, 1–22.

Bandiera, S.M., Torok, S.M., Lin, S., Ramsay, M.A., Norstrom, R.J., 1995. Catalytic and immunological characterization of hepatic and lung cytochromes p450 in the polar bear. Biochemical Pharmacology 49, 1135–1146.

Basu, N., Scheuhammer, A.M., Sonne, C., Letcher, R.J., Born, E.W., Dietz, R., 2009. Is dietary mercury of neurotoxicological concern to wild polar bears (*Ursus maritimus*)? Environmental Toxicology and Chemistry 28, 133–140.

Bergman, A., Olsson, M., 1985. Pathology of Baltic grey seal and ringed seal females with special reference to adrenocortical hyperplasia: is environmental pollution the cause of a widely distributed disease syndrome? Finnish Game Research 44, 47–62.

Bernhoft, A., Wiig, Ø., Utne Skaare, J., 1997. Organochlorines in polar bears (*Ursus maritimus*) at Svalbard. Environmental Pollution 95, 159–175.

Bourgeon, S., Riemer, A.K., Tartu, S., Aars, J., Polder, A., Jenssen, B.M., Routti, H., 2017. Potentiation of ecological factors on the disruption of thyroid hormones by organo-halogenated contaminants in female polar bears (*Ursus maritimus*) from the Barents Sea. Environmental Research 158, 94–104.

Braathen, M., Derocher, A.E., Wiig, O., Sormo, E.G., Lie, E., Skaare, J.U., Jenssen, B.M., 2004. Relationships between PCBs and thyroid hormones and retinol in female and male polar bears. Environmental Health Perspectives 112, 826–833.

Braune, B.M., Norstrom, R.J., Wong, M.P., Collins, B.T., Lee, J., 1991. Geographical-distribution of metals in livers of polar bears from the Northwest-Territories, Canada. The Science of the Total Environment 100, 283–299.

Bromaghin, J.F., Mcdonald, T.L., Stirling, I., Derocher, A.E., Richardson, E.S., Regehr, E.V., Douglas, D.C., Durner, G.M., Atwood, T., Amstrup, S., 2015. Polar bear population dynamics in the southern Beaufort Sea during a period of sea ice decline. Ecological Applications 25, 634–651.

Brown, T.M., Ross, P.S., Reimer, K.J., Veldhoen, N., Dangerfield, N.J., Fisk, A.T., Helbing, C.C., 2014. PCB related effects thresholds as derived through gene transcript profiles in locally contaminated ringed seals (*Pusa hispida*). Environmental Science and Technology 48, 12952–12961.

Brown, T.M., Iverson, S.J., Fisk, A.T., Macdonald, R.W., Helbing, C.C., Reimer, K.J., 2015. Local contamination, and not feeding preferences, explains elevated PCB concentrations in Labrador ringed seals (*Pusa hispida*). The Science of the Total Environment 515, 188–197.

Brown, T.M., Fisk, A.T., Wang, X.W., Ferguson, S.H., Young, B.G., Reimer, K.J., Muir, D.C.G., 2016. Mercury and cadmium in ringed seals in the Canadian Arctic: influence of location and diet. The Science of the Total Environment 545, 503–511.

Butt, C.M., Mabury, S.A., Kwan, M., Wang, X., Muir, D.C.G., 2008. Spatial trends of perfluoroalkyl compounds in ringed seals (*Phoca hispida*) from the Canadian Arctic. Environmental Toxicology and Chemistry 27, 542–553.

Bytingsvik, J., Lie, E., Aars, J., Derocher, A.E., Wiig, O., Jenssen, B.M., 2012a. PCBs and OH-PCBs in polar bear mother-cub pairs: a comparative study based on plasma levels in 1998 and 2008. The Science of the Total Environment 417, 117–128.

Bytingsvik, J., Van Leeuwen, S.P.J., Hamers, T., Swart, K., Aars, J., Lie, E., Nilsen, E.M.E., Wiig, O., Derocher, A.E., Jenssen, B.M., 2012b. Perfluoroalkyl substances in polar bear mother-cub pairs: a comparative study based on plasma levels from 1998 and 2008. Environment International 49, 92–99.

Bytingsvik, J., Simon, E., Leonards, P.E.G., Lamoree, M., Lie, E., Aars, J., Derocher, A.E., Wiig, O., Jenssen, B.M., Hamers, T., 2013. Transthyretin-binding activity of contaminants in blood from polar bear (*Ursus maritimus*) cubs. Environmental Science and Technology 47, 4778–4786.

Callaghan, T.V., Björn, L.O., Chernov, Y., Chapin, T., Christensen, T.R., Huntley, B., Ims, R.A., Johansson, M., Jolly, D., Jonasson, S., Matveyeva, N., Panikov, N., Oechel, W., Shaver, G., Elster, J., Henttonen, H., Laine, K., Taulavuori, K., Taulavuori, E., Zöckler, C., 2004. Biodiversity, distributions and adaptations of Arctic species in the context of environmental change. Ambio 33, 404–417.

Casals-Casas, C., Desvergne, B., 2011. Endocrine disruptors: from endocrine to metabolic disruption. Annual Review of Physiology 73, 135–162.

Castelli, M.G., Rusten, M., Goksøyr, A., Routti, H., 2014. mRNA expression of genes regulating lipid metabolism in ringed seals (*Pusa hispida*) from differently polluted areas. Aquatic Toxicology 146, 239–246.

Cattet, M., 2000. Biochemical and Physiological Aspects of Obesity, High Fat Diet, and Prolonged Fasting in Free-Ranging Polar Bears (Ph.D. dissertation). University of Saskatchewan, Saskatchewan.

Cattet, M.R.L., Caulkett, N.A., Obbard, M.E., Stenhouse, G.B., 2002. A body-condition index for ursids. Canadian Journal of Zoology 80, 1156–1161.

Christensen, J.R., Macduffee, M., Yunker, M.B., Ross, P.S., 2007. Hibernation-associated changes in persistent organic pollutant (POP) levels and patterns in British Columbia grizzly bears (*Ursus arctos horribilis*). Environmental Science and Technology 41, 1834–1840.

Ciesielski, T.M., Hansen, I.T., Bytingsvik, J., Hansen, M., Lie, E., Aars, J., Jenssen, B.M., Styrishave, B., 2017. Relationships between POPs, biometrics and circulating steroid levels in male polar bears (*Ursus maritimus*) from Svalbard. Environmental Pollution 230, 598–608.

D'eon, J.C., Hurley, M.D., Wallington, T.J., Mabury, S.A., 2006. Atmospheric chemistry of *N*-methyl perfluorobutane sulfonamidoethanol, $C_4F_9SO_2N(CH_3)CH_2CH_2OH$: kinetics and mechanism of reaction with OH. Environmental Science and Technology 40, 1862–1868.

Deguise, S., Martineau, D., Beland, P., Fournier, M., 1995. Possible mechanisms of action of environmental contaminants on St-Lawrence beluga whales (*Delphinapterus-leucas*). Environmental Health Perspectives 103, 73–77.

Dehn, L.-A., Follmann, E.H., Thomas, D.L., Sheffield, G.G., Rosa, C., Duffy, L.K., O'hara, T.M., 2006. Trophic relationships in an Arctic food web and implications for trace metal transfer. The Science of the Total Environment 362, 103–123.

Derocher, A.E., Stirling, I., 1994. Age–specific reproductive performance of female polar bears (*Ursus maritimus*). Journal of Zoology 234, 527–536.

Derocher, A.E., Wiig, Ø., Andersen, M., 2002. Diet composition of polar bears in Svalbard and the western Barents Sea. Polar Biology 25, 448–452.

Descamps, S., Aars, J., Fuglei, E., Kovacs, K.M., Lydersen, C., Pavlova, O., Pedersen, Å.Ø., Ravolainen, V., Strøm, H., 2016. Climate change impacts on wildlife in a High Arctic archipelago – Svalbard, Norway. Global Change Biology 490–502.

Desforges, J.-P.W., Sonne, C., Levin, M., Siebert, U., De Guise, S., Dietz, R., 2016. Immunotoxic effects of environmental pollutants in marine mammals. Environment International 86, 126–139.

Dietz, R., Riget, F., Born, E.W., 2000. An assessment of selenium to mercury in Greenland marine animals. The Science of the Total Environment 245, 15–24.

Dietz, R., Riget, F.F., Sonne, C., Born, E.W., Bechshoft, T., Mckinney, M.A., Drimmie, R.J., Muir, D.C.G., Letcher, R.J., 2013a. Three decades (1983-2010) of contaminant trends in East Greenland polar bears (*Ursus maritimus*). Part 2: brominated flame retardants. Environment International 59, 494–500.

Dietz, R., Sonne, C., Basu, N., Braune, B., O'hara, T., Letcher, R.J., Scheuhammer, T., Andersen, M., Andreasen, C., Andriashek, D., Asmund, G., Aubail, A., Baagoe, H., Born, E.W., Chan, H.M., Derocher, A.E., Grandjean, P., Knott, K., Kirkegaard, M., Krey, A., Lunn, N., Messier, F., Obbard, M., Olsen, M.T., Ostertag, S., Peacock, E., Renzoni, A., Riget, F.F., Skaare, J.U., Stern, G., Stirling, I., Taylor, M., Wiig, O., Wilson, S., Aars, J., 2013b. What are the toxicological effects of mercury in Arctic biota? The Science of the Total Environment 443, 775–790.

Dietz, R., Gustavson, K., Sonne, C., Desforges, J.P., Riget, F.F., Pavlova, V., Mckinney, M.A., Letcher, R.J., 2015. Physiologically-based pharmacokinetic modelling of immune, reproductive and carcinogenic effects from contaminant exposure in polar bears (*Ursus maritimus*) across the Arctic. Environmental Research 140, 45–55.

Dobricic, S., Vignati, E., Russo, S., 2016. Large-scale atmospheric warming in winter and the Arctic Sea ice retreat. Journal of Climate 29, 2869–2888.

Durner, G.M., Douglas, D.C., Nielson, R.M., Amstrup, S.C., Mcdonald, T.L., Stirling, I., Mauritzen, M., Born, E.W., Wiig, O., Deweaver, E., Serreze, M.C., Belikov, S.E., Holland, M.M., Maslanik, J., Aars, J., Bailey, D.A., Derocher, A.E., 2009. Predicting 21st-century polar bear habitat distribution from global climate models. Ecological Monographs 79, 25–58.

Ellingsen, I.H., Dalpadado, P., Slagstad, D., Loeng, H., 2008. Impact of climatic change on the biological production in the Barents Sea. Climatic Change 87, 155–175.

Ellis, D.A., Martin, J.W., De Silva, A.O., Mabury, S.A., Hurley, M.D., Andersen, M.P.S., Wallington, T.J., 2004. Degradation of fluorotelomer alcohols: a likely atmospheric source of perfluorinated carboxylic acids. Environmental Science and Technology 38, 3316–3321.

Falk-Petersen, S., Hop, H., Budgell, W.P., Hegseth, E.N., Korsnes, R., Løyning, T.B., Børre Ørbæk, J., Kawamura, T., Shirasawa, K., 2000. Physical and ecological processes in the marginal ice zone of the northern Barents Sea during the summer melt period. Journal of Marine Systems 27, 131–159.

Falk-Petersen, S., Haug, T., Hop, H., Nilssen, K.T., Wold, A., 2009. Transfer of lipids from plankton to blubber of harp and hooded seals off East Greenland. Deep Sea Research Part II: Topical Studies in Oceanography 56, 2080–2086.

Fisk, A.T., Hobson, K.A., Norstrom, R.J., 2001. Influence of chemical and biological factors on trophic transfer of persistent organic pollutants in the northwater polynya marine food web. Environmental Science and Technology 35, 732–738.

Frank, W., Dugani, C.B., 2003. Assessing the long-range transport potential of polybrominated diphenyl ethers: a comparison of four multimedia models. Environmental Toxicology and Chemistry 22, 1252–1261.

Fratev, F., Tsakovska, I., Al Sharif, M., Mihaylova, E., Pajeva, I., 2015. Structural and dynamical Insight into PPARγ antagonism: in silico study of the ligand-receptor interactions of non-covalent antagonists. International Journal of Molecular Sciences 16, 15405.

Gabrielsen, K.M., Krokstad, J.S., Villanger, G.D., Blair, D.A.D., Obregon, M.-J., Sonne, C., Dietz, R., Letcher, R.J., Jenssen, B.M., 2015. Thyroid hormones and deiodinase activity in plasma and tissues in relation to high levels of organohalogen contaminants in East Greenland polar bears (*Ursus maritimus*). Environmental Research 136, 413–423.

Gajdosechova, Z., Lawan, M.M., Urgast, D.S., Raab, A., Scheckel, K.G., Lombi, E., Kopittke, P.M., Loeschner, K., Larsen, E.H., Woods, G., Brownlow, A., Read, F.L., Feldmann, J., Krupp, E.M., 2016. In vivo formation of natural HgSe nanoparticles in the liver and brain of pilot whales. Scientific Reports 6, 11.

Galicia, M.P., Thiemann, G.W., Dyck, M.G., Ferguson, S.H., Higdon, J.W., 2016. Dietary habits of polar bears in Foxe Basin, Canada: possible evidence of a trophic regime shift mediated by a new top predator. Ecology and Evolution 6, 6005–6018.

Gebbink, W.A., Sonne, C., Dietz, R., Kirkegaard, M., Riget, F.F., Born, E., Muir, D.C.G., Letcher, R., 2008. Tissue-specific congener composition of organohalogen and metabolite contaminants in East Greenland polar bears (*Ursus maritimus*). Environmental Pollution 152, 621–629.

Gebbink, W.A., Bossi, R., Riget, F.F., Rosing-Asvid, A., Sonne, C., Dietz, R., 2016. Observation of emerging per- and polyfluoroalkyl substances (PFASs) in Greenland marine mammals. Chemosphere 144, 2384–2391.

Grebmeier, J.M., Overland, J.E., Moore, S.E., Farley, E.V., Carmack, E.C., Cooper, L.W., Frey, K.E., Helle, J.H., McLaughlin, F.A., McNutt, S.L., 2006. A major ecosystem shift in the Northern Bering Sea. Science 311, 1461–1464.

Grønnestad, R., Villanger, G.D., Polder, A., Kovacs, K.M., Lydersen, C., Jenssen, B.M., Borgå, K., 2017. Maternal transfer of perfluoroalkyl substances in hooded seals. Environmental Toxicology and Chemistry 36, 763–770.

Grün, F., Blumberg, B., 2009. Endocrine disrupters as obesogens. Molecular and Cellular Endocrinology 304, 19–29.

Gustavson, L., Ciesielski, T.M., Bytingsvik, J., Styrishave, B., Hansen, M., Lie, E., Aars, J., Jenssen, B.M., 2015. Hydroxylated polychlorinated biphenyls decrease circulating steroids in female polar bears (*Ursus maritimus*). Environmental Research 138, 191–201.

Gutleb, A.C., Cenijn, P., Van Velzen, M., Lie, E., Ropstad, E., Skaare, J.U., Malmberg, T., Bergman, A., Gabrielsen, G.W., Legler, J., 2010. In vitro assay shows that PCB metabolites completely saturate thyroid hormone transport capacity in blood of wild polar bears (*Ursus maritimus*). Environmental Science and Technology 44, 3149–3154.

Hakk, H., Letcher, R.J., 2003. Metabolism in the toxicokinetics and fate of brominated flame retardants—a review. Environment International 29, 801–828.

Hallanger, I.G., Warner, N.A., Ruus, A., Evenset, A., Christensen, G., Herzke, D., Gabrielsen, G.W., Borgå, K., 2011. Seasonality in contaminant accumulation in Arctic marine pelagic food webs using trophic magnification factor as a measure of bioaccumulation. Environmental Toxicology and Chemistry 30 (5), 1026–1035.

Hallanger, I.G., Sagerup, K., Evenset, A., Kovacs, K.M., Leonards, P., Fuglei, E., Routti, H., Aars, J., Strøm, H., Lydersen, C., Gabrielsen, G.W., 2015. Organophosphorous flame retardants in biota from Svalbard, Norway. Marine Pollution Bulletin 101, 442–447.

Heindel, J.J., Newbold, R., Schug, T.T., 2015. Endocrine disruptors and obesity. Nature Reviews Endocrinology 11, 653–661.

Helgason, L.B., Wolkers, H., Fuglei, E., Ahlstrom, O., Muir, D., Jorgensen, E.H., 2013. Seasonal emaciation causes tissue redistribution and an increased potential for toxicity of lipophilic pollutants in farmed arctic fox (*Vulpes lagopus*). Environmental Toxicology and Chemistry 32, 1784–1792.

Helland-Hansen, B., Nansen, F., 1909. The Norwegian Sea - its Physical Oceanography Based upon the Norwegian Researches 1900-1904, p. 422.

Helle, E., Olsson, M., Jensen, S., 1976. PCB levels correlated with pathological changes in seal uteri. Ambio 5, 261–263.

Hobson, K.A., Fisk, A., Karnovsky, N., Holst, M., Gagnon, J.-M., Fortier, M., 2002. A stable isotope ($\delta^{13}C$, $\delta^{15}N$) model for the North Water food web: implications for evaluating trophodynamics and the flow of energy and contaminants. Deep Sea Research Part II: Topical Studies in Oceanography 49, 5131–5150.

Hoegh-Guldberg, O., Bruno, J.F., 2010. The impact of climate change on the world's marine ecosystems. Science 328, 1523–1528.

Hoekstra, P.F., Letcher, R.J., O'hara, T.M., Backus, S.M., Solomon, K.R., Muir, D.C.G., 2003. Hydroxylated and methylsulfone-containing metabolites of polychlorinated biphenyls in the plasma and blubber of bowhead whales (*Balaena mysticetus*). Environmental Toxicology and Chemistry 22, 2650–2658.

Hop, H., Poltermann, M., Lønne, O.J., Falk-Petersen, S., Korsnes, R., Budgell, W.P., 2000. Ice amphipod distribution relative to ice density and under-ice topography in the northern Barents Sea. Polar Biology 23, 357–367.

Ikemoto, T., Kunito, T., Tanaka, H., Baba, N., Miyazaki, N., Tanabe, S., 2004. Detoxification mechanism of heavy metals in marine mammals and seabirds: interaction of selenium with mercury, silver, copper, zinc, and cadmium in liver. Archives of Environmental Contamination and Toxicology 47, 402–413.

Jenssen, B.M., Sormo, E.G., Baek, K., Bytingsvik, J., Gaustad, H., Ruus, A., Skaare, J.U., 2007. Brominated flame retardants in North-East Atlantic ecosystems. Environmental Health Perspectives 115 (Suppl. 1), 35–41.

Jenssen, B.M., Villanger, G.D., Gabrielsen, K.M., Bytingsvik, J., Bechshøft, T.Ø., Ciesielski, T.M., Sonne, C., Dietz, R., 2015. Anthropogenic flank attack on polar bears: interacting consequences of climate warming and pollutant exposure. Frontiers in Ecology and Evolution 3 (16), 1–7.

Kanerva, M., Routti, H., Tamuz, Y., Nyman, M., Nikinmaa, M., 2012. Antioxidative defense and oxidative stress in ringed seals (*Pusa hispida*) from differently polluted areas. Aquatic Toxicology 114–115, 67–72.

Kelly, B.C., Ikonomou, M.G., Blair, J.D., Morin, A.E., Gobas, F.A.P.C., 2007. Food web specific biomagnification of persistent organic pollutants. Science 317, 236–239.

Kelly, B.C., Ikonomou, M.G., Blair, J.D., Surridge, B., Hoover, D., Grace, R., Gobas, F., 2009. Perfluoroalkyl contaminants in an Arctic marine food web: trophic magnification and wildlife exposure. Environmental Science and Technology 43, 4037–4043.

Kliewer, S.A., Moore, J.T., Wade, L., Staudinger, J.L., Watson, M.A., Jones, S.A., Mckee, D.D., Oliver, B.B., Willson, T.M., Zetterström, R.H., Perlmann, T., Lehmann, J.M., 1998. An orphan nuclear receptor activated by pregnanes defines a novel steroid signaling pathway. Cell 92, 73–82.

Koenigk, T., Brodeau, L., Graversen, R.G., Karlsson, J., Svensson, G., Tjernström, M., Willén, U., Wyser, K., 2013. Arctic climate change in 21st century CMIP5 simulations with EC-Earth. Climate Dynamics 40, 2719–2743.

Kovacs, K.M., Lydersen, C., Overland, J.E., Moore, S.E., 2011. Impacts of changing sea-ice conditions on Arctic marine mammals. Marine Biodiversity 41, 181–194.

Krey, A., Kwan, M., Chan, H.M., 2014. In vivo and in vitro changes in neurochemical parameters related to mercury concentrations from specific brain regions of polar bears (*Ursus maritimus*). Environmental Toxicology and Chemistry 33, 2463–2471.

Krey, A., Ostertag, S.K., Chan, H.M., 2015. Assessment of neurotoxic effects of mercury in beluga whales (*Delphinapterus leucas*), ringed seals (*Pusa hispida*), and polar bears (*Ursus maritimus*) from the Canadian Arctic. The Science of the Total Environment 509–510, 237–247.

Krieger, L.K., Szeitz, A., Bandiera, S.M., 2016. Evaluation of hepatic biotransformation of polybrominated diphenyl ethers in the polar bear (*Ursus maritimus*). Chemosphere 146, 555–564.

Kuzyk, Z.A., Stow, J.P., Burgess, N.M., Solomon, S.M., Reimer, K.J., 2005. PCBs in sediments and the coastal food web near a local contaminant source in Saglek Bay, Labrador. The Science of the Total Environment 351–352, 264–284.

Laidre, K.L., Stirling, I., Lowry, L.F., Wiig, Ø., Heide-Jørgensen, M.P., Ferguson, S.H., 2008. Quantifying the sensitivity of arctic marine mammals to climate-induced habitat change. Ecological Applications 18, 97–125.

Laidre, K.L., Stern, H., Kovacs, K.M., Lowry, L., Moore, S.E., Regehr, E.V., Ferguson, S.H., Wiig, Ø., Boveng, P., Angliss, R.P., Born, E.W., Litovka, D., Quakenbush, L., Lydersen, C., Vongraven, D., Ugarte, F., 2015. Arctic marine mammal population status, sea ice habitat loss, and conservation recommendations for the 21st century. Conservation Biology 29, 724–737.

Lee, D.-H., Porta, M., Jacobs Jr., D.R., Vandenberg, L.N., 2014. Chlorinated persistent organic pollutants, obesity, and type 2 diabetes. Endocrine Reviews 35, 557–601.

Leesnitzer, L.M., Parks, D.J., Bledsoe, R.K., Cobb, J.E., Collins, J.L., Consler, T.G., Davis, R.G., Hull-Ryde, E.A., Lenhard, J.M., Patel, L., Plunket, K.D., Shenk, J.L., Stimmel, J.B., Therapontos, C., Willson, T.M., Blanchard, S.G., 2002. Functional consequences of cysteine modification in the ligand binding sites of peroxisome proliferator activated receptors by GW9662. Biochemistry 41, 6640–6650.

Lefterova, M.I., Haakonsson, A.K., Lazar, M.A., Mandrup, S., 2014. PPARγ and the global map of adipogenesis and beyond. Trends in Endocrinology and Metabolism 25, 293–302.

Legler, J., Fletcher, T., Govarts, E., Porta, M., Blumberg, B., Heindel, J.J., Trasande, L., 2015. Obesity, diabetes, and associated costs of exposure to endocrine-disrupting chemicals in the European Union. Journal of Clinical Endocrinology and Metabolism 100, 1278–1288.

Letcher, R.J., Norstrom, R.J., Lin, S., Ramsay, M.A., Bandiera, S.M., 1996. Immunoquantitation and microsomal monooxygenase activities of hepatic cytochromes P4501A and P4502B and chlorinated hydrocarbon contaminant levels in polar bear (*Ursus maritimus*). Toxicology and Applied Pharmacology 137, 127–140.

Letcher, R.J., Norstrom, R.J., Muir, D.G.G., 1998. Biotransformation versus bioaccumulation: sources of methyl sulfone PCB and 4,4′-DDE metabolites in the polar bear food chain. Environmental Science and Technology 32, 1656–1661.

Letcher, R.J., Klasson-Wehler, E., Bergman, Å., Paasivirta, J., 2000. Methyl sulfone and hydroxylated metabolites of polychlorinated biphenyls. In: Paasivirta, J. (Ed.), The Handbook of Environmental Chemistry. Springer-Verlag, Berlin, Heidelberg.

Letcher, R.J., Gebbink, W.A., Sonne, C., Born, E.W., Mckinney, M.A., Dietz, R., 2009. Bioaccumulation and biotransformation of brominated and chlorinated contaminants and their metabolites in ringed seals (*Pusa hispida*) and polar bears (*Ursus maritimus*) from East Greenland. Environment International 35, 1118–1124.

Letcher, R.J., Bustnes, J.O., Dietz, R., Jenssen, B.M., Jorgensen, E.H., Sonne, C., Verreault, J., Vijayan, M.M., Gabrielsen, G.W., 2010. Exposure and effects assessment of persistent organohalogen contaminants in arctic wildlife and fish. The Science of the Total Environment 408, 2995–3043.

Letcher, R.J., Chu, S., Mckinney, M.A., Tomy, G.T., Sonne, C., Dietz, R., 2014. Comparative hepatic in vitro depletion and metabolite formation of major perfluorooctane sulfonate precursors in arctic polar bear, beluga whale, and ringed seal. Chemosphere 112, 225–231.

Li, Y.F., Macdonald, R.W., 2005. Sources and pathways of selected organochlorine pesticides to the Arctic and the effect of pathway divergence on HCH trends in biota: a review. The Science of the Total Environment 342, 87–106.

Lie, E., Bernhoft, A., Riget, F., Belikov, S.E., Boltunov, A.N., Derocher, A.E., Garner, G.W., Wiig, O., Skaare, J.U., 2003. Geographical distribution of organochlorine pesticides (OCPs) in polar bears (*Ursus maritimus*) in the Norwegian and Russian Arctic. The Science of the Total Environment 306, 159–170.

Lie, E., Larsen, H.J.S., Larsen, S., Johansen, G.M., Derocher, A.E., Lunn, N.J., Norstrom, R.J., Wiig, Ø., Skaare, J.U., 2004. Does high organochlorine (OC) exposure impair the resistance to infection in polar bears (*Ursus maritimus*)? Part I: effect of OCs on the humoral immunity. Journal of Toxicology and Environmental Health, Part A: Current Issues 67, 555–582.

Lie, E., Larsen, H.J.S., Larsen, S., Johansen, G.M., Derocher, A.E., Lunn, N.J., Norstrom, R.J., Wiig, O., Skaare, J.U., 2005. Does high organochlorine (OC) exposure impair the resistance to infection in polar bears (*Ursus maritimus*)? Part II: Possible effect of OCs on mitogen- and antigen-induced lymphocyte proliferation. Journal of Toxicology and Environmental Health, Part A: Current Issues 68, 457–484.

Lille-Langøy, R., Goldstone, J.V., Rusten, M., Milnes, M.R., Male, R., Stegeman, J.J., Blumberg, B., Goksøyr, A., 2015. Environmental contaminants activate human and polar bear (*Ursus maritimus*) pregnane X receptors (PXR, NR1I2) differently. Toxicology and Applied Pharmacology 284, 54–64.

Liu, S., Lorenzen, E.D., Fumagalli, M., Li, B., Harris, K., Xiong, Z., Zhou, L., Korneliussen, T.S., Somel, M., Babbitt, C., Wray, G., Li, J., He, W., Wang, Z., Fu, W., Xiang, X., Morgan, C.C., Doherty, A., O'Connell, M.J., McInerney, J.O., Born, E.W., Dalén, L., Dietz, R., Orlando, L., Sonne, C., Zhang, G., Nielsen, R., Willerslev, E., Wang, J., 2014. Population genomics reveal recent speciation and rapid evolutionary adaptation in polar bears. Cell 157, 785–794.

Lockhart, W.L., Stern, G.A., Wagemann, R., Hunt, R.V., Metner, D.A., Delaronde, J., Dunn, B., Stewart, R.E.A., Hyatt, C.K., Harwood, L., Mount, K., 2005. Concentrations of mercury in tissues of beluga whales (*Delphinapterus leucas*) from several communities in the Canadian Arctic from 1981 to 2002. The Science of the Total Environment 351, 391–412.

Lone, K., Merkel, B., Lydersen, C., Kovacs, K.M., Aars, J., 2017. Sea ice resource selection models for polar bears in the Barents Sea subpopulation. Ecography. https://doi.org/10.1111/ecog.03020.

Lønø, O., 1970. The Polar Bear (*Ursus maritimus* Phipps) in the Svalbard Area (Ph.D. dissertation).

Lunn, N.J., Servanty, S., Regehr, E.V., Converse, S.J., Richardson, E., Stirling, I., 2016. Demography of an apex predator at the edge of its range: impacts of changing sea ice on polar bears in Hudson Bay. Ecological Applications 26, 1302–1320.

Lydersen, C., Wolkers, H., Severinsen, T., Kleivane, L., Nordøy, E.S., Skaare, J.U., 2002. Blood is a poor substrate for monitoring pollution burdens in phocid seals. The Science of the Total Environment 292, 193–203.

Ma, J., 2010. Atmospheric transport of persistent semi-volatile organic chemicals to the Arctic and cold condensation in the mid-troposphere – part 1: 2-D modeling in mean atmosphere. Atmospheric Chemistry and Physics 10, 7303–7314.

Macdonald, R.W., Harner, T., Fyfe, J., 2005. Recent climate change in the Arctic and its impact on contaminant pathways and interpretation of temporal trend data. The Science of the Total Environment 342, 5–86.

Martin, J.W., Ellis, D.A., Mabury, S.A., Hurley, M.D., Wallington, T.J., 2006. Atmospheric chemistry of perfluoroalkanesulfonamides: kinetic and product studies of the OH radical and Cl atom initiated oxidation of *N*-ethyl perfluorobutanesulfonamide. Environmental Science and Technology 40, 864–872.

Martineau, D., Deguise, S., Fournier, M., Shugart, L., Girard, C., Lagace, A., Beland, P., 1994. Pathology and toxicology of beluga whales from the St-Lawrence estuary, Quebec, Canada - past, present and future. The Science of the Total Environment 154, 201–215.

McKinney, M.A., Arukwe, A., De Guise, S., Martineau, D., Beland, P., Dallaire, A., Lair, S., Lebeuf, M., Letcher, R.J., 2004. Characterization and profiling of hepatic cytochromes P450 and phase II xenobiotic-metabolizing enzymes in beluga whales (*Delphinapterus leucas*) from the St. Lawrence River Estuary and the Canadian Arctic. Aquatic Toxicology 69, 35–49.

McKinney, M.A., De Guise, S., Martineau, D., Béland, P., Arukwe, A., Letcher, R.J., 2006a. Biotransformation of polybrominated diphenyl ethers and polychlorinated biphenyls in beluga whale (*Delphinapterus leucas*) and rat mammalian model using an in vitro hepatic microsomal assay. Aquatic Toxicology 77, 87–97.

McKinney, M.A., De Guise, S., Martineau, D., Beland, P., Lebeuf, M., Letcher, R., 2006b. Organohalogen contaminants and metabolites in beluga whale (*Delphinapterus leucas*) liver from two Canadian populations. Environmental Toxicology and Chemistry 25, 1246–1257.

McKinney, M.A., Peacock, E., Letcher, R.J., 2009. Sea Ice-associated diet change increases the levels of chlorinated and brominated contaminants in polar bears. Environmental Science and Technology 43, 4334–4339.

McKinney, M.A., Stirling, I., Lunn, N.J., Peacock, E., Letcher, R.J., 2010. The role of diet on long-term concentration and pattern trends of brominated and chlorinated contaminants in western Hudson Bay polar bears, 1991-2007. The Science of the Total Environment 408, 6210–6222.

McKinney, M.A., Dietz, R., Sonne, C., De Guise, S., Skirnisson, K., Karlsson, K., Steingrimsson, E., Letcher, R.J., 2011a. Comparative hepatic microsomal biotransformation of selected PBDEs, including decabromodiphenyl ether, and decabromodiphenyl ethane flame retardants in arctic marine-feeding mammals. Environmental Toxicology and Chemistry 30, 1506–1514.

McKinney, M.A., Letcher, R., Aars, J., Born, E., Branigan, M., Dietz, R., Evans, T., Gabrielsen, G.W., Peacock, E., Sonne, C., 2011b. Flame retardants and legacy contaminants in polar bears from Alaska, Canada, East Greenland and Svalbard, 2005-2008. Environmental International 37, 365–374.

McKinney, M.A., Letcher, R.J., Aars, J., Born, E.W., Branigan, M., Dietz, R., Evans, T.J., Gabrielsen, G.W., Muir, D.C.G., Peacock, E., Sonne, C., 2011c. Regional contamination versus regional dietary differences: understanding geographic variation in brominated and chlorinated contaminant levels in polar bears. Environmental Science and Technology 45, 896–902.

McKinney, M.A., Mcmeans, B.C., Tomy, G.T., Rosenberg, B., Ferguson, S.H., Morris, A., Muir, D.C.G., Fisk, A.T., 2012. Trophic transfer of contaminants in a changing Arctic marine food web: Cumberland Sound, Nunavut, Canada. Environmental Science and Technology 46, 9914–9922.

McKinney, M.A., Iverson, S.J., Fisk, A.T., Sonne, C., Rigét, F.F., Letcher, R.J., Arts, M.T., Born, E.W., Rosing-Asvid, A., Dietz, R., 2013. Global change effects on the long-term feeding ecology and contaminant exposures of East Greenland polar bears. Global Change Biology 19, 2360–2372.

McKinney, M.A., Pedro, S., Dietz, R., Sonne, C., Fisk, A.T., Roy, D., Jenssen, B.M., Letcher, R.J., 2015. A review of ecological impacts of global climate change on persistent organic pollutant and mercury pathways and exposures in arctic marine ecosystems. Current Zoology 61, 617–628.

McKinney, M.A., Atwood, T.C., Iverson, S.J., Peacock, E., 2017. Temporal complexity of southern Beaufort Sea polar bear diets during a period of increasing land use. Ecosphere 8, e01633.

Messier, F., Taylor, M.K., Ramsay, M.A., 1992. Seasonal activity patterns of female polar bears (*Ursus maritimus*) in the Canadian Arctic as revealed by satellite telemetry. Journal of Zoology 226, 219–229.

Moore, S.E., Huntington, H.P., 2008. Arctic marine mammals and climate change: impacts and resilience. Ecological Applications 18, 157–165.

Morris, A.D., Muir, D.C.G., Solomon, K.R., Letcher, R.J., Mckinney, M.A., Fisk, A.T., Mcmeans, B.C., Tomy, G.T., Teixeira, C., Wang, X.W., Duric, M., 2016. Current-use pesticides in seawater and their bioaccumulation in polar bear-ringed seal food chains of the Canadian Arctic. Environmental Toxicology and Chemistry 35, 1695–1707.

Muir, D., Riget, F., Cleemann, M., Skaare, J., Kleivane, L., Nakata, H., Dietz, R., Severinsen, T., Tanabe, S., 2000. Circumpolar trends of PCBs and organochlorine pesticides in the arctic marine environment inferred from levels in ringed seals. Environmental Science and Technology 34, 2431–2438.

Muir, D.C.G., Backus, S., Derocher, A.E., Dietz, R., Evans, T.J., Gabrielsen, G.W., Nagy, J., Norstrom, R.J., Sonne, C., Stirling, I., Taylor, M.K., Letcher, R.J., 2006. Brominated flame retardants in polar bears (*Ursus maritimus*) from Alaska, the Canadian Arctic, East Greenland, and Svalbard. Environmental Science and Technology 40, 449–455.

Muir, D.C.G., Kurt-Karakus, P., Stow, J., 2013. Chapter 4 — Occurrence and Trends in the Biological Environment. Canadian Arctic Contaminants Assessment Report On Persistent Organic Pollutants. Aboriginal Affairs and Northern Development Canada, Ottawa, ON.

Nelson, R., Jones, J., Wahner, H., McGill, D., Code, C., 1975. Nitrogen metabolism in bears: urea metabolism in summer starvation and in winter sleep and role of urinary bladder in water and nitrogen conservation. Mayo Clinic Proceedings 50, 141–146.

Niwa, T., Yabusaki, Y., Honma, K., Matsuo, N., Tatsuta, K., Ishibashi, F., Katagiri, M., 1998. Contribution of human hepatic cytochrome P450 isoforms to regioselective hydroxylation of steroid hormones. Xenobiotica 28, 539–547.

Noel, M., Loseto, L.L., Helbing, C.C., Veldhoen, N., Dangerfield, N.J., Ross, P.S., 2014. PCBs are associated with altered gene transcript profiles in Arctic beluga whales (*Delphinapterus leucas*). Environmental Science and Technology 48, 2942–2951.

Norstrom, R.J., Muir, D.C.G., 1994. Chlorinated-hydrocarbon contaminants in arctic marine mammals. The Science of the Total Environment 154, 107–128.

Norstrom, R.J., Schweinsberg, R.E., Collins, B.T., 1986. Heavy metals and essential elements in livers of the polar bear (*Ursus Maritimus*) in the Canadian Arctic. The Science of the Total Environment 48, 195–212.

Nuijten, R.J.M., Hendriks, A.J., Jenssen, B.M., Schipper, A.M., 2016. Circumpolar contaminant concentrations in polar bears (*Ursus maritimus*) and potential population-level effects. Environmental Research 151, 50–57.

Nyman, M., Raunio, H., Taavitsainen, P., Pelkonen, O., 2001. Characterization of xenobiotic-metabolizing cytochrome P450 (CYP) forms in ringed and grey seals from the Baltic Sea and reference sites. Comparative Biochemistry and Physiology Part C 128, 99–112.

Nyman, M., Koistinen, J., Fant, M.L., Vartiainen, T., Helle, E., 2002. Current levels of DDT, PCB and trace elements in the Baltic ringed seals (*Phoca hispida baltica*) and grey seals (*Halichoerus grypus*). Environmental Pollution 119, 399–412.

Nyman, M., Bergknut, M., Fant, M.L., Raunio, H., Jestoi, M., Bengs, C., Murk, A., Koistinen, J., Backman, C., Pelkonen, O., Tysklind, M., Hirvi, T., Helle, E., 2003. Contaminant exposure and effects in Baltic ringed and grey seals as assessed by biomarkers. Marine Environmental Research 55, 73–99.

Obbard, M.E., Cattet, M.R.L., Howe, E.J., Middel, K.R., Newton, E.J., Kolenosky, G.B., Abraham, K.F., Greenwood, C.J., 2016. Trends in body condition in polar bears (*Ursus maritimus*) from the Southern Hudson Bay subpopulation in relation to changes in sea ice. Arctic Science 2, 15–32.

Ohashi, M., Gamo, K., Tanaka, Y., Waki, M., Beniyama, Y., Matsuno, K., Wada, J., Tenta, M., Eguchi, J., Makishima, M., Matsuura, N., Oyama, T., Miyachi, H., 2015. Structural design and synthesis of arylalkynyl amide-type peroxisome proliferator-activated receptor γ (PPARγ)-selective antagonists based on the helix12-folding inhibition hypothesis. European Journal of Medicinal Chemistry 90, 53–67.

Olsen, G.H., Mauritzen, M., Derocher, A.E., Sormo, E.G., Skaare, J.U., Wiig, O., Jenssen, B.M., 2003. Space-use strategy is an important determinant of PCB concentrations in female polar bears in the barents sea. Environmental Science and Technology 37, 4919–4924.

Olsson, M., Karlsson, B., Ahnland, E., 1994. Diseases and environmental contaminants in seals from the Baltic and the Swedish west coast. The Science of the Total Environment 154, 217–227.

Oskam, I.C., Ropstad, E., Dahl, E., Lie, E., Derocher, A.E., Wiig, Ø., Larsen, S., Wiger, R., Skaare, J.U., 2003. Organochlorines affect the major androgenic hormone, testosterone, in male polar bears (*Ursus maritimus*) at Svalbard. Journal of Toxicology and Environmental Health, Part A: Current Issues 66A, 2119–2139.

Ostertag, S.K., Shaw, A.C., Basu, N., Chan, H.M., 2014. Molecular and neurochemical biomarkers in Arctic beluga whales (*Delphinapterus leucas*) were correlated to brain mercury and selenium concentrations. Environmental Science and Technology 48, 11551–11559.

Outridge, P.M., Macdonald, R.W., Wang, F., Stern, G.A., Dastoor, A.P., 2008. A mass balance inventory of mercury in the Arctic Ocean. Environmental Chemistry 5, 89–111.

Pavlova, O., Pavlov, V., Gerland, S., 2014. The impact of winds and sea surface temperatures on the Barents Sea ice extent, a statistical approach. Journal of Marine Systems 130, 248–255.

Pavlova, V., Nabe-Nielsen, J., Dietz, R., Sonne, C., Grimm, V., 2016. Allee effect in polar bears: a potential consequence of polychlorinated biphenyl contamination. Proceedings of the Royal Society of London B: Biological Sciences 283, 9.

Pedersen, K.E., Basu, N., Letcher, R., Greaves, A.K., Sonne, C., Dietz, R., Styrishave, B., 2015. Brain region-specific perfluoroalkylated sulfonate (PFSA) and carboxylic acid (PFCA) accumulation and neurochemical biomarker Responses in east Greenland polar Bears (*Ursus maritimus*). Environmental Research 138, 22–31.

Pedersen, K.E., Letcher, R.J., Sonne, C., Dietz, R., Styrishave, B., 2016. Per- and polyfluoroalkyl substances (PFASs) - new endocrine disruptors in polar bears (*Ursus maritimus*)? Environment International 96, 180–189.

Pereira-Fernandes, A., Demaegdt, H., Vandermeiren, K., Hectors, T.L., Jorens, P.G., Blust, R., Vanparys, C., 2013. Evaluation of a screening system for obesogenic compounds: screening of endocrine disrupting compounds and evaluation of the PPAR dependency of the effect. PLoS One 8, e77481.

Pilfold, N.W., Hedman, D., Stirling, I., Derocher, A.E., Lunn, N.J., Richardson, E., 2016. Mass loss rates of fasting polar bears. Physiological and Biochemical Zoology 89, 377–388.

Pilfold, N.W., McCall, A., Derocher, A.E., Lunn, N.J., Richardson, E., 2017. Migratory response of polar bears to sea ice loss: to swim or not to swim. Ecography 40, 189–199.

Pilsner, J.R., Lazarus, A.L., Nam, D.-H., Letcher, R.J., Sonne, C., Dietz, R., Basu, N., 2010. Mercury-associated DNA hypomethylation in polar bear brains via the LUminometric Methylation Assay: a sensitive method to study epigenetics in wildlife. Molecular Ecology 19, 307–314.

Polischuk, S.C., Norstrom, R.J., Ramsay, M.A., 2002. Body burdens and tissue concentrations of organochlorines in polar bears (*Ursus maritimus*) vary during seasonal fasts. Environmental Pollution 118, 29–39.

Post, E., 2017. Implications of earlier sea ice melt for phenological cascades in arctic marine food webs. Food Webs 13, 60–66.

Prop, J., Aars, J., Bårdsen, B.-J., Hanssen, S.A., Bech, C., Bourgeon, S., de Fouw, J., Gabrielsen, G.W., Lang, J., Noreen, E., Oudman, T., Sittler, B., Stempniewicz, L., Tombre, I., Wolters, E., Moe, B., 2015. Climate change and the increasing impact of polar bears on bird populations. Frontiers in Ecology and Evolution 3, 1–12.

Ramsay, M., Stirling, I., 1988. Reproductive biology and ecology of female polar bears (*Ursus maritimus*). Journal of Zoology 214, 601–633.

Regehr, E.V., Laidre, K.L., Akçakaya, H.R., Amstrup, S.C., Atwood, T.C., Lunn, N.J., Obbard, M., Stern, H., Thiemann, G.W., Wiig, Ø., 2016. Conservation status of polar bears (*Ursus maritimus*) in relation to projected sea-ice declines. Biology Letters 12, 20160556.

Rigét, F., Bignert, A., Braune, B., Stow, J., Wilson, S., 2010. Temporal trends of legacy POPs in Arctic biota, an update. The Science of the Total Environment 408, 2874–2884.

Rigét, F., Braune, B., Bignert, A., Wilson, S., Aars, J., Born, E., Dam, M., Dietz, R., Evans, M., Evans, T., Gamberg, M., Gantner, N., Green, N., Gunnlaugsdottir, H., Kannan, K., Letcher, R., Muir, D., Roach, P., Sonne, C., Stern, G., Wiig, Ø., 2011. Temporal trends of Hg in Arctic biota, an update. The Science of the Total Environment 409, 3520–3526.

Rigét, F., Bossi, R., Sonne, C., Vorkamp, K., Dietz, R., 2013. Trends of perfluorochemicals in Greenland ringed seals and polar bears: indications of shifts to decreasing trends. Chemosphere 93, 1607–1614.

Rigét, F., Vorkamp, K., Bossi, R., Sonne, C., Letcher, R.J., Dietz, R., 2016. Twenty years of monitoring of persistent organic pollutants in Greenland biota. A review. Environmental Pollution 217, 114–123.

Routti, H., Letcher, R.J., Arukwe, A., Van Bavel, B., Yoccoz, N.G., Chu, S.G., Gabrielsen, G.W., 2008. Biotransformation of PCBs in relation to phase I and II xenobiotic-metabolizing enzyme activities in ringed seals (*Phoca hispida*) from Svalbard and the Baltic Sea. Environmental Science and Technology 42, 8952–8958.

Routti, H., Letcher, R.J., Van Bavel, B., Arukwe, A., Chu, S., Gabrielsen, G., 2009a. Concentrations, patterns and metabolites of organochlorine pesticides in relation to xenobiotic phase I and II enzyme activities in ringed seals (*Phoca hispida*) from Svalbard and the Baltic Sea. Environmental Pollution 157, 2428–2434.

Routti, H., Letcher, R.J., Van Bavel, B., Chu, S., Gabrielsen, G.W., 2009b. Polybrominated diphenyl ethers and their hydroxylated analogues in ringed seals (*Phoca hispida*) from Svalbard and the Baltic Sea. Environmental Science and Technology 43, 3494–3499.

Routti, H., Arukwe, A., Jenssen, B.M., Letcher, R.J., Nyman, M., Bäckman, C., Gabrielsen, G.W., 2010a. Comparative endocrine disruptive effects of contaminants in ringed seals (*Phoca hispida*) from Svalbard and the Baltic Sea. Comparative Biochemistry and Physiology Part C 152, 306–312.

Routti, H., Jenssen, B.M., Lydersen, C., Bäckman, C., Arukwe, A., Nyman, M., Kovacs, K.M., Gabrielsen, G.W., 2010b. Hormone, vitamin and contaminant status during moulting/fasting period in ringed seals (*Phoca* [Pusa] *hispida*) from Svalbard. Comparative Biochemistry and Physiology Part A 155, 70–76.

Routti, H., Letcher, R., Born, E.W., Branigan, M., Dietz, R., Evans, T.J., Fisk, A.T., Peacock, E., Sonne, C., 2011. Spatio-temporal trends of selected trace elements in liver tissue from polar bears (*Ursus maritimus*) from Alaska, Canada and Greenland. Journal of Environmental Monitoring 13, 2260–2267.

Routti, H., Letcher, R.J., Born, E.W., Branigan, M., Dietz, R., Evans, T.J., Mckinney, M.A., Peacock, E., Sonne, C., 2012. Influence of carbon and lipid sources on variation of mercury and other trace elements in polar bears (*Ursus maritimus*). Environmental Toxicology and Chemistry 31, 2739–2747.

Routti, H., Lydersen, C., Hanssen, L., Kovacs, K.M., 2014. Contaminant levels in the world's northernmost harbor seals (*Phoca vitulina*). Marine Pollution Bulletin 87, 140–146.

Routti, H., Gabrielsen, G.W., Herzke, D., Kovacs, K.M., Lydersen, C., 2016a. Spatial and temporal trends in perfluoroalkyl substances (PFASs) in ringed seals (*Pusa hispida*) from Svalbard. Environmental Pollution 214, 230–238.

Routti, H., Lille-Langøy, R., Berg, M.K., Fink, T., Harju, M., Kristiansen, K., Rostkowski, P., Rusten, M., Sylte, I., Øygarden, L., Goksøyr, A., 2016b. Environmental chemicals modulate polar bear (*Ursus maritimus*) peroxisome proliferator-activated receptor gamma (PPARG) and adipogenesis in vitro. Environmental Science and Technology 50, 10708–10720.

Routti, H., Aars, J., Fuglei, E., Hanssen, L., Lone, K., Polder, A., Pedersen, Å.Ø., Tartu, S., Welker, J.M., Yoccoz, N.G., 2017. Emission changes dwarf the influence of feeding habits on temporal trends of per- and polyfluoroalkyl substances in two Arctic top predators. Environmental Science and Technology 51, 11996–12006.

Rush, S.A., Borgå, K., Dietz, R., Born, E.W., Sonne, C., Evans, T., Muir, D.C.G., Letcher, R.J., Norstrom, R.J., Fisk, A.T., 2008. Geographic distribution of selected elements in the livers of polar bears from Greenland, Canada and the United States. Environmental Pollution 153, 618–626.

Savinov, V., Muir, D.C.G., Svetochev, V., Svetocheva, O., Belikov, S., Boltunov, A., Alekseeva, L., Reiersen, L.-O., Savinova, T., 2011. Persistent organic pollutants in ringed seals from the Russian Arctic. The Science of the Total Environment 409, 2734–2745.

Shen, L., Jobst, K.J., Helm, P.A., Reiner, E.J., Mccrindle, R., Tomy, G.T., Backus, S., Brindle, I.D., Marvin, C.H., 2012. Identification and determination of the dechlorination products of Dechlorane 602 in Great Lakes fish and Arctic beluga whales by gas chromatography–high resolution mass spectrometry. Analytical and Bioanalytical Chemistry 404, 2737–2748.

Simon, E., Bytingsvik, J., Jonker, W., Leonards, P.E.G., De Boer, J., Jenssen, B.M., Lie, E., Aars, J., Hamers, T., Lamoree, M.H., 2011. Blood plasma sample preparation method for the assessment of thyroid hormone-disrupting potency in effect-directed analysis. Environmental Science and Technology 45, 7936–7944.

Simon, E., Van Velzen, M., Brandsma, S.H., Lie, E., Loken, K., De Boer, J., Bytingsvik, J., Jenssen, B.M., Aars, J., Hamers, T., Lamoree, M.H., 2013. Effect-directed analysis to explore the polar bear exposome: identification of thyroid hormone disrupting compounds in plasma. Environmental Science and Technology 47, 8902–8912.

Skaare, J.U., Bernhoft, A., Wiig, O., Norum, K.R., Haug, E., Eide, D.M., Derocher, A.E., 2001. Relationships between plasma levels of organochlorines, retinol and thyroid hormones from polar bears (*Ursus maritimus*) at Svalbard. Journal of Toxicology and Environmental Health, Part A: Current Issues 62, 227–241.

Smedsrud, L.H., Esau, I., Ingvaldsen, R.B., Eldevik, T., Haugan, P.M., Li, C., Lien, V.S., Olsen, A., Omar, A.M., Otterå, O.H., Risebrobakken, B., Sandø, A.B., Semenov, V.A., Sorokina, S.A., 2013. The role of the Barents Sea in the Arctic climate system. Reviews of Geophysics 51, 415–449.

Smith, T.G., Lydersen, C., 1991. Availability of suitable land-fast ice and predation as factors limiting ringed seal populations, *Phoca hispida*, in Svalbard. Polar Research 10, 585–594.

Smithwick, M., Mabury, S.A., Solomon, K.R., Sonne, C., Martin, J.W., Born, E.W., Dietz, R., Derocher, A.E., Letcher, R.J., Evans, T.J., Gabrielsen, G.W., Nagy, J., Stirling, I., Taylor, M.K., Muir, D.C.G., 2005. Circumpolar study of perfluoroalkyl contaminants in polar bears (*Ursus maritimus*). Environmental Science and Technology 39, 5517–5523.

Sonne, C., Dietz, R., Born, E.W., Riget, F.F., Kirkegaard, M., Hyldstrup, L., Letcher, R.J., Muir, D.C.G., 2004. Is bone mineral composition disrupted by organochlorines in East Greenland polar bears (*Ursus maritimus*)? Environmental Health Perspectives 112, 1711–1716.

Sonne, C., Leifsson, P.S., Dietz, R., Born, E.W., Letcher, R.J., Hyldstrup, L., Riget, F.F., Kirkegaard, M., Muir, D.C.G., 2006. Xenoendocrine pollutants may reduce size of sexual organs in East Greenland polar bears (*Ursus maritimus*). Environmental Science and Technology 40, 5668–5674.

Sonne, C., Iburg, T., Leifsson, P.S., Born, E.W., Letcher, R.J., Dietz, R., 2011. Thyroid gland lesions in organohalogen contaminated East Greenland polar bears (*Ursus maritimus*). Toxicological and Environmental Chemistry 93, 789–805.

Sonne, C., Letcher, R., Bechshoft, T.O., Riget, F., Muir, D., Leifsson, P., Born, E., Hyldstrup, L., Basu, N., Kirkegaard, M., Dietz, R., 2012a. Two decades of biomonitoring polar bear health in Greenland: a review. Acta Veterinaria Scandinavica 54, S15.

Sonne, C., Letcher, R.J., Leifsson, P.S., Riget, F.F., Bechshoft, T.O., Bossi, R., Asmund, G., Dietz, R., 2012b. Temporal monitoring of liver and kidney lesions in contaminated East Greenland polar bears (*Ursus maritimus*) during 1999-2010. Environment International 48, 143–149.

Sonne, C., Bechshoft, T.O., Riget, F.F., Baagoe, H.J., Hedayat, A., Andersen, M., Bech-Jensen, J.E., Hyldstrup, L., Letcher, R.J., Dietz, R., 2013. Size and density of East Greenland polar bear (*Ursus maritimus*) skulls: valuable bio-indicators of environmental changes? Ecological Indicators 34, 290–295.

Sonne, C., Dyck, M., Riget, F.F., Jensen, J.E.B., Hyldstrup, L., Letcher, R.J., Gustavson, K., Gilbert, M.T.P., Dietz, R., 2015. Penile density and globally used chemicals in Canadian and Greenland polar bears. Environmental Research 137, 287–291.

Sørmo, E.G., Salmer, M.P., Jenssen, B.M., Hop, H., Baek, K., Kovacs, K.M., Lydersen, C., Falk-Petersen, S., Gabrielsen, G.W., Lie, E., Skaare, J.U., 2006. Biomagnification of polybrominated diphenyl ether and hexabromocyclododecane flame retardants in the polar bear food chain in Svalbard, Norway. Environmental Toxicology and Chemistry 25, 2502–2511.

Stempniewicz, L., 2006. Polar bear predatory behaviour toward molting barnacle geese and nesting glaucous gulls on Spitsbergen. Arctic 59, 247–251.

Stempniewicz, L., Kidawa, D., Barcikowski, M., Iliszko, L., 2014. Unusual hunting and feeding behaviour of polar bears on Spitsbergen. Polar Record 50, 216–219.

Stern, G.A., Macdonald, C.R., Armstrong, D., Dunn, B., Fuchs, C., Harwood, L., Muir, D.C.G., Rosenberg, B., 2005. Spatial trends and factors affecting variation of organochlorine contaminants levels in Canadian Arctic beluga (*Delphinapterus leucas*). The Science of the Total Environment 351–352, 344–368.

Stockholm Convention, 2001. In: Final Act of the Conference of Plenipotentiaries on the Stockholm Convention on Persistent Organic Pollutants. UNEP/POPS/CONF/4. UNEP, Stockholm.

Stockholm Convention, 2011. C.N.703.2011.TREATIES-8.

Stroeve, J., Notz, D., 2015. Insights on past and future sea-ice evolution from combining observations and models. Global and Planetary Change 135, 119–132.

Taniyasu, S., Yamashita, N., Moon, H.B., Kwok, K.Y., Lam, P.K.S., Horii, Y., Petrick, G., Kannan, K., 2013. Does wet precipitation represent local and regional atmospheric transportation by perfluorinated alkyl substances? Environment International 55, 25–32.

Tartu, S., Bourgeon, S., Aars, J., Andersen, M., Ehrich, D., Thiemann, G.W., Welker, J.M., Routti, H., 2016. Geographical area and life history traits influence diet in an Arctic marine predator. PLoS One 11, e0155980.

Tartu, S., Bourgeon, S., Aars, J., Andersen, M., Polder, A., Thiemann, G.W., Welker, J.M., Routti, H., 2017a. Sea ice-associated decline in body condition leads to increased concentrations of lipophilic pollutants in polar bears (*Ursus maritimus*) from Svalbard, Norway. The Science of the Total Environment 576, 409–419.

Tartu, S., Bourgeon, S., Aars, J., Andersen, M., Lone, K., Jenssen, B.M., Polder, A., Thiemann, G.W., Torget, V., Welker, J.M., Routti, H., 2017b. Diet and metabolic state are the main factors determining concentrations of perfluoroalkyl substances in female polar bears from Svalbard. Environmental Pollution 229, 146–158.

Tartu, S., Lille-Langøy, R., Størseth, T.R., Bourgeon, S., Brunsvik, A., Aars, J., Goksøyr, A., Jenssen, B.M., Polder, A., Thiemann, G.W., Torget, V., Routti, H., 2017c. Multiple-stressor effects in an apex predator: combined influence of pollutants and sea ice decline on lipid metabolism in polar bears. Scientific Reports 7, 16487.

Thiemann, G.W., Iverson, S.J., Stirling, I., 2008. Polar bear diets and arctic marine food webs: insights from fatty acid analysis. Ecological Monographs 78, 591–613.

Tomy, G.T., Muir, D.C.G., Stern, G.A., Westmore, J.B., 2000. Levels of C_{10}–C_{13} polychloro-n-alkanes in marine mammals from the Arctic and the St. Lawrence River Estuary. Environmental Science and Technology 34, 1615–1619.

Tomy, G.T., Pleskach, K., Arsenault, G., Potter, D., Mccrindle, R., Marvin, C.H., Sverko, E., Tittlemier, S., 2008. Identification of the novel cycloaliphatic brominated flame retardant 1,2-dibromo-4-(1,2-dibromoethyl)cyclohexane in Canadian Arctic beluga (*Delphinapterus leucas*). Environmental Science and Technology 42, 543–549.

Tomy, G.T., Pleskach, K., Ferguson, S.H., Hare, J., Stern, G., Macinnis, G., Marvin, C.H., Loseto, L., 2009. Trophodynamics of some PFCs and BFRs in a Western Canadian Arctic marine food web. Environmental Science and Technology 43, 4076–4081.

Verreault, J., Gabrielsen, G.W., Chu, S., Muir, D.C., Andersen, M., Hamaed, A., Letcher, R.J., 2005a. Flame retardants and methoxylated and hydroxylated polybrominated diphenyl ethers in two Norwegian Arctic top predators: glaucous gulls and polar bears. Environmental Science and Technology 39, 6021–6028.

Verreault, J., Muir, D.C., Norstrom, R.J., Stirling, I., Fisk, A.T., Gabrielsen, G.W., Derocher, A.E., Evans, T.J., Dietz, R., Sonne, C., Sandala, G.M., Gebbink, W., Riget, F.F., Born, E.W., Taylor, M.K., Nagy, J., Letcher, R.J., 2005b. Chlorinated hydrocarbon contaminants and metabolites in polar bears (*Ursus maritimus*) from Alaska, Canada, East Greenland, and Svalbard: 1996-2002. The Science of the Total Environment 351–352, 369–390.

Villa, S., Migliorati, S., Monti, G.S., Holoubek, I., Vighi, M., 2017. Risk of POP mixtures on the Arctic food chain. Environmental Toxicology and Chemistry 36, 1181–1192.

Villanger, G.D., Jenssen, B.M., Fjeldberg, R.R., Letcher, R.J., Muir, D.C.G., Kirkegaard, M., Sonne, C., Dietz, R., 2011a. Exposure to mixtures of organohalogen contaminants and associative interactions with thyroid hormones in East Greenland polar bears (*Ursus maritimus*). Environment International 37, 694–708.

Villanger, G.D., Lydersen, C., Kovacs, K.M., Lie, E., Skaare, J.U., Jenssen, B.M., 2011b. Disruptive effects of persistent organohalogen contaminants on thyroid function in white whales (*Delphinapterus leucas*) from Svalbard. The Science of the Total Environment 409, 2511–2524.

Villanger, G.D., Gabrielsen, K.M., Kovacs, K.M., Lydersen, C., Lie, E., Karimi, M., Sormo, E.G., Jenssen, B.M., 2013. Effects of complex organohalogen contaminant mixtures on thyroid homeostasis in hooded seal (*Cystophora cristata*) mother-pup pairs. Chemosphere 92, 828–842.

Vinje, T., Kvambekk, Å.S., 1991. Barents Sea drift ice characteristics. Polar Research 10, 59–68.

Vorkamp, K., Riget, F.F., 2014. A review of new and current-use contaminants in the Arctic environment: evidence of long-range transport and indications of bioaccumulation. Chemosphere 111, 379–395.

Vorkamp, K., Rigét, F.F., Glasius, M., Muir, D.C.G., Dietz, R., 2008. Levels and trends of persistent organic pollutants in ringed seals (*Phoca hispida*) from Central West Greenland, with particular focus on polybrominated diphenyl ethers (PBDEs). Environment International 34, 499–508.

Vorkamp, K., Riget, F.F., Bossi, R., Dietz, R., 2011. Temporal trends of hexabromocyclododecane, polybrominated diphenyl ethers and polychlorinated biphenyls in ringed seals from East Greenland. Environmental Science and Technology 45, 1243–1249.

Vorkamp, K., Bossi, R., Rigét, F.F., Skov, H., Sonne, C., Dietz, R., 2015. Novel brominated flame retardants and dechlorane plus in Greenland air and biota. Environmental Pollution 196, 284–291.

Wagemann, R., Innes, S., Richard, P.R., 1996. Overview and regional and temporal differences of heavy metals in Arctic whales and ringed seals in the Canadian Arctic. The Science of the Total Environment 186, 41–66.

Wania, F., 2007. A global mass balance analysis of the source of perfluorocarboxylic acids in the Arctic ocean. Environmental Science and Technology 41, 4529–4535.

Wania, F., Mackay, D., 1993. Global fractionation and cold condensation of low volatility organochlorine compounds in polar-regions. Ambio 22, 10–18.

Ware, J.V., Rode, K.D., Bromaghin, J.F., Douglas, D.C., Wilson, R.R., Regehr, E.V., Amstrup, S.C., Durner, G.M., Pagano, A.M., Olson, J., Robbins, C.T., Jansen, H.T., 2017. Habitat degradation affects the summer activity of polar bears. Oecologia 1–13.

Weber, J., Halsall, C.J., Muir, D., Teixeira, C., Small, J., Solomon, K., Hermanson, M., Hung, H., Bidleman, T., 2010. Endosulfan, a global pesticide: a review of its fate in the environment and occurrence in the Arctic. The Science of the Total Environment 408, 2966–2984.

Welch, H.E., Bergmann, M.A., Siferd, T.D., Martin, K.A., Curtis, M.F., Crawford, R.E., Conover, R.J., Hop, H., 1992. Energy-flow through the marine ecosystem of the Lancaster Sound region, Arctic Canada. Arctic 45, 343–357.

White, R.D., Hahn, M.E., Lockhart, W.L., Stegeman, J.J., 1994. Catalytic and immunochemical characterization of hepatic-microsomal cytochromes p450 in beluga whale (*Delphinapterus leucas*). Toxicology and Applied Pharmacology 126, 45–57.

Wilson, R.R., Regehr, E.V., Rode, K.D., Martin, M.S., 2016. Invariant polar bear habitat selection during a period of sea ice loss. Proceedings of the Royal Society of London B: Biological Sciences 283, 20160380.

Wolkers, J., Burkow, I.C., Lydersen, C., Dahle, S., Monshouwer, M., Witkamp, R.F., 1998a. Congener specific PCB and polychlorinated camphene (toxaphene) levels in Svalbard ringed seals (*Phoca hispida*) in relation to sex, age, condition and cytochrome P450 enzyme activity. The Science of the Total Environment 216, 1–11.

Wolkers, J., Witkamp, R.F., Nijmeijer, S.M., Burkow, I.C., De Groene, E.M., Lydersen, C., Dahle, S., Monshouwer, M., 1998b. Phase I and phase II enzyme activities in Ringed seals (*Phoca hispida*): characterization of hepatic cytochrome P450 by activity patterns, inhibition studies, mRNA analyses, and Western blotting. Aquatic Toxicology 44, 103–115.

Wolkers, H., Burkow, I.C., Lydersen, C., Witkamp, R.F., 2000. Chlorinated pesticide concentrations, with an emphasis on polychlorinated camphenes (toxaphenes), in relation to cytochrome P450 enzyme activities in harp seals (*Phoca groenlandica*) from the Barents Sea. Environmental Toxicology and Chemistry 19, 1632–1637.

Wolkers, H., Burkow, I.C., Hammill, M.O., Lydersen, C., Witkamp, R.F., 2002. Transfer of polychlorinated biphenyls and chlorinated pesticides from mother to pup in relation to cytochrome P450 enzyme activities in harp seals (*Phoca groenlandica*) from the Gulf of St. Lawrence, Canada. Environmental Toxicology and Chemistry 21, 94–101.

Wolkers, H., Lydersen, C., Kovacs, K.M., 2004a. Accumulation and lactational transfer of PCBs and pesticides in harbor seals (*Phoca vitulina*) from Svalbard, Norway. The Science of the Total Environment 319, 137–146.

Wolkers, H., Van Bavel, B., Derocher, A.E., Wiig, Ø., Kovacs, K.M., Lydersen, C., Lindstöm, G., 2004b. Congener-specific accumulation and food chain transfer of polybrominated diphenyl ethers in two Arctic food chains. Environmental Science and Technology 38, 1667–1674.

Wolkers, H., Lydersen, C., Kovacs, K.M., Burkow, I., Bavel, B., 2006. Accumulation, metabolism, and food-chain transfer of chlorinated and brominated contaminants in subadult white whales (*Delphinapterus leucas*) and narwhals (*Monodon monoceros*) from Svalbard, Norway. Archives of Environmental Contamination and Toxicology 50, 69–78.

Wolkers, H., Corkeron, P.J., Van Parijs, S.M., Simila, T., Van Bavel, B., 2007. Accumulation and transfer of contaminants in killer whales (*Orcinus orca*) from Norway: indications for contaminant metabolism. Environmental Toxicology and Chemistry 26, 1582.

Wolkers, H., Krafft, B.A., Van Bavel, B., Helgason, L.B., Lydersen, C., Kovacs, K.M., 2008. Biomarker responses and decreasing contaminant levels in ringed seals (*Pusa hispida*) from Svalbard, Norway. Journal of Toxicology and Environmental Health, Part A 71, 1009–1018.

Wolkers, H., Boily, F., Fink-Gremmels, J., Van Bavel, B., Hammill, M.O., Primicerio, R., 2009. Tissue-specific contaminant accumulation and associated effects on hepatic serum analytes and cytochrome P450 enzyme activities in hooded seals (*Cystophora cristata*) from the Gulf of St. Lawrence. Archives of Environmental Contamination and Toxicology 56, 360–370.

Yoneda, S., Suzuki, K.T., 1997. Detoxification of mercury by selenium by binding of equimolar Hg-Se complex to a specific plasma protein. Toxicology and Applied Pharmacology 142, 274–280.

Zhao, Z., Xie, Z.Y., Moller, A., Sturm, R., Tang, J.H., Zhang, G., Ebinghaus, R., 2012. Distribution and long-range transport of polyfluoroalkyl substances in the Arctic, Atlantic Ocean and Antarctic coast. Environmental Pollution 170, 71–77.

Chapter 14

Toxicological Risks and Considerations Associated With Lipophilic Contaminant Burdens of Southern Ocean Mysticetes

Susan M. Bengtson Nash
Griffith University, Brisbane, QLD, Australia

SYNTHETIC CHEMICAL CONTAMINATION OF THE ANTARCTIC REGION

Synthetic chemicals were first detected in Antarctic biota in the 1960s (George and Frear, 1966). The detection of dichlorodiphenyltrichloroethane (DDT) evidenced the long-range environmental transport capabilities of this agricultural pesticide, now identified as a persistent organic pollutant (POP). POPs share the four defining characteristics of toxicity: persistence, the ability to accumulate in organisms and magnify between trophic levels, and the capacity to undergo long-range environmental dispersal. Recognition of the threat posed by POPs to human and environmental health led to the promulgation of the Stockholm Convention, on POPs, in 2004 (UNEP, 2001).

The majority of POPs reach the polar regions of the Earth via air, as a function of their persistence and semivolatility. As volatility is temperature dependent, they deposit out of their volatilized state at higher latitudes, where cold temperatures further extend their lifetime (Wania and Mackay, 1993). They favor accumulation in the lipid reserves of organisms due to their lipophilicity and hydrophobicity. The exception to these chemical properties and environmental behaviors came with the listing of perfluoroalkylated substances (PFAS) as POPs in 2011, hereby emphasizing the diversity of chemical structures that were capable of fulfilling the four POP criteria. Rather than accumulating in the lipid reserves of organisms like their chlorinated and brominated counterparts, ionic PFAS are proteinophilic (Prevedouros et al., 2006). Similarly, as opposed to undergoing long-range dispersal through atmospheric transport, ionic PFAS are hydrophilic, favoring accumulation in surface waters and dispersal via oceanic pathways (Yamashita et al., 2008).

Marine Mammal Ecotoxicology. https://doi.org/10.1016/B978-0-12-812144-3.00014-0

In addition to the two major abiotic environmental transport pathways of water and air, Antarctica supports a number of highly migratory species that introduce the potential for animals to serve as biological "vectors" of chemical transport to the region (Wild, 2016). For example, south polar skua (*Catharacta maccormicki*), observed to overwinter as far north as North America (Kopp et al., 2011), carry more diverse chemical profiles than those of endemic bird species (Wild, 2016), reflecting feeding in closer proximity to pollution sources. These body burdens may in turn be delivered to the Antarctic environment through excrement, failed eggs, or carcasses.

Increasing human polar activity also introduces a heightened risk of in-situ release of chemical contamination to the local environment. Indeed, Antarctic research stations have repeatedly been flagged as local emitters of modern POPs (Bengtson Nash et al., 2008; Hale et al., 2008; Wild et al., 2014), representing a present-day challenge for the Antarctic Treaty Agreement related to harmful substances (UNEP, 2002).

Finally, it should be acknowledged that the Stockholm Convention lists but a tiny fraction of approximately 30,000 high-use (>1 ton/year) commercial chemicals (Muir and Howard, 2006). Any investigation of chemical exposure and risk must necessarily work on the assumption that there are other, currently unidentified POPs (Burton et al., 2017), and as such a hyperconservative approach to wildlife exposure evaluations as part of species-specific risk assessments must be assumed.

THE ANTARCTIC SEA ICE ECOSYSTEM

The central role of Antarctic krill (*Euphausia superba*) in food web interactions of the Southern Ocean has often been described (e.g., Knox, 1994). Almost all higher order consumers rely either directly or indirectly on the seasonal productivity of this circumpolar species. Recognition of the keystone role of krill was behind adoption of the "ecosystem approach" to management by the Convention for the Conservation of Antarctic Marine Living Resources (Eversen, 2000).

Antarctic krill undergo 12 larval stages before assuming their adult pelagic life stage (George, 1982). The first three of these are nonfeeding stages that undertake the larval ascent to reach the underside of the sea ice where they commence feeding upon ice algae–associated microbes to fuel further development (Nicol, 2006). This sympagic species therefore relies on the sea ice ecosystem as a nursery ground, with the implication that larval recruitment has often been described as being closely tied to sea ice extent (Loeb et al., 1997). Under this hypothesis, a decrease in suitable sea ice habitat in a warming climate would result in reduced krill recruitment success, which in turn would translate to reduced prey biomass available to higher trophic level Antarctic consumers in subsequent years (Flores et al., 2012). More recently, the proximal role of sea ice extent in cyclical krill biomass fluctuations has been found to be secondary to intercohort competition for food resources (Ryabov et al., 2017). Nonetheless, climatic conditions defined by cryosphere loss and changing ocean pH are predicted to have a devastating impact on this sympagic species, with ocean acidification forecasted to result in a complete collapse of Antarctic krill populations by 2300 if current CO_2 emissions are left unmitigated (Kawaguchi et al., 2013).

FEEDING ECOLOGY OF SOUTHERN OCEAN MYSTICETES

Six species of mysticetes occur in the Southern Ocean (Knox, 1994): blue whales (*Balaenoptera musculus*), fin whales (*B. physalus*), sei whales (*B. borealis*), Antarctic minke whales (*B. bonaerensis*), southern right whales (*B. australis*), and humpback whales (*Megaptera novaeangliae*).

Mysticetes or baleen whales are identified by their keratinous baleen plates extending from their upper jaw, which facilitate "sieving" or filter feeding on small food items from the water column. Baleen whales foraging in the Southern Ocean (termed "Southern Ocean mysticetes" throughout this chapter) exploit the annual swarms of Antarctic krill that form through summer months. Lunge feeding in rorqual whales (Balaenopteridae) has been posed as only being energetically viable at certain prey densities (Goldbogen et al., 2011; Keen, 2017), pointing to apparent coevolution of Southern Ocean mysticetes with Antarctic krill, through positive iron feedback loops, also known as the "manuring mechanism," in the iron-limited Southern Ocean ecosystem (Nicol et al., 2010; Smetacek, 2008).

Adaptation to the Polar Environment

The seasonal productivity of high-latitude environments has necessitated that species inhabiting these regions evolve life-history strategies that allow them to cope with extended periods of low food availability. For mysticetes, this has led to adaptive obesity, permitting a capital breeding and migratory life history. Seasonal migrations between summer Antarctic feeding grounds and lower latitude winter breeding grounds allow these populations to exploit the seasonal productivity of Antarctic krill, while also satisfying the physiological needs of newborn calves, born without a substantial blubber layer, thus impacting their ability to thermoregulate and energy investment directed to this process.

While the general notion of migratory dispersal during winter months is thought to be true for all Southern Ocean foraging mysticetes, our understanding of extra-Antarctic migratory range, distribution, timing, and regularity of migration events varies significantly between the species, and it is far from complete. For example, Antarctic blue whales are found largely south of the Antarctic Convergence in summer months; however, there remains little evidence of known overwintering grounds (Samaran et al., 2013). For other species, such as humpback whales, distinct calving grounds define identified breeding stocks (e.g., Smith et al., 2012) and their management by the International Whaling Commission. Confounding existing knowledge gaps is new evidence, obtained on the basis of acoustic recordings and historical whaling data, of the year-round presence of minke, humpback, and blue whales in Antarctica (Samaran et al., 2013; Van Opzeeland et al., 2013). It is postulated that a complete annual migration may indeed not be energetically favorable for all individuals, most notably, immature individuals or mature females rebuilding their energetic reserves following weaning of a calf.

While the length of the migratory journey for Southern Ocean foraging mysticetes may vary between species, and the regularity of these events may vary between individuals of a species, all have adopted a capital breeding strategy, that is, servicing extended offspring dependence, and periods of a negative energy balance, with energy stores accumulated during intensive summer feeding. The breeding season represents a period of intensive energy utilization. Males undertake competitive breeding behavior, while migrating females are predominantly pregnant and/or nursing young calves. Prolonged negative energy balance in turn has direct implications for circulating lipophilic contaminant exposure.

BIOGEOCHEMICAL CYCLING OF POPs IN THE POLAR ENVIRONMENT

Polar Regions have long been established as receiving or "sink" environments for POPs, due to their position at the end of the global distillation pathway (Wania and Mackay, 1993). As such, chemicals have been retained by polar ice caps and oceans, effectively "charging" these environmental compartments (Dachs, 2011).

Ice and snow characterize the polar landscape and play a major role in POP partitioning at high latitudes. For example, sea ice and associated hydrologic cycles greatly influence both the delivery of chemicals to the marine environment as well as the seasonal onset of biological productivity. Sea ice acts as a lid at the air–water interface and a reservoir for POPs during winter months. Melting of sea ice may therefore result in concentration peaks in receiving water bodies and pulse exposure to aquatic organisms (Bigot et al., 2017; Fuoco et al., 1991; Halsall, 2004; Bigot et al., 2016) (Fig. 14.1).

The majority of POPs are hydrophobic and lipophilic. Once released into the water column, they will preferentially adsorb to particulate organic matter (Fig. 14.1). In turn, seasonal phytoplankton blooms ensure efficient transfer of contaminants to higher trophic levels, while also playing a key role in chemical "drawdown" from the atmosphere (Dachs et al., 2002).

The effects of climate change are expected to be most pronounced in polar regions due to snow and ice albedo feedbacks as the polar cryosphere is lost (Box, 2012). Rising temperatures will serve to not only change the physical "icescape" but also chemical equilibria (Ma et al., 2011). Similarly, altered cryosphere dynamics, primary productivity, and food web interactions carry

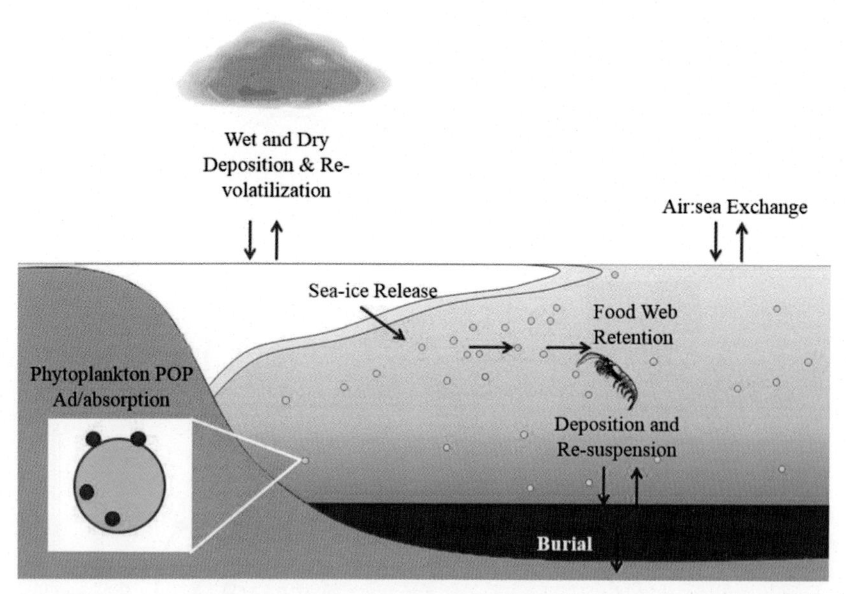

FIGURE 14.1 POP transport pathways in the Antarctic seasonal ice zone as indicated by arrows and red markers.

direct implications for chemical pathways, bioavailability, and fate through the Antarctic ecosystem.

MYSTICETE POP EXPOSURE AND UPTAKE

Like other mammals, the primary route of POP exposure of Southern Ocean foraging mysticetes is via ingestion of contaminated prey. Polar biota are characterized by slow metabolism and extended lifespans (King and Riddle, 2001). For example, Antarctic krill may live for up to 7 years (Kilada et al., 2017), significantly longer than temporal or tropical counterparts. Their extended lifetime enhances their potential for chemical bioaccumulation, while a dependence in polar food webs on a lipid-rich diet favors efficient transfer and assimilation of lipophilic POPs (Borgå et al., 2004).

Krill perform an important ecological role as "grazers" of the ecosystem, assimilating organic carbon from the water column, mainly in the form of phytoplankton (Eversen, 2000). Unsurprisingly, this grazing role of krill, combined with the central role of krill in the Antarctic sea ice ecosystem, also makes them primary vectors for POP transfer and dispersal to the remaining ecosystem (Bengtson Nash et al. 2014). Recently, a regional ecosystem-coupled fugacity model demonstrated this dominant pelagic transfer pathway for the modeled POP compound, hexachlorobenzene (HCB), to higher trophic levels (Bates et al., 2017) (Fig. 14.2).

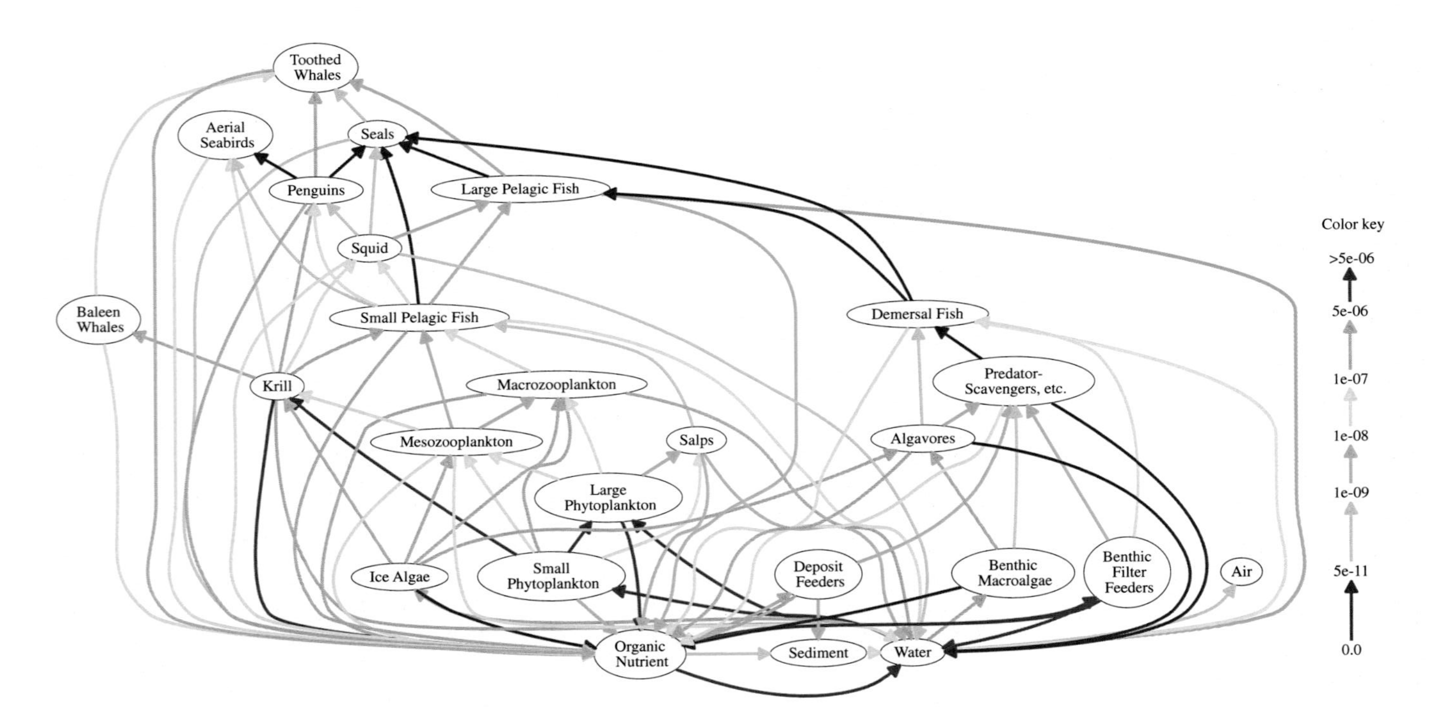

FIGURE 14.2 A graphic representation of the fluxes of HCB (mol/h) through an Antarctic ecosystem. *(Image reproduced from Bates, M., Bengtson Nash, S.M., Hawker, D., Cropp, R., 2017. The distribution of persistent organic pollutants in a trophically complex Antarctic ecosystem model. Journal of Marine Systems 170, 103–114.)*

To date, the POP burdens of Antarctic krill around the circum-Antarctic region have been observed to be relatively low and homogenous, with only some evidence of "hot spots" for the modern brominated POPs, polybromindated diphenylethers (PBDEs) (Bengtson Nash et al., 2008; Chiuchiolo et al., 2004). The recent, common application of PBDEs in firefighting foams, building materials, and clothing (Alaee et al., 2003) has led researchers to attribute localized, elevated PBDE levels in krill to local polar research station emissions, particularly those of higher molecular weight and with limited potential for Long Range Atmospheric Transport (Wild et al., 2014).

Comparatively low, homogenous levels and profiles of legacy POPs at the base of the Antarctic food web reflect a number of human activities and Earth system processes. Historically, the southern hemisphere has experienced lower usage of industrial chemicals compared to the northern hemisphere (Bengtson Nash, 2011). Further, the diffuse nature of hemispheric sources of contamination and the length of the global distillation pathway for chemicals to reach the Antarctic region ensure a relatively well-mixed chemical profile across the region. The required lifetime of a chemical to reach the continent is finally also extended by the heightened protection of the Antarctic region from direct atmospheric and hydrospheric advective chemical transport from lower latitudes, due to natural hydrospheric and atmospheric circulation barriers surrounding the Antarctic continent (Bengtson Nash et al., 2010, 2017; Hunt et al., 2016).

The principal energetic provisioning of Southern Ocean foraging mysticetes via a single, circumpolar Antarctic prey species greatly simplifies exposure assessments and bioenergetics calculations. The importance of "supplementary" extra-Antarctic feeding, and hence the relevance of exposure via alternate prey, however, remains to be characterized at species and population levels. In recent years, there have been increasing observations of temperate feeding by migrating humpback whales (e.g., Stamation, 2007; Stockin and Burgess, 2005), while fecal analysis of southern right whales in late spring, off the Peninsula of Valdes, revealed primarily copepod remains (D'Agostino et al., 2015). Extra-Antarctic feeding by humpback and right whales has further been evidenced through biochemical investigation of feeding records contained in baleen plates (Best and Schell, 1996; Eisenmann et al., 2016, 2017).

Extra-Antarctic feeding may be either a routine or opportunistic behavior for a species or population (Owen et al., 2017), or indeed a sign of present-day ecological change (Bengtson Nash et al., 2018). Regardless, ingestion of temperate and/or higher trophic level prey items will result in disproportionately high contaminant exposure compared to that of a diet of Antarctic krill. This is due to the higher levels, and more diverse chemical profiles, observed closer to pollution source regions, as well as biomagnification processes associated with feeding at higher trophic levels. Monitoring of temporal trends of biomagnification factors (Kelly et al., 2007) between Southern Ocean foraging mysticetes and Antarctic krill may provide valuable insight into the current and changing nature of mysticete trophodynamics and supplementary feeding.

TEMPORAL TRENDS OF SOUTHERN OCEAN MYSTICETE POP BURDENS

One of the greater difficulties in evaluating contaminant exposure of Antarctic mysticetes is the low number of studies that have historically been performed evaluating POP burdens, and the limited overlap of targeted species, chemical analytes, and sampling regions between studies (Bengtson Nash, 2011). This same gap prevents robust temporal trend evaluation. One study, derived from individuals harvested under the Japanese Research Program in Antarctica (JARPA) (Yasunaga et al., 2015), provides some insight. Yasunaga et al. report on POP burdens in Antarctic minke whales between 1987 and 2005. This publication supersedes earlier reports from the JARPA program by constraining the dataset to only male individuals, between 20 and 25 years of age. Only a declining temporal trend in hexachlorocyclohexane (ΣHCH) burdens was found to be significant in this study, across the study timeframe, which corresponds to findings of atmospheric contamination of this compound during the same interim (Bengtson Nash et al., 2017; Bigot et al., 2016; Dickhut et al., 2005) (Fig. 14.3).

When Antarctic minke whale POP burdens are compared to levels reported in humpback whales (Bengtson Nash et al., 2013; Das et al., 2017; Dorneles et al., 2015) (Fig. 14.3), the only other Southern Ocean foraging mysticete species for which POP data is available, there appears to be relatively good consistency of findings despite the species difference. This is to be expected given the shared low trophic level feeding niche and the diffuse and well-mixed nature of contamination in the region.

Differences in sampling season (stage of fasting), genders, and breeding stock foraging grounds between studies prevent any evaluation of POP temporal trends within Southern Ocean foraging humpback whales. Outer blubber levels of some POPs have been shown to fluctuate by up to 500 times across just 4 months of the migration journey in males of the same breeding stock (Bengtson Nash et al., 2013). This dramatic effect of life-history and nutritional state on tissue-specific chemical concentrations is expected to cloud any evaluation of differences between breeding stocks, or indeed elucidation of trends within a breeding stock, necessitating larger sample numbers and longer monitoring periods. Indeed, even when annual sampling was constrained to males of the same breeding stock, sampled within the same 2-week period each year, the coefficient of variation was found to be such that it would take 12, 15, and 18 years of annual sampling to detect a 5% change in HCB, ΣDDE, and ΣPCB levels respectively, per year with 80% confidence (Bengtson Nash et al., 2018).

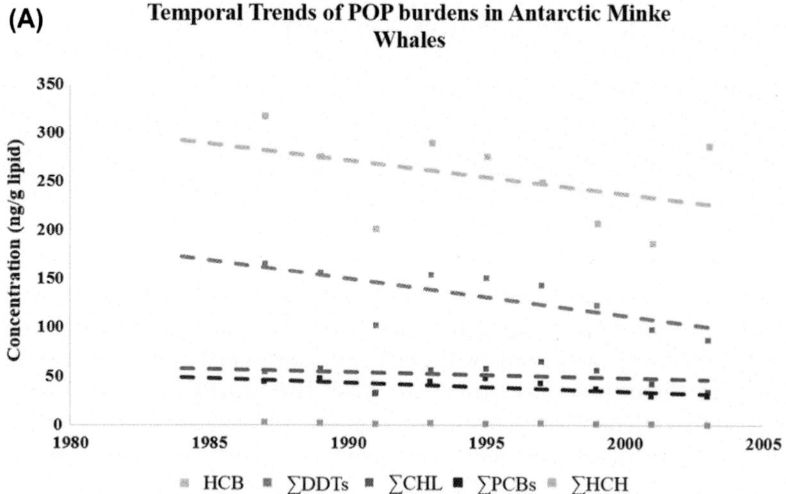

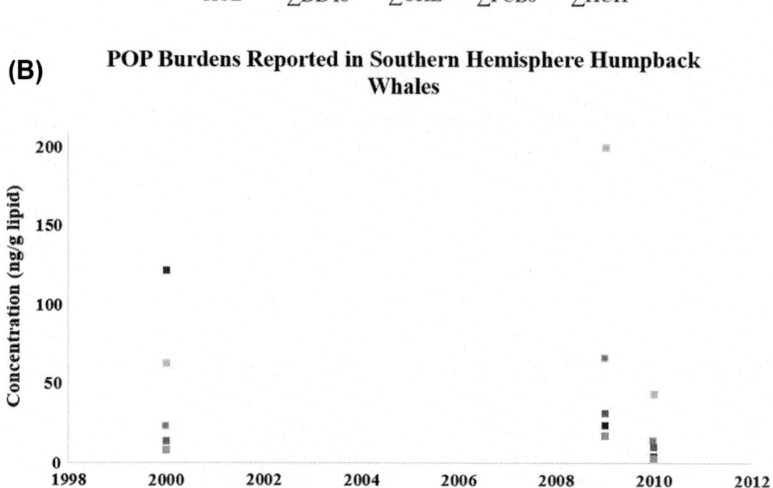

FIGURE 14.3 Blubber POP burdens of Southern Ocean foraging minke (A) and humpback (B) whales (ng/g lipid). *CHL*, chlordanes; *DDTs*, dichlorodiphenyltrichloroethanes; *HCHs*, hexacyclohexanes; *PCBs*, polychlorinated biphenyls. *(Data for minke whales derived from Yasunaga et al. (2015). Data for humpback whales derived from Bengtson Nash et al. (2013), Das et al. (2017), and Dorneles et al. (2008).)*

Future efforts must seek to provide standardized intrapopulation temporal data for robust trend evaluation and should seek to expand chemical screening to modern POPs and POP candidates, to reflect the rapid progress of the global chemical industry, and thereby the current and changing landscape of environmental chemical contamination (Burton et al., 2017).

POP TOXICOKINETICS AND THE TOXICOLOGICAL RISK POSED TO SOUTHERN OCEAN MYSTICETES

Seasonal Cycles

Seasonality has been shown to influence contaminant burdens in Antarctic biota (Bates et al., 2017; Bengtson Nash et al., 2013; Cropp et al., 2014). The major driver of this seasonality are chemical toxicokinetics. The majority of POPs are lipophilic, and their toxicokinetics are governed by lipid dynamics (Bengtson Nash et al., 2013; Yordy et al., 2010). The seasonal winter fast represents a time of negative energy balance for the whales. As the animals draw on accumulated lipid stores for energy, an apparent rise in lipophilic chemical burdens occurs, as rapid lipid depletion is not associated with corresponding chemical metabolism and elimination (Chevrier et al., 2000; Imbeault et al., 2002; Pelletier et al., 2002; Tremblay et al., 2004). Whole-body lipid depletion is thus reflected through the redistribution and concentration of lipophilic chemical burdens among the body's remaining lipid stores (Fig. 14.4).

Life Stages

Such periods of lipophilic contaminant redistribution must be considered a time of "re-exposure." Although the role of adipose tissue as an endocrine active tissue (Kershaw and Flier, 2004), and therefore as a site of potential toxicological action (La Merrill et al., 2013), is receiving increasing attention, the traditional view of contaminant storage in adipose tissue has been that the chemicals

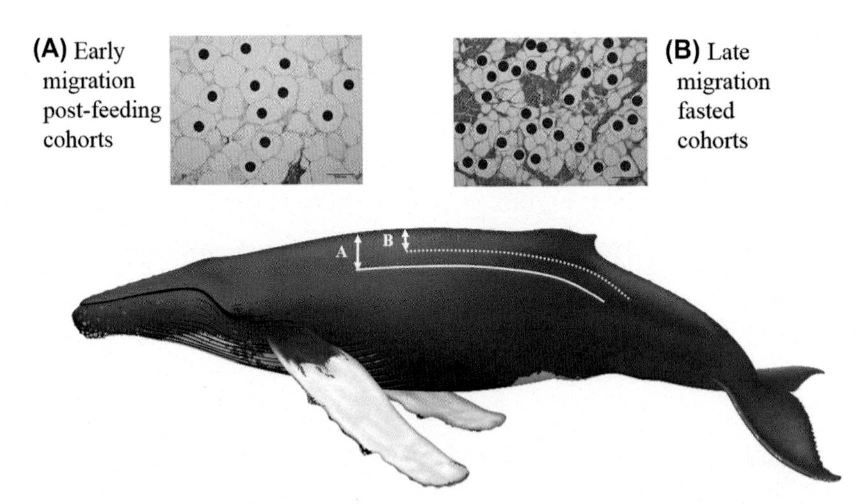

FIGURE 14.4 Schematic representation of lipophilic POP redistribution and concentration within fat cells of the thinning blubber layer, between early migration (A) and late migration (B) time points, where red markers depict POP molecules.

remain "inert" or "dormant" in the adipocyte. By contrast, once mobilized back into circulation, the chemical's potential for both incurring detoxification action as well as intercepting target sites of toxicity increases dramatically. Notably, the seasonal timing of remobilization of POP burdens in Southern Ocean foraging mysticetes occurs when females are predominantly pregnant or nursing young calves. Transplacental transfer and "offloading" of lipophilic chemicals through lipid-rich breastmilk is well documented in mammals, including cetaceans (Borrell and Aguilar, 2005; Wolkers et al., 2007). The size difference between the mother and embryo or calf leads to inherent differences in chemical exposure on a body size basis (Nielsen et al., 2008), with proportionally elevated exposure experienced by the young. To date, our understanding of the toxicological impacts of seasonal re-exposure of adults to accumulated body burdens is unknown, let alone the comparatively elevated exposure experienced by younger developmental stages. A significant toxicological threat of some POPs is their ability to act as endocrine disrupters. Through disruption of normal endocrine function, they carry the capacity to interfere with a multitude of hormonally driven processes, from reproduction, metabolism, and neurological function to fetal and pubertal development (Meeker, 2012). Early developmental stages are therefore particularly vulnerable to the harmful effects of chemicals, both as a function of their small size, maternal metabolic concentration and offloading, as well as the rate at which the young is developing, and therefore the number of critical, hormonally governed processes that present points of potential chemical interference and adverse outcomes.

Effects Assessment

Despite the fact that cetaceans are among the most susceptible organisms to the accumulation of elevated levels of POPs on account of their large lipid mass and extended lifespans, few approaches are available for toxicological effect assessments on these species. Reliable assessment is complicated by the clear logistical and ethical considerations when working with large, free-roaming species. While tissues obtained via remote biopsy have provided a basis for nonlethal exposure assessment (Bengtson Nash et al., 2013; Das et al., 2017; Dorneles et al., 2008), very limited data exists regarding species-specific toxicological sensitivity.

As with humans, chemical risk assessment for cetaceans relies largely on effect assessment data derived from studies on mammalian laboratory models. Species differences can, however, be significant, even between species of the same taxonomic group (Nielsen et al., 2006). As such, there remains great interest in the continued development of non-lethal, species-specific toxicity evaluation approaches (Bengtson Nash et al., 2011).

Enzyme biomarkers of effect, derived from epidermal tissues, have been identified and tested in southern hemisphere humpback whales (Bengtson Nash et al., 2014; Waugh et al., 2011). Similar enzyme biomarkers have been previously utilized for toxicological evaluation of mysticetes in other global regions, (e.g., Marsili et al., 1998). In the former study, the analysis of glutathione-s-transferase (GST) and cytochrome P450 1A1 were found not to correlate with each other, nor dominant POP groups identified in the blubber of the same animals. Rather, levels of particularly GST were found to be low and relatively steady between individuals, indicating basal activity, possibly impeded by fasting. These findings did not support future application of these common enzymes as effective biomarkers of POP exposure in the species.

Developments in the use of cell lines for cell-based in vitro toxicity evaluation have shown great promise in facilitating mechanistic understanding of cellular processes and the identification of molecular and cellular chemical targets (Allen et al., 2005). The first humpback whale fibroblast cell line was developed (Burkard et al., 2015) and has since been immortalized within the author's team. The dose-response relationship between fibroblast cell viability and *p,p′*-DDE exposure (a priority contaminant accumulating in southern hemisphere populations) was derived alongside a human fibroblast cell line. Direct experimental comparison found that the human fibroblasts both grew more rapidly and were significantly more sensitive to *p,p′*-DDE exposure than the humpback whale cells.

In the Northern Hemisphere, several cell lines have been developed from different tissues of the North Atlantic right whale (*Eubalaena glacialis*), and chemical dose-response relationships have been derived (Godard et al., 2004). Despite intrinsic inter-species differences, these contributions significantly advance our understanding of mysticete toxicological sensitivity.

Once toxicity thresholds are established, the application of physiologically based pharmacokinetic or toxicokinetic (PBPK or PBTK) models can offer valuable extension of findings and insight into potential toxicological impacts across species and populations. PBTK models use the physiology of the species of interest and the physicochemical properties of the chemical of interest to estimate the absorption, distribution, metabolism, and elimination of a chemical in the body of an organism over time (Lipscomb et al., 2012). A pharmacokinetic model was, for example, used to further explore the impact of fasting on accumulated POP loads in Southern Hemisphere humpback whales (Cropp et al., 2014). The value of such adaptable risk evaluation tools is underscored in a rapidly changing global chemical environment, with inherent implications for wildlife exposure pathways, trophodynamics and nutritional state.

Adjusted Risk Categories

A generic approach to risk "ranking" for whales and dolphins has been proposed, based on chemical emission sources, environmental POP chemodynamics, and consequent exposure potential of a species or population in its given habitat (Bengtson Nash et al., 2011). The approach assumes that Northern Hemisphere, high trophic level, coastal species should experience the highest level of chemical exposure through the dietary pathway, and therefore they should be placed in the highest risk category. By contrast, Southern Ocean foraging mysticetes, feeding in remote locations far from contaminant sources, at a lower trophic level to toothed cetaceans, should be at the lowest risk category. Critically, the focus on dietary exposure of this risk categorization approach does not account for periods of negative energy balance and the potentially significant increases in contaminant risk, particularly to young individuals, experienced at this time. It is therefore proposed that chemical risk evaluation prioritizes closer investigation of periods of negative energy balance, which stand to play a role of increasing importance, not only for Southern Ocean foraging mysticetes, but numerous species globally impacted by climate change–driven habitat degradation and resulting food limitation and starvation.

ANOMALIES AND OPPORTUNITIES IN A CHANGING POLAR CLIMATE

Lipophilic Contaminants as Tracers of Energetic Health: A Case Study on Humpback Whales

The potential application of lipophilic contaminant burdens in defined populations, or individuals, to serve as tracers of whole-body lipid gain or loss over time has recently been validated on humpback whales (Bengtson Nash et al., 2018). In 2013, Bengtson Nash et al. showed that lipophilic contaminant burdens measured in outer blubber biopsies of humpback whales rose rapidly across 4 months of the migration journey. This increase did not show a linear relationship with the corresponding reduction in blubber lipid percent. As the outer blubber layer of cetaceans plays an important role in a number of physiological functions, aside from lipid storage, such as buoyancy and thermoregulation (Koopman et al., 2002), it is likely that there exists a threshold under which lipid depletion cannot occur from this blubber layer without compromising these ancillary functions, and therefore individual survival. As such, these lipid depots are to some extent conserved and may be considered “maintenance lipids”. Consequently, as whole-body lipid depletion occurs, the outer blubber layer starts to represent an increasing proportion of the individual’s remaining lipid stores and accumulates proportionately greater pollutant loads (Fig. 14.4). This led the authors to propose that standardized temporal monitoring of outer blubber POP burdens, in a defined population (males only to avoid the confounding factors of pregnancy and lactation), may provide insight into fluctuating population adiposity and therefore seasonal foraging success in Antarctica. This potential was validated through the observation that outer blubber POP burdens in a defined population, targeted during the same 2-week window each year, revealed a direct correlation with another established adiposity measure in the species, the adipocyte index (Bengtson Nash et al., 2018; Castrillon et al., 2017). These two metrics in turn showed synchronous, interannual oscillations with climate parameters in the corresponding Antarctic feeding ground (Bengtson Nash et al., 2018) (Fig. 14.5).

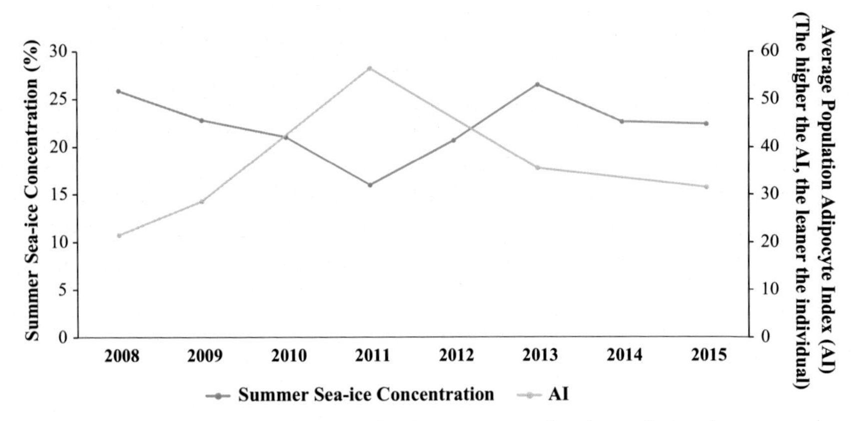

FIGURE 14.5 Summer sea ice concentration in the corresponding Antarctic foraging ground during the summer preceding population sampling, displayed together with the average winter adipocyte index (AI) of migrating individuals.

In the aforementioned study, the authors applied fluctuating lipophilic POP burdens as a proxy for variable adiposity by focusing on only the males of a distinct migrating breeding stock. The approach would be equally valid for individuals tracked over time or similarly defined populations. The approach could not, however, be used to compare two diverse populations with different diets, nor to track populations with a high level of individual exchange with other populations.

CONCLUSION

In the past, Southern Ocean foraging mysticetes have undoubtedly experienced comparatively lower levels of dietary POP exposure relative to temperate, tropical, or Arctic foraging counterparts. Chemical contamination of the Antarctic system, however, continues to evolve in line with both increasing levels of local human activity and pollution, as well as the rapid, and largely unregulated, expansion of chemical production globally (Burton et al., 2017). Confounding species chemical risk calculations, based upon direct dietary exposure, is the life history adaptation to a capital breeding strategy adopted by mysticetes feeding in the Southern Ocean. The seasonal lipid dynamics and associated toxicokinetics associated with this life history places them at an elevated risk of toxicological effects associated with their accumulated contaminant burdens, due to repeated exposure of sensitive sites of toxicity (Pelletier et al., 2002). Further, a high dependence upon a single prey item, vulnerable to the direct effects of climate change, places these species in a precarious ecological position in a changing polar climate. It is hypothesized that those species adapted to physiological extremes, that is, operating at the limits of energetic viability, may be the most responsive to environmental change (Bengtson Nash et al., 2018; Bengtson Nash, 2011). The expression of toxicological effects, such as immunosuppression, often manifest at times when animals are weakened through, for example, extended periods of negative energy balance associated with elevated levels of circulating POPs (Ross et al., 2003). At these times, immunotoxic impacts of circulating burdens may render animals vulnerable to pathogenic attack. Polar capital breeders are therefore likely to be particularly susceptible to the effects of multiple stressors in a rapidly changing polar climate.

REFERENCES

Alaee, M., Arias, P., Sjodin, A., Bergman, Å., 2003. An overview of commercially used brominated flame retardants, their applications, their use patterns in different countries/regions and possible modes of release. Environment International 29, 683–689.

Allen, D.D., Caviedes, R., Cárdenas, A.M., Shimahara, T., Segura-Aguilar, J., Caviedes, P.A., 2005. Cell Lines as In Vitro Models for Drug Screening and Toxicity Studies. Drug Development and Industrial Pharmacy 31, 757–768.

Bates, M., Bengtson Nash, S.M., Hawker, D., Cropp, R., 2017. The distribution of persistent organic pollutants in a trophically complex Antarctic ecosystem model. Journal of Marine Systems 170, 103–114.

Bengtson Nash, S., Castrillon, J., Eisenmann, P., Fry, B., Shuker, J., Cropp, R., Dawson, A., Bignert, A., Bohlin-Nizzetto, P., Waugh, C.A., Polkinghorne, B., Dalle Luche, G., McLagan, D., 2018. Signals from the south; humpback whales carry messages of Antarctic sea-ice ecosystem variability. Global Change Biology 24, 1500–1510.

Bengtson Nash, S., Wild, S., Hawker, D.W., Cropp, R., Hung, H., Wania, F., Xiao, H., Bohlin Nizzetto, P., Bignert, A., Broomhall, S., 2017. Persistent organic pollutants in the East Antarctic atmosphere: inter-annual observations from 2010-2015 using high-flow-through passive sampling. Environmental Science and Technology.

Bengtson Nash, S.M., 2011. Persistent organic pollutants in Antarctica; current and future research priorities. Journal of Environmental Monitoring 13, 497–504.

Bengtson Nash, S.M., Bolton, J., Brownell, R., Collier, T., Dorneles, P.R., Godard-Codding, C., Gulland, F., Kucklick, J., Schwacke, L., Venn-Watson, S., Burkhardt-Holm, P., Donovan, G., Fossi, C., Hall, A., Rosa, C., Simmonds, M., Rowles, T., Ylitalo, G., 2011. Report of the IWC pollution 2000+ phase II workshop. Journal of Cetacean Research and Management 12.

Bengtson Nash, S.M., Poulsen, A.H., Kawaguchi, S., Vetter, W., Schlabach, M., 2008. Persistent organohalogen contaminant burdens in Antarctic krill (*Euphausia superba*) from the Eastern Antarctic sector: a baseline study. The Science of the Total Environment 407, 304–314.

Bengtson Nash, S., Schlabach, M., Nichols, P., 2014. A Nutritional-Toxicological Assessment of Antarctic Krill Oil versus Fish Oil Dietary Supplements. Nutrients 6, 3382–3402.

Bengtson Nash, S.M., Rintoul, S.R., Kawaguchi, S., Staniland, I., Van den Hoff, J., Tierney, M., Bossi, R., 2010. Perfluorinated compounds in the Antarctic region: ocean circulation provides prolonged protection. Environmental Pollution 158, 1985–1991.

Bengtson Nash, S.M., Waugh, C.A., Burkard, M., Dawson, A., Huston, W., 2014. Detoxification enzyme activities (CYP1A1 and GST) in the skin of humpback whales as a function of organochlorine burdens and migration status. Aquatic Toxicology 155, 207–212.

Bengtson Nash, S.M., Waugh, C.A., Schlabach, M., 2013. Metabolic concentration of lipid soluble organochlorine burdens in humpback whales through migration and fasting. Environmental Science and Technology 47, 9404–9413.

Best, P.B., Schell, D.M., 1996. Stable isotopes in southern right whales (*Eubalaena australis*) baleen as indicators of seasonal movements, feeding and growth. Marine Biology 124, 483–494.

Bigot, M., Hawker, D., Cropp, R., Muir, D., Jensen, B., Bossi, R., Bengtson Nash, S., 2017. Spring melt and the redistribution of organochlorine pesticides in the sea-ice environment: a comparative study between Arctic and Antarctic regions. Environmental Science and Technology 51, 8944–8952.

Bigot, M., Muir, D., Hawker, D., Cropp, R., Dachs, J., Bengtson Nash, S.M., 2016. Air-seawater exchange of organochlorine pesticides in the Southern Ocean between Australia and Antarctica. Environmental Science and Technology 50, 8001–8009.

Borgå, K., Fisk, A.T., Hoekstra, P.F., Muir, D.C.G., 2004. Biological and chemical factors of importance in the bioaccumulation and trophic transfer of persistent organochlorine contaminants in Arctic marine food webs. Environmental Toxicology and Chemistry 23, 2367–2385.

Borrell, A., Aguilar, A., 2005. Mother-calf transfer of organochlorine compounds in the common dolphin (*Delphinus delphis*). Bulletin of Environmental Contamination and Toxicology 75, 149–156.

Box, J.E., 2012. Climate science: melting triggers more melting. Nature 488, 433.

Burkard, M., Schirmer, K., Whitworth, D., Bengtson Nash, S.M., 2015. Establishment of the first humpback whale fibroblast cell lines and their application in chemical risk assessment. Aquatic Toxicology 167, 240–247.

Burton, G.A., Di Giulio, R., Costello, D., Rohr, J.R., 2017. Slipping through the cracks: why is the U.S. Environmental Protection Agency not funding extramural research on chemicals in our environment? Environmental Science and Technology 51, 755–756.

Castrillon, J., Huston, W., Bengtson Nash, S., 2017. The blubber adipocyte index: a nondestructive biomarker of adiposity in humpback whales (*Megaptera novaeangliae*). Ecology and Evolution 7, 5131–5139.

Chevrier, J., Dewailly, E., Ayotte, P., Mauriege, P., Despres, J.P., Tremblay, A., 2000. Body weight loss increases plasma and adipose tissue concentrations of potentially toxic pollutants in obese individuals. International Journal of Obesity 24, 1272–1278.

Chiuchiolo, A.L., Dickhut, R.M., Cochran, M.A., Ducklow, H.W., 2004. Persistent organic pollutants at the base of the Antarctic Marine food web. Environmental Science and Technology 38, 3551–3557.

Cropp, R., Bengtson Nash, S.M., Hawker, D., 2014. A model to resolve the dynamics of organochlorine pharmacokinetics in migrating humpback whales. Environmental Toxicology and Chemistry 33, 1638–1649.

D'Agostino, V.C., Hoffmeyer, M.S., Degrati, M., 2015. Faecal analysis of southern right whales (*Eubalaena australis*) in Península Valdés calving ground, Argentina: *Calanus australis*, a key prey species. Journal of the Marine Biological Association of the United Kingdom 96, 859–868.

Dachs, J., 2011. Coming in from the cold. Nature 1, 247–248.

Dachs, J., Lohmann, R., Ockenden, W.A., Mejanelle, L., Eisenreich, S.J., Jones, K.C., 2002. Oceanic biogeochemical controls on global dynamics of persistent organic pollutants. Environmental Science and Technology 36, 4229–4237.

Das, K., Malarvannan, G., Dirtu, A., Dulau, V., Dumont, M., Lepoint, G., Mongin, P., Covaci, A., 2017. Linking pollutant exposure of humpback whales breeding in the Indian Ocean to their feeding habits and feeding areas off Antarctica. Environmental Pollution 220 (Part B), 1090–1099.

Dickhut, R.M., Cincinelli, A., Cochran, M.A., Ducklow, H.W., 2005. Atmospheric concentrations and air - water flux of organochlorine pesticides along the western Antarctic peninsula. Environmental Science and Technology 39, 465–470.

Dorneles, P.R., Lailson-Brito, J., Azeredo, A., Meyer, J., Vidal, L., Fragoso, A.B., Torres, J.P., Malm, O., Blust, R., Das, K., 2008. High accumulation of perfluorooctane sulfonate (PFOS) in marine tucuxi dolphins (*Sotalia guianensis*) from the Brazillian coast. Environmental Science and Technology 42, 5368–5373.

Dorneles, P.R., Lailson-Brito, J., Secchi, E., Dirtu, A., Weijs, L., Dalla Rosa, L., Bassoi, M., Cunha, H., Azevedo, A., Covaci, A., 2015. Levels and profiles of chlorinated and brominated contaminants in southern hemisphere humpback whales, *Megaptera novaeangliae*. Environmental Research 138, 49–57.

Eisenmann, P., Fry, B., Holyoake, C., Coughran, D., Nicol, S., Bengtson Nash, S., 2016. Isotopic evidence of a wide spectrum of feeding strategies in southern hemisphere humpback whale baleen records. PLoS One 11.

Eisenmann, P., Fry, B., Mazumder, D., Jacobsen, G., Holyoake, C., Coughran, D., Bengtson Nash, S.M., 2017. Radiocarbon as a tracer of migration in southern hemisphere humpback whales. Scientific Reports 7, 4366.

Eversen, I., 2000. Krill: Biology, Ecology and Fisheries. Blackwell Sciences Ltd., Oxford.

Flores, H., Atkinson, A., Kawaguchi, S., Krafft, B., Milinevsky, G., Nicol, S., Reiss, C., Tarling, G., Werner, R., Rebolledo, B., Cirelli, V., Cuzin-Roudy, J., Fielding, S., Groenveld, J., Haraldsson, M., Lombana, A., Marschoff, E., Meyer, B., Pakhomov, E.A., Rombola, E., Schmidt, K., Siegel, V., Tesche, M., Tonkes, H., Toullec, J., Trathan, P.N., Tremblay, N., Van de Putte, A., Van Franeker, J., Werner, T., 2012. Impact of climate change on Antarctic krill. Marine Ecology Progress Series 458, 1–19.

Fuoco, R., Colombini, M.P., Abete, C., 1991. Evaluation of pack melting effect on polychlobiphenyl content in sea water samples from Terra Nova Bay - Ross Sea (Antarctica). Societa Chimica Italiana 81, 383–394.

George, R.Y., 1982. Crustacean Biology: The biology of the Antarctic krill (Euphausia superba), The First International Symposium on Krill. Wilmington, North Carolina: Allen Press Inc.

George, J.L., Frear, D.E.H., 1966. Pesticides in Antarctica. Journal of Applied Ecology 3, 155–167.

Godard, C.A.J., Smolowitz, R.M., Wilson, J.Y., Payne, R.S., Stegeman, J.J., 2004. Induction of cetacean cytochrome P4501A1 by β-naphthoflavone exposure of skin biopsy slices. Toxicological Sciences 80, 268–275.

Goldbogen, J.A., Calambokidis, J., Oleson, E., Potvin, J., Pyenson, N.D., Schorr, G., Shadwick, R.E., 2011. Mechanics, hydrodynamics and energetics of blue whale lunge feeding: efficiency dependence on krill density. The Journal of Experimental Biology 214, 131–146.

Hale, R.C., Kim, S.L., Harvey, E., La Guardia, M.J., Bush, E.O., Jacobs, E.M., 2008. Antarctic research bases: local sources of polybrominated diphenyl ether (PBDE) flame retardants. Environmental Science and Technology 42, 1452–1457.

Halsall, C., 2004. Investigating the occurrence of persistent organic pollutants in the Arctic: their atmospheric behaviour and interaction with the seasonal snow pack. Environmental Pollution 128.

Hunt Jr., G.L., Drinkwater, K.F., Arrigo, K., Berge, J., Daly, K.L., Danielson, S., Daase, M., Hop, H., Isla, E., Karnovsky, N., Laidre, K., Mueter, F.J., Murphy, E.J., Renaud, P.E., Smith Jr., W.O., Trathan, P., Turner, J., Wolf-Gladrow, D., 2016. Advection in polar and sub-polar environments: impacts on high latitude marine ecosystems. Progress in Oceanography 149, 40–81.

Imbeault, P., Chevrier, J., Dewailly, E., Ayotte, P., Despres, J.P., Mauriege, P., Tremblay, A., 2002. Increase in plasma pollutant levels in response to weight loss is associated with the reduction of fasting insulin levels in men but not in women. Metabolism 51, 482–486.

Kawaguchi, S., Ishida, A., King, R., Raymond, B., Waller, N., Constable, A., Nicol, S., Wakita, M., Ishimatsu, A., 2013. Risk maps for Antarctic krill under projected Southern Ocean acidification. Nature Climate Change 3, 842–847.

Keen, E.M., 2017. Aggregative and feeding thresholds of sympatric rorqual whales within a fjord system. Ecosphere 8 e01702–n/a.

Kelly, B.C., Ikonomou, M.G., Blair, J.D., Morin, A.E., Gobas, F.A.P.C., 2007. Food web-specific biomagnification of persistent organic pollutants. Science 317, 236–239.

Kershaw, E.E., Flier, J., 2004. Adipose tissue as an endocrine organ. The Journal of Endocrinology and Metabolism 86, 2548–2556.

Kilada, R., Reiss, C.S., Kawaguchi, S., King, R.A., Matsuda, T., Ichii, T., 2017. Validation of band counts in eyestalks for the determination of age of Antarctic krill, *Euphausia superba*. PLoS One 12, e0171773.

King, C., Riddle, M.J., 2001. Effects of metal contamination on the development of the common Antarctic sea urchin *Sterechinus neumayeri* and comparisons of sensitivity with tropical and temperate echinoids. Marine Ecology Progress Series 215, 143–154.

Knox, G.A., 1994. The Biology of the Southern Ocean. Cambridge University Press, Cambridge.

Koopman, H., Pabst, D.A., McLellan, W.A., Dillaman, R.M., Read, A.J., 2002. Changes in blubber distribution and morphology associated with starvation in the harbour porpoise (*Phocoena phocoena*): evidence for regional differences in blubber structure and function. Physiological and Biochemical Zoology 75, 498–512.

Kopp, M., Peter, H.U., Mustafa, O., Lisovski, S., Ritz, M.S., Phillips, R.A., Hahn, S., 2011. South polar skuas from a single breeding population overwinter in different oceans though show similar migration patterns. Marine Ecology Progress Series 435, 263–267.

La Merrill, M., Emond, C., Kim, M., Antignac, J., Le Bizec, B., Clement, K., Birnbaum, L., Barouki, R., 2013. Toxicological function of adipose tissue: focus on persistent organic pollutants. Environmental Health Perspectives 121, 162–168.

Lipscomb, J.C., Haddad, S., Poet, T., Krishnan, K., 2012. Physiologically-based pharmacokinetic (PBPK) models in toxicity testing and risk assessment. Advances in Experimental Medicine and Biology 745, 76–95.

Loeb, V., Siegel, V., Holm-Hansen, O., Hewitt, R.P., Fraser, W., Trivelpiece, W., Trivelpiece, S., 1997. Effects of sea-ice extent and krill or salp dominance on the Antarctic food web. Nature 387, 897–900.

Ma, J., Hung, C.L.H., Tian, C., Kallenborn, R., 2011. Revolatilization of persistent organic pollutants in the Arctic induced by climate change. Nature Climate Change 1, 255–258.

Marsili, L., Fossi, M.C., Notarbartolo di Sciara, G., Zanardelli, M., Nani, B., Panigada, S., Focardi, S., 1998. Relationship between organochlorine contaminants and mixed-function oxidase activity in skin biopsy specimens of mediterranean fin whales (*Balaenoptera physalus*). Chemosphere 37, 1501–1510.

Meeker, J.D., 2012. Exposure to environmental endocrine disruptors and child development. Archives of Pediatrics and Adolescent Medicine 166, E1–E7.

Muir, D.C.G., Howard, P.H., 2006. Are there other persistent organic pollutants? A challenge for environmental chemists. Environmental Science and Technology 40, 7157–7166.

Nicol, S., 2006. Krill, Currents, and Sea Ice: Euphausia superba and Its Changing Environment. Bioscience 56.

Nicol, S., Bowie, A., Jarman, S., Lannuzel, D., Meiners, K.M., Van Der Merwe, P., 2010. Southern Ocean iron fertilization by baleen whales and Antarctic krill. Fish and Fisheries 11, 203–209.

Nielsen, E., Larsen, J.C., Ladefoged, O., 2006. Risk Assessment of Contaminant Intake from Traditional Greenland Food Items. Danish Veterinary and Food Administration, p. 178.

Nielsen, E., Østergaard, G., Larsen, J.C., 2008. Toxicological Risk Assessments of Chemicals. Informa, New York, London.

Owen, K., Kavanagh, A.S., Warren, J.D., Noad, M.J., Donnelly, D., Goldizen, A.W., Dunlop, R.A., 2017. Potential energy gain by whales outside of the Antarctic: prey preferences and consumption rates of migrating humpback whales (*Megaptera novaeangliae*). Polar Biology 40, 277–289.

Pelletier, C., Doucet, E., Imbeault, P., Tremblay, A., 2002. Associations between weight loss induced changes in plasma organochlorine concentrations, serum T_3 concentrations and resting metabolic rate. Toxicological Sciences 67, 46–51.

Prevedouros, K., Cousins, I., Buck, R.C., Korzeniowski, S.H., 2006. Sources, fate and transport of perfluorocarboxylates. Environmental Science and Technology 40, 32–44.

Ross, P.S., Vos, J.G., Osterhaus, A.D.M.E., 2003. The immune system, environmental contaminants and virus-associated mass mortalities among pinnipeds. In: Vos, J.G., Bossart, G.D., Fournier, M., O'Shea, J.D. (Eds.), Toxicology of Marine Mammals (New Perspectives: Toxicology and Environment). Taylor and Francis, London, pp. 534–557.

Ryabov, A.B., de Roos, A.M., Meyer, B., Kawaguchi, S., Blasius, B., 2017. Competition-induced starvation drives large-scale population cycles in Antarctic krill. 1, 0177.

Samaran, F., Stafford, K., Branch, T., Gedamke, J., Royer, J., Dziak, R., Guinet, C., 2013. Seasonal and geographic variation of southern blue whale sub-species in the Indian Ocean. PLoS One 8.

Smetacek, V., 2008. Are declining Antarctic krill stocks a result of global warming or the decimation of whales? In: Duarte, C. (Ed.), Impacts of Global Warming on Polar Ecosystems, pp. 47–83.

Smith, J.N., Grantham, H.S., Gales, N., Double, M.C., Noad, M.J., Paton, D., 2012. Identification of humpback whale breeding and calving habitat in the Great Barrier Reef. Marine Ecology Progress Series 447, 259–272.

Stamation, A.K., 2007. Observations of humpback whales (*Megaptera novaeangliae*) feeding during their southward migration along the coast of southeastern New South Wales, Australia: identification of a possible supplemental feeding ground. Aquatic Mammals 33, 165–174.

Stockin, K.A., Burgess, A., 2005. Opportunistic feeding of an adult male humpback whale (*Megaptera novaeangliae*) migrating along the coast of southeastern Queensland, Australia. Aquatic Mammals 31, 120–123.

Tremblay, A., Pelletier, C., Doucet, E., Imbeault, P., 2004. Thermogenesis and weight loss in obese individuals: a primary association with organochlorine pollution. International Journal of Obesity 28, 936–939.

UNEP, 2001. Stockholm Convention on Persistent Organic Pollutants. United Nations Environment Program, Stockholm.

UNEP, 2002. Antarctic Regional Report: Regionally Based Assessment of Persistent Toxic Substances. United Nations Environment Programme, pp. 1–77.

Van Opzeeland, I., Van Parijs, S., Kindermann, L., Burkhardt, E., Boebel, O., 2013. Calling in the cold: pervasive acoustic presence of humpback whales (*Megaptera novaeangliae*) in Antarctic coastal waters. PLoS One 8, e73007.

Wania, F., Mackay, D., 1993. Global fractionation and cold condensation of low volatility organochlorine compounds in polar regions. Ambio 22, 10–18.

Waugh, C.A., Huston, W., Noad, M., Bengtson Nash, S.M., 2011. Cytochrome P450 isoenzyme protein verified in the skin of southern hemisphere humpback whales (*Megaptera novaeangliae*); Implications for biochemical biomarker assessment. Marine Pollution Bulletin 62, 758–761.

Wild, S., 2016. Investigating Input Pathways of Persistent Organic Pollutants to Eastern Antarctica. School of Environment, Griffith University, Brisbane, Australia.

Wild, S., McLagan, D., Schlabach, M., Bossi, R., Hawker, D., Cropp, R., King, C., Stark, J., Mondon, J., Bengtson Nash, S.M., 2014. An Antarctic research station as a source of brominated and perfluorinated persistent organic pollutants to the local environment. Environmental Science and Technology 49, 103–112.

Wolkers, H., Corkeron, P.J., van Parijs, S., Simila, T., van Bavel, B., 2007. Accumulation and transfer of contaminants in killer whales (*Orcinus orca*) from Norway: indications for contaminant metabolism. Environmental Toxicology and Chemistry 26, 1582–1590.

Yamashita, N., Taniyasu, S., Petrick, G., Wei, S., Gamo, T., Lam, P.K.S., Kannan, K., 2008. Perfluorinated acids as novel chemical tracers of global circulation of ocean waters. Chemosphere 70, 1247–1255.

Yasunaga, G., Fujise, Y., Zenitani, R., Tanabe, S., Kato, H., 2015. Spatial and temporal variation of PCBs and organochlorine pesticides in the Antarctic minke whales, *Balaenoptera bonaerensis*, in the period 1987–2005. Chemosphere 126, 11–17.

Yordy, J.E., Wells, R., Balmer, B.C., Schwacke, L.H., Rowles, T.K., Kucklick, J.R., 2010. Partitioning of persistent organic pollutants between blubber and blood of wild bottlenose dolphins: implications for biomonitoring and health. Environmental Science and Technology 44, 4789–4795.

Chapter 15

Emerging Pathogens and Stress Syndromes of Cetaceans in European Waters: Cumulative Effects

Sandro Mazzariol[1], Manuel Arbelo[2], Cinzia Centelleghe[1], Giovanni Di Guardo[3], Antonio Fernandez[2], Eva Sierra[2]
[1]University of Padua, Padova, Italy; [2]University of Las Palmas de Gran Canaria, Las Palmas, Spain; [3]University of Teramo, Teramo, Italy

INTRODUCTION

There is a growing awareness among the veterinary and human health disciplines of the importance of the link between human and animal health and environmental conditions, particularly in the context of species conservation (Deem et al., 2001; Aguirre et al., 2012). Connections between the health of humans, animals, and the environments in which they live are well recognized and have been referred to as "one health, one medicine."

Pathogens may use many different methods to disperse from an infected to an uninfected host. As a consequence, factors that affect these pathogens, hosts, and vectors, and their probability of close contact, are fundamentally crucial to disease dynamics. Several ecologic processes influence risk and disease incidence. Environmental changes due to anthropogenic activity such as climate change and pollution have been associated with disease states in human and animal populations. Pathogens, along with habitat loss, overexploitation, human disturbance, and pollution, are becoming more important factors in the conservation of species (Lafferty and Gerber, 2002; Smith et al., 2006). In some marine mammal mass mortality events (Heide-Jørgensen et al., 1992; Houde et al., 2005) and epidemics of infectious diseases in amphibians and reptiles (Herbst, 1994; Berger et al., 1998), environmental pollution has been hypothesized to be a relevant causative factor. To evaluate impacts of pollution on cetaceans' health and its role on the epidemiology of infectious diseases, it is important to investigate routinely stranding events. In fact, cetacean strandings are an important source

Marine Mammal Ecotoxicology. https://doi.org/10.1016/B978-0-12-812144-3.00015-2

of information on cetacean population health status, allowing determination of not only the causes of mortality but also the threats to these populations, including anthropogenic and natural risks (Peltier et al., 2014).

Global contamination has become a great concern, especially for cetaceans because they are one of the populations exposed to highest concentrations of persistent organic pollutants (POPs) arising out of an alarming anthropogenic pressure (Tanabe, 2002). Although the use of the majority of the toxic chemicals detected in cetaceans is currently banned, relevant levels still persist in the environment, accumulate in lipid-rich tissue, and build up along trophic levels, therefore affecting populations of cetaceans all over the world. Besides, various authors have attempted to establish thresholds for toxicity in different tissues (Kannan et al., 2000; AMAP, 2002; Letcher et al., 2010), and numerous studies have shown that these endpoints are commonly exceeded in marine mammals.

After the 1990–92 morbillivirus epizootic occurred in striped dolphins (*Stenella coeruloalba*) of the Mediterranean Sea (Domingo et al., 1992), many publications stressed the role of contaminant levels in facilitating infectious diseases (Ross, 2002). For instance, it has been reported that PCBs were implicated in the development of the morbillivirus epidemic in this area (Kannan et al., 1993; Marsili and Focardi, 1996; Storelli and Marcotrigiano, 2003; Aguilar and Borrell, 1994, 2005; Fossi et al., 2006; Wafo et al., 2005; Borrell and Aguilar, 2007; Shoham-Frider et al., 2009; Castrillón et al., 2010; Storelli et al., 2012). Although direct effects of POPs are difficult to assess, many studies have related the pollutant load with adverse health effects, such as immune suppression, endocrine disruption, reproductive impairment, and carcinogenic effects (Martineau et al., 1994; De Swart et al., 1995; Lahvis et al., 1995; Schwacke et al., 2002, 2012; Wells et al., 2005; Yap et al., 2012), but conclusive evidence of the influences of pollution in the development of an infectious disease are not clearly evident during postmortem investigations of cetaceans. For these reasons, many studies have been conducted by many scientists to understand the role of spontaneous diseases in cetaceans' conservation, focusing in particular on the interaction between the pathogen and its natural host.

Newly recognized and evolving infectious diseases showing either a recent increase in incidence or changes in pathogenicity or a geographic expansion are gathered under the common definition of "emerging infectious diseases" (EIDs) (Bengis et al., 2004). The modification in the host and pathogen ecology as well as in their interaction dynamics, largely influenced by anthropogenic factors (i.e., environmental chemical pollution and global climate change) frequently underlie EID development (Daszak et al., 2000a,b; Bengis et al., 2004; Cunningham, 2005; Munson et al., 2008; Datta et al., 2009; Johnson et al., 2009). Mass die-offs, population declines, biodiversity loss, and increased risk of extinction may be the result of EID occurrence within wild animal communities (Plowright, 1982; Thorne and Williams, 1988; Daszak et al., 2000a,b; Johnson et al., 2009). EIDs have been reported in several cetacean species and populations throughout the world, being responsible for dramatic mortality outbreaks, affecting their reproductive function and biology, with associated/related population declines (Van Bressem et al., 2009; Di Guardo et al., 2010, 2011a,b; Van Bressem et al., 2014). Emerging and reemerging viral, bacterial, protozoal, and fungal diseases are being increasingly described in cetaceans. Additionally, complex diseases involving emerging infectious and neoplastic components have been reported, and these diseases may provide important information on aquatic ecosystem and public health (Bossart, 2011).

We herein review the most significant cetacean EIDs, namely *Morbillivirus*, *Herpesvirus*, *Brucella ceti*, and *Toxoplasma gondii* infections, along with a number of additional emerging infectious pathogens that have gained progressive importance, providing thereafter a brief overview of the epidemiological situation regarding the aforementioned agents both in the Canary Islands archipelago (Atlantic Ocean) and in the Mediterranean Sea.

MORBILLIVIRUS

Dolphin morbillivirus (DMV) (family *Morbillivirus*, genus *Pasamyxovirus*) is one of the three more common strains of cetacean morbillivirus (CeMV), which also includes the porpoise morbillivirus (PMV) (Barrett et al., 1993; Kennedy, 1998) and the pilot whale morbillivirus (PWMV) (Taubenberger et al., 2000). It has been recognized for at least 25 years as a biologic disease agent of great concern for free-ranging cetaceans (Van Bressem et al., 2009), being responsible for several outbreaks in marine mammals worldwide in the last 25–30 years (Van Bressem et al., 2014; Duignan et al., 2014). Focusing on cetacean species, the most dramatic episodes affected bottlenose dolphins (*Tursiops truncatus*) along the Atlantic coast of the United States in 1987–88 and in 2013–15 (Lipscomb et al., 1994; Schulman et al., 1997; NOAA, 2013), and striped dolphins in the Mediterranean Sea between 1990 and 1992 (Domingo et al., 1990, 1992) and in 2007–08 (Raga et al., 2008). Interestingly, before this latter event, at the end of 2006, a morbilliviral epidemic was also reported in long-finned pilot whales (*Globicephala melas*) around the Strait of Gibraltar (Fernández et al., 2008) and in the following months, it was reported in striped dolphins and pilot whales along the Spanish Mediterranean coast, involving a pilot whale and a bottlenose dolphin found stranded on the French Mediterranean coast (Keck et al., 2010), as well as striped dolphins (Di Guardo et al., 2011a,b). In the following years, other smaller episodes were reported mainly in Italian waters (Casalone et al., 2014), affecting not only small odontocetes but also larger ones (Mazzariol et al., 2017; Centelleghe et al., 2017) and mysticetes (Mazzariol et al., 2016). In most of these episodes, the dolphin morbillivirus strain was involved, supporting the hypothesis of the capability of this virus to affect different cetacean species, as previously suggested by several cases in the Pacific Ocean (Van Bressem et al., 2014). In specific conditions, this strain was also able to infect pinnipeds, as in the case of a monk seal (Monachus) mortality episode in 2000 (Van de Bildt et al., 2000) and in a single captive harbor seal (*Phoca vitulina*) (Mazzariol et al., 2012).

DMV causes either a systemic infection, characterized by bronchointerstitial pneumonia, lymphoid depletion with germinal center necrosis, and nonsuppurative encephalitis (Fig. 15.1) (Domingo et al., 1992; Duignan et al., 1992; Raga et al., 2008; Soto et al., 2011) or a localized infection restricted to the central nervous system (Domingo et al., 1995; Di Guardo et al., 2011a,b), resembling human "subacute sclerosing panencephalitis" and "old dog encephalitis" (Di Guardo and Mazzariol, 2013). Typical microscopic findings observed mainly in lungs and lymph nodes are the syncytial cells (Warthin-Finkeldey type) (Fig. 15.1C). In case of the animal's recovery, as occur for other morbilliviral species, the immune function of the affected animal is impaired, and the capability to resist other diseases is consequently reduced, so these animals could die from subsequent complication of the diseases (Van Bressem et al., 2014).

HERPESVIRUS

Herpesvirus (family *Herpesviridae*, including three subfamilies, *Alphaherpesvirinae*, *Betaherpesvirinae*, and *Gammaherpesvirinae*) causes

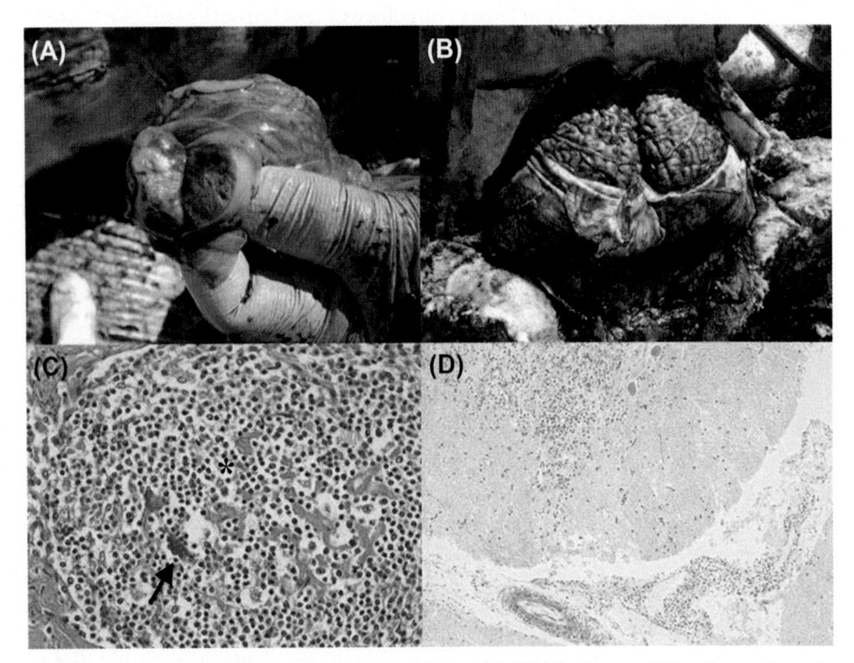

FIGURE 15.1 In this figure, main lesions related to DMV infection could be observed: on the left figures, lymph node involvement is shown with the enlargement of pulmonary organs in a fin whale (A) and the lymphocytic depletion areas in the lymph nodes of a harbor seal (*Phoca vitulina*) (*asterisk*) along with one syncytial cell (Warthin-Finkeldey type), microscopically evident as a multinucleated giant cell (*arrow*) (EE, 20x) (C) On the right, cerebral findings could be observed with a severe and diffuse cerebral hyperemia (B) and the chronic meningeal inflammation (EE, 4x) (D) in a fin whale (*Balaenoptera physalus*).

disease of variable severity in many species, including cetaceans. Nevertheless, little is known about the distribution and the pathogenic effects of these viral agents in dolphins and whales (Arbelo et al., 2012; Lecis et al., 2014; Melero et al., 2015). Phylogenetic analysis of their partial DNA polymerase coding sequence allows classification of cetacean herpesviruses into *Alphaherpesvirinae* and *Gammaherpesvirinae* (van Elk et al., 2009).

Infections induced by alphaherpesviruses, the least specific within the *Herpesviridae* family (Esperón et al., 2008), have been reported in bottlenose dolphins (Blanchard et al., 2001), in which they have been also found associated with proliferative dermatitis lesions (Manire et al., 2006). Similar infections have been additionally described in a Cuvier's beaked whale (*Ziphius cavirostris*) with severe lymphoid tissue necrosis (Arbelo et al., 2010) and in a Blainville's beaked whale (*Mesoplodon densirostris*) with membranous glomerulonephritis and nonsuppurative, interstitial nephritis (Arbelo et al., 2012). Reports of primary (alpha)herpesviral infection in free-ranging cetaceans include cases of nonsuppurative encephalitis in bottlenose dolphins (Esperón et al., 2008) as well as in harbor porpoises (*Phocoena phocoena*) (Kennedy et al., 1992; van Elk and van de Bildt, 2016). However, infection often cooccurs with *Toxoplasma gondii* (Van Bressem et al., 2009). The relationship between herpesviruses and immunocompromised hosts has been largely described, including the presence of systemic herpesviral lesions in a striped dolphin, probably secondary to immunosuppression caused by morbillivirus coinfection (Soto et al., 2012).

BRUCELLA AND OTHER BACTERIA

Cetaceans living close to the shores, such as bottlenose dolphin populations, can be exposed to pathogens normally associated with humans or domestic animals, especially in urbanized areas. Bacteria such as *Salmonella* spp., *Escherichia coli.* and *Listeria monocytogenes* have been found in some stranded animals. Furthermore, methicillin-resistant *Staphylococcus aureus*, a bacterial species responsible for several nosocomial infections both in human beings and in farm animals, has been isolated from free-ranging cetaceans close to the Florida shores. These findings, even if they are generally detected in single animals, support the idea of a telluric biologic pollution with these bacteria being carried by sewage waters or flooding from the land facilities to the marine environment. Among these bacteria, *Brucella* spp. and *Erysipelothrix rusiopathiae* are considered among the most worrying ones.

Brucella spp. have been isolated from free-living cetaceans in northern European, Mediterranean (including Italian), and US waters. While these infections are probably not fatal, they can lead to a number of chronic disease conditions that make animals more susceptible to other pathogens or prevent them from feeding in an effective manner.

Brucella spp. infections have been described in different marine mammal worldwide (Nymo et al., 2011) since its first reference in these species in 1994

(Ewalt et al., 1994; Ross et al., 1994), related to placentitis, abortion (Miller et al., 1999), and nonsuppurative meningoencephalitis (Fig. 15.2) (González et al., 2002; Davison et al., 2009). Several cases of Brucella infection in bottlenose dolphins have been previously described with a wide range of induced lesions, such as pulmonary abscesses (Cassle et al., 2013), vertebral osteomyelitis (Goertz et al., 2011), and abortion and placentitis (Miller et al., 1999).

Genetic studies have revealed that these bacterial agents probably originated from *B. ovis*, meaning that they distinguished themselves in ancient times, thus following the evolution of their respective host (seals, porpoises, and dolphins) (Bourg et al., 2007) Two species of *Brucella* spp. have been recognized on the basis of their biologic and molecular characteristics: the *Brucella* strains infecting cetaceans, classified as *B. ceti*, and those infecting pinnipeds, classified as *B. pinnipedialis*. *Brucella* spp. isolates obtained from cetaceans were later distinguished into *B. delphini* and *B. phocoenae*. In some populations, as in small cetaceans from Peruvian waters, the infection appears to be endemic, in particular infecting sexually mature young individuals, thus representing a potential threat for the growth of resident cetacean populations. The zoonotic potential of these microorganisms must be also carefully evaluated, and some cases of human infection have been reported: *Brucella*-infected patients complain about sinusitis, weakness, and headache, symptoms that last for a week if treated with antibiotics. Two cases of neurobrucellosis associated with osteomyelitis were also reported in human patients with a *Brucella* spp. genotype ST27 infection.

Erysipelothrix rhusiopathiae is the only pathogenic species of the genus *Erysipelothrix*. This bacterium is ubiquitous and can persist for long periods in the environment, including marine locations (Wang et al., 2010). *E. rhusiopathiae* is the causative agent of erysipelas, a disease of many mammalian and

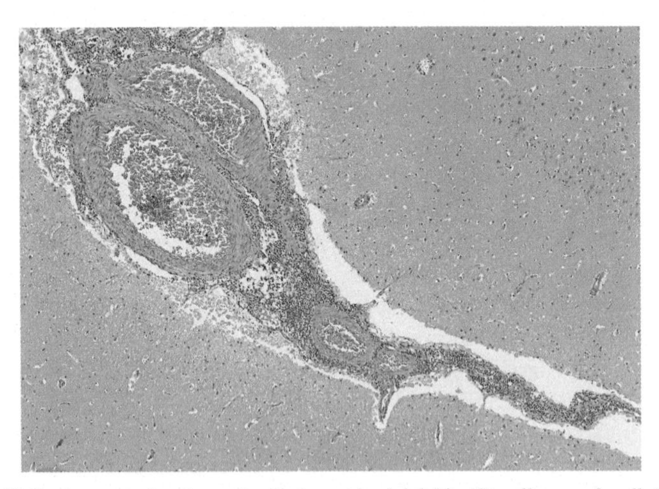

FIGURE 15.2 A severe chronic meningitis in a striped dolphin (*Stenella coeruleoalba*) stranded along the Ionian shores: a diffuse population of lymphocytes massively infiltrates and thickens the meningeal tissues (EE, 10x).

avian species, mainly swine and turkeys (Kinsel et al., 1997). In humans, is considered an occupational zoonosis caused by contact with contaminated animals, specially handling fish, their products, their wastes, or soil (Wang et al., 2010). The dermatologic and acute septicemic forms of this disease have been reported in several cetacean species, including free-range bottlenose dolphins (Melero et al., 2011).

Regarding cetacean parasites, normally, these agents cause little damage to their hosts; although, sometimes they produce pathologic processes harmful enough to inflict substantial morbidity and mortality, usually related to great parasitic loads associated with a general weakening or immunosuppression of the animals (Dailey, 1985; Geraci and St Aubin, 1987; Cornaglia et al., 2000).

TOXOPLASMA GONDII

Toxoplasma gondii, an apicomplexan protozoan parasite, infects a range of hosts worldwide, including several marine mammal species, in which it may cause abortion, lethal systemic disease (Dubey et al., 2003), and nonsuppurative encephalitis (Resendes et al., 2002; Dubey et al., 2009; Di Guardo et al., 2010, 2011a,b). Free-ranging bottlenose dolphins rank among *T. gondii*–susceptible hosts inhabiting the Mediterranean Sea and many other marine ecosystems worldwide. *T. gondii* is believed to be a pathogen of concern for this and other cetaceans species, with documented potential to affect their already threatened health and conservation status, as clearly highlighted by the prominent subacute-to-chronic, nonsuppurative meningoencephalitis lesions reported in several striped dolphins found stranded between 2007 and 2008 along the Ligurian Sea coast of Italy (Di Guardo et al., 2010). The presence of parasitic bodies and zoites was, moreover, documented in earlier studies at the encephalic with the simultaneous finding of a mild inflammatory reaction (Fig. 15.3), lung and lymph node levels; in the latter site a moderate to serious necrotizing lymphadenitis was also apparent.

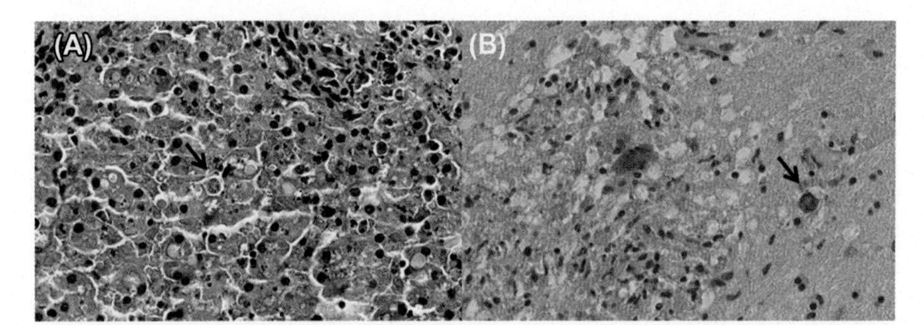

FIGURE 15.3 (A) *Toxoplasma gondii* oocysts observed in the liver of a striped dolphin (*Stenella coeruleoalba*) (*arrow*, EE, 40x) and (B) in the brain of a bottlenose dolphin (*Tursiops truncatus*) along with a mild focal granulomatous reaction (immunohistochemistry against goat *T. gondii* SLD-IFA-TOXO, avidin-biotin-peroxidase with hematoxylin counterstain; 40x).

This protozoa has been also documented in pelagic species of large dimensions such as the sperm whale (*Physeter macrocephalus*) and the fin whale (Mazzariol et al., 2011, 2012), generally as the consequence of either virus- or pollutant-induced/related immunosuppression or cachexia/starvation. The number and the nature of infections caused by *T. gondii* underlines how this agent has spread in coastline waters likely affected by anthropic pressure and by coastline changes, along with the prolonged resistance of protozoan oocysts even in sea water.

CANARY ISLANDS

In the Canary Islands archipelago, 30 of the 87 described species of cetaceans on the planet have been sighted, making the Canary Islands one of the most diverse regions for North Atlantic cetaceans (MAGRAMA, 2012). Most are oceanic species, little known and representative of the Macaronesian region. Nevertheless, their presence and frequency of presentation vary. The set of strandings and sightings analysis allows describing the year-round presence of some species, while others show a marked seasonality. Furthermore, some species are frequently observed, while others are considered rare or occasional.

Since 1992, the Unit of Cetacean Research of the Division of Histology and Veterinary Pathology of Institute for Animal Health (University of Las Palmas de Gran Canaria) has been carrying out a health monitoring program of cetacean populations through the study of their pathologies (Fernández et al., 2005; Jaber et al., 2005; Arbelo et al., 2013; Sierra et al., 2013) and pollutants burdens (Carballo et al., 2008; García-Alvarez et al., 2014a,b) found in the stranded animals in these islands. The research projects carried out by this unit are designed with the intention of providing a multidisciplinary knowledge based on the biology, physiology, and pathology of cetaceans protocols of laboratory analysis, including histopathology, virology, parasitology, bacteriology, toxicology, and chromatographic analysis, performed at the IUSA, ULPGC, and specialized laboratories. In addition, a cetacean tissue bank has been generated, containing samples from more than 600 animals of more than 20 different species. The corresponding analyses are carried out to determine the pathologies and etiological agents present in stranded cetaceans. Currently, IUSA has all the laboratories for etiological analysis. More specific or special cases, in which very specialized support is needed, are carried out through the national and international collaborations in which our Institute is currently involved.

Natural causes of death have so far accounted for the majority of the cases in the cetaceans stranded along the coasts of the Canary Islands, including infectious and noninfectious diseases, neonatal pathology, and intra- and interspecific interactions (Arbelo et al., 2013).

Dolphin Morbillvirus

To date, 177 specimens stranded in the Canary Islands have been tested for the presence of CeMV. From them, 15/177 (8.5%) resulted positive in one or more tissues (by means of immunohistochemistry and/or molecular methods). The majority of the specimens belonged to the striped dolphin species ($n=6$), followed by the short-finned pilot whale (*Globicephala macrorynchus*) ($n=4$). The common dolphin (*Delphinus delphis*) and the Risso's dolphin (*Grampus griseus*) were represented by two specimens, respectively, and the bottlenose dolphin for one animal (Belliére et al., 2011; Sierra et al., 2014a,b). Two different strains of CeMV have been detected among the positive cases: DMV and PWMV. The latter was exclusively detected in the short-finned pilot whale species: in the brain of a case that showed lesions compatible with morbilliviral disease (Fig. 15.4) (Belliére et al., 2011) and in an unusual mortality event (UME) affecting three cases with the systemic infection presentation form of the disease (Sierra et al., 2016), indicating that lethal infections by this viral strain are not as rare as previously believed. It is also worthy of mention that the Canary Islands is the only place in the world where the PWMV has been detected in the short-finned pilot whale species (Fig. 15.5).

Regarding the presentation form of the disease, the chronic systemic infection and the chronic localized CeMV encephalitis were represented by four cases, respectively; and the acute and the subacute systemic diseases presented two cases each one. In one case, due to the advanced autolysis of the carcass, it was not possible to associated one of these presentation forms to the case. Concerning the dates of stranding, positive cases were detected in 1999 ($n=1$), 2002 ($n=1$), 2005 ($n=1$), 2007 ($n=2$), 2008 ($n=2$), 2009 ($n=1$), 2011 ($n=1$), 2012 ($n=1$), 2015 ($n=4$), and 2016 ($n=1$). These data reveal that despite that no epidemics have been described in the Canary Islands, sporadic mortalities associated with the morbillivirus infection have been described along the last 18 years, with some of them coinciding temporarily with some of the outbreaks and/or UME that occurred in the in

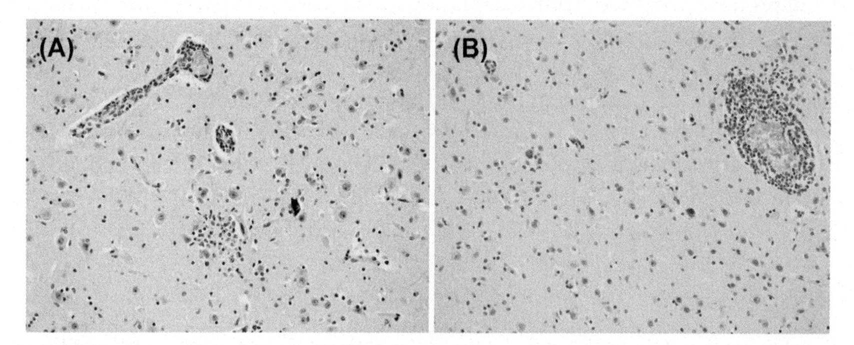

FIGURE 15.4 Localized chronic morbilliviral infection of the central nervous system in a juvenile male short-finned pilot whale stranded in the Canary Islands in 1996. Immunohistochemistry against canine distemper virus (nucleoprotein), avidin-biotin-peroxidase with Harris hematoxylin counterstain; 20x. (A) Gliosis nodules and occasional immunopositive neurons are present. (B) A severe perivascular cuffing is also present (5–10 layers of lymphoplasmacytic cells) and scattered immunopositive neurons.

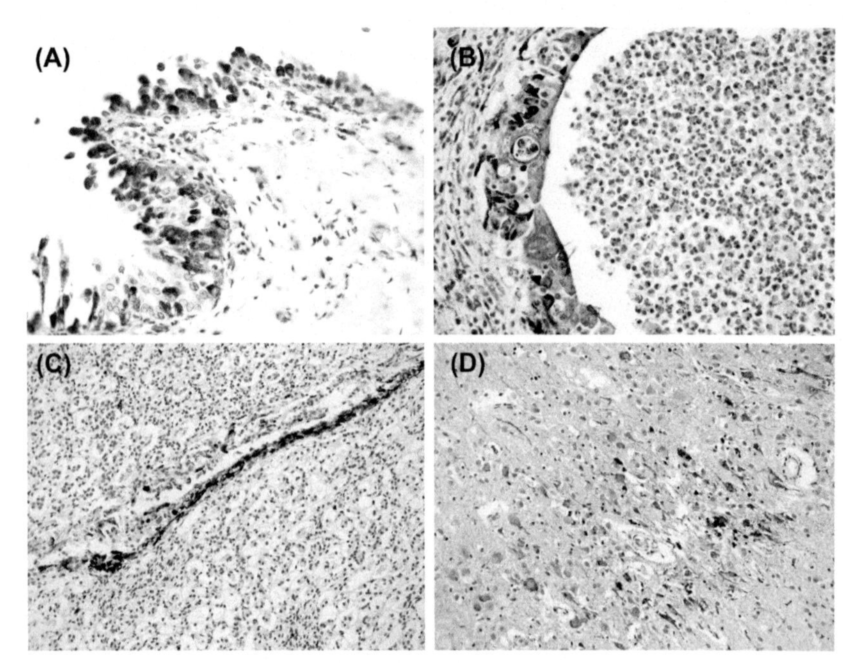

FIGURE 15.5 Subacute systemic disease in an adult female of short-finned pilot whale stranded in the Canary Islands in 2015. Immunohistochemistry against canine distemper virus (nucleoprotein), avidin-biotin-peroxidase with Harris hematoxylin counterstain. Intense immunoperoxidase staining of morbillivirus was observed in cytoplasm and nuclei of urinary bladder epithelial cells, (A) fibroblasts and laryngeal tonsil epithelial cells (B), kidney ureotelia (C), and neurons and glial cells (D).

Northwest Atlantic Ocean and Mediterranean Sea (2006–08; 2011). It has been proposed that the strain circulating in the Central Eastern (CE) Atlantic Ocean was introduced in the Mediterranean Sea in 2006 (Van Bressem et al., 2014). The Canary Islands have a geographically strategic position since they are located between three continents: Europe, America, and Africa. Thus, a regular monitoring for the presence of CeMV in free-ranging cetaceans in this particular point of the CE Atlantic is essential to detect the occurrence and distribution of this infectious disease and to predict novel worldwide epidemics.

Herpesvirus

As already stated, in the Canary Islands, some cases of herpesvirus infection have been described (by means of immunohistochemistry and/or molecular methods). More specifically, herpesviruses have been responsible for serious diseases such as encephalitis in a striped dolphin (Sierra et al., 2014a,b) and a bottlenose dolphin (Esperón et al., 2008). Molecular studies also described alphaherpesviruses associated with a localized infection of the lymphoid system in a Cuvier's beaked whale (Arbelo et al., 2010), and acute multifocal necrotizing tubulointerstitial nephritis in a Blainville's beaked whale (Arbelo et al., 2012).

A gammaherpesvirus was detected within the penile lesions (two masses composed of hyperplastic epithelial cells with ballooning degeneration and margination of chromatin at the stratum corneum of the epidermis) of an adult male striped dolphin stranded alive at Arico, Tenerife, Canary Islands, Spain (Sierra et al., 2015).

Brucella spp., *Erysipelothrix rhusiopathiae*, and *Toxoplasma gondii*

We have detected two cases of neurobrucellosis in a striped and a bottlenose dolphin stranded in the Canary Islands in 2004 and 2005, respectively (unpublished data). Furthermore, two cases of fatal *Erysipelothrix rhusiopathiae* septicemia have been reported in the Canary Islands in two different species: Atlantic spotted and bottlenose dolphin species (Díaz-Delgado et al., 2015). In the Canary Islands, several cases of encephalitis associated with *Toxoplasma gondii* have been diagnosed (by means of immunohistochemistry and molecular methods) in the *Stenella frontalis* species in the absence of concurrent morbillivirus infection, suggesting that this protozoan parasite could act as a primary etiological agent in the development of the observed severe lesions in these cases (Arbelo et al., 2013).

CENTRAL MEDITERRANEAN SEA: ITALY

In Italian waters, eight of the species regularly documented in the Mediterranean Sea, mainly bottlenose and striped dolphins, occur. Most of these species can be found within the Pelagos Sanctuary a special marine protected area dedicated to the protection of these species, extending in the northwestern Mediterranean Sea between Italy, France, and Sardinia, encompassing Corsica and the Archipelago Toscano. Since 2011, the Italian Ministry for the Environment and the Ministry for Health decided to establish a national stranding network including technical experts: veterinary pathologists and diagnosticians working in the regional public laboratories (Istituti Zooprofilattici Sperimentali) and in the veterinary schools of Padova and Teramo, working in cooperation with ecotoxicologists and biologists of other universities and museums, ensuring a constant monitoring of the about 7400 km of shores. Data related to stranding events are reported to a national stranding database, while collected tissues are stored in the Mediterranean Marine Mammals Tissue Bank, available for further research. The cooperation of all these scientists and Institutions ensure a multidisciplinary approach to the stranding event (Mazzariol et al., 2016), in particular during UMEs and mass strandings (Casalone et al., 2014; Mazzariol et al., 2011, 2016). Data obtained from this monitoring are resumed here as follows.

Morbillivirus

After the two epidemics occurred in the Mediterranean Sea in 1990–92 and 2006–08 related to DMV, another three mortality outbreaks were reported in the Italian waters in the following years: in 2010–11, two striped dolphins, one bottlenose dolphins, and two fin whales showed morbilliviral infections with related lesions (Di Guardo et al., 2012), and during this mortality outbreak a harbor seal kept under human care (Fig. 15.1C) was found affected by the same virus (Mazzariol et al., 2012); in 2013, in a larger UME, 24/57 (42%) cetaceans of different species were found to be molecularly positive to DMV antigens, and during the postmortem investigations the evaluation of the toxicologic stress using the model proposed by Marsili et al. (2004) estimated hazardous levels of organochlorine pollutants in the examined tissues, higher in those affected by the viral disease (Casalone et al., 2014); a third outbreak was reported in the summer of 2016, affecting mainly young striped dolphins in the Ionian Sea (unpublished data). Besides these small mortality events the stranding program showed a constant presence of DMV-positive animals (19.5% in 2015, Diagnostic Report of the Italian Stranding Network, 2015 and Centelleghe et al., 2017; 23.6% in 2016, Diagnostic Report of the Italian Stranding Network, 2016), suggesting an endemic circulation of the virus in the Mediterranean Sea.

This hypothesis was supported also by an increasing number of species sensitive to the virus, as fin whales, beaked whales, and sperm whales. More in details, a mortality outbreak was supposed for fin whales, since five animals out of the nine examined between 2011 and 2013 were found to be positive with DMV-related lesions (Mazzariol et al., 2016), while the same virus was deemed to be one of the causative factors in a sperm whale mass stranding that occurred along the Italian coastline of the Adriatic basin (Centelleghe et al., 2016; Mazzariol et al., 2013). Regarding Cuvier's beaked whales, a single newborn individual was reported in the Southern Tyrrhenian Sea (Centelleghe et al., 2017). The lower density of striped dolphins and bottlenose dolphins reported by recent monitoring efforts in the Mediterranean basin (Panigada et al., 2011), considered the classic hosts for DMV (Sato et al., 2012), could have play a role in the cross-species infection, along with other potential factors as an intermediate and amplifier hosts with the critical role of bringing DMV in close contact with alternative hosts. In fact, the rate and intensity of contact may be relevant as well as the evolutionary relatedness (Parrish et al., 2008).

Similarly to wild carnivores for canine distemper virus, the species reported to be sensitive to morbilliviral infection, as that herein described, could act as reservoir for DMV being a possible cause of spillback events involving a naive population of classic hosts and supporting the presence and circulation of the virus in this limited basin (Beineke et al., 2015; Mazzariol et al., 2016). These recent reports, furthermore, support the relevance of vertical transmission as the likely route of entry, as well as the possible survival strategy of the virus for these rarely affected species (West et al., 2015; Mazzariol et al., 2013, 2016;

Centelleghe et al., 2016). The involvement of newborns, young, and pregnant females could represent a serious menace for the conservation of these already threatened species.

An in-depth investigation conducted by Beffagna et al. (2017) on the whole genome of DMV recovered from a fin whale stranded in 2013, aimed to investigate genomic, aminoacidic, and structural changes, showed two relevant substitutions (Ile21Thr on the N protein and Asn52Ser on the F antigen), already reported in the viral strain isolated from striped dolphins and long-finned pilot whales during the 2006–08 DMV outbreak in Spanish waters. An additional four nonsynonymous aminoacidic substitutions arose after the 2006–08 outbreak, which could have favored the different DMV pathogenicity in fin whale. A relevant proportion of the amino acids that differentiate this strain from those recovered from the two main epidemics in 1990–92 and 2006–08) were proven to be under diversifying selection, thus supporting their potential role in DMV host switch. Besides the apparent change of DMV in fin whales, the reported mutations support the hypothesis of a constant and prolonged morbilliviral infection's presence among Mediterranean cetaceans: the phylogenetic trees obtained comparing the viral sequence characterized in the fin whale with those detected in striped dolphins stranded along the Atlantic coastlines of Portugal and Spain between 2012 and 2014 show a relevant distance, suggestive of a possible divergence between the viral strains circulating in the Mediterranean and in the adjacent Atlantic waters (Fig. 15.6). On the contrary, the virus found in animals stranded along the Italian coastlines during the summer 2016 is more closely to the viral strain found in the Atlantic waters (Casalone, personal communication), underlining periodical DMV entries through the Strait of Gibraltar, as previously suggested (Van Bressem et al., 2001; Raga et al., 2008; Sierra et al., 2014a).

Herpesvirus

These viruses have been less investigated in the Mediterranean Sea and, as already reported, most of the literature describes a herpesviral agent found as coinfection during morbilliviral outbreaks, often without any evident pathologic changes (Belliére et al., 2010; Casalone et al., 2014; Lecis et al., 2014). Interestingly Melero and colleagues reported molecular evidences of alphaherpesviruses in tissues of two mysticetes, namely one fin whale and one minke whale (*Balaenoptera acutorostrata*), stranded along the Valencian coast. In the last 2 years of constant monitoring, herpesvirus was isolated in three dolphins in 2015 (one striped dolphin, one bottlenose dolphin, and one Risso's dolphin out of 77 examined cetaceans) and four striped dolphins in 2016 (out of 76), with one animal showing severe ulcerative changes to the skin related to this infection (Diagnostic Report of the Italian Stranding Network, 2015, 2016). All these animals were clustered in the Central and Northern Tyrrhenian Sea. Finally, an interesting finding of herpesviral infection should be underlined,

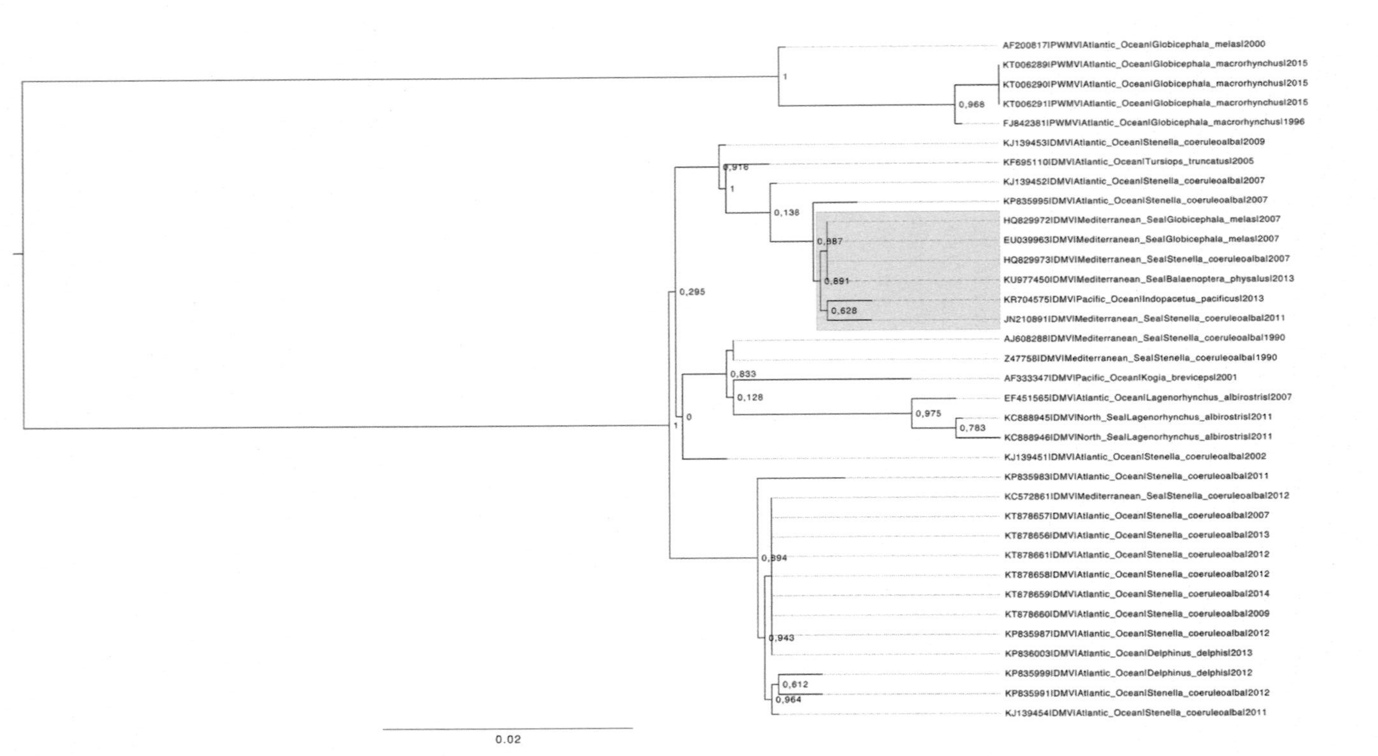

FIGURE 15.6 Maximum likelihood (ML) phylogenetic tree reconstructed on the basis of phosphoprotein (P) sequences of DMV and pilot whale morbillivirus (PWMV) isolated over time in different seas. The bootstrap support is reported close to the corresponding node. The clade including the strain detected in fin whale (*Balaenoptera physalus*) is highlighted in *yellow*.

that in the renal tissues of a fetus harbored by a pregnant female sperm whale involved in a mass stranding that occurred along the Italian Adriatic shores in 2014 (unpublished data).

Brucella ceti and *Erysipelothrix rhusiopathiae*

Reports of brucellosis in Mediterranean cetaceans are quite rare. *Brucella ceti* was reported for the first time in this basin by Alba et al. (2013), who reported a striped dolphin showing nonpurulent meningoencephalitis and a coinfection with *Toxoplasma gondii*. Another two striped dolphins that showed a severe inflammatory reaction to the brain and in pulmonary tissues were reported in Apulia Region in 2012 (Fig. 15.2), along the Ionian shores (Garofolo et al., 2014). In both the reports, the sequence type was ST26. Isidoro-Ayza et al. described in 2014 brucellosis in three dolphin with different lesions stranded along the Catalonian coast: one striped dolphin with evidences of meningoencephalitis (Fig. 15.2), a bottlenose dolphin with discospondylitis, and another striped dolphin without pathologic changes related to brucellosis. A detailed characterization of the three *B. ceti* isolates performed by several approaches showed that they cluster together in a distinct phylogenetic clade, close to that formed by *B. ceti* isolates from cetaceans living in the Atlantic Ocean. A more recent description of meningoencephalitis in a striped dolphin reported *B. ceti*, interestingly, associated to *T. gondii* and *Listeria monocytogenes*. Despite the low number of cases reported and the absence of outbreaks related to this bacterium, the infection due to *B. ceti* arouses a great concern for its zoonotic potential, in particular for personnel working during the stranding events.

A similar consideration could be done also for *Erysipelothrix rhusiopatiae*, which was reported by Melero et al. in 2011 in a free-ranging bottlenose dolphin. Similar cases characterized by an acute septicemia and by a bacteriological isolation were previously reported in a common dolphin stranded in Cefalù (1994), in rough-toothed dolphins stranded alive in Ragusa (2002), and finally in 2008 in a striped dolphin stranded along the Campania coastline in 2008 (Maio et al., 2012).

Toxoplasma gondii

As stated earlier, this protozoon has been frequently encountered during postmortem investigations in cetaceans stranded along the Mediterranean shores and, in particular, in the Italian waters. *Toxoplasma gondii* is an opportunistic pathogen often related to morbilliviral infection. In the case of multiple strandings of striped dolphins that occurred in Ligurian waters in 2008, this pathogen was deemed as the likely cause of death for these animals due to the severe chronic meningoencephalitis evident at microscopic examination without any other relevant agent responsible for such severe changes (Di Guardo et al., 2010). Besides the Ligurian Sea, *T. gondii* has been regularly found in specific areas, possibly related to runoff waters during flooding episodes (Shapiro et al., 2015), as other areas of the Pelagos Sanctuary (Northern Sardinia and Tuscany), and in the bottlenose dolphins inhabiting the Northern Adriatic. The most recent surveys confirm a prevalence of 6.6% in 2015 ($n=5$) and 11.7% in 2016 ($n=9$). As already mentioned, interesting cases of *T. gondii* evidences were reported in pelagic species as sperm whales and fin whales (Mazzariol et al., 2011, 2012). Genotyping confirms that most of the isolates belong to type II or X, confirming a terrestrial source from domestic cats and wild carnivores. Even if this protozoon is not always lethal for cetaceans, its presence in the marine environment mirrors how the human development is rapidly reshaping landscapes, changing pathogen transmission cycles with significant impacts on public health and wildlife conservation. The continued development of the terrestrial landscape may therefore increase the number *T. gondii* oocysts and other pathogens flowing into freshwater systems and the near-shore marine environment, where they pose a risk to marine mammals and humans (VanWormer et al., 2014).

CONCLUDING REMARKS

In spite of the large amount of high-quality research that has been carried out throughout the last 30 years on cetacean EIDs, with special emphasis on *Morbillivirus* infections, the origin, evolutionary phylogeny, host range, pathogenicity, ecology, and epidemiology of the herein dealt infectious pathogens warrant further investigation. Furthermore, these infections may also represent valuable comparative pathology models in the study of similar disease conditions affecting man and terrestrial mammals (Di Guardo, 2012; Di Guardo and Mazzariol, 2013).

Environmental anthropogenic changes may increase the prevalence and severity of infectious diseases in free-ranging dolphins and whales worldwide, with coastal, estuarine, and river cetaceans appearing to be especially at risk, due to the severe degradation frequently experienced by coastal and fluvial ecosystems, which are more consistently polluted both biologically and chemically, being more prone to the effects of climate change (Van Bressem et al., 2009).

In this respect, a crucial issue is the potential synergistic effect, if any, exerted by a number of persistent environmental contaminants, with special emphasis on organochlorines (PCBs, DDTs, dioxins, etc.) and heavy metals (Hg, Pb, Cd, etc.) in modulating the pathogenicity of emerging infectious agents, along with their impact on cetacean health and conservation. In consideration of their "top predator"-related ecologic behavior, in fact, odontocete cetaceans are known to harbor inside their body, with subsequent "bioaccumulation" and "biomagnification," high tissue loads of the aforementioned and other pollutants (Fossi et al., 2007), such as "flame retardants" (PBDEs) and microplastic polymers.

REFERENCES

Aguilar, A., Borrell, A., 1994. Abnormally high polychlorinated biphenyl levels in striped dolphins (*Stenella coeruleoalba*) affected by the 1990–1992 Mediterranean epizootic. Science of the Total Environment 154, 237–247.

Aguilar, A., Borrell, A., 2005. DDT and PCB reduction in the western Mediterranean from 1987 to 2002, as shown by levels in striped dolphins (*Stenella coeruleoalba*). Marine Environment Research 59, 391–404.

Aguirre, A.A., Ostfeld, R.S., Daszak, P., 2012. New Directions in Conservation Medicine, Applied Cases of Ecological Health. Oxford University Press, New York.

Alba, P., Terracciano, G., Franco, A., Lorenzetti, S., Cocumelli, C., Fichi, G., Eleni, C., Zygmunt, M.S., Cloeckaert, A., Battisti, A., 2013. The presence of *Brucella ceti* ST26 in a striped dolphin (*Stenella coeruleoalba*) with meningoencephalitis from the Mediterranean Sea. Veterinary Microbiology 164, 158–163.

AMAP, 2002. AMAP assessment 2002: persistent organic pollutants in the arctic. In: Arctic Monitoring and Assessment Programme (AMAP) Oslo, Norway.

Arbelo, M., Bellière, E.N., Sierra, E., Sacchinni, S., Esperón, F., Andrada, M., Rivero, M., Diaz-Delgado, J., Fernández, A., December 13, 2012. Herpes virus infection associated with interstitial nephritis in a beaked whale (*Mesoplodon densirostris*). BMC Veterinary Research 8, 243. https://doi.org/10.1186/1746-6148-8-243.

Arbelo, M., Sierra, E., Esperón, F., Watanabe, T.T., Belliere, E.N., Espinosa de los Monteros, A., Fernández, A., 2010. Herpesvirus infection with severe lymphoid necrosis affecting a beaked whale stranded in the Canary Islands. Diseases of Aquatic Organism 89, 261–264.

Arbelo, M., Los Monteros, A.E., Herráez, P., Andrada, M., Sierra, E., Rodríguez, F., Jepson, P.D., Fernández, A., 2013. Pathology and causes of death of stranded cetaceans in the Canary Islands (1999–2005). Diseases of Aquatic Organism 103, 87–99.

Barrett, T., Visser, I.K., Mamaev, L., Goatley, L., Van Bressem, M.F., Osterhaust, A.D., 1993. Dolphin and porpoise Morbilliviruses are genetically distinct from phocine distemper virus. Virology 193, 1010–1012.

Beffagna, G., Centelleghe, C., Franzo, G., Di Guardo, G., Mazzariol, S., 2017. Genomic and structural investigation on dolphin Morbillivirus (DMV) in Mediterranean fin whales (*Balaenoptera physalus*). Scientific Report 7, 41554.

Beineke, A., Baumgärtner, W., Wohlsein, P., 2015. Cross-species transmission of canine distemper virus-an update. One Health 1, 49–59.

Bellière, E.N., Esperón, F., Arbelo, M., Muñoz, M.J., Fernández, A., Sánchez-Vizcaíno, J.M., 2010. Presence of Herpesvirus in striped dolphins stranded during the cetacean Morbillivirus epizootic along the Mediterranean Spanish coast in 2007. Archives of Virology 155, 1307–1311.

Bellière, E.N., Esperon, F., Fernandez, A., Arbelo, M., Munoz, M.J., Sanchez-Vizcaino, J.M., 2011. Phylogenetic analysis of a new Cetacean Morbillivirus from a short-finned pilot whale stranded in the Canary Islands. Research in Veterinary Science 90, 324–328.

Bengis, R.G., Leighton, F.A., Fischer, J.R., Artois, M., Mörner, T., Tate, C.M., 2004. The role of wildlife in emerging and re-emerging zoonoses. Revue Scientifique et Technique-Office International des Epizooties 23 (2), 497–511.

Berger, L., Speare, R., Daszak, P., Green, D.E., Cunningham, A.A., Goggin, C.L., Slocombe, R., Ragan, M.A., Hyatt, A.D., McDonald, K.R., Hines, H.B., Lips, K.R., Marantelli, G., Parkes, H., 1998. Chytridiomycosis causes amphibian mortality associated with population declines in the rain forests of Australia and Central America. Proceedings of the National Academy of Sciences of the United States of America 95, 9031–9036.

Blanchard, T.W., Santiago, N.T., Lipscomb, T.P., Garber, R.L., McFee, W.E., Knowles, S., 2001. Two novel alphaherpesviruses associated with fatal disseminated infections in Atlantic bottlenose dolphins. Journal of Wildlife Diseases 37, 297–305.

Borrell, A., Aguilar, A., 2007. Organochlorine concentrations declined during 1987–2002 in western Mediterranean bottlenose dolphins, a coastal top predator. Chemosphere 66, 347–352.

Bossart, G.D., 2011. Marine mammals as sentinel species for oceans and human health. Veterinary Pathology 48 (3), 676–690.

Bourg, G., O'Callaghan, D., Boschiroli, M.L., 2007. The genomic structure of Brucella strains isolated from marine mammals gives clues to evolutionary history within the genus. Veterinary Microbiology 125, 375–380.

Carballo, M., Arbelo, M., Esperón, F., Méndez, M., De la Torre, A., Muñoz, M.J., 2008. Organochlorine residues in the blubber and liver of bottlenose dolphins (*Tursiops truncatus*) stranded in the Canary Islands, North Atlantic Ocean. Environmental Toxicology 23 (2), 200–210.

Casalone, C., Mazzariol, S., Pautasso, A., Di Guardo, G., Di Nocera, F., Lucifora, G., Ligios, C., Franco, A., Fichi, G., Cocumelli, C., Cersini, A., Guercio, A., Puleio, R., Goria, M., Podestà, M., Marsili, L., Pavan, G., Pintore, A., De Carlo, E., Eleni, C., Caracappa, S., 2014. Cetacean strandings in Italy: an unusual mortality event along the Tyrrhenian Sea coast in 2013. Diseases of Aquatic Organism 109 (1), 81–86.

Cassle, S.E., Jensen, E.D., Smith, C.R., Meegan, J.M., Johnson, S.P., Lutmerding, B., Ridgway, S.H., Francis-Floyd, R., 2013. Diagnosis and successful treatment of a lung abscess associated with Brucella species infection in a bottlenose dolphin (*Tursiops truncatus*). Journal of Zoo and Wildlife Medicine 44, 495–499.

Castrillón, J., Gómez-Campos, E., Aguilar, A., Berdie, L., Borrell, A., 2010. PCB and DDT levels do not appear to have enhanced the mortality of striped dolphins (*Stenella coeruleoalba*) in the 2007 Mediterranean epizootic. Chemosphere 81, 459–463.

Centelleghe, C., Beffagna, G., Palmisano, G., Franzo, G., Casalone, C., Pautasso, A., Giorda, F., Di Nocera, F., Iaccarino, D., Santoro, M., Di Guardo, G., Mazzariol, S., 2017. Dolphin morbillivirus in a Cuvier's beaked whale (*Ziphius cavirostris*), Italy. Frontiers in Microbiology 8, 111.

Centelleghe, C., Beffagna, G., Zanetti, R., Zappulli, V., Di Guardo, G., Mazzariol, S., 2016. Molecular analysis of dolphin Morbillivirus: a new sensitive detection method based on nested RT-PCR. Journal of Virological Methods 235, 85–91.

Cornaglia, E., Rebora, L., Gili, C., Di Guardo, G., 2000. Histopathological and immunohistochemical studies on cetaceans found stranded on the coast of Italy between 1990 and 1997. Journal of Veterinary Medicine Series A 47 (3), 129–142.

Cunningham, A.A., 2005. A walk on the wild side–emerging wildlife diseases. BMJ 331 (7527), 1214–1215.

Dailey, M.D., 1985. In: Kinne, O. (Ed.), Diseases of Marine Animals, vol. IV. Biologishe Anstalt Helgoland, Hamburg.

Daszak, P., Cunningham, A.A., 2000a. More on the ecological impact of fungal infections on wildlife populations. Parasitology Today 16 (9), 404–405.

Daszak, P., Cunningham, A.A., Hyatt, A.D., 2000b. Emerging infectious diseases of wildlife–threats to biodiversity and human health. Science 287 (5452), 443–449.

Datta, K., Bartlett, K.H., Marr, K.A., 2009. *Cryptococcus gattii*: emergence in western North America: exploitation of a novel ecological niche. Interdisciplinary Perspectives on Infectious Diseases 2009, 176532.

Davison, N.J., Cranwell, M.P., Perrett, L.L., Dawson, C.E., Deaville, R., Stubberfield, E.J., Jarvis, D.S., Jepson, P.D., 2009. Meningoencephalitis associated with Brucella species in a live-stranded striped dolphin (*Stenella coeruleoalba*) in south-west England. Veterinary Record 165, 86–89.

Deem, S.L., Karesh, W.B., Weisman, W., 2001. Putting theory into practice: wildlife health in conservation. Conservation Biology 15, 1224–1233.

Di Guardo, G., Mazzariol, S., 2013. Dolphin Morbillivirus: a lethal but valuable infection model. Emerging Microbes and Infection 2, e74.

Di Guardo, G., Cocumelli, C., Scholl, F., Di Francesco, C.E., Speranza, R., Pennelli, M., Eleni, C., 2011a. Morbilliviral encephalitis in a striped dolphin *Stenella coeruleoalba* calf from Italy. Diseases of Aquatic Organism 95, 247–251.

Di Guardo, G., Di Cesare, A., Otranto, D., Casalone, C., Iulini, B., Mignone, W., Tittarelli, C., Meloni, S., Castagna, G., Forster, F., Kennedy, S., Traversa, D., 2011b. Genotyping of *Toxoplasma gondii* isolates in meningo-encephalitis affected striped dolphins (*Stenella coeruleoalba*) from Italy. Veterinary Parasitology 183, 31–36.

Di Guardo, G., Di Francesco, C.E., Eleni, C., Cocumelli, C., Scholl, F., Casalone, C., Peletto, S., Mignone, W., Tittarelli, C., Di Nocera, F., Leonardi, L., Fernández, A., Marcer, F., Mazzariol, S., 2013. Morbillivirus infection in cetaceans stranded along the Italian coastline: pathological, immunohistochemical and biomolecular findings. Research in Veterinary Science 94 (1), 132–137.

Di Guardo, G., 2012. Morbillivirus-host interaction: lessons from aquatic mammals. Frontiers in Microbiology 3, 431.

Di Guardo, G., Proietto, U., Di Francesco, C.E., Marsilio, F., Zaccaroni, A., Scaravelli, D., Mignone, W., Garibaldi, F., Kennedy, S., Forster, F., Iulini, B., Bozzetta, E., Casalone, C., 2010. Cerebral toxoplasmosis in striped dolphins (*Stenella coeruleoalba*) stranded along the Ligurian Sea coast of Italy. Veterinary Pathology 47, 245–253.

De Swart, R.L., Ross, P.S., Timmerman, H.H., Vos, H.W., Reijnders, P.J., Vos, J.G., Osterhaus, A.D., 1995. Impaired cellular immune response in harbour seals (*Phoca vitulina*) feeding on environmentally contaminated herring. Clinical and Experimental Immunology 101 (3), 480–486.

Díaz-Delgado, J., Arbelo, M., Sierra, E., Vela, A., Domínguez, M., Paz, Y., Andrada, M., Domínguez, L., Fernández, A., 2015. Fatal *Erysipelothrix rhusiopathiae* septicemia in two Atlantic dolphins (*Stenella frontalis* and *Tursiops truncatus*). Diseases of Aquatic Organism 116 (1), 75–81.

Domingo, M., Visa, J., Pumarola, M., Marco, A.J., Ferrer, L., Rabanal, R., Kennedy, S., 1992. Pathologic and immunocytochemical studies of Morbillivirus infection in striped dolphins (*Stenella coeruleoalba*). Veterinary Pathology 29, 1–10.

Domingo, M., Ferrer, L., Pumarola, M., Marco, A., Plana, J., Kennedy, S., McAlisey, M., Rima, B.K., 1990. Morbillivirus in dolphins. Nature 384, 21.

Domingo, M., Vilafranca, M., Visa, J., Prats, N., Trudgett, A., Visser, I., 1995. Evidence for chronic Morbillivirus infection in the Mediterranean striped dolphin (*Stenella coeruleoalba*). Veterinary Microbiology 44, 229–239.

Dubey, J.P., Mergl, J., Gehring, E., Sundar, N., Velmurugan, G.V., Kwok, O.C., Grigg, M.E., Su, C., Martineau, D., 2009. Toxoplasmosis in captive dolphins (*Tursiops truncatus*) and walrus (*Odobenus rosmarus*). Journal of Parasitology 95, 82–85.

Dubey, J.P., Zarnke, R., Thomas, N.J., Wong, S.K., Van Bonn, W., Briggs, M., Davis, J.W., Ewing, R., Mense, M., Kwok, O.C., Romand, S., Thulliez, P., 2003. Toxoplasma gondii, *Neospora caninum*, *Sarcocystis neurona*, and Sarcocystis canis-like infections in marine mammals. Veterinary Parasitology 116, 275–296.

Duignan, P.J., Geraci, J.R., Raga, J.A., Calzada, N., 1992. Pathology of Morbillivirus infection in striped dolphins (*Stenella coeruleoalba*) from Valencia and Murcia, Spain. Journal of Veterinary Research 56, 242–248.

Duignan, P.J., Van Bressem, M.F., Baker, J.D., Barbieri, M., Colegrove, K.M., De Guise, S., de Swart, R.L., Di Guardo, G., Dobson, A., Duprex, W.P., Early, G., Fauquier, D., Goldstein, T., Goodman, S.J., Grenfell, B., Groch, K.R., Gulland, F., Hall, A., Jensen, B.A., Lamy, K., Matassa, K., Mazzariol, S., Morris, S.E., Nielsen, O., Rotstein, D., Rowles, T.K., Saliki, J.T., Siebert, U., Waltzek, T., Wellehan, J.F., 2014. Phocine distemper virus: current knowledge and future directions. Viruses 6 (12), 5093–5134.

Esperón, F., Fernández, A., Sánchez-Vizcaíno, J.M., 2008. Herpes simplex-like infection in a bottlenose dolphin stranded in the Canary Islands. Diseases of Aquatic Organism 81, 73–76.

Ewalt, D.R., Payeur, J.B., Martin, B.M., Cummins, D.R., Miller, W.G., 1994. Characteristics of a Brucella species from a bottlenose dolphin (*Tursiops truncatus*). Journal of Veterinary Diagnostical Investigation 6, 448–452.

Fernández, A., Edwards, J.F., Rodríguez, F., Espinosa de los Monteros, A., Herráez, P., Castro, P., Jaber, J.R., Martín, V., Arbelo, M., 2005. "Gas and fat embolic syndrome" involving a mass stranding of beaked whales (*family Ziphiidae*) exposed to anthropogenic sonar signals. Veterinary Pathology 42, 446–457.

Fernández, A., Esperon, F., Herraez, P., de Los Monteros, A.E., Clavel, C., Bernabé, A., Sánchez-Vizcaino, J.M., Verborgh, P., DeStephanis, R., Toledano, F., Bayón, A., 2008. Morbillivirus and pilot whale deaths, Mediterranean Sea. Emerging Infectious Diseases 14, 792–794.

Fossi, M.C., Casini, S., Marsili, L., 2006. Endocrine Disruptors in Mediterranean top marine predators. Environmental Science and Pollution Research 13, 204–207.

Fossi, M.C., Casini, S., Marsili, L., 2007. Potential toxicological hazard due to endocrine-disrupting chemicals on Mediterraneantop predators: state of art, gender differences and methodological tools. Environmental Research 104, 174–182.

García-Alvarez, N., Boada, L.D., Fernández, A., Zumbado, M., Arbelo, M., Sierra, E., Xuriach, A., Almunia, J., Camacho, M., Luzardo, O.P., 2014a. Assessment of the levels of polycyclic aromatic hydrocarbons and organochlorine contaminants in bottlenose dolphins (*Tursiops truncatus*) from the Eastern Atlantic Ocean. Marine Environmental Research 100, 48–56.

García-Alvarez, N., Martín, V., Fernández, A., Almunia, J., Xuriach, A., Arbelo, M., Tejedor, M., Boada, L.D., Zumbado, M., Luzardo, O.P., 2014b. Levels and profiles of POPs (organochlorine pesticides, PCBs, and PAHs) in free-ranging common bottlenose dolphins of the Canary Islands, Spain. Science of the Total Environment 493, 22–31.

Garofolo, G., Zilli, K., Troiano, P., Petrella, A., Marotta, F., Di Serafino, G., Ancora, M., Di Giannatale, E., 2014. *Brucella ceti* from two striped dolphins stranded on the Apulia coastline, Italy. Journal of Medical Microbiology 63, 325–329.

Geraci, J.R., St. Aubin, D.J., 1987. Effects of parasites on marine mammals. International Journal for Parasitology 17, 407–414.

Goertz, C.E., Frasca Jr., S., Bohach, G.A., Cowan, D.F., Buck, J.D., French, R.A., De Guise, S., Maratea, J., Hinckley, L., Ewalt, D., Schlievert, P.M., Karst, S.M., Deobald, C.F., St Aubin, D.J., Dunn, J.L., 2011. Brucella sp. vertebral osteomyelitis with intercurrent fatal *Staphylococcus aureus* toxigenic enteritis in a bottlenose dolphin (*Tursiops truncatus*). Journal of Veterinary Diagnostic Investigation 23, 845–851.

González, L., Patterson, I.A., Reid, R.J., Foster, G., Barberan, M., Blasco, J.M., Kennedy, S., Howie, F.E., Godfroid, J., MacMillan, A.P., Schock, A., Buxton, D., 2002. Chronic meningoencephalitis associated with Brucella sp. infection in live-stranded striped dolphins (*Stenella coeruleoalba*). Journal of Comparative Pathology 126, 147–152.

Heide-Jørgensen, M.P., Härkönen, T., Dietz, R., Thompson, P.M., 1992. Retrospective of the 1988 European seal epizootic. Diseases of Aquatic Organisms 13, 37–62.

Herbst, L.H., 1994. Fibropapillomatosis of marine turtles. Annual Review of Fish Diseases 4, 389–425.

Houde, M., Hoekstra, P.F., Solomon, K.R., Muir, D.C., 2005. Organohalogen contaminants in delphinoid cetaceans. Reviews of Environmental Contamination and Toxicology 184, 1–57.

Isidoro-Ayza, M., Ruiz-Villalobos, N., Pérez, L., Guzmán-Verri, C., Muñoz, P.M., Alegre, F., Barberán, M., Chacón-Díaz, C., Chaves-Olarte, E., González-Barrientos, R., Moreno, E., Blasco, J.M., Domingo, M., 2014. *Brucella ceti* infection in dolphins from the Western Mediterranean Sea. BMC Veterinary Research 10, 206.

Italian Diagnostic Report on Stranded Cetaceans, 2015. Cre.Di.Ma. Istituto Zooprofilattico Sperimentale del Piemonte. Liguria e Val d'Aosta (IZSPLVA), Torino, Italy.

Italian Diagnostic Report on Stranded Cetaceans, 2016. Cre.Di.Ma. Istituto Zooprofilattico Sperimentale del Piemonte. Liguria e Val d'Aosta (IZSPLVA), Torino, Italy.

Jaber, J.R., Pérez, J., Carballo, M., Arbelo, M., Espinosa de los Monteros, A., Herráez, P., Muñoz, J., Andrada, M., Rodríguez, F., Fernández, A., 2005. Hepatosplenic large cell immunoblastic lymphoma in a bottlenose dolphin (*Tursiops truncatus*) with high levels of polychlorinated biphenyl congeners. Journal of Comparative Pathology 132, 242–247.

Johnson, C.K., Tinker, M.T., Estes, J.A., Conrad, P.A., Staedler, M., Miller, M.A., Jessup, D.A., Mazet, J.A., 2009. Prey choice and habitat use drive sea otter pathogen exposure in a resource-limited coastal system. Proceedings of the National Academy of Sciences of the United States of America 106 (7), 2242–2247.

Kannan, K., Blankenship, A.L., Jones, P.D., Giesy, J.P., 2000. Toxicity reference values for the toxic effects of polychlorinated biphenyls to aquatic mammals. Human and Ecological Risk Assessment 6, 181–201.

Kannan, K., Tanabe, S., Borrell, A., Aguilar, A., Focardi, S., Tatsukawa, R., 1993. Isomer-specific analysis and toxic evaluation of polychlorinated biphenyls in striped dolphins affected by an epizootic in the Western Mediterranean Sea. Archive of Environmental Contaminant and Toxicology 25, 227–233.

Keck, N., Kwiatek, O., Dhermain, F., Dupraz, F., Boulet, H., Danes, C., Laprie, C., Perrin, A., Godenir, J., Micout, L., Libeau, G., 2010. Resurgence of Morbillivirus infection in Mediterranean dolphins off the French coast. Veterinary Record 166, 654–655.

Kennedy, S., Lindstedt, I.J., McAliskey, M.M., McConnell, S.A., McCullough, S.J., 1992. Herpesviral encephalitis in a harbor porpoise (*Phocoena phocoena*). Journal of Zoo and Wildlife Medicine 23, 374–379.

Kennedy, S., 1998. Morbillivirus infections in aquatic mammals. Journal of Comparative Pathology 119, 201–225.

Kinsel, M.J., Boehm, J.R., Harris, B., Murnane, R.D., 1997. Fatal *Erysipelothrix rhusiopathiae* septicemia in a captive Pacific white-sided dolphin (*Lagenorhyncus obliquidens*). Journal of Zoo and Wildlife Medicine 28, 494–497.

Lafferty, K.D., Gerber, L.R., 2002. Good medicine for conservation biology: the intersection of epidemiology and conservation theory. Conservation Biology 16, 593–604.

Lahvis, G.P., Wells, R.S., Kuehl, D.W., Stewart, J.L., Rhinehart, H.L., Via, C.S., 1995. Decreased lymphocyte responses in free-ranging bottlenose dolphins (*Tursiops truncatus*) are associated with increased concentrations of PCBs and DDT in peripheral blood. Environmental Health Perspectives 4, 67–72.

Lecis, R., Tocchetti, M., Rotta, A., Naitana, S., Ganges, L., Pittau, M., Alberti, A., 2014. First Gammaherpesvirus detection in a free-living Mediterranean bottlenose dolphin. Journal of Zoo and Wildlife Medicine 45, 922–925.

Letcher, R.J., Bustnes, J.O., Dietz, R., Jenssen, B.M., Jorgensen, E.H., Sonne, C., Verreault, J., Vijayan, M.M., Gabrielsen, G.W., 2010. Exposure and effects assessment of persistent organohalogen contaminants in arctic wildlife and fish. Science of the Total Environment 408, 2995–3043.

Lipscomb, T.P., Schulman, F.Y., Moffett, D., Kennedy, S., 1994. Morbilliviral disease in Atlantic bottlenose dolphins (*Tursiops truncatus*) from the 1987–1988 epizootic. Journal of Wildlife Diseases 30, 567–571.

MAGRAMA, 2012. In: Evaluación inicial y buen estado ambiental del Grupo Mamíferos Marinos para las Estrategias Marinas. Ministerio de Agricultura, Alimentación y Medio Ambiente. Secretaría General Técnica. Centro de Publicaciones.

Maio, N., Pollaro, F., Di Nocera, F., De Carlo, E., Galiero, G., 2012. Cetaceans stranded along the coasts of Campania region (Mammalia: Cetacea). Report 2006–2011. In: Atti Italian Society of Natural Sciencet. Milan Natural History Museum, vol. 153, pp. 241–255.

Manire, C.A., Smolarek, K.A., Romero, C.H., Kinsel, M.J., Clauss, T.M., Byrd, L., 2006. Proliferative dermatitis associated with a novel alphaherpesvirus in an Atlantic bottlenose dolphin (*Tursiops truncatus*). Journal of Zoo and Wildlife Medicine 37 (2), 174–181.

Marsili, L., Focardi, S., 1996. Organochlorine levels in subcutaneous blubber biopsies of fin whales (*Balaenoptera physalus*) and striped dolphins (*Stenella coeruleoalba*) from the Mediterranean Sea. Environmental Pollution 91, 1–9.

Marsili, L., D'Agostino, A., Bucalossi, D., Malatesta, T., Fossi, M.C., 2004. Theoretical models to evaluate hazard due to organochlorine compounds (OCs) in Mediterranean striped dolphin (*Stenella coeruleoalba*). Chemosphere 56 (8), 791–801.

Martineau, D., De Guise, S., Fournier, M., Shugart, L., Girard, C., Lagace, A., Beland, P., 1994. Pathology and toxicology of beluga whales from the St. Lawrence Estuary, Quebec, Canada. Past, present and future. Science of the Total Environment 154, 201–215.

Mazzariol, S., Di Guardo, G., Petrella, A., Marsili, L., Fossi, C.M., Leonzio, C., Zizzo, N., Vizzini, S., Gaspari, S., Pavan, G., Podestà, M., Garibaldi, F., Ferrante, M., Copat, C., Traversa, D., Marcer, F., Airoldi, S., Frantzis, A., Quirós, Y., Cozzi, B., Fernández, A., 2011. Sometimes sperm whales (*Physeter macrocephalus*) cannot find their way back to the high seas: a multidisciplinary study on a mass stranding. PLoS One 6 (5), e19417.

Mazzariol, S., Marcer, F., Mignone, W., Serracca, L., Goria, M., Marsili, L., Di Guardo, G., Casalone, C., 2012. Dolphin Morbillivirus and *Toxoplasma gondii* coinfection in a Mediterranean fin whale (*Balaenoptera physalus*). BMC Veterinary Research 7, 8–20.

Mazzariol, S., Centelleghe, C., Beffagna, G., Povinelli, M., Terracciano, G., Cocumelli, C., Pintore, A., Denurra, D., Casalone, C., Pautasso, A., Di Francesco, C.E., Di Guardo, G., 2016. Mediterranean fin whales (*Balaenoptera physalus*) threatened by dolphin Morbillivirus. Emerging Infectious Diseases 22 (2).

Mazzariol, S., Centelleghe, C., Di Provvido, A., Di Renzo, L., Cardeti, G., Cersini, A., Fichi, G., Petrella, A., Di Francesco, C.E., Mignone, W., Casalone, C., Di Guardo, G. 2017. Dolphin Morbillivirus Associated with a Mass Stranding of Sperm Whales, Italy. Emerging Infectious Diseases 23 (1), 144–166.

Mazzariol, S., Peletto, S., Mondin, A., Centelleghe, C., Di Guardo, G., Di Francesco, C.E., Casalone, C., Acutis, P.L., February, 2013. Dolphin Morbillivirus infection in a captive harbor seal (*Phoca vitulina*). Journal of Clinical Microbiology 51 (2), 708–711.

Melero, M., Crespo-Picazo, J.L., Rubio-Guerri, C., García-Párraga, D., Sánchez-Vizcaín, J.M., 2015. First molecular determination of herpesvirus from two mysticete species stranded in the Mediterranean Sea. BMC Veterinary Research 11, 283.

Melero, M., Rubio-Guerri, C., Crespo, J.L., Arbelo, M., Vela, A.I., García-Párraga, D., Sierra, E., Domínguez, L., Sánchez-Vizcaíno, J.M., 2011. First case of erysipelas in a free-ranging bottlenose dolphin (*Tursiops truncatus*) stranded in the Mediterranean Sea. Diseases of Aquatic Organism 97, 167–170.

Miller, W.G., Adams, L.G., Ficht, T.A., Cheville, N.F., Payeur, J.P., et al., 1999. Brucella-induced abortions and infection in bottlenose dolphins (*Tursiops truncatus*). Journal of Zoo and Wildlife Medicine 30, 100–110.

Munson, L., Terio, K.A., Kock, R., Mlengeya, T., Roelke, M.E., Dubovi, E., Summers, B., Sinclair, A.R., Packer, C., 2008. Climate extremes promote fatal co-infections during canine distemper epidemics in African lions. PLoS One 3 (6), e2545.

NOAA, 2013. What's Causing the Bottlenose Dolphin Deaths along the Mid-Atlantic? http://ner-onoaagov/stories/2013/bottlenoseDolphinAnnouncementhtml Consulted on September 20th.

Nymo, I.H., Tryland, M., Godfroid, J., 2011. A review of Brucella infection in marine mammals, with special emphasis on *Brucella pinnipedialis* in the hooded seal (*Cystophora cristata*). Veterinary Research 42, 1297–9716.

Panigada, S., Lauriano, G., Burt, L., Pierantonio, N., Donovan, G., 2011. Monitoring winter and summer abundance of cetaceans in the Pelagos Sanctuary (Northwestern Mediterranean Sea) through aerial surveys. PLoS One 6 (7), e22878.

Parrish, C.R., Holmes, E.C., Morens, D.M., Park, E.C., Burke, D.S., Calisher, C.H., Laughlin, C.A., Saif, L.J., Daszak, P., 2008. Cross-species virus transmission and the emergence of new epidemic diseases. Microbiology and Molecular Biology Reviews 72, 457–470.

Peltier, H., Jepson, P.D., Dabin, W., Deaville, R., Daniel, P., Van Canneyt, O., Ridoux, V., 2014. The contribution of stranding data to monitoring and conservationstrategies for cetaceans: developing spatially explicit mortalityindicators for common dolphins (*Delphinus delphis*) in the eastern North-Atlantic. Ecological Indicators 39, 203–214.

Plowright, W., 1982. The effects of rinderpest and rinderpest control on wildlife in Africa. In: Edwards, M.A., McDonnel, U. (Eds.), Symposia of the Zoological Society of London No. 50. Academic press, London, UK, pp. 1–28.

Raga, J.A., Banyard, A., Domingo, M., Corteyn, M., Van Bressem, M.F., Fernández, M., Aznar, F.J., Barrett, T., 2008. Dolphin Morbillivirus epizootic resurgence, Mediterranean Sea. Emerging Infectious Diseases 14, 471–473.

Resendes, A.R., Almería, S., Dubey, J.P., Obón, E., Juan-Sallés, C., Degollada, E., Alegre, F., Cabezón, O., Pont, S., Domingo, M., 2002. Disseminated toxoplasmosis in a Mediterranean pregnant Risso's dolphin (*Grampus griseus*) with transplacental fetal infection. Journal of Parasitology 88, 1029–1032.

Ross, H.M., Foster, G., Reid, R.J., Jahans, K.L., MacMillan, A.P., 1994. Brucella species infection in sea-mammals. Veterinary Record 134, 359.

Ross, P.S., 2002. The role of immunotoxic environmental contaminants in facilitating the emergence of infectious diseases in marine mammals. Human and Ecological Risk Assessment 8, 277–292.

Schulman, F.Y., Lipscomb, T.P., Moffett, D., Krafft, A.E., Lichy, J.H., Tsai, M.M., Taubenberger, J.K., Kennedy, S., 1997. Histologic, immunohistochemical, and polymerase chain reaction studies of bottlenose dolphins from the 1987–1988 United States Atlantic Coast epizootic. Veterinary Pathology 34, 288–295.

Schwacke, L.H., Voit, E.O., Hansen, L.J., Wells, R.S., Mitchum, G.B., Hohn, A.A., Fair, P.A., 2002. Probabilistic risk assessment of reproductive effects of polychlorinated biphenyls on bottlenose dolphins (*Tursiops truncatus*) from the Southeast United States Coast. Environmental Toxicology and Chemistry 21, 2752–2764.

Schwacke, L.H., Zolman, E.S., Balmer, B.C., De Guise, S., George, R.C., Hoguet, J., Hohn, A.A., Kucklick, J.R., Lamb, S., Levin, M., Litz, J.A., McFee, W.E., Place, N.J., Townsend, F.I., Wells, R.S., Rowles, T.K., 2012. Anaemia, hypothyroidism and immune suppression associated with polychlorinated biphenyl exposure in bottlenose dolphins (*Tursiops truncatus*). Proceeding of the Royal Society B 279, 48–57.

Shoham-Frider, E., Kress, N., Wynne, D., Scheinin, A., Roditi-Elsar, M., Kerem, D., 2009. Persistent organochlorine pollutants and heavy metals in tissues of common bottlenose dolphin (*Tursiops truncatus*) from the Levantine Basin of the Eastern Mediterranean. Chemosphere 77, 621–627.

Sierra, E., Diaz-Delgado, J., Arbelo, M., Andrada, M., Sacchini, S., Fernandez, A., 2015. Herpesvirus-associated genital lesions in a stranded striped dolphin (*Stenella coeruleoalba*) in the Canary Islands, Spain. Journal of Wildlife Diseases 51 (3), 696–702.

Sierra, E., Fernández, A., de los Monteros, A.E., Arbelo, M., de Quirós, Y.B., Herráez, P., 2013. Muscular senescence in cetaceans: adaptation towards a slow muscle fibre phenotype. Scientific Report 3, 1795.

Sierra, E., Fernández, A., Suárez-Santana, C., Xuriach, A., Zucca, D., Bernaldo de Quirós, Y., García-Álvarez, N., De la Fuente, J., Sacchini, S., Andrada, M., Díaz-Delgado, J., Arbelo, M., 2016. Morbillivirus and pilot whale deaths, Canary Islands, Spain, 2015. Emerging Infectious Disease 22 (4).

Sierra, E., Sánchez, S., Saliki, J.T., Blas-Machado, U., Arbelo, M., Zucca, D., Fernandez, A., 2014a. Retrospective study of etiologic agents associated with nonsuppurative meningoencephalitis in stranded cetaceans in the Canary Islands. Journal of Clinical Microbiology 52, 2390–2397.

Sierra, E., Zucca, D., Arbelo, M., García-Alvarez, N., Andrada, M., Déniz, S., Fernández, A., 2014b. Fatal systemic Morbillivirus infection in bottlenose dolphin, Canary Islands, Spain. Emerging Infectious Diseases 20, 269–271.

Smith, K.F., Sax, D.F., Lafferty, K.D., 2006. Evidence for the role of infectious disease in species extinction and endangerment. Conservation Biology 20, 1349–1357.

Sato, H., Yoneda, M., Honda, T., Kai, C., 2012. Morbillivirus receptors and tropism: multiple pathways for infection. Frontiers in Microbiology 3, 1–9.

Shapiro, K., VanWormer, E., Aguilar, B., Conrad, P.A., 2015. Surveillance for Toxoplasma gondii in California mussels (*Mytilus californianus*) reveals transmission of atypical genotypes from land to sea. Environmental Microbiology 17 (11), 4177–4188.

Soto, S., González, R., Alegre, F., González, B., Medina, P., Raga, J.A., Marco, A., Domingo, M., 2011. Epizootic of dolphin Morbillivirus on the Catalonian Mediterranean coast in 2007. Veterinary Record 169, 22.

Soto, S., González, B., Willoughby, K., Maley, M., Olvera, A., Kennedy, S., Marco, A., Domingo, M., 2012. Systemic herpesvirus and Morbillivirus co-infection in a striped dolphin (*Stenella coeruleoalba*). Journal of Comparative Pathology 146, 269–273.

Storelli, M.M., Marcotrigiano, G.O., 2003. Levels and congener pattern of polychlorinated biphenyls in the blubber of the Mediterranean bottlenose dolphins (*Tursiops truncatus*). Environment International 28, 559–565.

Storelli, M.M., Barone, G., Giacominelli-Stuffler, R., Marcotrigiano, G.O., 2012. Contamination by polychlorinated biphenyls (PCBs) in striped dolphins (*Stenella coeruleoalba*) from the Southeastern Mediterranean Sea. Environmental Monitoring and Assessment 184, 5797–5805.

Tanabe, S., 2002. Contamination and toxic effects of persistent endocrine disrupters in marine mammals and birds. Marine Pollution Bulletin 45, 69–77.

Taubenberger, J.K., Tsai, M.M., Atkin, T.J., Fanning, T.G., Krafft, A.E., Moeller, R.B., Kodsi, S.E., Mense, M.G., Lipscomb, T.P., 2000. Molecular genetic evidence of a novel Morbillivirus in a long-finned pilot whale (*Globicephalus melas*). Emerging Infectious Diseases 6, 42–45.

Thorne, E.T., Williams, E.S., Spraker, T.R., Helms, W., Segerstrom, T., 1988. Bluetongue in free-ranging pronghorn antelope (*Antilocapra americana*) in Wyoming: 1976 and 1984. Journal of Wildlife Diseases 24 (1), 113–119.

Van Bressem, M.F., Raga, J.A., Di Guardo, G., Jepson, P.D., Duignan, P.J., Siebert, U., Barrett, T., Santos, M.C., Moreno, I.B., Siciliano, S., Aguilar, A., Van Waerebeek, K., 2009. Emerging infectious diseases in cetaceans worldwide and the possible role of environmental stressors. Diseases of Aquatic Organism 86, 143–157.

Van Bressem, M.F., Duignan, P.J., Banyard, A., Barbieri, M., Colegrove, K.M., De Guise, S., Di Guardo, G., Dobson, A., Domingo, M., Fauquier, D., Fernandez, A., Goldstein, T., Grenfell, B., Groch, K.R., Gulland, F., Jensen, B.A., Jepson, P.D., Hall, A., Kuiken, T., Mazzariol, S., Morris, S.E., Nielsen, O., Raga, J.A., Rowles, T.K., Saliki, J., Sierra, E., Stephens, N., Stone, B., Tomo, I., Wang, J., Waltzek, T., Wellehan, J.F., 2014. Cetacean Morbillivirus: current knowledge and future directions. Viruses 6 (12), 5145–5181.

Van Bressem, M., Waerebeek, K.V., Jepson, P.D., Raga, J.A., Duignan, P.J., Nielsen, O., Di Beneditto, A.P., Siciliano, S., Ramos, R., Kant, W., Peddemors, V., Kinoshita, R., Ross, P.S., López-Fernandez, A., Evans, K., Crespo, E., Barrett, T., 2001. An insight into the epidemiology of dolphin morbillivirus worldwide. Veterinary Microbiology 81 (4), 287–304.

Van de Bildt, M.W., Martina, B.E., Vedder, E.J., Androukaki, E., Kotomatas, S., Komnenou, A., Sidi, B.A., Jiddou, A.B., Barham, M.E., Niesters, H.G., Osterhaus, A.D., 2000. Identification of Morbilliviruses of probable cetacean origin in carcases of Mediterranean monk seals (*Monachus monachus*). Veterinary Record 146 (24), 691–694.

van Elk, C.E., van de Bildt, M.W., de Jong, A.A., Osterhaus, A.D., Kuiken, T., 2009. Herpesvirus in bottlenose dolphins (*Tursiops truncatus*): cultivation, epidemiology, and associated pathology. Journal of Wildlife Diseases 45 (4), 895–906.

van Elk, C., van de Bildt, M., van Run, P., de Jong, A., Getu, S., Verjans, G., Osterhaus, A., Kuiken, T., 2016. Central nervous system disease and genital disease in harbor porpoises (*Phocoena phocoena*) are associated with different herpesviruses. Veterinary Research 47, 28. https://doi.org/10.1186/s13567-016-0310-8.

VanWormer, E., Miller, M.A., Conrad, P.A., Grigg, M.E., Rejmanek, D., Carpenter, T.E., Mazet, J.A., 2014. Using molecular epidemiology to track *Toxoplasma gondii* from terrestrial carnivores to marine hosts: implications for public health and conservation. PLoS Neglected Tropical Diseases 8, e2852.

Wafo, E., Sarrazin, L., Diana, C., Dhermain, F., Schembri, T., Lagadec, V., Pecchia, M., Rebouillon, P., 2005. Accumulation and distribution of organochlorines (PCBs and DDTs) in various organs of *Stenella coeruleoalba* and a *Tursiops truncatus* from Mediterranean littoral environment (France). Science of the Total Environment 348, 115–127.

Wang, Q., Chang, B.J., Riley, T.V., 2010. Erysipelothrix rhusiopathiae. Veterinary Microbiology 140, 405–417.

Wells, R.S., Tornero, V., Borrell, A., Aguilar, A., Rowles, T.K., Rhinehart, H.L., Hofmann, S., Jarman, W.M., Hohn, A.A., Sweeney, J.C., 2005. Integrating life-history and reproductive success data to examine potential relationships with organochlorine compounds for bottlenose dolphins (*Tursiops truncatus*) in Sarasota Bay, Florida. Science of the Total Environment 349, 106–119.

West, K.L., Levine, G., Jacob, J., Jensen, B., Sanchez, S., Colegrove, K., Rotstein, D., 2015. Coinfection and vertical transmission of Brucella and Morbillivirus in a neonatal sperm whale (*Physeter macrocephalus*) in Hawaii, USA. Journal of Wildlife Diseases 51, 227–232.

Yap, X., Deaville, R., Perkins, M.W., Penrose, R., Law, R.J., Jepson, P.D., 2012. Investigating links between polychlorinated biphenyl (PCB) exposure and thymic involution and thymic cysts in harbour porpoises (*Phocoena phocoena*). Marine Pollutant Bulletin 64, 2168–2176.

FURTHER READING

BOE-A-2011-14661, 2011. Orden ARM/2417/2011, de 30 de agosto, por la que se declaran zonas especiales de conservación los lugares de importancia comunitaria marinos de la región biogeográfica Macaronésica de la Red Natura 2000 y se aprueban sus correspondientes medidas de conservación. Ministerio de Medio Ambiente, y Medio Rural y Marino.

Camacho, M., Boada, L.D., Orós, J., Calabuig, P., Zumbado, M., Luzardo, O.P., 2012. Comparative study of polycyclic aromatic hydrocarbons (PAHs) in plasma of Eastern Atlantic juvenile and adult nesting loggerhead sea turtles (*Caretta caretta*). Marine Pollutant Bulletin 64, 1974–1980.

Camacho, M., Calabuig, P., Luzardo, O.P., Boada, L.D., Zumbado, M., Orós, J., 2013. Crude oil as a stranding cause among loggerhead sea turtles (*Caretta caretta*) in the Canary Islands, Spain (1998–2011). Journal of Wildlife Diseases 49, 637–640.

Culik, B.M., 2004. Review of Small Cetaceans. Distribution, Behaviour, Migration and Threats. United Nations Environment Programme and the Convention on the Conservation of Migratory Species of Wild Animals (UNEP/CMS).

Dailey, M.D., 2001. Parasitic diseases. In: Dierauf, L.A., Gulland, F.M. (Eds.), CRC Handbook of Marine Mammal Medicine, second ed. CRC Press, pp. 357–379.

De Guise, S., Bisaillon, A., Seguin, B., Lagace, A., 1994. The anatomy of the male genital system of the beluga whale, *Delphinapterus leucas*, with special reference to the penis. Anatomia Histologia Embryologia 23, 207–216.

Di Guardo, G., Agrimi, U., Amaddeo, D., McAliskey, M., Kennedy, S., 1992. Morbillivirus infection in a striped dolphin (*Stenella coeruleoalba*) from the coast of Italy. Veterinary Record 130, 579–580.

Di Guardo, G., Marruchela, G., Agrimi, M., Kennedy, S., 2005. Morbillivirus infections in aquatic mammals: a brief overview. Journal of Veterinary Medicine A 52, 88–93.

Di Guardo, G., Di Francesco, C.E., Eleni, C., Cocumelli, C., Scholl, F., Casalone, C., Peletto, S., Mignone, W., Tittarelli, C., Di Nocera, F., Leonardi, L., Fernández, A., Marcer, F., Mazzariol, S., 2013. Morbillivirus infection in cetaceans stranded along the Italian coastline: pathological, immunohistochemical and biomolecular findings. Research in Veterinary Science 94, 132–137.

Dubey, J.P., Fair, P.A., Bossart, G.D., Hill, D., Fayer, R., Sreekumar, C., Kwok, O.C.H., Thulliez, P.A., 2005. Comparison of several serologic tests to detect antibodies to *Toxoplasma gondii* in naturally exposed bottlenose dolphins (*Tursiops truncatus*). Journal of Parasitology 91, 1074–1081.

Edvinsson, B., Lappalainen, M., Evengard, B., 2006. Real-time PCR targeting a 529-bp repeat element for diagnosis of toxoplasmosis. Clinical Microbiology and Infection 12, 131–136.

Foster, G., Osterman, B.S., Godfroid, J., Jacques, I., Cloeckaert, A., 2007. *Brucella ceti* sp. nov. and *Brucella pinnipedialis* sp. nov for Brucella strains with cetaceans and seals as their preferred hosts. International Journal of Systematic and Evolutionary Microbiology 57, 2688–2693.

Geraci, J.R., Lounsbury, V.J., 2005. Marine Mammals Ashore: A Field Guide for Strandings, second ed. National Aquarium Baltimore, Baltimore, MD.

Groussaud, P., Shankster, S.J., Koylass, M.S., Whatmore, A.M., 2007. Molecular typing divides marine mammal strains of Brucella into at least three groups with distinct host preferences. Journal of Medical Microbiology 56, 1512–1518.

Hernandez-Mora, G., Gonzalez-Barrientos, R., Morales, J.A., Chaves-Olarte, E., Guzman-Verri, C., Baquero-Calvo, E., De-Miguel, M.J., Marin, C.M., Blasco, J.M., Moreno, E., 2008. Neurobrucellosis in stranded dolphins, Costa Rica. Emerging Infectious Diseases 14, 1430–1433.

Jackson, J.B.C., Kirby, M.X., Berger, W.H., Bjorndal, K.A., Botsford, L.W., Bourque, B.J., Bradbury, R.H., Cooke, R., Erlandson, J., Estes, J.A., Hughes, T.P., Kidwell, S., Lange, C.B., Lenihan, H.S., Pandolfi, J.M., Peterson, C.H., Steneck, R.S., Tegner, M.J., Warner, R.R., 2001. Historical overfishing and the recent collapse of coastal ecosystems. Science 293, 629–637.

Kucklick, J., Schwacke, L., Wells, R., Hohn, A., Guichard, A., Yordy, J., Hansen, L., Zolman, E., Wilson, R., Litz, J., Nowacek, D., Rowles, T., Pugh, R., Balmer, B., Sinclair, C., Rosel, P., 2011. Bottlenose dolphins as indicators of persistent organic pollutants in the western North Atlantic Ocean and northern Gulf of Mexico. Environmental Science and Technology 45, 4270–4277.

Kuiken, T., García-Hartmann, M., 1993. Cetacean pathology: dissection techniques and tissue sampling. ECS Newsletter 17 (Special Issue).

Lamere, S.A., St Leger, J.A., Schrenzel, M.D., Anthony, S.J., Rideout, B.A., Salomon, D.R., 2009. Molecular characterization of a novel gammaretrovirus in killer whales (*Orcinus orca*). Journal of Virology 83, 12956–12967.

Law, R.J., Barry, J., Barber, J.L., Bersuder, P., Deaville, R., Reid, R.J., Brownlow, A., Penrose, R., Barnett, J., Loveridge, J., Smith, B., Jepson, P.D., 2012. Contaminants in cetaceans from UK waters: status as assessed within the Cetacean Strandings Investigation Programme from 1990 to 2008. Marine Pollutant Bulletine 64, 1485–1494.

Martineau, D., Lagace, A., Beland, P., Higgins, R., Armstrong, D., Shugart, L.R., 1988. Pathology of stranded beluga whales (*Delphinapterus leucas*) from the St. Lawrence Estuary, Quebec, Canada. Journal of Comparative Pathology 98, 287–311.

Maucher, J.M., Briggs, L., Podmore, C., Ramsdell, J.S., 2007. Optimization of blood collection card method/enzyme-linked immunoassay for monitoring exposure of bottlenose dolphin to brevetoxin-producing red tides. Environmental Science and Technology 41 (2), 563–567.

Ohishi, K., Zenitani, R., Bando, T., Goto, Y., Uchida, K., Maruyama, T., Yamamoto, S., Miyazaki, N., Fujise, Y., 2003. Pathological and serological evidence of Brucella-infection in baleen whales (Mysticeti) in the western North Pacific. Comparative Immunology, Microbiology and Infectious Diseases 26, 125–136.

Pauly, D., Christensen, V., Dalsgaard, J., Froese, R., Torres Jr., F., 1998. Fishing down marine food webs. Science 279 (5352), 860.

Prenger-Berninghoff, E., Thoele, A., Siebert, U., Stede, M., Weiss, R., 2003. Phenotypic and molecular characterization of Brucella strains isolated from marine mammals of the German North Sea. International Journal of Medical Microbiology 293 (36), 320.

Probert, W.S., Schrader, K.N., Khuong, N.Y., Bystrom, S.L., Graves, M.H., 2004. Real-time multiplex PCR assay for detection of Brucella spp., *B. abortus*, and *B. melitensis*. Journal of Clinical Microbiology 42, 1290–1293.

Reddy, M.L., Dierauf, L.A., Gulland, F.M.D., 2001. Marine mammals as sentinels of ocean health. In: Dierauf, L.A., Gulland, F.M.D. (Eds.), CRC Handbook of Marine Mammal Medicine. CRC Press, Boca Raton, FL, pp. 3–14.

Reidarson, T.H., McBain, J., House, C., King, D.P., Stott, J.L., Krafft, A., Taubenberger, J.K., Heyning, J., Lipscomb, T.P., 1998. Morbillivirus infection in stranded common dolphins from the Pacific Ocean. Journal of Wildlife Diseases 34, 771–776.

Reynolds III, J.E., Wells, R.S., Eide, S.D., 2000. The bottlenose dolphin: biology and conservation. University Press of Florida, Gainesville, FL.

Saliki, J.T., Cooper, E.J., Rotstein, D.S., Caseltine, S.L., Pabst, D.A., McLellan, W.A., Govett, P., Harms, C., Smolarek, K.A., Romero, C.H., 2006. A novel gammaherpesvirus associated with genital lesions in a Blainville's beaked whale (*Mesoplodon densirostris*). Journal of Wildlife Diseases 42, 142–148.

Smolarek Benson, K.A., Manire, C.A., Ewing, R.Y., Saliki, J.T., Townsend, F.I., Ehlers, B., Romero, C.H., 2006. Identification of novel alpha- and gammaherpesviruses from cutaneous and mucosal lesions of dolphins and whales. Journal of Virological Methods 136, 261–266.

Tamura, K., Peterson, D., Peterson, N., Stecher, G., Nei, M., Kumar, S., 2011. MEGA5: molecular evolutionary genetics analysis using maximum likelihood, evolutionary distance, and maximum parsimony methods. Molecular Biology and Evolution 28, 2731–2739.

To, H., Koyama, T., Nagai, S., Tuchiya, K., Nunoya, T., 2009. Development of quantitative real-time polymerase chain reaction for detection of and discrimination between *Erysipelothrix rhusiopathiae* and other Erysipelothrix species. Journal of Veterinary Diagniostic Investigation 21, 701–706.

Tobeña, M., Escánez, A., Rodríguez, Y., López, C., Ritter, F., Aguilar, N., 2014. Inter-island movements of common bottlenose dolphins *Tursiops truncatus* among the Canary Islands: online catalogues and implications for conservation and management. African Journal of Marine Science 36, 137–141.

VanDevanter, D.R., Warrener, P., Bennett, L., Schultz, E.R., Coulter, S., Garber, R.L., Rose, T.M., 1996. Detection and analysis of diverse herpesviral species by consensus primer PCR. Journal of Clinical Microbiology 34, 1666–1671.

Wells, R.S., Scott, M.D., 1999. Bottlenose dolphin – *Tursiops truncatus* (Montagu, 1821). In: Ridgway, S.H., Harrison, R. (Eds.), Handbook of Marine Mammals, Volume 6, the Second Book of Dolphins and Porpoises. Academic Press, San Diego, CA, pp. 137–182.

Wells, R.S., Rhinehart, H.L., Hansen, L.J., Sweeney, J.C., Townsend, F.I., Stone, R., Casper, D.R., Scott, M.D., Hohn, A.A., Rowles, T.K., 2004. Bottlenose dolphins as marine ecosystem sentinels: developing a health monitoring system. EcoHealth 1, 246–254.

Whatmore, A.M., Dawson, C., Groussaud, P., Koylass, M.S., King, A., Shankster, S.J., 2008. Marine mammal Brucella genotype associated with zoonotic infection. Emerging Infectious Diseases 14, 517–518.

Yordy, J.E., Pabst, D.A., McLellan, W.A., Wells, R.S., Rowles, T.K., Kucklick, J.R., 2010a. Tissue-specific distribution and whole-body burden estimates of persistent organic pollutants in the bottlenose dolphin (*Tursiops truncatus*). Environmental Toxicology and Chemistry 29, 1263–1273.

Yordy, J.E., Wells, R.S., Balmer, B.C., Schwacke, L.H., Rowles, T.K., Kucklick, J.R., 2010b. Life history as a source of variation for persistent organic pollutant (POP) patterns in a community of common bottlenose dolphins (*Tursiops truncatus*) resident to Sarasota Bay, FL. Science of the Total Environmental 408, 2163–2172.

Yordy, J.E., Wells, R.S., Balmer, B.C., Schwacke, L.H., Rowles, T.K., Kucklick, J.R., 2010c. Partitioning of persistent organic pollutants between blubber and blood of wild bottlenose dolphins: implications for biomonitoring and health. Environmental Science Technology 44, 478.

Chapter 16

Ecotoxicology of the Sirenia in the Twenty-First Century

Thomas J. O'Shea[1], Noel Y. Takeuchi[2], Liesbeth Weijs[3], Helene Marsh[4]

[1]U.S. Geological Survey, retired, Glen Haven, CO, United States; [2]University of South Florida St. Petersburg and Fish and Wildlife Research Institute, Florida Fish and Wildlife Conservation Commission, St Petersburg, FL, United States; [3]The University of Queensland, Coopers Plains, QLD, Australia; [4]James Cook University, Townsville, QLD, Australia

THE SIRENIA

Sirenians are unique among marine mammals because of their distinct evolutionary histories and their unique positions in aquatic food webs. The Sirenia evolved from terrestrial herbivores about 60 million years ago and are more closely related to other Afrotherians such as aardvarks, golden moles, tenrecs, sengis, elephants, and hyraxes than they are to cetaceans or pinnipeds (see review in Marsh et al., 2012). As summarized in Table 16.1, there are three extant species in the family Trichechidae (the Amazonian, West African, and West Indian manatees) and a single species in the family Dugongidae, the dugong. The West Indian manatee has two subspecies, the Florida manatee and the Antillean manatee. Sirenians feed primarily on aquatic vascular plants. Thus because of the low trophic position of sirenians in food webs, their exposure to toxic substances is very different to that of most other medium-sized marine mammals. Indeed, many of the contaminants of traditional concern (such as organochlorine pesticides and metabolites, PCBs, and mercury) are found only at low concentrations in sirenians in contrast to the high concentrations in tissues of piscivorous marine mammals. Sirenians are hindgut digesters, and they have different nutritional metabolic capacities and digestive systems than other marine mammals, including long gut retention times (Marsh et al., 2012). These attributes may also mean that sirenians must face challenges from toxic substances in their environments in ways that are fundamentally different to most other marine mammals. These challenges are likely to be exacerbated by the fact that more than 70 sirenian range states are developing countries with limited pollution control (Marsh et al., 2012).

Marine Mammal Ecotoxicology. https://doi.org/10.1016/B978-0-12-812144-3.00016-4

TABLE 16.1 The Extant Species and Subspecies of the Sirenia and Their General Distribution

Species or Subspecies	General Distribution
Family Dugongidae	
Dugong *(Dugong dugon)*	Tropical and subtropical Indo-Pacific coastal marine waters from Australia through southern Asia and the Middle East to Mozambique
Family Trichechidae	
Amazonian manatee (*Trichechus inunguis*)	Amazon River and larger tributaries in Brazil, Colombia, Ecuador, Peru
West African manatee (*Trichechus senegalensis*)	Tropical and subtropical eastern Atlantic and associated large inland rivers and estuaries from Senegal to Angola
West Indian manatee (*Trichechus manatus*)	Tropical and subtropical western Atlantic and Caribbean Sea and associated large inland rivers and estuaries
Florida manatee (*Trichechus manatus latirostris*)	Southeastern United States, principally Florida
Antillean manatee (*Trichechus manatus manatus*)	Tropical and subtropical western Atlantic from Mexico to Alagoas state in Brazil, Caribbean Sea

After Marsh et al. (2012).

The two families and four species of sirenians are distributed globally in tropical and subtropical waters (Table 16.1), sometimes near burgeoning coastal populations of humans where they can be exposed to contaminants from mainland runoff as well as other direct inputs. All are categorized as vulnerable on the IUCN Red List. The principal known threats to sirenian populations are from direct human causes of mortality such as incidental take in artisanal net fisheries, hunting, and strikes from boats (reviewed in Marsh et al., 2012). Demographic features of sirenian population dynamics that make them particularly vulnerable to such additive sources of mortality are lengthy periods before first breeding in females (age at sexual maturity 6–14 years in dugongs), lengthy gestation periods and litter sizes of one, and a long period of lactation (up to 2 years). This combination of traits requires high adult survival to maintain stable or growing populations, which can in turn result in high maximum longevities (60–70 years in some individuals; see life history review in Marsh et al., 2012).

The objectives of our review are to summarize past findings pertinent to the ecotoxicology of sirenians during the 20th century, to provide an updated summary of information published since the year 2000, and to critically appraise current knowledge with suggestions for possible new directions for ecotoxicologic research. Twentieth-century research on ecotoxicology of persistent contaminants and biotoxins in dugongs and manatees has been reviewed in greater detail in several past summaries (e.g., O'Shea and Aguilar, 2001; Bossart et al., 2002; O'Shea, 2003; Marsh et al., 2012). Impacts of oil pollution on sirenians have not been clearly demonstrated but were reviewed in 2015 (Helm et al., 2015).

TWENTIETH-CENTURY CONCERNS AND FINDINGS

Concern for conservation of the Sirenia mounted during the 20th century because of their relative rarity, declining numbers, and lack of knowledge about threats to their populations. This situation led to special status designations and protective laws and regulations for dugongs and manatees throughout most of the world. Concurrently, the large growth in the chemical industry following World War II led to synthesis of persistent contaminants that entered aquatic ecosystems globally and with unforeseen consequences. Realization of these consequences for humans began to unfold soon thereafter in Japan during the 1950s and 1960s, with the recognition of methyl mercury as the source of tragic human poisonings at Minamata and Nagata Bays (Harada, 1995), cadmium poisoning, or "itai-itai disease," resulting in long-term renal dysfunction and bone deformities from the Jinzu River basin (Aoshima, 2016), and contamination of rice oils with PCBs and PCDD/Fs as the cause of "Yusho disease" (Yoshimura, 2003, 2012). Impacts of chemical contaminants on wildlife also became obvious with firm evidence for direct lethality of organochlorine pesticides in birds (e.g., Stickel et al., 1966), as well as more insidious effects on avian reproduction through eggshell thinning (e.g., Wiemeyer and Porter, 1970). Human health concerns and effects on wildlife ushered in a global era of environmental legislation and regulation.

Researchers turned their attention to marine mammals as possibly susceptible to contamination by persistent chemicals with recognition of contamination of the oceans on a global scale. DDT and metabolites were documented in Antarctic and European seals (George and Frear, 1966; Sladen et al., 1966; Koeman and van Genderen, 1966). Coastal populations of pinnipeds and small cetaceans were found to have blubber with some of the highest concentrations of organochlorines and mercury seen in any wildlife on Earth (e.g., Helle et al., 1976a,b; LeBouef and Bonnell, 1971; O'Shea et al., 1980). These observations were followed by correlative field studies suggestive of possible reproductive impairment and increased susceptibility to disease in pinnipeds (e.g., DeLong et al., 1973; Helle et al., 1976a,b; Ross et al., 2003), later augmented by captive feeding experiments with harbor seals (e.g., Reijnders, 1986; see reviews in Reijnders, 2003, Vos et al., 2003). A historical overview of contaminants in marine mammals during these times was provided along with additional details by O'Shea and Tanabe (2003), who noted that by the end of the 20th century, chemical residue surveys in marine mammal tissues had included over 20,000 individuals and 80 species globally.

Given the startling findings of the ubiquity and sometimes high concentrations of organochlorines and toxic elements in other marine mammals and the declining status of sirenian populations, investigators began asking questions about exposure of manatees and dugongs to persistent contaminants. Following, we give a brief synopsis of these findings, including the recognition of biotoxins as causes of direct mortality in manatees (see also Table 16.2).

Organochlorine Pesticides and Metabolites

Dugongs. In comparison with other marine mammals, low concentrations of organochlorine pesticides and metabolites were reported in dugongs. Two dugongs sampled at Sulawesi Island, Indonesia, in 1975 lacked detectable residues of six organochlorine pesticides or metabolites in muscle (Miyazaki et al., 1979). Very low concentrations of lindane and dieldrin were found in livers of four dugongs stranded in Queensland in 1977 (unpublished data of Heinsohn and Marsh cited in Haynes and Johnson, 2000). Analyses of blubber from seven dugong carcasses sampled from 1996 to 1999 in northern Queensland at detection levels of 2 ng/g (ppb) wet weight (ww) revealed metabolites of DDT (primarily *p,p′*-DDE) in six (maximum 173 ng/g lipid weight [lw] ΣDDT), dieldrin in one (14 ng/g lw), and no residues of eight other compounds (Vetter et al., 2001). Blubber samples were analyzed for nine organochlorine pesticides or metabolites in 52 dugong carcasses that had stranded between 1996 and 2000 in Queensland, Australia (Haynes et al., 2005). Organochlorine compounds were found at the limits of detection of 0.1–0.2 μg/kg (ppb) only in 59% of the samples, with the following maximum concentrations (ww): 9.2 μg/kg dieldrin, 59 μg/kg DDT, and 15 μg/kg DDE (Haynes et al., 2005). Lindane, heptachlor, and HCB were not detected, and one sample contained DDD at 6 μg/kg ww.

Manatees. Tissue samples from a single Florida manatee (*Trichechus manatus latirostris*) carcass recovered in 1974 had only trace amounts (<1.0 μg/g ww) of DDE, PCBs, and dieldrin (Forrester et al., 1975). Blubber of 26 Florida manatees found dead in 1977–81 were examined for organochlorine pesticides and metabolites as well as PCBs (quantified as Aroclor mixtures) at a level of sensitivity of 0.1 μg/g (ppm) using gas chromatography or combined gas chromatography-mass spectrometry (O'Shea et al., 1984). Compounds that were targeted included *p,p′*-DDE, *p,p′*-DDT, *p,p′*-TDE, dieldrin, heptachlor epoxide, oxychlordane, *cis*-chlordane, *trans*-nonachlor, *cis*-nonachlor, endrin, HCB, mirex, and toxaphene. However, just 5 of 26 manatees had detectable concentrations of DDT metabolites (range 0.14–0.28 μg/g ww), with dieldrin found in four samples (range 0.12–0.36 μg/g ww). No other organochlorines were detected.

These samples from the 1970s had very low concentrations compared to the hundreds or thousands of μg/g of organochlorines detected in the blubber of some other marine mammals sampled elsewhere during the same era (e.g., Helle, 1976a,b; LeBouef and Bonnell, 1971; O'Shea et al., 1980; for review, see O'Shea, 1999), a finding consistent with the low position of sirenians in marine

TABLE 16.2 Qualitative Summary of Principal 20th-Century Ecotoxicologic Findings on Sirenians

Substances	Species or Subspecies	Sampling Periods and Locations	General Findings
Organochlorine pesticides and metabolites	Dugong	1970s, Sulawesi, Indonesia; Queensland, Australia	Very low concentrations compared to pinnipeds and cetaceans, but low sample sizes, c. six dugongs
		1990s, Queensland, Australia	Very low concentrations compared to pinnipeds and cetaceans, c. 59 dugongs
	Florida manatees	1970s–early 1980s, Florida, U.S.	Very low concentrations compared to pinnipeds and cetaceans, c. 31 manatees
		1990s, Florida, U.S.	Very low concentrations, some compounds lower than previous studies in Florida, up to 19 manatees
	Antillean, Amazonian, and West African manatees	No research	No knowledge
Other organohalogens	Dugong	Late 1990s to early 2000s, Queensland, Australia; Thailand	Low concentrations of PCBs, individual congeners quantified; organobromine compounds detected; PCDD\Fs congeners quantified, some at high concentrations relative to other marine mammals, c. 30 dugongs
		1990s, Florida, U.S.	No PCBs detected, up to 19 manatees. No analyses for other organohalogens

Continued

TABLE 16.2 Qualitative Summary of Principal 20th-Century Ecotoxicologic Findings on Sirenians—cont'd

Substances	Species or Subspecies	Sampling Periods and Locations	General Findings
	Antillean, Amazonian, and West African manatees	No research	No knowledge
Metals	Dugong	1970s, Sulawesi, Indonesia; Queensland, Australia	Fifteen elements analyzed in c. 57 dugongs. Low concentrations of elemental mercury, cadmium increases in kidneys with age
		Late 1990s, Queensland, Australia	Twelve elements analyzed in c. 38 dugongs. Conclusions similar to previous studies, none of ecotoxicologic significance
	Florida manatees	1970s–early 1980s, Florida, U.S.	Six elements analyzed in up to 54 manatees. Very low concentrations of elemental mercury compared to other marine mammals; cadmium increases in kidneys with age. Copper concentrations in liver very high in areas with copper herbicide use, thought to be of possible ecotoxicologic significance
	Antillean manatees	Late 1980s–mid-1990s, Mexico	Up to 14 elements analyzed in bone of 19 manatees. Lead and cadmium higher than in some other marine mammals

TABLE 16.2 Qualitative Summary of Principal 20th-Century Ecotoxicologic Findings on Sirenians—cont'd

Substances	Species or Subspecies	Sampling Periods and Locations	General Findings
		1990–2013, Mexico	Up to seven elements analyzed in bone of 33 manatees. Higher lead in adults, no results of seeming ecotoxicologic significance
	Amazonian and West African manatees	No research	No knowledge
Biotoxins	Dugongs	No research	No knowledge
	Florida manatees	1963, 1982, Southwest Florida	Circumstantial evidence for mortality due to concurrence with *Karenia brevis* red tide and absence of evidence for other causes, c. 44 deaths
		1996, Southwest Florida	Presence of brevetoxin in tissues with associated pathology confirmed, c. 149 deaths
	Antillean, Amazonian, and West African manatees	No research	No knowledge

See chapter narrative for greater details and references.

food webs. As a result of an epizootic of unknown causes during 1982, an additional four blubber and two brain tissue samples from the Florida manatees were analyzed for organochlorines with similar results (O'Shea et al., 1991). Additional chemical residue surveys for organochlorines were carried out on 11 blubber, 15 liver, and 19 kidney samples (limits of detection 0.1–1.0 ng/g ww) gathered from Florida manatee carcasses during 1990–93; no *p,p′*-DDE or dieldrin residues were found in these later samples, although *o,p′*-isomers of DDD and DDT were extracted from some livers and kidneys at low concentrations (Ames and Van Vleet, 1996).

These findings took the focus off organochlorine pesticides as possible threats to Florida manatees, whereas losses of habitat and direct human-related mortality due to strikes by boats, deaths in water control structures, cold-related die-offs, and a large fraction of deaths due to undetermined causes remained obvious major management concerns (e.g., O'Shea et al., 1985; Ackerman et al., 1995; see review in Marsh et al., 2012).

Other Organohalogens and Polycyclic Aromatic Hydrocarbons

Dugongs. Two dugongs sampled at Sulawesi Island, Indonesia, in 1975 lacked detectable residues of PCBs in muscle (Miyazaki et al., 1979). Analyses of blubber from seven dugong carcasses sampled from 1996 to 1999 in Queensland at detection levels of 2 ng/g (ppb) ww revealed the presence of PCBs 153 (maximum 171 ng/g lw) and 138 (maximum 127 ng/g lw) in all samples, PCBs 180 (maximum 35 ng/g lw) and 170 (maximum 29 ng/g lw) in six dugongs, and evidence for transplacental transfer to a fetus; measurable total PCBs reached 209 ng/g lw (Vetter et al., 2001). Unknown brominated organic compounds were also detected in some of these dugongs (Vetter et al., 2001). The concentration of total PCBs in blubber from a dugong sampled in Australia's Northern Territory at an unspecified time was 490 pg/g lw, with PCB 118 found at the highest concentration (Gaus et al., 2005). PCBs were quantified in muscle from two dugongs taken from coastal Thailand in 1999 and 2001, with PCB 118 found at the greatest concentration in one dugong and PCB 180 in the second (Kumar et al., 2002).

Considerable attention was given to PCDDs and PCDFs in dugongs in Queensland, Australia, in the context of concerns about pollution from terrestrial runoff in the Great Barrier Reef World Heritage Area. Fat from the blubber of three dugongs sampled during 1996 in northern Queensland had quantifiable residues of these compounds (Haynes et al., 1999). These residues consisted of five higher chlorinated congeners of PCDDs (penta to hepta range 5.8–72 pg/g [ppt] lw) and low concentrations of 2,3,7,8-tetrachlorodibenzo-*p*-dioxin (the most toxic dioxin known, range 2.3–6.0 pg/g [ppt] lw). In the same samples, octachlorodibenzodioxin (OCDD) occurred at the highest concentrations (range 170–250 pg/g lw, Müller et al., 1998; Haynes et al., 1999), higher than estimates for any odontocetes sampled globally at the time (reviewed in O'Shea, 2003). Ten polychlorinated dibenzofurans were also detected at lower concentrations, and a maximum concentration of 390 pg/g fat of total PCDD/Fs was reported (Müller et al., 1998; Haynes et al., 1999).

These findings were unusual compared to other marine mammals because PCDD/Fs contributed more to the total toxic equivalents (TEQs) than did PCBs. Exposure to PCDDs and PCDFs may have been through dugongs foraging in contaminated sediments in seagrass systems (perhaps with regional inputs from agriculture as the source), biotransformation from precursor compounds during hindgut digestion, or unique metabolic degradation pathways (Müller et al., 1998; Haynes et al., 1999). Additional research on PCDDs and PCDFs in sediments and seagrasses along the Queensland coast supported the hypothesis of exposure from contaminated sediments (McLachlan et al., 2001). As in the three dugongs discussed earlier, OCDD predominated in both seagrass and sediment samples, but PCDFs and lower chlorinated dioxins were not detected or were found at much lower concentrations (McLachlan et al., 2001). Samples of muscle tissue of two dugongs collected in 1999 and 2002 from coastal Thailand had similar patterns in PCDD/Fs and TEQs relative to PCBs (Kumar et al., 2002).

Blubber samples from additional dugongs (total 16), sediments, and seagrasses from northeastern Queensland verified the previously discussed findings and congener patterns (Gaus et al., 2001). Maximum concentrations of ΣPCDD/Fs reached 2400 pg/g lw, and maximum TEQs were 135 pg/g lw; regional differences in concentrations of OCDD occurred in male dugongs that corresponded with differences in sediments, and evidence was presented that PCDD/Fs may be taken up from sediments by plant tissues (Gaus et al., 2001). Later, more samples were added from an unspecified time period, resulting in analyses of blubber from 25 dugongs in total from the region (Gaus et al., 2004). Bioconcentration factors were calculated for different congeners, and previous patterns were further confirmed, including the correlation between ΣPCDD/F concentrations in blubber and OCDD in sediments, and higher concentrations of PCDD/Fs in adult males compared to adult females, which can excrete some of these compounds via pregnancy and lactation (Gaus et al., 2004). Assuming that the lowest-observed-adverse-effect level (LOAEL) obtained for experimental laboratory animals applied to dugongs, a theoretical LOAEL for average daily intake of TEQs would be exceeded by dugongs in some of the Australian sampling regions (Gaus et al., 2004).

Although there have been no analyses of PAHs in dugongs (PAHs would likely be easily metabolized), there is circumstantial evidence of direct mortality of dugongs during oil spills in the Middle East. An estimated 150 dugongs died during the Norwuz oil spill in the Arabian Gulf during 1983–84, and 14 dead dugongs were observed in 1991 during the Gulf War oil spill (Preen, 1988, 2004; Preen et al., 2012).

Manatees. PCBs were detected in blubber of 13 of 26 Florida manatees (range 0.50–4.6 μg/g) sampled during 1977–81 (O'Shea et al., 1984). All manatees with detectable PCB residues were recovered from locations in relatively urbanized areas. Additional chemical residue surveys for PCBs were carried out on 11 blubber, 15 liver, and 19 kidney samples gathered from Florida manatee carcasses during 1990–93, but no PCBs were detected (Ames and Van Vleet, 1996).

We are unaware of any information concerning PAHs or congener-specific studies of PCBs, PCDD/Fs, or PBDEs in manatees during the 20th century, nor of any published results on other organohalogens in the Antillean subspecies (*Trichechus m. manatus*) of the West Indian manatee, Amazonian manatees, or West African manatees from this period.

Heavy Metals and Other Toxic Elements

Dugongs. The death of a captive dugong in northern Queensland, Australia, was attributed to exposure to copper sulfate added to its tank as an algicide (Oke, 1967), but this claim was not substantiated with toxicologic or pathologic evidence. Fifteen metals and other toxic elements were reported in muscle, livers, and kidneys of about 57 dugongs in eight publications, nearly all based on samples from Australia, including very remote regions (Torres Strait between northern Queensland and Papua New Guinea and the Gulf of Carpentaria). Detailed findings and references were tabulated in the review by O'Shea (2003). The most extensive study involved samples from 42 dugongs collected during 1974–78 in northern Queensland, including 28 dugongs from the very remote regions listed before (Denton et al., 1980). As in manatees and other marine mammals, cadmium in kidneys increased with age, with a maximum of 309 μg/g dry weight (dw) in a dugong estimated to be 30 years old. Maximum concentrations of some elements (iron, zinc, copper, cobalt, and silver in livers) were high for marine mammals as known at the time (Denton et al., 1980). Results were most pertinent to dugong nutrition rather than toxicology. Elemental composition of seagrasses was examined and related to the nutritional status of the dugongs, as revealed in element concentrations of their tissues. High iron in seagrasses, for example, was related to storage of excess dietary iron in livers, as histologically observed hemosiderin granules. No pathologic findings were evident (Denton et al., 1980). Mercury analyses in dugongs were limited to two muscle samples from Indonesia (Miyazaki et al., 1979), and muscle and liver samples ($n = 11$ dugongs each organ) and kidney ($n = 2$ individuals) of dugongs taken from the mid-1970s to early 1990s at Torres Strait, northern Queensland, and the Northern Territory of Australia (summarized in O'Shea, 2003 and reports cited therein). Concentrations of elemental mercury from dugongs sampled during this period were very low, ranging from undetected to 0.15 μg/g dw. Muscle samples from the two dugongs from Indonesia were also analyzed for methyl mercury, with 0.004 μg/g dw found in one animal, none in the second (Miyazaki et al., 1979).

Similar concentrations of 12 elements were reported in the livers of 38 dugongs sampled in Queensland from 1996 to 2000 (Haynes et al., 2005). Mature dugongs had higher concentrations of aluminum, cadmium, iron, lead, mercury, and zinc than immature dugongs. Maximum concentrations in livers for elements of general concern were 32.5 μg/g ww cadmium, 1.1 μg/g ww mercury, 3.1 μg/g ww lead, and 7.7 μg/g ww arsenic, none of seeming toxicologic significance (Haynes et al., 2005). With analysis of individual trace metal concentrations, chemical speciation of metals became of interest. In one dugong liver sample from Okinawa, Japan, from 1982, arsenobetaine concentrations were low compared to those of methylarsonic acid and dimethylarsinic acid, unlike in other marine mammals and presumably due to the dugongs' herbivorous diet (Kubota et al., 2002).

Manatees. Quantitative determination of concentrations of cadmium, copper, iron, lead, mercury, and selenium in tissues was conducted on Florida manatee carcasses found during 1977–82 (O'Shea et al., 1984, 1991). Elemental mercury (limit of detection 0.02 μg/g dw) was either not detected or found at very low concentrations (maximum of 0.5 μg/g dw) in livers (but not in muscle) from 27 individuals. These concentrations were far lower than in cetaceans and pinnipeds globally, in which mercury can reach 3–4 orders of magnitude higher concentrations but is detoxified in complexes involving positive correlations with selenium (studies summarized in O'Shea, 1999). Selenium also was at low concentrations (maximum of 1 μg/g dw) in Florida manatee livers.

Concentrations of cadmium and lead in tissues of Florida manatees was of interest because these metals can occur at elevated concentrations in tissues of aquatic plants in contaminated environments (O'Shea et al., 1984). Cadmium concentrations in kidneys of 38 manatees sampled during 1977–81 increased with total body length (a proxy for age), as previously known in other marine mammals, reaching a maximum of 190 μg/g dw. Lead concentrations in livers and kidneys of 20 manatees reached maxima (7.1 μg/g dw in kidney) similar to maxima in other marine mammals (as summarized in O'Shea, 1999). These concentrations were not thought to be indicative of toxicity (O'Shea et al., 1984).

Copper was of great interest in the early Florida manatee studies because copper-based herbicides were being used to control aquatic plants in an area that had large seasonal aggregations of manatees, and copper concentrations in treated plants can reach high levels (O'Shea et al., 1984). Copper is well known to be highest in livers of young mammals and to decrease with age. Maximum concentrations in 54 manatees sampled from 1977 to 1981 were found in those individuals from the region subject to copper herbicide treatment, with a maximum of 1200 μg/g dw reached in a young animal (O'Shea et al., 1984). An excess or deficiency in iron intake can influence copper concentrations in livers, but no relationship between iron and copper was found in the Florida study. Although no pathologic evidence of toxic effects of copper on manatees was available, investigators suggested that evidence of accumulation and comparative studies of copper toxicity in other mammals were sufficient to recommend restricted use of copper herbicides in areas frequented by manatees (O'Shea et al., 1984).

Bones of 19 Antillean manatees sampled during 1987–96 in Chetumal Bay, Mexico, were examined for up to 14 elements by atomic absorption spectrophotometry (Rojas-Mingüer and Morales-Vela, 2002). Concentrations of lead (maximum 128 μg/g ww) and cadmium (maximum 5 μg/g ww) were higher than maxima in bone of several other species of marine mammals. A comparative study of Antillean manatees from Quintana Roo (Chetumal Bay), Tabasco, and Campeche, Mexico, focused on concentrations of seven elements (arsenic, cadmium, chromium, copper, lead, nickel, zinc) in bones of 33 stranded carcasses sampled from 1990 to 2013 (Romero-Calderón et al., 2016). Lead concentrations were higher in adults than in calves, and differences in some metal concentrations were found among regions, but no toxic effects could be ascribed to the concentrations observed (Romero-Calderón et al., 2016).

Biotoxins and Harmful Algal Blooms

Manatees. The term "red tide" is applied to various dinoflagellate blooms worldwide, but in Florida, it usually refers to blooms of *Karenia brevis* and its associated brevetoxins (PbTx), which consist of multiple polycyclic polyether compounds. These blooms occur with greatest frequency on the Gulf Coast of Florida. The first reported deaths of marine mammals associated with Florida red tides involved bottlenose dolphins during a bloom in 1946–47 (Gunter et al., 1948), and seven dead manatees concurrent with a bloom in 1963 (Layne, 1965). Stronger circumstantial evidence supported the likelihood of associated mortality of at least 37 Florida manatees during a red tide event on the southwestern Gulf Coast in 1982 (O'Shea et al., 1991). A definitive link between PbTx poisoning and manatee mortality took place in a 6-month event in 1996, during which at least 149 Florida manatees died (Bossart et al., 1998). Deaths were accompanied by characteristic gross and histopathologic lesions of the respiratory system, liver, kidneys, meninges, and choroid plexus with immunohistochemical confirmation of PbTx in multiple tissues, as well as in tissues of manatees archived from the 1982 event; strong support was provided for proposed cellular mechanisms of toxicosis (Bossart et al., 1998, 2002).

TWENTY-FIRST-CENTURY FINDINGS AND CONCERNS

During the 21st century, ecotoxicologic research continued to focus on dugongs and Florida manatees, but with increased attention given to sampling of the Antillean subspecies of manatee (see Table 16.3 for summary). Concentrations of contaminants in tissues remained the focus, but sampling was expanded. Additional substances, blood, and biomarkers were investigated in living manatees, and the depth of knowledge concerning biotoxins was increased. PBDEs and organotins were discovered in dugongs, and studies of PCDD/Fs were expanded to include in vitro bioassays. As in the prior century, no studies were conducted with West African or Amazonian manatees.

Organochlorine Pesticides and Metabolites

Dugongs. We are unaware of any recent developments concerning dugongs and organochlorine pesticides.

Manatees. Blood samples from 16 captive Antillean manatees from coastal Brazil were analyzed for metabolites of DDT, chlordanes, HCHs, cyclodienes, HCB, mirex, and 47 PCB congeners at a level of sensitivity of 0.1 ng/g: none of these compounds were detected (Anzolin et al., 2012). We are unaware of other published literature on organochlorine concentrations in samples of manatees taken during this century. Unpublished data from tissues of Florida manatees sampled in 2003 referenced by Wetzel et al. (2012) apparently continue to show minor and declining concentrations of the organochlorines as reported in previous studies, consistent with curtailment of their usage. Antillean manatees sampled in Chetumal Bay, Mexico, during 2006–07 seemed to have concentrations of organochlorines in tissues similar to those seen in Florida 25–30 years earlier, but including two compounds not previously reported in manatees (lindane and endosulfan), as well as heptachlor, dieldrin, and mirex (Wetzel et al., 2012).

Other Organohalogens and PAHs

Dugongs. Low concentrations of polybrominated diphenyl ether (PBDEs) flame retardants were found in adipose tissue from a single adult female dugong sampled in about 2005 in coastal Queensland (Hermanussen et al., 2008). Five PBDE congeners (28, 47, 49, 99, and 100) were detected at ranges of 0.02–0.69 ng/g lw, whereas two were not detected (153 and 154); ΣPBDEs were estimated at 1.4 ng/g lw, "among the lowest levels reported to date in marine megafauna studies from across the globe" (Hermanussen et al., 2008). Data from dugongs sampled in Australia for PCDD/Fs were mostly taken during the late 1990s, but with some overlap in the early 2000s. See summary for these data earlier in the text.

Seven different PCB congeners (PCBs 28, 52, 101, 118, 153, 138, 180) were analyzed in liver, muscle, and blubber of dugongs from two different Queensland coastal locations sampled from 1998 to 2005 (Weijs et al., 2016). Higher proportions of lower chlorinated PCBs were found in the liver and muscle compared to the blubber, whereas higher chlorinated PCBs were more abundant in the blubber. PCBs were highest in muscle (range 1486–42,284 pg/g lw), followed by blubber (range 1008–12,537 pg/g lw) and liver (range 1620–10,920 pg/g lw) (Weijs et al., 2016). The higher levels in muscle compared to blubber and liver were attributed to the low sample mass available, as well as a generally low lipid content and potential associated error in gravimetric lipid measurements. Generally, levels of total PCBs found in these dugongs ($n = 10$) were orders of magnitude lower than toxicity thresholds reported in other marine mammal species, thereby suggesting that PCBs were a negligible threat to the health of dugongs during the early 2000s in Queensland.

Novel recent research concerning PCDD/Fs and other organohalogens in dugongs has focused on the development of tools to assess exposure to and potential associated risks from contaminant mixtures. Passive sampling and bioanalytical screening are routinely used to detect contaminants in abiotic matrices (e.g., wastewater). However, these two approaches, especially in combination, have only been applied to marine biota in recent years. Attention has also shifted toward investigating pollutant mixtures, triggered by the need to establish cause-effect relationships, given that wildlife are routinely exposed to pollutant mixtures. Effect studies reporting interactions among pollutant

TABLE 16.3 Qualitative Summary of Principal 21st Century Ecotoxicologic Findings on Sirenians

Substances	Species or Subspecies	Sampling Periods and Locations	General Findings
Organochlorine pesticides and metabolites	Dugong	None published	No recent findings
	Florida manatees	None published	No recent findings
	Antillean manatees	Captive manatees held in states of Pernambuco, Paraíba and Alagoas, Brazil; dates unspecified	No residues of six compounds and metabolites detected in blood of 16 captive manatees
	Amazonian and West African manatees	No research	No knowledge
Other organohalogens	Dugong	2005, Queensland, Australia	Very low concentrations of PBDEs found in one dugong compared to other marine mammals
		Early 2000s, Queensland, Australia	Low concentrations of 7 PCB congeners detected in tissues of 10 dugongs, thought to be of negligible significance
		2015, Queensland, Australia	In vitro bioassay methods developed to assess mixture toxicity, with PCDDs responsible for most AhR induction but only 5% of oxidative stress.
	Antillean manatees	Captive manatees held in states of Pernambuco, Paraíba and Alagoas, Brazil; dates unspecified	No residues of PCB congeners detected in blood of 16 captive manatees

TABLE 16.3 Qualitative Summary of Principal 21st Century Ecotoxicologic Findings on Sirenians—cont'd

Substances	Species or Subspecies	Sampling Periods and Locations	General Findings
	Amazonian and West African manatees	No research	No knowledge
Metals	Dugong	Presumed 2000s, Torres Strait, Australia	Women who consumed dugongs and turtles had higher cadmium in urine.
		1998–2003, Thailand	Organotins found in tissues of 10 dugongs at concentrations higher than most marine mammals, thought to be of possible ecotoxicologic significance
	Florida manatees	2000s, Florida	Concentrations of up to 22 trace metals determined in blood and skin of eight manatees sampled 2007, blood of 31 manatees sampled 2005–06, and blood of 137 manatees sampled 2007–2011. Concern expressed about higher concentrations of zinc, aluminum, copper. Lead and copper in livers declined since 1970s
	Antillean manatees	2008–2011, Belize	Concentrations of up to 22 trace metals determined in blood of 33 manatees, some elements vary by location, arsenic in wild manatees higher than in captives
	Amazonian and West African manatees	No research	No knowledge

Continued

TABLE 16.3 Qualitative Summary of Principal 21st Century Ecotoxicologic Findings on Sirenians—cont'd

Substances	Species or Subspecies	Sampling Periods and Locations	General Findings
		Captive manatees held in states of Pernambuco, Paraíba and Alagoas, Brazil; dates unspecified	Concentrations of nine trace elements determined in blood of 16 captive manatees along with biomarkers of potential toxic exposure. Aluminum, tin, lead and cadmium appeared high in comparison to the literature on Florida manatees and in some cases other marine mammals. Higher biomarker levels found in manatees from Paraíba, where higher lead was found (although below suggested thresholds for harm)
Biotoxins	Dugongs	2000–04, Queensland, Australia	No okadaic acid found in tissues of 41 dugongs
	Florida manatees	2000s, Florida	Significant increases in knowledge concerning factors that *promote Karenia brevis* red tides, mechanisms of brevetoxin poisoning in manatees, and sublethal effects on manatee immune function and susceptibility to other sources of mortality. Discovery of manatee exposure to okadaic acid, saxitoxin
	Antillean, Amazonian, and West African manatees	No research	No knowledge

See chapter narrative for greater details and references.

mixtures even if present at low levels (reviewed in Altenburger et al., 2013) are also receiving attention. As a result, multitarget and nontarget methods have been increasingly employed. In vitro bioassays are nontarget methods that offer the possibility to screen mixture toxicity in tissue samples based on cell responses. Such nontarget screening approaches can only work if tissue samples are extracted and prepared in a nontarget or nonselective manner.

Following this approach, Jin et al. (2013) used passive equilibrium sampling with polydimethylsiloxane (PDMS) for unbiased sampling of mixtures from dugong blubber samples. Resulting extracts were dosed into a cell-based bioassay indicative for AhR-receptor induction (i.e., CAFLUX). PDMS extraction was found to be a suitable nontarget way to extract bioaccumulative pollutants with varying degrees of hydrophobicity, without a need for a thorough preparation of the samples (thereby further increasing the likelihood of being nonselective). PCDDs were not solely responsible for the bioassay outcomes for all dugongs tested, despite PCDDs being known triggers of the AhR-receptor toxicity pathway. The same PDMS sampling approach was repeated by Jin et al. (2015), but dugong blubber extracts were utilized, in addition to CAFLUX, in a battery of bioassays with different endpoints including DNA damage, oxidative stress, and inflammation. Exposure to blubber extracts from dugongs did not yield significant responses indicative for DNA damage and inflammation. In contrast, significant but dissimilar responses were found for bioassays indicative for AhR-receptor induction and oxidative stress, with quantified PCDD levels responsible for the majority of AhR-mediated activity, but only 5% of the cell responses in the bioassays were indicative for oxidative stress. From this work, it is clear that the pollutant loads in dugongs include more relevant compounds than PCDDs per se, a result that gives new impetus to more studies focusing on chemicals other than PCDDs.

Manatees. Blood samples from 16 captive Antillean manatees from coastal Brazil were analyzed for 47 PCB congeners at a level of sensitivity of 0.1 ng/g: none were detected (Anzolin et al., 2012). We are unaware of other published literature on other organohalogens in manatees taken during this century. Unpublished data referenced by Wetzel et al. (2012) on PCBs in tissues of Antillean manatees sampled in Chetumal Bay, Mexico, during 2006–07 seemed to indicate concentrations of PCBs similar to those seen in Florida 25–30 years earlier; PAHs were not detected in these samples (Wetzel et al., 2012).

Heavy Metals and Other Toxic Elements

Dugongs. Organotin compounds are widely used in antifouling paints on hulls of ships. Ten dugongs found dead in coastal Thailand during 1998–2003 were sampled for organotins in blubber, kidney, liver, lung, heart, and muscle (Harino et al., 2007). Maximum concentrations of butyltins (14,468 μg/kg in kidney) were higher than those reported for multiple species of cetaceans and pinnipeds in other studies, with liver more typically containing the highest amounts.

Monobutyltins and dibutyltins dominated. Phenyltins were also found in some dugongs, but at lower concentrations than butyltins (Harino et al., 2007). Concern was expressed about possible toxicity of these compounds to dugongs, which are sometimes observed grazing on algae on ships bottoms and wharf pilings (Marsh, unpublished observation).

Elevated levels of cadmium have been a concern to the Indigenous peoples of the Torres Strait region, who consume the meat of dugongs and green turtles as traditional foods (Marsh et al., 2015). Sixty women between the ages of 30–50 years old participated in a questionnaire survey, urine collection, and health examinations. Cadmium concentrations in their urine were elevated in people consuming traditional food such as dugong kidney at least monthly, with cadmium concentrations increasing with age (Haswell-Elkins et al., 2007).

Trace metals were analyzed for forensic crime applications in several protected mammals in Thailand using a handheld X-ray fluorescence analyzer (Nganvongpanit et al., 2017). Up to 22 elements were analyzed in 60 dugong (molar) teeth and 43 tusk (incisor) samples with results compared with similar material from other marine mammals, Asian elephants (*Elephas maximus*), and tigers *(Panthera tigris tigris*). Aluminum was detected only in dugongs, which also had higher Ca/P ratios, and elevated levels of silver in male tusks (which erupt and are exposed to the sediment) compared to female tusks (which generally remain inside the body). This study was of importance to law enforcement officers for distinguishing among species for confiscated tusks and other teeth (Nganvongpanit et al., 2017).

Manatees. Trace element concentrations, including some metals and toxic elements, were reported for whole blood and skin samples of eight Florida manatees sampled in 2007, along with concentrations of elements in sediments and plants (Stavros et al., 2008). This study was the first to utilize whole blood as a biomarker for circulating levels of trace metals in live manatees rather than examining contaminant loads in tissue samples from carcasses. Most elements were at concentrations not indicative of concern, although aluminum in skin may warrant additional investigation. In a separate study, concentrations of 30 trace elements were determined in whole blood and serum of 14 Antillean manatees from Belize and 31 Florida manatees sampled in 2005–06 (Siegal-Willott et al., 2013). Trace metal concentrations varied according to sample type (whole blood vs. serum), location, age, and sex. Although some metal concentrations were found at levels toxic to domestic terrestrial species, manatees lacked clinical signs of toxicosis. In a larger study, up to 22 trace elements were analyzed in whole blood from 170 free-ranging and managed manatees from Florida and Belize sampled during 2007–11 (Takeuchi et al., 2016). High levels of zinc (11.20 ± 0.30 ppm, ww) were found compared to other species of marine mammals, suggesting a physiologic need for an elevated level of zinc. This result warrants further investigation into uses of zinc in wound healing, the role of metallothionein in detoxification (Takeuchi, 2012), and bone remodeling (Anné et al., 2016). Selenium concentrations (0.18 ± 0.09 ppm, ww) were lower in long-term and short-term captive manatees compared to those in wild manatees and other marine mammal species, which suggests needed monitoring for dietary supplementation. In Florida manatees, whole blood copper levels were the highest in urban areas, as well as in calves, whereas arsenic may be of concern to the Antillean manatee in Belize (Takeuchi et al., 2016).

In addition to manatees in Florida and Belize, concentrations of nine trace elements (including cadmium and lead) were determined in blood of 16 captive Antillean manatees from three states in coastal Brazil, along with three biochemical and seven hematological biomarkers of potential toxic exposure (Anzolin et al., 2012). Aluminum, tin, lead, and cadmium concentrations appeared high in comparison to the literature on Florida manatees, and in some cases of other marine mammals. Higher mean corpuscular volume and smaller mean corpuscular hemoglobin concentration was found in manatees from Paraíba, where higher lead (range 0.05–0.19 μg/g ww) was found in blood compared to two other areas, although lead concentrations were lower than suggested thresholds for harm. Two enzymes (catalase and glutathione-S-transferase) did not appear to be at levels indicating problems, but inhibition of butyrilcholinesterase activity was seen in blood of manatees kept at two areas, perhaps indicating exposure to carbamate or organophosphate pesticides rather than effects of metals (Anzolin et al., 2012).

When comparing hepatic and renal metal burdens in Florida manatees from the late 1970s to the 21st century, there was a significant decrease in lead concentrations and a lower level of copper (Takeuchi, 2012). This comparison displays the importance of continual monitoring of manatees and other sirenians to understand potential anthropogenic influences and the possible beneficial effect of increasing environmental protection efforts. Although metal concentrations in Florida manatees do not seem to be of major concern today, consistent monitoring of metal loads in the context of the age of the animals is highly recommended.

Biotoxins and Harmful Algal Blooms

Dugongs. Certain dinoflagellates (*Prorocentrum* spp.) are epiphytic on seagrasses in dugong habitat and produce the toxin okadaic acid, which can be a tumor promoter (Takahashi et al., 2008). No okadaic acid or tumors were detected in 41 dugongs sampled during 2000–04 in Moreton Bay, Queensland, an area where these dinoflagellates are found, and no associated morbidity or mortality was observed (Takahashi et al., 2008). It is not known whether these animals avoided eating the affected seagrasses, but it is known that dugongs tend to avoid eating seagrasses with high epiphyte loads (Marsh et al., 2012).

Manatees. Manatee mortality events involving dozens of animals were seen in Florida during *Karenia brevis* blooms in 2002, 2003, 2005–07 (Landsberg et al., 2009; Fire et al., 2015), 2012–13, and 2016 (Florida Fish and Wildlife Conservation Commission, 2017). Events were typically located on the Gulf

Coast, with brevetoxicosis in manatees first reported on the Atlantic coast in 2007–08 (Fire et al., 2015). Increased understanding as a result of these and prior events led to further generalizations and insights into PbTx poisoning in Florida manatees. Manatee mortality events seem to be most likely during red tides where blooms come inshore in the winter–spring, inshore salinities are high (>24 ppt; sometimes due to droughts), and blooms persist for weeks or months (Landsberg et al., 2009). Exposure can occur through ingestion of PbTx that remains associated with seagrasses and their epiphytes, even after red tides dissipate (Flewelling et al., 2005; Landsberg et al., 2009; Fire et al., 2015), as well as through inhalation as indicated by pathology (Bossart et al., 2002; Bossart, 2011) or, perhaps, through maternal transfer during lactation (based on presence of PbTx in milk; Flewelling et al., 2005; Fire et al., 2015). Mechanisms of lethality may involve inflammatory mediators that culminate in fatal toxic shock (Bossart, 2011).

With mass mortality events in marine mammals increasing, there is an interest in synergistic effects of biotoxins and effects of multiple other stressors (Gulland and Hall, 2007). Although manatees may not be as sensitive to the immunotoxic effects of PbTx as dolphins (Gebhard et al., 2015), the immune health of manatees can be affected and exposure to PbTx can occur to some degree throughout the year (Landsberg et al., 2014). Between 2000 and 2003, blood samples ($n=71$) were collected in Florida from healthy wild manatees, cold-stressed manatees, and red tide–exposed manatees. Using peripheral blood mononuclear cells and in vitro exposures, lymphocyte proliferation decreased in red tide–exposed animals, and they were even further depleted in cold-stressed manatees exposed to PbTx. Thus, red tide exposure may impact immunocompetence, as well as increase the immune sensitivity to additional environmental stressors affecting the overall health of manatees (Walsh et al., 2005). Thirty-seven blood samples were collected from nonexposed manatees and manatees sublethally exposed to red tide during 2006–13 (Walsh et al., 2015). Various immune function parameters were analyzed, which included lymphocyte proliferation, lysozyme activity, inflammation, and oxidative stress. In red tide–exposed manatees, lymphocyte proliferation decreased, while oxidative stress and inflammation increased, also providing evidence that sublethal exposure to red tide can have a detrimental impact on immune function and overall health (Walsh et al., 2015), as well as possible vulnerability to incidental causes of death, such as boat strikes while disoriented or listing, and other possible physiologic effects of exposure (Wetzel et al., 2010).

In addition to brevetoxicosis, manatees have been exposed to additional biotoxins, such as debromoaplysiatoxin (also a tumor promoter) produced by *Lyngbya* sp., collected on the skin of manatees in 2006, which may be associated with dermal ulcerations (Harr et al., 2008). Capper et al. (2013) examined 14 manatees sampled in 2003–06 for brevetoxin (PbTx), okadaic acid (OA), lyngbyatoxin-A, and saxitoxin (STX). This report was the first time that manatee tissues had been reported as testing positive for OA and STX and as testing positive for more than one biotoxin (Capper et al., 2013).

CONCLUSIONS AND SUGGESTIONS FOR FUTURE RESEARCH

Sirenians differ from other species of marine mammals in many fundamental ways, and this is reflected in their status in relation to toxic substances. Exposure through dermal or inhalation pathways has not been demonstrated other than brevetoxin poisoning. However, dietary intake seems the most likely exposure pathway for many toxic substances, particularly considering their daily food intake of about 7%–16% of their body mass (adult sirenians range 450–1600 kg; Marsh et al., 2012). Unlike piscivorous marine mammals that are exposed to biomagnified lipophilic substances in their prey, concentrations of organochlorine pesticides and metabolites, PCBs, and methyl mercury in tissues appear uniformly low in manatees and dugongs, presumably because they feed largely on aquatic plants. Toxic substances that can accumulate in sediments and tissues of aquatic plants are thus of greatest concern, particularly in nearshore and riverine habitats close to human urban, industrial, or agricultural sources. These substances include toxic elements such as copper, cadmium, aluminum, and lead, as well as PCDD/Fs and biotoxins.

Biotoxins have been well demonstrated to cause death, as well as sublethal effects that may increase the susceptibility of Florida manatees to other causes of death. Global climate change may exacerbate this and other threats, with an increase in harmful algal blooms, diseases, and severe storms, which can impact food supplies (Edwards, 2013). Although none of the persistent contaminants identified in sirenian tissues thus far have been definitively associated with mortality or morbidity of manatees or dugongs, the possibility of such effects continues to warrant suspicion based on elevated concentrations in tissues in contaminated locales. Newly applied bioassay techniques suggest effects of PCDD/Fs, but also reveal effects of yet unknown substances as well as mixtures of contaminants present in Queensland dugongs.

Sampling of sirenians for toxic substances remains rudimentary and incomplete. There are no data for any contaminants in two of the four living species. Exploratory analyses of sirenian tissues, food plants, and sediments should also be undertaken in attempts to identify yet unknown substances to which sirenians may be exposed. For example, based on the work of Jin et al. (2015) the measured PCDDs only explained 5% of the toxicity observed in the bioassay indicative for oxidative stress, suggesting that important toxic chemicals are being overlooked.

Many biochemical, molecular genetic, and immunological biomarkers are obtainable by biopsy (e.g., Fossi et al., 1999; Hahn, 2002). Examples of applications of such tools are provided by the blood chemistry and hematological biomarkers used for Antillean manatees by Anzolin et al. (2012) and biomarkers of immune function used in Florida manatees (Walsh et al., 2015). Anatomic and developmental endpoints of endocrine disruptor exposure (e.g., hypospadias, cleft clitorine urethra, reduced anogenital distances, sex organ size, retained nipples; Gray et al., 1998; Working Group on Endocrinology and Reproduction, 1999) also can be routinely measured. There is significant opportunity to apply other technologic advances in ecotoxicology to the study of sirenians. Examples include the transition from organ sampling for metals during necropsy to use of blood sampled from living manatees using specific trace metal vacutainer blood tubes. Analytical techniques, such as atomic absorption spectroscopy and inductively coupled plasma spectroscopy, have been used in the 20th and 21st century with potential for analytical equipment with higher specificity. Fatty acid signature analysis may also be a potential biomarker for red tide exposure in manatees (Wetzel et al., 2010).

We recommend the following approaches be taken in future ecotoxicologic studies of sirenians to more fully understand the threats.

1. expanded sampling of all species, particularly:
   a. the Amazonian and West African manatees for which no data are available, with priority given to accessible populations potentially exposed to high levels of chemical contaminants;
   b. the populations of dugongs and Florida manatees for which demographic patterns are reasonably well-known and health assessments are regularly performed to assist in interpreting the population implications of findings (for references, see Marsh et al., 2012).
2. periodic monitoring of persistent contaminants in the more accessible sirenian populations as indices of changes to the quality of their environments in the face of climate change and habitat degradation;
3. greater emphasis on the following:
   a. the poorly known PCDD/Fs, PBDEs, or organotins in manatees;
   b. the chemical forms of organically bound metals (e.g., Kubota et al., 2002; Harino et al., 2007; Takeuchi et al., 2016);
   c. novel organic pollutants (chlorinated paraffins, perfluorinated compounds, new emerging flame retardants), which have never been investigated in any sirenian species;
   d. the synergistic effects of multiple toxic substances on sirenian health by expanded use of in vitro bioassays and passive equilibrium sampling such as PDMS for sampling of mixtures of persistent contaminants from tissue samples of sirenians, building on the approaches of Jin et al. (2013, 2015) and the in vitro immunological assays used by Walsh et al. (2005) in Florida manatees exposed to brevetoxins;
   e. corroborative biomarker studies to improve interpretation of possible effects of substances detected by chemical analyses during carcass necropsies.

FINAL REMARKS

The four extant species of sirenians are very different from the other species of medium-sized marine mammals because of their herbivorous diet and consequently unique digestive anatomy, physiology, and metabolism. Many sirenian populations are vulnerable to extinction and have coastal and riverine distributions in range states that are mostly developing countries with limited pollution controls. These factors indicate unique potential exposures to environmental contaminants and biotoxins that warrant greater research. Long-term monitoring of persistent contaminants in accessible sirenian populations has the potential to offer important insights into the changing quality of such environments in the face of climate change and ongoing habitat degradation.

REFERENCES

Ackerman, B.B., Wright, S.D., Bonde, R.K., Odell, D.K., Banowetz, D.J., 1995. Trends and patterns in mortality of manatees in Florida, 1974-1992. In: O'Shea, T.J., Ackerman, B.B., Percival, H.F. (Eds.), Population Biology of the Florida Manatee, U.S. Department of the Interior, National Biological Service, Information and Technology Report 1, Washington, DC, pp. 223–258.

Altenburger, R., Backhaus, T., Boedeker, W., Faust, M., Scholze, M., 2013. Simplifying complexity: mixture toxicity assessment in the last 20 years. Environmental Toxicology and Chemistry 32, 1685–1687.

Ames, A.L., Van Vleet, E.S., 1996. Organochlorine residues in the Florida manatee, *Trichechus manatus latirostris*. Marine Pollution Bulletin 32, 374–377.

Anné, J., Wogelius, R.A., Edwards, N.P., van Veelen, A., Ignatyev, K., Manning, P.L., 2016. Chemistry of bone remodeling preserved in extant and fossil Sirenia. Metallomics 8, 508–513.

Anzolin, D.G., Sarkis, J.E.S., Diaz, E., Soares, D.G., Serrano, I.L., Borges, J.C.G., Souto, A.S., Taniguchi, S., Montone, R.C., Bainy, A.C.D., Carvalho, P.S.M., 2012. Contaminant concentrations, biochemical and hematological biomarkers in blood of West Indian manatees *Trichechus manatus* from Brazil. Marine Pollution Bulletin 64, 1402–1408.

Aoshima, K., 2016. Itai-itai disease: renal tubular osteomalacia induced by environmental exposure to cadmium—historical review and perspectives. Soil Science and Plant Nutrition 62 (4), 319–326.

Bossart, G.D., 2011. Marine mammals as sentinel species for oceans and human health. Veterinary Pathology 48, 676–690.

Bossart, G.D., Baden, D.G., Ewing, R.Y., Roberts, B., Wright, S.D., 1998. Brevetoxicosis in manatees (*Trichechus manatus latirostris*) from the 1996 epizootic: gross, histologic, and immunohistochemical features. Toxicologic Pathology 26, 276–282.

Bossart, G.D., Baden, D.G., Ewing, R.Y., Wright, S.D., 2002. Manatees and brevetoxicosis. In: Pfeiffer, C.J. (Ed.), Molecular and Cell Biology of Marine Mammals. Krieger Publishing Company, Melbourne, Florida, pp. 205–212.

Capper, A., Flewelling, L.J., Arthur, K., 2013. Dietary exposure to harmful algal bloom (HAB) toxins in the endangered manatee (*Trichechus manatus latirostris*) and green sea turtle (*Chelonia mydas*) in Florida, USA. Harmful Algae 28, 1–9.

DeLong, R.L., Gilmartin, W.G., Simpson, J.G., 1973. Premature births in California sea lions: association with high organochlorine pollutant residue levels. Science 181, 1168–1170.

Denton, G.R.W., Marsh, H., Heinsohn, G.E., Burdon-Jones, C., 1980. The unusual metal status of the dugong, *Dugon dugon*. Marine Biology 57, 201–219.

Edwards, H.H., 2013. Potential impacts of climate change on warm water megafauna: the Florida manatee example (*Trichechus manatus latirostris*). Climatic Change 121, 727–738.

Fire, S.E., Flewelling, L.J., Stolen, M., Durden, W.N., de Wit, M., Spellman, A.C., Wang, Z., 2015. Brevetoxin-associated mass mortality event of bottlenose dolphins and manatees along the east coast of Florida, USA. Marine Ecology Progress Series 526, 241–251.

Flewelling, L.J., Naar, J.P., Abbott, J.P., Baden, D.G., Barros, N.B., Bossart, G.D., Bottein, M.Y.D., Hammond, D.G., Haubold, E.M., Heil, C.A., Henry, M.S., Jacocks, H.M., Leighfield, T.A., Pierce, R.H., Pitchford, T.D., Rommel, S.A., Scott, P.S., Steidinger, K.A., Truby, E.W., Van Dolah, F.M., Landsberg, J.H., 2005. Brevetoxicosis: red tides and marine mammal mortalities. Nature 435, 755–756.

Florida Fish and Wildlife Conservation Commission, Red Tide Manatee Mortalities. 2017. Retrieved from: http://myfwc.com/research/manatee/rescue-mortality-response/mortality-statistics/red-tide/.

Forrester, D.J., White, F.H., Woodard, J.C., Thompson, N.P., 1975. Intussusception in a Florida manatee. Journal of Wildlife Diseases 11, 566–568.

Fossi, M.C., Casini, S., Marsili, L., 1999. Nondestructive biomarkers of exposure to endocrine disrupting chemicals in endangered species of wildlife. Chemosphere 39, 1273–1285.

Gaus, C., Päpke, O., Blanchard, W., Haynes, D., Connell, D.W., Müller, J., 2001. Bioaccumulation and pathways of PCDDs in the lower trophic marine system. Organohalogen Compounds 52, 95–99.

Gaus, C., O'Donohue, M., Connell, D., Müller, J., Haynes, D., Päpke, O., 2004. Exposure and potential risks of dioxins to the marine mammal dugong. Organohalogen Compounds 66, 1559–1566.

Gaus, C., Correll, R., Müller, J., Holt, E., Ellis, D., Prange, J., Shaw, M., Bauer, U., Symons, R., Burniston, D., 2005. Dioxins and dioxin-like PCBs in marine mammals from Australia. Organohalogen Compounds 67, 1271–1275.

Gebhard, E., Levin, M., Bogomolni, A., De Guise, S., 2015. Immunomodulatory effects of brevetoxin (PbTx-3) upon in vitro exposure in bottlenose dolphins (*Tursiops truncatus*). Harmful Algae 44, 54–62.

George, J.L., Frear, D.E.H., 1966. Pesticides in the Antarctic. In: Moore, N.W. (Ed.), Pesticides in the Environment and Their Effects on Wildlife. In: Journal of Applied Ecology, vol. 3 (Suppl.), 155–167.

Gray, L. E., Ostby, J., Wolf, C., Lambright, C., Kelce, W.R., 1998. The value of mechanistic studies in laboratory animals for the prediction of reproductive effects in wildlife. Environmental Toxicology and Chemistry 17, 109–118.

Gulland, F.M.D., Hall, A.J., 2007. Is marine mammal health deteriorating? Trends in the global reporting of marine mammal disease. EcoHealth 4, 135–150.

Gunter, G., Williams, R.H., Davis, C.C., Walton Smith, F.G., 1948. Catastrophic mass mortality of marine animals and coincident phytoplankton bloom on the west coast of Florida, November 1946 to August 1947. Ecological Monographs 18, 309–324.

Hahn, M.E., 2002. Biomarkers and bioassays for detecting dioxin-like compounds in the marine environment. The Science of the Total Environment 289, 49–69.

Harada, M., 1995. Minamata disease: methylmercury poisoning in Japan caused by environmental pollution. Critical Reviews in Toxicology 25, 1–24.

Harino, H., Ohji, M., Wattayakorn, G., Adulyanukosol, K., Arai, T., Miyazaki, N., 2007. Concentrations of organotin compounds in tissues and organs of dugongs from Thai coastal waters. Archives of Environmental Contamination and Toxicology 53, 495–502.

Harr, K.E., Szabo, N.J., Cichra, M., Philips, E.J., 2008. Debromoaplysiatoxin in *Lyngbya*-dominated mats on manatees (*Trichechus manatus latirostris*) in the Florida King's Bay ecosystem. Toxicon 52, 385–388.

Haswell-Elkins, M., McGrath, V., Moore, M., Satarug, S., Walmby, M., Ng, J., 2007. Exploring potential dietary contributions including traditional seafood and other determinants of urinary cadmium levels among indigenous women of a Torres Strait Island (Australia). Journal of Exposure Science and Environmental Epidemiology 17, 298–306.

Haynes, D., Johnson, J.E., 2000. Organochlorine, heavy metal and polyaromatic hydrocarbon pollutant concentrations in the Great Barrier Reef (Australia) environment: a review. Marine Pollution Bulletin 41, 267–278.

Haynes, D., Müller, J.F., McLachlan, M.S., 1999. Polychlorinated dibenzo-*p*-dioxins and dibenzofurans in Great Barrier Reef (Australia) dugongs (*Dugong dugon*). Chemosphere 38, 255–262.

Haynes, D., Carter, S., Gaus, C., Müller, J., Dennison, W., 2005. Organochlorine and heavy metal concentrations in blubber and liver tissue collected from Queensland (Australia) dugong (*Dugong dugon*). Marine Pollution Bulletin 51, 361–369.

Helle, E., Olsson, M., Jensen, S., 1976a. DDT and PCB [polychlorinated biphenyls] levels and reproduction in ringed seal from the Bothnian Bay. Ambio 5, 188–189.

Helle, E., Olsson, M., Jensen, S., 1976b. PCB levels correlated with pathological changes in seal uteri. Ambio 5, 261–262.

Helm, R.C., Costa, D.P., O'Shea, T.J., Wells, R.S., Williams, T.M., 2015. Overview of effects of oil spills on marine mammals. In: Fingas, M.V. (Ed.), Handbook of Oil Spill Science and Technology. John Wiley & Sons, Inc., Hoboken, New Jersey, pp. 455–475.

Hermanussen, S., Matthews, V., Päpke, O., Limpus, C.J., Gaus, C., 2008. Flame retardants (PBDEs) contamination of marine turtles, dugongs and seafood from Queensland, Australia. Marine Pollution Bulletin 57, 409–418.

Jin, L., Gaus, C., van Mourik, L., Escher, B.I., 2013. Applicability of passive sampling to bioanalytical screening of bioaccumulative chemicals in marine wildlife. Environmental Science and Technology 47, 7982–7988.

Jin, L., Gaus, C., Escher, B.I., 2015. Adaptive stress response pathways induced by environmental mixtures of bioaccumulative chemicals in dugongs. Environmental Science and Technology 49, 6963–6973.

Koeman, J.H., van Genderen, H., 1966. Some preliminary notes on residues of chlorinated hydrocarbon insecticides in birds and mammals in the Netherlands. In: Moore, N.W. (Ed.), Pesticides in the Environment and Their Effects on Wildlife. In: Journal of Applied Ecology, vol. 3 (Suppl.), pp. 99–106.

Kubota, R., Kunito, T., Tanabe, S., 2002. Chemical speciation of arsenic in the livers of higher trophic marine animals. Marine Pollution Bulletin 45, 218–223.

Kumar, K.S., Yamamuro, M., Pitaksintorn, S., Takasuga, T., Masunaga, S., 2002. Dioxins/furans and dioxin-like polychlorinated biphenyls in dugongs from Thailand coast. Organohalogen Compounds 57, 177–180.

Landsberg, J.H., Flewelling, L.J., Naar, J., 2009. *Karenia brevis* red tides, brevetoxins in the food web, and impacts on natural resources: decadal advancements. Harmful Algae 8, 598–607.

Landsberg, J.L., Lefebvre, K.A., Flewelling, L.J., 2014. Effects of toxic microalgae on marine organisms. In: G.P. Rossini (Ed.), Toxins and Biologically Active Compounds from Microalgae . Biological Effects and Risk Management, vol. 2. CRC Press, Boca Raton, pp. 379–449.

Layne, J.N., 1965. Observations on marine mammals in Florida waters. Bulletin of the Florida State Museum 9, 131–181.

LeBouef, B.J., Bonnell, M.L., 1971. DDT in California sea lions. Nature 234, 108–109.

Marsh, H., O'Shea, T.J., Reynolds III, J.E., 2012. Ecology and Conservation of the Sirenia: Dugongs and Manatees. Cambridge University Press, Cambridge, UK. 538 p.

Marsh, H., Grayson, J., Grech, A., Hagihara, R., Sobtzick, S., 2015. Re-evaluation of the sustainability of a marine mammal harvest by indigenous people using several lines of evidence. Biological Conservation 192, 324–330.

McLachlan, M.S., Haynes, D., Müller, J.F., 2001. PCDDs in the water-sediment-seagrass-dugong (*Dugong dugon*) food chain on the Great Barrier Reef (Australia). Environmental Pollution 113, 129–134.

Miyazaki, N., Itano, K., Fukushima, M., Kawai, S.-I., Honda, K., 1979. Metals and organochlorine compounds in the muscle of dugong from sulawesi island. Scientific Reports of the Whales Research Institute, vol. 31, pp. 125–128.

Müller, J.F., Haynes, D., McLachlan, M.S., 1998. PCDD/Fs in the Great Barrier Reef environment of Australia. Organohalogen Compounds 39, 105–108.

Nganvongpanit, K., Buddhachat, K., Piboon, P., Euppayo, T., Kaewmong, P., Cherdsukjai, P., Kittiwatanawong, K., Thitaram, C., 2017. Elemental classification of the tusks of dugong (*Dugong dugong*) by HH-XRF analysis and comparison with other species. Scientific Reports 7, 1–12.

Oke, V.R., 1967. A brief note on the dugong *Dugong dugon* at Cairns Oceanarium. International Zoo Yearbook, vol. 7, pp. 220–221.

O'Shea, T.J., 1999. Environmental contaminants and marine mammals. In: Reynolds III, J.E., Rommel, S.A. (Eds.), Biology of Marine Mammals. Smithsonian Institution Press, Washington, DC, pp. 485–564.

O'Shea, T.J., 2003. Toxicology of sirenians. In: Vos, J., Bossart, G., Fournier, M., O'Shea, T.J. (Eds.), Toxicology of Marine Mammals. Taylor & Francis Publishers, London, pp. 270–287.

O'Shea, T.J., Aguilar, A., 2001. Cetaceans and sirenians. In: Shore, R.F., Rattner, B.A. (Eds.), Ecotoxicology of Wild Mammals. John Wiley and Sons, N.Y., pp. 427–496.

O'Shea, T.J., Tanabe, S.A., 2003. Marine mammals and persistent ocean contaminants: a retrospective overview. In: Vos, J., Bossart, G., Fournier, M., O'Shea, T.J. (Eds.), Toxicology of Marine Mammals. Taylor & Francis Publishers, London, pp. 99–134.

O'Shea, T.J., Beck, C.A., Bonde, R.K., Kochman, H.I., Odell, D.K., 1985. An analysis of manatee mortality patterns in Florida, 1976-1981. Journal of Wildlife Management 49, 1–11.

O'Shea, T.J., Rathbun, G.B., Bonde, R.K., Buergelt, C.D., Odell, D.K., 1991. An epizootic of Florida manatees associated with a dinoflagellate bloom. Marine Mammal Science 7, 165–179.

O'Shea, T.J., Moore, J.F., Kochman, H.I., 1984. Contaminant concentrations in manatees in Florida. Journal of Wildlife Management 48, 741–748.

O'Shea, T.J., Brownell Jr., R.L., Clark Jr., D.R., Walker, W.A., Gay, M.L., Lamont, T.G., 1980. Organochlorine pollutants in small cetaceans from the Pacific and South Atlantic Oceans, November 1968-June 1976. Pesticides Monitoring Journal 14, 35–46.

Preen, A., 1988. The Status and Conservation of the Dugong in the Arabian Region, Final Report, Meteorology and Environmental Protection Administration, Saudi Arabia, vols. 1–2.

Preen, A., 2004. Distribution, abundance and conservation status of dugongs and dolphins in the southern and western Arabian Gulf. Biological Conservation 118, 205–218.

Preen, A., Das, H., Al-Rumaidh, M., Hodgson, A., 2012. Dugongs in Arabia. In: Hines, E.M., Reynolds III, J.E., Aragones, L.M., Mignucci-Giannoni, A.A., Marmontel, M. (Eds.), Sirenian Conservation: Issues and Strategies in Developing Countries. University Press of Florida, Gainesville, pp. 91–98.

Reijnders, P.J.H., 1986. Reproductive failure in common seals feeding on fish from polluted coastal waters. Nature 324, 456–457.

Reijnders, P.J.H., 2003. Reproductive and developmental effects of environmental organochlorines on marine mammals. In: Vos, J., Bossart, G., Fournier, M., O'Shea, T.J. (Eds.), Toxicology of Marine Mammals. Taylor & Francis Publishers, London, pp. 55–66.

Romero-Calderón, A.G., Morales-Vela, B., Rosíles-Martínez, R., Olivera-Gómez, L.D., Delgado-Estrella, A., 2016. Metals in bone tissue of Antillean manatees from the Gulf of Mexico and Chetumal Bay, Mexico. Bulletin of Environmental Contamination and Toxicology 96, 9–14.

Rojas-Mingüer, A., Morales-Vela, B., 2002. Metales en hueso y sangre de manatíes de (*Trichechus manatus manatus*) de la Bahía de Chetumal, Quintana Roo, México. In: Rosado-May, F.J., Romero Mayo, R., De Jesús Navarrete, A. (Eds.), Contribuciones de la Ciencia al Manejo Costero Integrado de la Bahía de Chetumal y su Área de Influencia. Universidad de Quintana Roo, Chetumal, Mexico, pp. 133–142.

Ross, P.S., Vos, J.G., Osterhaus, A.D.M.E., 2003. The immune system, environmental contaminants and virus-associated mass mortalities among pinnipeds. In: Vos, J., Bossart, G., Fournier, M., O'Shea, T.J. (Eds.), Toxicology of Marine Mammals. Taylor & Francis Publishers, London, pp. 534–557.

Siegal-Willott, J.L., Harr, K.E., Hall, J.O., Hayek, L.-A.C., Auil-Gomez, N., Powell, J.A., Bonde, R.K., Heard, D., 2013. Blood mineral concentrations in manatees (*Trichechus manatus latirostris* and *T. manatus manatus*). Journal of Zoo and Wildlife Medicine 44, 285–294.

Sladen, W.J.L., Menzie, C.M., Reichel, W.L., 1966. DDT residues in Adelie penguins and a crabeater seal from Antarctica. Nature 210, 670–673.

Stickel, L.F., Stickel, W.H., Christensen, R., 1966. Residues of DDT in brains and bodies of birds that died on dosage and survivors. Science 151, 1549–1551.

Stavros, H.C.W., Bonde, R.K., Fair, P.A., 2008. Concentrations of trace elements in blood and skin of Florida manatees (*Trichechus manatus latirostris*). Marine Pollution Bulletin 56, 1221–1225.

Takahashi, E.M., Arthur, K.E., Shaw, G.R., 2008. Occurrence of okadaic acid in the feeding grounds of dugongs (*Dugong dugon*) and green turtles (*Chelonia mydas*) in Moreton Bay, Australia. Harmful Algae 7, 430–437.

Takeuchi, N.Y., 2012. Trace Metal Concentrations and the Physiological Role of Zinc in the West Indian Manatee (*Trichechus manatus*) (Dissertation). University of Florida.

Takeuchi, N.Y., Walsh, M.T., Bonde, R.K., Powell, J.A., Bass, D.A., Gaspard III, J.C., Barber, D.S., 2016. Baseline reference range for trace metal concentrations in whole blood of wild and managed West Indian manatees (*Trichechus manatus*) in Florida and Belize. Aquatic Mammals 42, 440–453.

Vetter, W., Scholz, E., Gaus, C., Müller, J.F., Haynes, D., 2001. Anthropogenic and natural organohalogen compounds in blubber of dolphins and dugongs (*Dugong dugon*) from northeastern Australia. Archives of Environmental Contamination and Toxicology 41, 221–231.

Vos, J.G., Ross, P.S., De Swart, R.L., Van Loveren, H., Osterhaus, A.D.M.E., 2003. The effects of chemical contaminants on immune functions in harbour seals: results of a semi-field study. In: Vos, J., Bossart, G., Fournier, M., O'Shea, T.J. (Eds.), Toxicology of Marine Mammals. Taylor & Francis Publishers, London, pp. 558–570.

Walsh, C.J., Butawan, M., Yordy, J., Ball, R., Flewelling, L., de Wit, M., Bonde, R.K., 2015. Sublethal red tide toxin exposure in free-ranging manatees (*Trichechus manatus*) affects the immune system through reduced lymphocyte proliferation responses, inflammation, and oxidative stress. Aquatic Toxicology 161, 73–84.

Walsh, C.J., Luer, C.A., Noyes, D.R., 2005. Effects of environmental stressors on lymphocyte proliferation in Florida manatees, *Trichechus manatus* latirostris. Veterinary Immunology and Immunopathy 103 (3–4), 247–256.

Weijs, L., Vijayasarathy, S., Gaus, C., 2016. PCB levels in tissues of the strictly herbivorous Australian dugong (*Dugong dugon*). Organohalogen Compounds 78, 1119–1122.

Wetzel, D.L., Pulster, E., Reynolds III, J.E., 2012. Organic contaminants and sirenians. In: Hines, E., Reynolds III, J.E., Aragones, L.V., Mignucci-Giannoni, A.A., Marmontel, M. (Eds.), Sirenian Conservation: Issues and Strategies in Developing Countries. University of Florida Press, Gainesville, pp. 196–203.

Wetzel, D.L., Reynolds, J.E., Sprinkel, J.M., Schwacke, L., Mercurio, P., Rommel, S.A., 2010. Fatty acid profiles as a potential lipidomic biomarker of exposure to brevetoxin for endangered Florida manatees (*Trichechus manatus latirostris*). The Science of the Total Environment 408, 6124–6133.

Wiemeyer, S.N., Porter, R.D., 1970. DDE thins eggshells of captive American kestrels. Nature 227, 737–738.

Working Group on Endocrinology and Reproduction, 1999. Indices of reproductive and developmental functional endocrine toxicology in marine mammals. In: O'Shea, T.J., Reeves, R.R., Long, A.K. (Eds.), Marine Mammals and Persistent Ocean Contaminants: Proceedings of the Marine Mammal Commission Workshop. U.S. Marine Mammal Commission, Bethesda, Maryland, pp. 35–40.

Yoshimura, T., 2003. Yusho in Japan. Industrial Health 41, 139–148.

Yoshimura, T., 2012. Yusho: 43 years later. The Kaohsiung Journal of Medical Sciences 28, S49–S52.

Section III

Implication for Monitoring and Conservation of Marine Mammals

Chapter 17

Marine Mammals and Multiple Stressors: Implications for Conservation and Policy

Mark P. Simmonds[1,2]

[1]*Humane Society International, London, United Kingdom;* [2]*University of Bristol, School of Veterinary Sciences, Bristol, United Kingdom*

INTRODUCTION

For many centuries, in many maritime countries, human interest in marine mammals was limited to consideration of them as a resource to be exploited for human consumption and then for profit. For example, whales were regarded as having such value that King Edward II of England made a formal claim to their ownership, followed by several other heads of state (Brakes and Simmonds, 2011). Widespread commercial whaling in the 19th and 20th centuries, eventually involving diesel-driven fleets including factory vessels, led to decimation of populations. Attitudes changed in the 1960s and 1970s when the animals started to be valued and appreciated in other ways, including aesthetically and for their entertainment value in captivity.

Considerable knowledge has been gained in recent decades about both the biology of the animals and the fast-evolving threats that they face, but increasing knowledge does not automatically lead to improved protection, and some species and populations are still heading toward extinction (Campagna, 2015). At the root of this is a complex and evolving array of factors that can impact on these animals. For example, the endangered North Atlantic right whale, *Eubalaena glacialis*, population was initially devastated by whaling. Now, as this much diminished population struggles to recover, ship strikes and entanglement in fishing gear are regarded as the primary threats (Reilly et al., 2012). Looking to the future, it seems likely that climate change will cause the species yet more problems (Greene and Pershing, 2004).

Another example of populations being affected by multiple threats might be found in the case of delphinids in the Northeast Atlantic where pollution, in the form of PCBs, has recently been recognized again as a major threat

Marine Mammal Ecotoxicology. https://doi.org/10.1016/B978-0-12-812144-3.00017-6

(see, for example, Jepson et al., 2016). These are the same populations that, in many cases, are also being affected by deaths in fishing nets and other factors.

To conserve wildlife populations, we need to address not one but the multiple factors that are affecting them simultaneously, and this is not a new realization. Nor is the notion that some factors act synergistically, creating greater harm together than when acting on their own. For example, enhanced exposure to pathogens from discharges into cetacean habitat combined with enhanced exposure to immunosuppressive contaminants might be expected to create more disease and even, potentially, drive mass mortalities (Simmonds and Mayer, 1997).

However, marine mammal science tends to focus on particular classes of threat, rather than trying to address their multiplicity and the consequences of the interactions between them for the species and populations being affected. There have been good reasons for this. Typically, scientists have had to specialize to be effective (and successful in their careers), and natural sciences and veterinary sciences (including animal welfare science) have tended to follow separate paths. Perhaps, as argued subsequently, the time may have come for a reunification of these specializations, as we struggle to address the realities of multiple stressors in wildlife conservation. Indeed, how to sensibly address this complexity is arguably now one of the "holy grails" of modern conservation. Inherent in this is understanding how the factors interact to cause outcomes for the animals concerned and also how multiple exposures to stressors over a lifetime might best be considered. None of this is easy. Indeed it has recently been suggested that assessing "cumulative effects" is "a problem that has proven nearly impossible to solve" (Tyack, 2016). Nonetheless, it is also argued that to discern the factors contributing to population trends, scientists must consider the full complement of threats faced by marine mammals (NAS, 2016). Only with such knowledge can effective decisions be made about which stressors to reduce, to bring the population back to a more favorable state, and this kind of assessment can also provide the environmental context for evaluating whether an additional activity could threaten it. However, this view of science driving policy, while eminently logical, may not be fully realistic.

AN INVENTORY OF THREATS

There is a wide and growing range of potential stressors that affect marine mammals, and Table 17.1 provides a list. These stressors are not static over time, as new ones continue to be created by human activities (take, for example, the evolution of marine noise pollution as a threat, as described in Simmonds et al., 2014) and populations may be exposed to new stressors as conditions change. In fact, novel technologies (combined with retreating ice at the poles) now allow us to access even the deepest and previously most inaccessible regions. In the Arctic, in particular, we are witnessing an influx of activities new to the region, including large-scale fishing, fossil fuel exploration, and shipping, all presenting new threats to wildlife (Simmonds, 2016).

TABLE 17.1 Factors That May Adversely Affect Cetacean and Other Marine Mammal Populations and Their Habitats

Climate change	Storm intensity changes
	Sea ice changes
	Changes in runoff water circulations
	Ozone depletion
	Climate change–driven *changes in human activities, e.g.,* • *increased shipping and fishing in Arctic waters* • *increased directed take of marine mammals*
Pollution	Nutrient pollution/eutrophication
	Harmful algal blooms
	Oil spills
Persistent organic pollutants, especially PCBs (but also potentially including brominated flame retardants and perfluorinated compounds)	
Heavy metals	
Nonfishery-derived marine debris, including microdebris	
Fisheries/related activities	Overfishing and prey-culling *and depletion*
	Mariculture
	Marine debris, including ghost nets
	Bycatch
Noise pollution	Seismic surveys
	Boat traffic (*also causing ship strikes*)
	Military sonar
	Construction
Pathogen emergent disease	
Physical habitat degradation	Bottom trawling
	Dredging
	Other destructive fishing techniques
	Reclamation
	Coastal construction
	Wind farms
	Dams and barrages
	Marine fossil fuel exploration/extraction

Continued

TABLE 17.1 Factors That May Adversely Affect Cetacean and Other Marine Mammal Populations and Their Habitats–cont'd

Tourism	Whale watching
	"Swim with" programs
War-related activities	Mines
	Munitions dumps
Introduced species	
Intentional takes	*Commercial whaling*
	Other marine mammal takes for profit or food.

After International Whaling Commission (2006), with additional factors from Brakes and Simmonds (2011).

Simmonds and Brakes (2011) compared a review of threats to cetaceans made in 1996, with their understanding in 2011, and suggested the following key developments:

- There had been a general acceptance of noise pollution as a substantive threat and some movement to address this.
- Climate change had also become an accepted phenomenon, with implications for cetaceans.
- Levels of some of the more infamous pollutants had fallen.
- There was much recent new research into marine mammal diseases and a growing awareness of the vulnerability of marine mammal populations to disease events and the potential of human activities to contribute to them.

A few years further on (I am now writing in mid-2017), it is now possible to recognize the reemergence of the threat posed by PCBs as a significant issue for the survival of some populations. Likewise, the growing number of harmful algal blooms (e.g., Anderson, 2009), possibly boosted by nutrient discharges, combined with changing climate, seems to be coming more clearly to the fore as a pressing issue (IWC, 2017). It is also now much more clearly recognized that intense sounds from human activities—such as seismic air guns—can have direct physiologic effects on marine mammals and that naval sonar triggers behavioral reactions that can lead to death by stranding (NAS, 2016).

Emerging threats at this time include the growing amounts of macro- and microdebris in the seas and oceans and, as noted before, rapidly changing human activities in the Arctic. Factors impacting marine mammals populations can be lethal (e.g., a ship strike or a launched harpoon) or sublethal, and when describing "stressors" here, it is a sublethal impact that is being primarily considered. For example, while loud noise can be lethal, the most common effect of noise on marine mammals is behavioral disturbance. From a population perspective, rather subtle behavioral changes affecting very large numbers of marine mammals may have greater consequences than occasional lethal events affecting a few (NAS, 2016).

AN EXAMPLE OF A COMPLEXITY: CLIMATE CHANGE

To help more fully comprehend the complex natures of the situations that marine mammal populations are facing, it may be worth considering further the various mechanisms through which climate change may come to impact them. Simmonds (2016) reviewed this, and it is apparent from the scientific literature that the primary concerns are not so much about a direct effect upon the individual marine mammals themselves (e.g., thermal stress) but more focused upon changes in prey and, to some extent, on changes in human activities (including their changing locations as highlighted for the Arctic earlier and discussed more broadly in Alter et al., 2010). This is not to say that there might not be direct responses from marine mammal populations to changing physical conditions in the sea. For example, cetacean population distribution is closely related to temperature, and it has long been theorized that there will be a general movement toward the poles as waters warm. There is already evidence that this is starting to happen. Prey may also change and shift distribution, so trying to separate out one effect from another in the future may be difficult.

Fig. 17.1 illustrates the various ways in which climate change–driven factors may come to affect marine mammals. It also highlights potential interactions with other factors. For example, access to prey might also be affected by competition with species that have changed distribution. And the fitness of the marine mammals (both as individuals and populations) might also be undermined by exposure to new pathogens, chemical and noise pollution, and so forth.

ENGAGING WITH MULTIPLE STRESSORS

The first serious attempt to try to address the issue of the multiple factors affecting marine mammals may have come from the International Whaling Commission (IWC). By the early 2000s, the member nations of the IWC had become concerned about the broad range of factors then known to be affecting cetaceans. It initiated an ambitious piece of work to look at this via a "Workshop on Habitat Degradation." While the workshop title indicates a focus on habitat, it was ultimately concerned with how to take an integrated approach to stressors/threats. The workshop was informed by an earlier smaller "scoping group" meeting of experts, and it is worth noting that this identified several potential ways forward, including consideration of individual health and body condition, "vital rates" (i.e., survival and fecundity and other life history parameters), population changes, and community-level changes (IWC, 2006). The scoping group suggested that the principal tools for linking habitat changes to these response variables were (1) correlative analyses comparing response variables across habitats with very different levels and patterns of impact; (2) "analogy

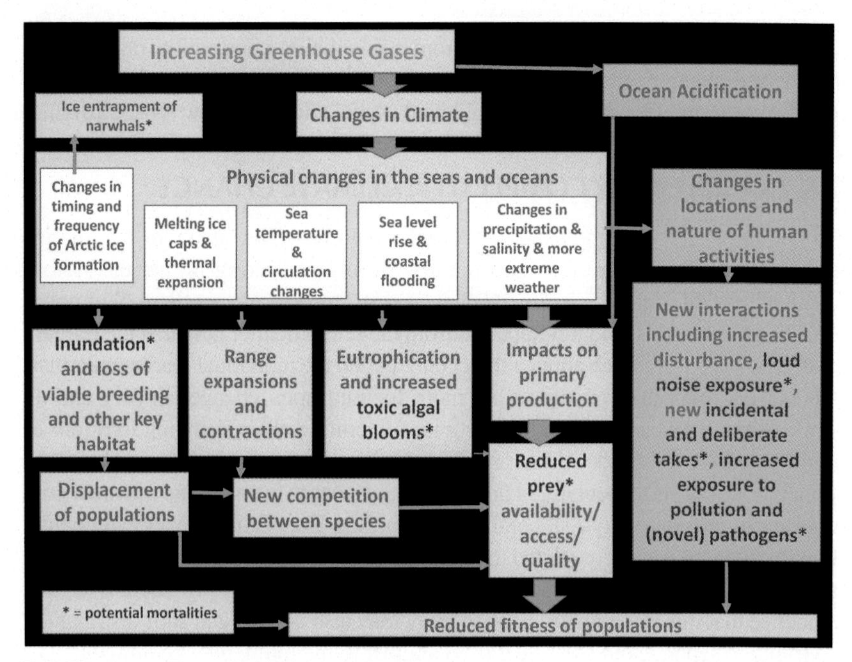

FIGURE 17.1 Climate change–driven factors and associated stressors and linkages. *(Modified from Simmonds, M.P., 2016. Impacts and effects of ocean warming on marine mammals. In: Laffoley, D., Baxter, J.M. (Eds.), Explaining Ocean Warming: Causes, Scale, Effects and Consequences. IUCN, pp. 305–322.)*

from more detailed mechanistic studies on model species"; and (3) modeling of population responses to changes in vital rates as a result of habitat degradation.

The IWC Workshop on Habitat Degradation met in 2004 and noted in its report that the IWC has been concerned about the influence of environmental changes on cetacean populations for many years, signified by various resolutions requesting that its Scientific Committee progress understanding of this issue (IWC, 2006). In response, the Scientific Committee had identified eight environmental priority topics:

- climate/environment change;
- physical and biologic habitat degradation;
- chemical pollution;
- direct and indirect effects of fisheries;
- impact of noise;
- disease and mortality events;
- ozone and UV-B radiation;
- Arctic issues.

The workshop's general conclusions stressed the importance of undertaking research relating habitat condition to cetacean status in the context of

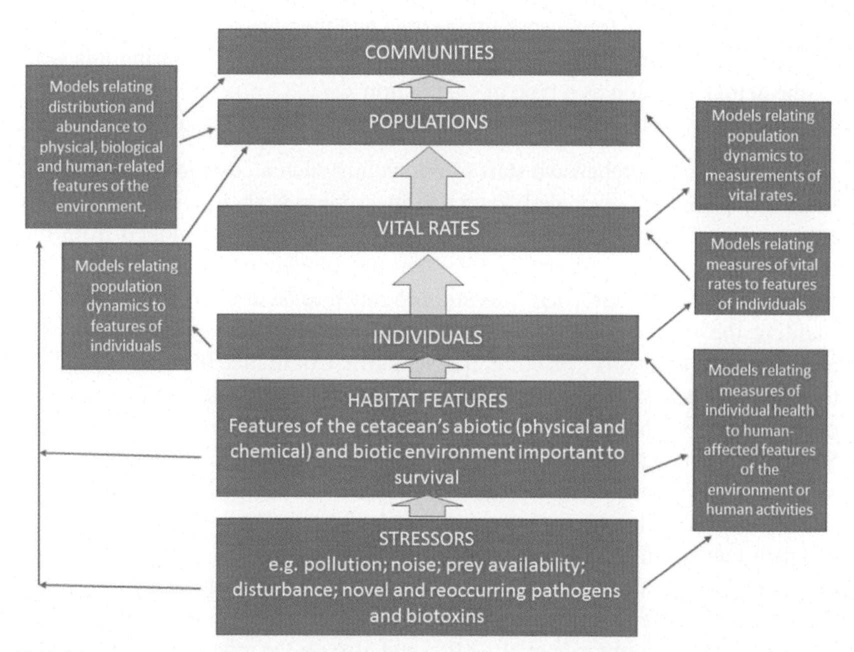

FIGURE 17.2 Framework for modeling the links between environmental stressors that degrade habitat and population effects. *(After IWC, 2006. Report of the IWC scientific committee workshop on habitat degradation. Journal of Cetacean Research and Management 8 (Suppl.), 313–335.)*

conservation and management. However, it also commented that "this is a particularly complex area of study, requiring both theoretical developments in modelling approaches and a commitment to long-term interdisciplinary data collection programmes." To help make progress, the workshop produced and strongly recommended a new framework for further investigation, which is shown in Fig. 17.2.

The workshop also commented that any general application of the framework would require that management and research bodies take a longer-term view and described the present ad hoc processes (giving "Environmental Impact Assessments," based on short-term limited datasets as an example) as unsatisfactory. In terms of further research, the workshop identified several cetacean populations with sufficiently broad sampling programs, covering sufficiently long time frames, which could be the focus of studies: Florida bottlenose dolphins; European harbor porpoises; and resident killer whales from the northwest coast of North America.

The workshop also proposed a workplan to develop the framework (as shown in Fig. 17.2) and that this should include:

1. application to specific case studies;
2. further development of approaches to distinguish the relative effects of different stressors via population and spatial modeling approaches;

3. application of the framework to one area and then using the results to make predictions for the same species in a different area and comparing this with the actual situation as a type of "validation";
4. a follow-up workshop to review the progress of this workplan.

Sadly, this comprehensive start to unraveling such a complex issue has not obviously positively resonated down the intervening years in terms of research either under the jurisdiction of the IWC or, as far as can be judged from the scientific literature, anywhere else! Perhaps the inherent problems were just too complicated, or perhaps, there was still too much to be done in terms of understanding the various stressors or developing the necessary models. However, most recently, at its 2017 meeting, the Scientific Committee of the IWC agreed to prepare for a workshop on cumulative threats, and it took note of the relevance of the outputs of the 2004 Habitat Degradation workshop to this (IWC, 2017). So, it may be hoped that there may yet be some further development and elaboration of the approaches and recommendations made by the 2004 workshop.

Certainly, there has been a lot of work on the factors affecting marine mammals and their habitats in the intervening years, and increasingly, this considers interactions with more than one stressor. The relevant scientific literature is too voluminous to review here, but examples include the copious amount of recent research on marine noise (Simmonds et al., 2014) and also on the effects of whale watching on cetacean populations (see, for example, New et al., 2015; Higham et al., 2014). Effort has also gone into modeling approaches, leading, for example, to the Population Consequences of Disturbance model (New et al., 2014).

THE LATEST WORK ON CUMULATIVE EFFECTS

Animals and populations of animals may be exposed to particular stressors once or many times. A good example is exposure to a loud noise, and multiple, frequent exposures might be more significant than rare exposures over a longer time. "Cumulative effect" has been defined as the combined effect of exposures to multiple stressors integrated over a defined relevant period: a day, a season, year, or lifetime (NAS, 2016).

In the United States, the National Academies of Sciences, Engineering, and Medicine has been looking at cumulative effects on marine mammals. The results of its deliberations were delivered in a substantive and substantial (250-page) report published in 2016 (NAS, 2016). The topic of cumulative effects was chosen by the federal agency sponsors because assessing cumulative effects has been an important part of US regulations protecting marine mammals since the 1970s, but "the approaches used have little predictive value." If cumulative effects cannot be accounted for, "then unexpected adverse impacts from interactions between stressors pose a risk to marine mammal populations and the marine ecosystems on which people and marine mammals depend" (Tyack, 2016).

Because quantitative prediction of cumulative effects of stressors on marine mammals is not currently possible, the authors of the NAS report have developed a conceptual framework for assessing the population consequences of multiple stressors (NAS, 2016). They call this the "Population Consequences of Multiple Stressors" model, and it uses indicators of health that integrate the short-term effects of different stressors that affect survival and reproduction, and the report explores a variety of methods to estimate health, stressor exposure, and responses to stressors. (For a full explanation of this approach and the study's full and detailed recommendations, readers are directed to the full report.)

Importantly, the authors concluded that scientific knowledge is not up to the task of predicting the cumulative effects of different combinations of stressors on marine mammal populations (NAS, 2016) and comment that "even though exposure to multiple stressors is an unquestioned reality for marine mammals, the best current approach for management and conservation is to identify which stressor combinations cause the greatest risk."

CONCLUSIONS AND RECOMMENDATIONS

This short review cannot do justice to the investigations that have been made into the effects of stressors on marine mammals and their habitats, alone, in combination, or cumulatively. However, what is emerging from these studies is that this is a very complex sphere of endeavor. Clearly, much research is ongoing, and inherent in this is information that will help to inform those seeking to conserve marine mammal populations. However, the integration of research into effective conservation policy is itself far from being straightforward.

Claudio Campagna, in an inspiring keynote address at the 2015 Conference of the Society for Marine Mammalogy, challenged his audience with a bleak but well-informed view of modern conservation (Campagna, 2015). He opined that the continuing crisis of imminent extinctions is being driven by a paradigm that he summarized as

> *",,,provide me with a good economic reason or I do nothing… or I will make small adjustments of no consequence".*

He argued that the current approach to species conservation is flawed as it is based on the notion that science informs policy and policy informs conservation and sustainable economic growth. However, in practice, he argued new information is used to intervene only when it is no more costly than doing nothing! Valuing nature only in economic terms avoids recognizing the disastrous consequences of what Campagna called "the species crisis."

Sadly, my own experience of conservation work aligns closely with this, and while scientists may work hard to understand matters and give advice, including in the complex context of the multiple stressors now affecting marine mammals, this does not necessarily mean that any effective action will follow.

Related to this is that many conservation approaches require a good understanding and ongoing monitoring of the populations concerned. This is rare for many marine mammal populations (which is why many remain "data deficient" on the International Union for Conservation of Nature Red List). What is clear, however, is that chemical pollution, noise pollution, disturbance (leading, for example, to displacement from important habitats), and other factors can substantially impact populations, and there are some instances where we know or can reasonably deduce which populations are being impacted to such an extent that their future is imperiled (for example, in the case of PCBs, certain populations in the Northeast Atlantic, including the Mediterranean and Black Sea areas). This then provides a case for action.

Pollution by PCBs and climate change are clearly difficult issues to address. There is no simple "off-tap" for either. However, it should be noted that various actions are being promoted, especially in a European context, to address PCBs (see Law and Jepson, 2017; Stuart-Smith and Jepson, 2017). However, in situations where we believe such intransigent stressors as these may be the primary cause of problems, addressing other more easily resolvable factors likely to be adversely affecting the population would seem at least precautionary and, indeed, sensible (e.g., taking action to stop or lessen incidental removals in fishing nets or death by ship strikes).

Such precautionary action—reducing stressors where this is possible—should not wait on perfect proof of impact or be inhibited by the knowledge that these stressors are not the primary causal factors in declines, but it should proceed to make populations as robust as possible to the multiple stressors they are facing. Sanctuaries or marine protected areas, wherein stressors are reduced or removed, will play an important role in this, and there is an ambitious program of work on this going forward at this time led by the Marine Mammal Protected Areas Task Force. The Task Force was created in 2013 and has been setting up regional workshops to identify Important Marine Mammal Areas, beginning with the Mediterranean in 2016, followed by the South Pacific, the Northeast Indian, the Northwest Indian and the Southeast Pacific oceans, and the waters of Oceania surrounding Australia and New Zealand (ICMMPA, 2017).

Another innovation (as hinted at in the introduction) is the use of animal health considerations to help pinpoint and better understand problems. Monitoring marine mammal population trends may not always be practical, and a measurable decline in a population should not necessarily be taken as the only possible cue for action. Welfare science and health assessments offer another set of tools. This idea is not entirely novel. While the 2004 IWC workshop did not formally include health assessments in its guiding framework (Fig. 17.2), the possible development and use of health parameters was certainly discussed there (IWC, 2006). Thirteen years later, the National Academies of Sciences, Engineering, and Medicine puts monitoring health at the center of its approach and recommendations.

More generally, monitoring the health of wild populations offers a new way to identify when significant problems are developing; perhaps providing a kind of early warning system. This relationship between welfare science and conservation now deserves to be further developed from the perspective of improving both conservation and welfare responses, and interestingly, the IWC, with its growing interest in whale welfare outside of the hunting context (IWC, 2016), may prove to be the crucible in which such things productively come to mix.

Finally, one of the biggest problems faced by those who want to conserve and protect marine mammals (or for that matter address pressing threats, including climate change) is convincing those in power and the public more generally that this actually matters: specifically that the survival of marine mammals has relevance to our own species.

Somehow, it appears that the human race has become detached from the natural environment that supports it by maintaining functioning ecosystems of which wild animals (including marine mammals) are components. This detachment is so profound that we do not recognize the threat to ourselves as our activities disrupt and damage ecosystems. Part of the response to this has to be in education (in the broadest sense) and explaining how we inherently fit into—and are supported by—something much bigger than ourselves. Without a better informed and sympathetic public, and policy makers, we have little hope of effectively addressing the complex issues besetting marine and other ecosystems.

ACKNOWLEDGMENTS

With thanks to the editors for the opportunity to contribute here, to my anonymous reviewer for guidance, and to Mike Archer for his review. The views expressed are my own and do not necessarily reflect those of any organization that I am or have been affiliated with.

REFERENCES

Alter, S., Simmonds, M.P., Brandon, J.R., 2010. Forecasting the consequences of climate-driven shifts in human behaviour on cetaceans. Marine Policy 34 (5). https://doi.org/10.1016/j.marpool.2010.01.02.

Anderson, D.M., 2009. Approaches to monitoring, control and management of harmful algal blooms (HABs). Ocean and Coastal Management 52 (7), 342. https://doi.org/10.1016/j.ocecoaman.2009.04.006.

Brakes, P., Simmonds, M.P., 2011. Whales and Dolphins: Cognition, Conservation and Human Perceptions. Earthscan, London.

Campagna, C., 2015. Language is the dress of thought. In: Keynote Presentation at the Society of Marine Mammalogy Conference. https://www.marinemammalscience.org/conference-schedule/program-highlights-and-special-events/.

Greene, C.H., Pershing, A.J., 2004. Climate and the conservation biology of North Atlantic right whales: the right whale at the wrong time? Frontiers in Ecology and the Environment 2, 29–34. https://doi.org/10.1890/15409295(2004)002[0029:CATCBO]2.0.CO;2.

Higham, J., Bejder, L., Williams, R., 2014. Whale-Watching – Sutainable Tourism and Ecological Management. Cambridge University Press, Cambridge, UK.

ICMMPA, 2017. Website of the Marine Mammal Protected Areas Task Force. https://www.marinemammalhabitat.org/.

IWC, 2006. Report of the IWC scientific committee workshop on habitat degradation. Journal of Cetacean Research and Management 8 (Suppl.), 313–335.

IWC, 2016. Report of the Workshop to Support the IWC's Consideration of Non-Hunting Related Aspects of Cetacean Welfare. IWC/66/WKM&WI Rep 01 34 pages.

IWC, 2017. Report of the Scientific Committee of the International Whaling Commission. (in press).

Jepson, P.D., et al., 2016. PCB pollution continues to impact populations of orcas and other dolphins in European waters. Scientific Reports 6. https://doi.org/10.1038/srep18573Article number: 18573.

Law, R.J., Jepson, P.D., 2017. Europe's insufficient pollutant remediation. Science 356 (6334), 148. https://doi.org/10.1126/science.aam6274.

NAS, 2016. Approaches to Understanding the Cumulative Effects of Stressors on Marine Mammals. National Academies of Sciences, Engineering, and Medicine. The National Academies Press, Washington, DC. https://doi.org/10.17226/23479.

New, L.F., Hall, A.J., Harcourt, R., Kaufman, G., Parsons, E.C.M., Pearson, H.C., Cosentino, A.M., Schick, R.S., 2015. The modeling and assessment of whale-watching impacts. Ocean and Coastal Management 115, 10–16.

New, L.F., Clark, J.S., Costa, D.P., Fleishman, E., Hindell, M.A., Klanjšček, T., Lusseau, D., Kraus, S., Mcmahon, C.R., Robinson, P.W., Schick, R.S., Schwartz, L.K., Simmons, S.E., Thomas, L., Tyack, P., Harwood, J., 2014. Using short-term measures of behaviour to estimate long-term fitness of southern elephant seals. Marine Ecology Progress Series 496, 99–108.

Reilly, S.B., Bannister, J.L., Best, P.B., Brown, M., Brownell Jr., R.L., Butterworth, D.S., Clapham, P.J., Cooke, J., Donovan, G., Urbán, J., Zerbini, A.N., 2012. Eubalaena glacialis. In: The IUCN Red List of Threatened Species 2012: e.T41712A17084065 https://doi.org/10.2305/IUCN.UK.2012.RLTS.T41712A17084065.en.

Simmonds, M.P., 2016. Impacts and effects of ocean warming on marine mammals. In: Laffoley, D., Baxter, J.M. (Eds.), Explaining Ocean Warming: Causes, Scale, Effects and Consequences. IUCN, pp. 305–322.

Simmonds, M.P., Brakes, P., 2011. Whales and dolphins on a rapidly changing planet. In: Brakes, P., Simmonds, M.P. (Eds.), Whales and Dolphins: Cognition, Conservation and Human Perceptions. Earthscan, London, pp. 169–178.

Simmonds, M.P., Dolman, S.J., Jasny, M., Parsons, E.C.M., Weilgart, L., Wright, A.J., Leaper, R., 2014. Marine noise pollution – increasing recognition but need for more practical action. The Journal of Ocean Technology 9, 70–90.

Simmonds, M.P., Mayer, S.J., 1997. An evaluation of enviornmental and other factors in some recent marine mammal mortalities in Europe: implications for conservation and management. Environmental Reviews 5, 89–98.

Stuart-Smith, S., Jepson, P.D., 2017. Persistent threats need persistent counteraction: responding to PCB pollution in marine mammals. Marine Policy 84, 69–75.

Tyack, P.L., 2016. Preface. In: NAS (2016) Approaches to Understanding the Cumulative Effects of Stressors on Marine Mammals. National Academies of Sciences, Engineering, and Medicine. The National Academies Press, Washington, DC, pp. vii–viii. https://doi.org/10.17226/23479.

Index

Note: 'Page numbers followed by "f" indicate figures, "t" indicate tables.'

C

F

G

H

I

P

Q

R

S

T

U

V

W

Y

Z

Printed in the United States
by Baker & Taylor Publisher Services